Aircraft Design Concepts

Aircraft Design Concepts

An Introductory Course

James DeLaurier

CRC Press is an imprint of the
Taylor & Francis Group, an **informa** business

MATLAB® is a trademark of The MathWorks, Inc. and is used with permission. The MathWorks does not warrant the accuracy of the text or exercises in this book. This book's use or discussion of MATLAB® software or related products does not constitute endorsement or sponsorship by The MathWorks of a particular pedagogical approach or particular use of the MATLAB® software.

First edition published 2022

by CRC Press
6000 Broken Sound Parkway NW, Suite 300, Boca Raton, FL 33487-2742

and by CRC Press
4 Park Square, Milton Park, Abingdon, Oxon, OX14 4RN

© 2022 James DeLaurier
CRC Press is an imprint of Taylor & Francis Group, LLC

Reasonable efforts have been made to publish reliable data and information, but the author and publisher cannot assume responsibility for the validity of all materials or the consequences of their use. The authors and publishers have attempted to trace the copyright holders of all material reproduced in this publication and apologize to copyright holders if permission to publish in this form has not been obtained. If any copyright material has not been acknowledged please write and let us know so we may rectify in any future reprint.

Except as permitted under U.S. Copyright Law, no part of this book may be reprinted, reproduced, transmitted, or utilized in any form by any electronic, mechanical, or other means, now known or hereafter invented, including photocopying, microfilming, and recording, or in any information storage or retrieval system, without written permission from the publishers.

For permission to photocopy or use material electronically from this work, access www.copyright.com or contact the Copyright Clearance Center, Inc. (CCC), 222 Rosewood Drive, Danvers, MA 01923, 978-750-8400. For works that are not available on CCC please contact mpkbookspermissions@tandf.co.uk

Trademark Notice: Product or corporate names may be trademarks or registered trademarks and are used only for identification and explanation without intent to infringe.

ISBN: 978-1-138-03339-9 (hbk)
ISBN: 978-1-032-11104-9 (pbk)
ISBN: 978-1-315-22816-7 (ebk)

DOI: 10.1201/9781315228167

Typeset in Times LT Std
by KnowledgeWorks Global Ltd.

Dedication

To my Susan

Contents

Preface xi
Acknowledgements xiii
Author xv

Chapter 1   Introduction 1
                   Design Features 1
                   Different Materials in Column Buckling 6
                   Illustrative Examples of Wood, Metal and Composite Airplanes 11
                   Examples Wing Structures 17

Chapter 2   Aerodynamic Review 25
                   Airfoils 25
                   Wings 36
                   Bodies 40
                   Undercarriage 46
                   Wing-Body Combination 48
                   Complete-Aircraft Aerodynamics (for a Tail-Aft Monoplane) 49
                         Lift 50
                         Drag 53
                         Pitching Moment (with non-extended undercarriage) 55
                         Pitching Moment (with extended undercarriage) 60
                   Numerical Example 60
                         Wing 61
                         Tail 64
                         Body 70
                         Propeller 74
                         Undercarriage 74
                         Complete Airplane Lift and Drag Coefficients 75
                         Complete Airplane Moment Characteristics 76
                   A Final Comment 78

Chapter 3   Propeller Analysis 79
                   Simple Blade-Element Analysis 79
                   Actuator-Disk Momentum Theory 82
                   Extensions to the Simple Blade-Element Analysis 84
                   Method of Calculation 87
                   Numerical Example 89
                   Application of the Analysis 94
                   Propeller Airfoils 97
                   Matlab Program 97

Chapter 4   Flying Wings (or Tailless Airplanes) 101
                   "Flying Planks" 101
                   Swept Flying Wings 105

Paragliders 108
Rogallo-Type Hang Gliders 109
Span-Loader Flying Wings 110
A Canadian Flying-Wing Glider 114
An Approximate Method for Estimating the Aerodynamic Characteristics
of Wings with Variable Twist, Taper and Sweep 115
    Example 1, Straight-Tapered Linear-Twisted Wing... 118
    Example 2, Double-Swept and Double-Tapered Wing... 129
    Matlab Computer Program 143
"Delta" Tailless Aircraft 154
Final Observations 162

Chapter 5   Canard Airplanes and Biplanes... 167

Canard Airplanes 167
    An Approximate Method for Estimating the Aerodynamic Characteristics
    of Canard Airplanes 179
    Example, Rectangular Wing 183
    Example, Canard Glider 185
Biplanes 197
    An Approximate Method for Estimating the Aerodynamic Characteristics
    of Biplanes with Wings of Equal Spans and Areas... 210
    Example, "Slow SHARP" Biplane 212
    Further Considerations of Biplane Analysis... 229
    Example, Biplane Glider 232
    Final Biplane Comment... 243

Chapter 6   Flight Dynamics 245

Introduction 245
Aircraft Longitudinal Small-Perturbation Dynamic Equations 256
    Non-Dimensional Form of the Equations 259
    Estimation of the Longitudinal Stability Derivatives... 264
        The $(\ )_0$ Terms 264
        The $\hat{u}$ Derivatives 265
        The α Derivatives 266
        The $\hat{q}$ derivatives 267
        The $D\hat{u}$ Derivatives 271
        The $D\alpha$ Derivatives 273
        The Dq Derivatives 276
        The Propulsion Derivatives 277
    Longitudinal Numerical Example ("Scholar" Tail-Aft Monoplane).................... 279
        Example Values of $(C_X)_0$, $(C_Z)_0$ and $(C_M)_0$ 280
        Example Values of $C_{X\alpha}$, $C_{Z\alpha}$ and $C_{M\alpha}$ 281
        Example values of C_{Xq}, C_{Zq} and C_{Mq} 282
        Example Values of C_{XDU} and C_{MDU} 285
        Example Values of $C_{ZD\alpha}$ and $C_{MD\alpha}$ 287
        Example Values of C_{XDq}, C_{ZDq} and C_{MDq} 289
        Example Values of $\hat{T}_{Xu}$, $\hat{T}_{X\alpha}$ and $\hat{T}_{Z\alpha}$ 292
Aircraft Lateral Small-Perturbation Dynamic Equations.. 294

- Non-Dimensional Form of the Equations 295
- Estimation of the Lateral Stability Derivatives 297
  - The β Derivatives: 298
  - The $\hat{r}$ Derivatives 314
  - The $\hat{p}$ Derivatives: 319
  - The $D\beta$ Derivatives: 324
  - The $D\hat{r}$ Derivatives 326
  - The $D\hat{p}$ Derivatives 327
- Example Lateral Stability Derivatives for the "Scholar" Tail-Aft Monoplane 329
  - Example Value of $C_{Y\beta}$ 329
  - An Extension to the Analysis 331
  - Example Value of $C_{N\beta}$ 337
  - Example Value of $C_{\bar{L}\beta}$ 339
  - Example Value of C_{Yr} 340
  - Example Value of C_{Nr} 341
  - Example Value of $C_{\bar{L}r}$ 343
  - Example Value of C_{Yp} 344
  - Example Value of C_{Np} 345
  - Example Value of $C_{\bar{L}p}$ 346
  - Example Value of $C_{YD\beta}$ 348
  - Example Value of $C_{ND\beta}$ 349
  - Example Value of $C_{\bar{L}D\beta}$ 350
  - Example Value of C_{YDr} 351
  - Example Value of C_{NDr} 352
  - An Extension to the Analysis 352
  - Example Value of $C_{\bar{L}Dr}$ 354
  - Example Value of C_{YDp} 354
  - Example Value of C_{NDp} 355
  - Example Value of $C_{\bar{L}Dp}$ 356
  - An Extension to the Analysis 356
- Radii-of-Gyration Values for Representative Airplanes 359
- Definitions of Stability 360
- Longitudinal Dynamic Stability 362
  - Numerical Example 366
  - Comments on α and θ 368
  - Flight Paths 369
  - Approximate Equations 374
    - Short-Period Mode 374
    - Phugoid Mode 376
  - Roots-Locus Plots 378
- Lateral Dynamic Stability 382
  - Flight Paths 386
  - Approximate Equations 392
    - Spiral Mode 393
    - Rolling Mode 394
    - Dutch-Roll Mode 395
  - Roots-Locus Plots 396
  - Stability-Boundary Plot 398
- Addendum 399

Chapter 7	Performance	405
	Glide Tests	405
	Equilibrium Flight	408
	Trim State	408
	Full Solution	412
	Performance Parameters	419
	Take-Off Run	428
	Wood's Methodology	428
	Final Comments	430
Chapter 8	Balloons and Airships	431
	Free Balloons	431
	The Physics of Buoyancy	439
	Tethered Balloons	441
	Airships	446
	Aerodynamics of Finned Axisymmetric Bodies	461
	A Method for Calculating the Longitudinal Static Aerodynamic Coefficients	464
	Example 1: Small Aerostat	472
	Example 2: Airship ZRS-4 "Akron" 5.98m Wind-Tunnel Model	479
	Example 3: Goodyear "Wingfoot2" Airship (Zeppelin LZ N07)	485
	Example 4: TCOM CBV-71 Aerostat	493
	Summary of Example Parameters	494
	Aerodynamic Corrections for Inflated Fins	494
	Additional Observations about Aerostats	496
	Lateral Force and Yawing Moment Calculation	498
	Airship Aerodynamic Mystery	502
	Matlab Program	503

Appendix A: Multhropp Body-Moment Equation ... 517

Appendix B: Alternative Swept-Wing Analysis ... 519

Appendix C: Rigid-Body Equations of Motion ... 527

Appendix D: Apparent-Mass Effects ... 539

Appendix E: Lift of Finite Wings Due To Oscillatory Plunging Acceleration ... 551

Index ... 559

Preface

The purpose of this book is to provide an introduction to the principles and concepts of aircraft design. This will be done in a quantitative manner, such that basic analyses may be performed upon a candidate subsonic flight vehicle. The motivation for this book was provided by two decades of teaching aircraft design to fourth-year undergraduates and graduate students who came into the class with a good grounding in the basic engineering topics of aerodynamics, structures and analytical techniques. However, it was typical that most of these students had no exposure to the fundamentals of aircraft design, or even had any model-airplane design or fabrication experience. The challenge for this course was to not only provide the skill set required to allow three or four-person teams to design a viable airplane (subject to constraints set by the instructor), but in certain instances to build and fly a remotely-piloted model. Further, it had to be done in one semester. In that spirit, the methodologies are in a somewhat "cookbook" fashion, with charts and design equations that do not require derivation. Thus, the students may proceed quickly with the design development within the time restraints of a formal course. Nonetheless, when possible, the underlying physics behind these methodologies is explained.

The book has another aspect that should also prove educational. Namely, it describes various types of aircraft and why they were configured for a presumed performance advantage by their designers. The history of aircraft has seen some unique configurations, involving multiple wings, tail-first, flying wings, etc. These will be described and analyzed, and some surprising things will be learned about these ideas. In fact, one way in which the course was organized was to assign these types of configurations to the various teams.

It will also be seen that the book contains a strong historical theme. Some of the design ideas of the past (biplanes, fixed landing gear, airships, etc.) seem quaint by today's perspective. However, such aircraft and design features had a purpose that was appropriate for the time. These will be presented and explained. Most importantly, it is the goal of this book to give the students an appreciation of the rich history of aeronautical engineering. Although it is a relatively young discipline compared with other fields of engineering, it has had an amazing development in only a century. Also, there is an unprecedented rigor required for an aircraft to safely achieve its design goals. This has given the history of aeronautical engineering a cast of characters and visionaries who are as unique as their aircraft designs.

James DeLaurier, Professor Emeritus,
University of Toronto Institute for Aerospace Studies
April, 2021

MATLAB® is a registered trademark of The Math Works, Inc. For product information, please contact:

The Math Works, Inc.
3 Apple Hill Drive
Natick, MA 01760-2098
Tel: 508-647-7000
Fax: 508-647-7001
E-mail: info@mathworks.com
Web: http://www.mathworks.com

Acknowledgements

The author is indebted to his son, Bradley DeLaurier, for keeping his computer up-to-date and functioning with the latest software. This book could not have been written otherwise. Also, DVD copies of successive updates of this book were stored off-site by the author's daughter, Prof. April DeLaurier, who also designed the book's unique cover art.

There are several generous contributors of pictures and illustrations used in this book. In particular, the author wishes to acknowledge Jim Marske, who has designed many successful flying-wing gliders, some of which are in Chapter 4. An excellent book about his productive career has recently been published titled "The Wing and I". The author also appreciates free use of photographs by Robert Mudd, a colleague of Mr. Marske.

John Shupek, who runs the amazing website Skytamer.com, allowed the author free use of any of his numerous excellent photos of historic airplanes. In fact, the author has his photo of a B-17 bomber as his screen picture. Also, Gerald Balzer provided photos of the Northrop Flying Wings as well as that of the 1930's Northrop "Gamma". This was made possible by the good services of Richard Navarro and Tony Chong of Northrop Grumman.

Likewise, Kenneth Ross of Lockheed Martin freely provided three excellent photos of their historic airplanes. Also, a special thanks has to be given to Ed Cunningham of TCOM L.P. for free use of images of all TCOM products. Chapter 8 would have been difficult to write without these pictures.

Another major aerospace corporation that helped the author was Boeing, through Heather Anderson, in the free use of the concept drawing of the "Spanlifter". Likewise, Paola Cheung of Piaggio Aerospace provided three beautiful photos of the remarkable "Avanti". Photo permissions were also given by Duane Swing of Velocity Aircraft, which manufactures superbly-crafted canard general-aviation airplanes.

Individuals who have helped the author include Joseph Biviano, who gave open permission to use any illustrations from the well-known reference book "Fluid-Dynamic Drag" by Sighard Hoerner. All serious aerodynamicists have copies of this on their shelves. Also, the author's friend, Prof. Michael Selig, gave permission to use any of the data or graphs resulting from his extensive research on airfoil design. It should be mentioned that Prof. Selig designed the airfoil for the ornithopter described in Chapter 1.

The author owes a great debt to John H. Leinhard, who made the author aware of the fact that Wikipedia is an enormous resource of freely-available photos. This book could not have been written, without going broke, without this revelation. Also, Dave Klyde of Systems Technology Inc. generously gave permission to use several figures from their very-enlightening reports on the modern analyses of the historic Wright Flyer.

The author has long been intrigued by the innovative structures of the WW1 Fokker airplanes, and this curiosity was satisfied by visiting John Weatherseed at Brampton Ontario, who is building a remarkably-accurate reproduction of a Fokker D-VII fighter. Also, the author is very grateful to Koloman Mayrhofer and Ron Sands, Jr. who provided photos of their replica wing constructions that clearly show the uniqueness and advanced thinking by the Fokker Company at that time.

Excellent photos of the Hughes H-4 "Hercules" Flying Boat were freely provided by Grace Clabaugh of the Evergreen Aviation & Space Museum in Oregon. Likewise, John Ball and Prof. Victor Fleischer of the University of Akron provided a great in-flight photo of the unique Goodyear "Inflatoplane".

The author's friend and colleague, Steve Wallace, is an expert on fabric structures, and drew the image of the laminate fabric sample in Chapter 8. Steve also read the draft of Chapter 8, and made many valuable suggestions. Further, he arranged for Hannah Cameron, of Cameron Balloons, to provide the excellent images of modern free balloons. The author is very grateful to both Steve and Hannah.

Alexander von Gablenz, Cargo Lifter Management, gave generous permission to use a concept drawing of the proposed CargoLifter heavy-load airship. Also, Xavier Cotton gave permission to use the photo of the 1922 Simplex-Arnoux Racer from www.passionpourlaviation.fr. This aircraft is unique in that it is a powered version of the "flying-plank" configuration.

Hoken Colting of Newmarket Ontario developed a series of successful spherical airships, and he gave a photo of his "baseball" design to Jeremy Harris (the author's friend and research partner). The photo was provided to the author and used in this book. Also, Jerry read and critiqued Chapter 1, which benefitted from his insightful comments. The author is likewise grateful to William McKinney (the author's former graduate student) who spent time finding photos that he had taken of the Colting "Baseball Blimp" during its flights in Atlanta, Georgia.

The author had difficulty finding a good in-flight photo of the de Havilland "Mosquito" fighter/bomber. However, this was resolved by Neil Hutchinson who provided an excellent photo that he took of a restored example of this remarkable airplane. Also, Chad Slattery, a professional photographer, permitted the author to use his superb in-flight photo of the Beech "Starship".

Patrick Zdunich, the author's former graduate student, designed, built and flew the "Extendo-Wing" retracting-wing model, which was part of a DARPA contract for morphing aircraft; and the photos and performance graphs are generously provided by Patrick for this book.

A great source of general-arrangement drawings of historic aircraft is found at aerofred.com, and the owner of this site gave the author full permission to use any of these required for this book.

The author badly needed an illustration of the Zeppelin NT's internal structure, and Juergen Fecher, through his contacts at Zeppelin, was able to obtain an excellent picture that the author has been pleased to include in this book. Also for the NT, operational information of the Goodyear "Wingfoot2" version of this airship was generously provided by Michael Dougherty, chief pilot of Goodyear Airship Operations.

I also want to give a big thanks to my colleagues in the Airships, Dirigibles and Zeppelins Facebook group, who unfailingly helped me whenever I posted questions. In particular, Jens Schenkenberger and Hendrick Stoops identified figures that I wanted to include in the book. They are a very knowledgeable group of enthusiasts.

Finally, I very much appreciate the help and guidance (and patience!) provided by my two editors at Taylor & Francis Group: Jonathan Plant and Kyra Lindholm. Also Nishant Bhagat, their colleague in India, always promptly answered my many questions. Since this is the first full book that I have written, I had a lot to learn about the process of bringing such a publication to completion.

Author

Professor James DeLaurier received his Ph.D. in Aeronautics and Astronautics from Stanford University, during which time he also worked at the NASA Ames Research Center on the Apollo Program in the summers of 1965 and 1966.

After post-doctoral research at the von Karman Institute in Belgium in 1970, he worked at the Sheldahl Corporation to develop modern aerostats as well as designing a wing-shaped balloon that would later be the basis for a hybrid-airship configuration.

In 1974, Prof. DeLaurier established a research team at the University of Toronto Institute for Aerospace Studies and was in charge of the Low-Speed Aerodynamics Laboratory. This group tested wind-tunnel models and scaled flying prototypes for several new types of airships, as well as developing flight-dynamic simulations for aerostats and airships that were used in the certification of new commercial airships in the U.K., U.S. and Germany.

Other projects include the first successful airplane powered by microwave power transmission. This was the initiative of the Communications Research Centre (Canada) and was called SHARP (Stationary High-Altitude Relay Platform).

Also, fundamental research was performed on the challenging topic of flapping-wing flight, which lead to a successful remotely-piloted proof-of-concept ornithopter (1991), as well as two human-carrying ornithopters (2006 and 2010). All three award-winning aircraft are now at the Canadian Aviation and Space Museum.

In 2022, Prof. DeLaurier was inducted into the Canadian Aviation Hall of Fame.

Prof. DeLaurier continues to perform consulting work on aerodynamic and LTA topics, as well as writing books…like this.

1 Introduction

DESIGN FEATURES

It is instructive to consider why some aircraft have certain design features and why others don't. It will be seen that these decisions are not always based on aerodynamic or performance considerations alone, but may also involve marketing and public perception. For example, consider the 1933 Northrop "Gamma" monoplane.

Courtesy of Gerald Balzer

This plane was designed by Jack Northrop, who had a well-deserved reputation for innovative and advanced airplanes. The Gamma was constructed entirely of metal (aluminum) like modern airplanes, using riveted semi-monocoque fabrication.

> **ASIDE**
>
> A monocoque structure uses its skin, or shell, to support all applied loads. An example of this is an eggshell. A semi-monocoque structure has internal bulkheads and stringers to help support the bending loads, and the skin primarily resists the twisting loads.
>
> **END ASIDE**

The Gamma looks like a fairly modern airplane except for the undercarriage, which is fixed and streamlined with rather massive-looking "trousers". This was counter to the design trends of the time, which favored retractable-undercarriage.

Designers had long understood that a retractable undercarriage can considerably reduce an airplane's aerodynamic drag. An example of this is the highly innovative Dayton-Wright Racer of 1922.

The Henry Ford

It is seen to incorporate undercarriage that retracts into the fuselage, as well as a variable-camber wing. This was extremely advanced for that time and the designers deserve a great deal of credit for their effort. However, the aircraft was not a success because too many new things were tried at once, and the development process wasn't long enough to work out the problems step by step. This is a common failing for engineering projects attempting too many simultaneous innovations.

Another very advanced airplane was the Verville-Sperry R3 monoplane racer, from the mid-1920s, where the undercarriage was designed to completely retract into the wing.

NASM

Introduction

Also, observe that the wing had no external bracing (completely cantilevered), which was unusual for that time. This airplane was very aerodynamically "clean", with a streamlined solution for engine cooling (wing-skin radiators) as well as a spinner on the propeller hub.

These features clearly had a favorable effect because this airplane won the prestigious 1924 Pulitzer Trophy Race at 347 km/h (216 mph). In truth, this airplane was at least 10 years ahead of contemporary operational airplanes. So, one may ask why it didn't have a greater immediate influence on aircraft design.

The answer is that, in fact, retractable undercarriage offers no real advantage unless an airplane flies faster than about 480 km/h (300 mph). Below this speed, the extra weight of the retraction mechanism generally more than offsets the performance advantage of less drag. The author was made aware of this in a discussion by Professor Walter G. Vincenti at Stanford University. Clearly, Jack Northrop was also aware of this, which is why his 352 km/h (200 mph) Gamma had fixed undercarriage. Of course, this observation assumes that an attempt was made to streamline the fixed undercarriage in some way. Northrop's solution was the "trousers" mentioned earlier. Other approaches involved streamlining of the support strut with a housing over the wheel itself "wheel pants", or thin spring steel (or carbon fiber) undercarriage legs with skinny wheels.

At the time this chapter was written, the Brazilian CEA-308 airplane, with streamlined fixed undercarriage, achieved a record speed of 360.13 km/h (223.77 mph) with an 80 hp engine.

Regarding the Northrop Gamma, however, it looked "old fashioned" in 1933 compared with the Lockheed 9D-2 Orion, whose performance was no better than the Gamma.

Courtesy of Lockheed Martin

The point, however, is that aircraft design can be driven by marketing considerations as well as engineering ones. All things being equal, the Lockheed design looked more up-to-date. In truth, this sort of consideration is far from confined to aircraft design. Public perception of what is the latest and greatest plays a major role in any product design. The "prime directive" for aircraft, though, is that the aircraft's safety and suitability for its mission must never be compromised by any exercise to improve the appeal of the design.

Another advance in airplane design in the 1930s was a fabrication from sheet metal instead of wood, tubing, and fabric.

Such type of construction dated back to the World War I; but with the evolution of aluminum alloys well suited for aeronautical use and the development of the semi-monocoque method of construction, practical all-metal airplanes were now possible (such as the Northrop Gamma). In the public perception, this is comparable to today's attitudes regarding composite construction.

Introduction

Metal construction has several advantages as compared to wood. First of all, the material has consistent, isotropic, uniform properties. Also, the skill level for fabrication is less because riveting is easily inspected and corrected compared with gluing and nailing. Furthermore, mass production doesn't depend so greatly on a limited natural resource.

Having said that, it should be pointed out that wood was still a viable construction material for aircraft of the 1930s and beyond. A graphic example of this is the de Havilland "Mosquito" of World War II.

Courtesy of Neil Hutchinson

This was an extraordinary fighter/bomber, built almost entirely of wood in a semi-monocoque fashion. Its maximum speed was 669 km/h (415 mph). Also, its wood construction made it nearly invisible to German radars, making it suitable for fast surprise attacks against "pin-point" targets. Its fabrication, though, required a very skilled workforce; and some of its major components were built at specialized woodworking and furniture factories. A noted British structures-and-materials researcher, James Edward Gordon, wrote that "gluing is more a responsibility than a skill", meaning that the builder must feel the responsibility of preparing the wood surfaces carefully (dry and uncontaminated, sanded with fresh abrasives so as to not close the pores that the adhesive has to penetrate into) and applying just the right amount of fresh adhesive (wetting both sides of the surfaces). Finally, the surfaces must be clamped together with just the right amount of pressure (too loose and the surfaces don't fully adhere, too tight and the glue is forced out). So, it is seen why riveting sheet metal is more straightforward and predictable.

However, these same considerations for gluing now apply to composite construction:

1. Surface preparation
2. Adhesive application
3. Pressure while curing

Wood is still a popular material for smaller scale and radio-controlled airplanes, so its engineering suitability will be described by the following column buckling examples.

DIFFERENT MATERIALS IN COLUMN BUCKLING

Consider a pin-ended slender column subject to Euler buckling. So, the buckling load, P, is given by:

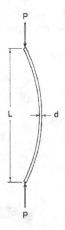

$$P = \pi^2 E I / L^2$$

where E is the modulus of elasticity, I is the area "moment of inertia", and L is column length.

Further, assume that the cross-section is square, so that:

$$I = d^4/12 = A d^2/12$$

where d is the side length of the square and A is the cross-sectional area.

Also, the mass of the column is given by:

$$m = \rho A L$$

where ρ is the material density.

Consider an example where the material is spruce, for which:

$$\rho_{\text{spruce}} = 0.513 \, \text{gm/cm}^3 \, (32.0 \, \text{lb/ft}^3)$$

$$E_{\text{spruce}} = 13.1 \, \text{GPa} = 13.1 \times 10^9 \, \text{N/m}^2 = 13.1 \times 10^5 \, \text{N/cm}^2 \, (1.9 \times 10^6 \, \text{psi})$$

Also, choose that:

$$L = 0.914 \, \text{m} = 91.4 \, \text{cm} \, (3.0 \, \text{ft} = 36.0 \, \text{in})$$

$$d_{\text{spruce}} = 1.27 \, \text{cm} = 0.0127 \, \text{m} \, (0.5 \, \text{in} = 0.0417 \, \text{ft})$$

One thus obtains:

$$m_{\text{spruce}} = 0.513 \times 1.27^2 \times 91.4 = 75.63 \, \text{g} \, (0.167 \, \text{lb})$$

$$I = 1.27^4/12 = 0.2168 \, \text{cm}^4 \, (0.005208 \, \text{in}^4)$$

which gives the following buckling load:

$$P_{\text{spruce}} = \frac{\pi^2 \times 13.1 \times 10^5 \times 0.2168}{91.4^2} = 335.5\,\text{N}\,(75.4\,\text{lb})$$

Now assume a column of the same mass and length, only made from aluminum:

$$\rho_{\text{al}} = 2.70\,\text{g/cm}^3\,(168.5\,\text{lb/ft}^3)$$

$$E_{\text{al}} = 73.0\,\text{GPa} = 73.0 \times 10^5\,\text{N/cm}^2\,(10.5 \times 10^6\,\text{psi})$$

One then obtains the following cross-sectional values:

$$d_{\text{al}} = 0.554\,\text{cm}\,(0.2181\,\text{in}); \quad I_{\text{al}} = 0.00785\,\text{cm}^4\,(0.000189\,\text{in}^4)$$

Thus, the buckling load for the aluminum column is:

$$P_{\text{al}} = \frac{\pi^2 \times 73.0 \times 10^5 \times 7.85 \times 10^{-3}}{91.4^2} = 67.70\,\text{N}\,(15.2\,\text{lb})$$

Note that for the same mass and length, the spruce column carries five times as much buckling load as the aluminum column. Therefore, one may conclude that for a frame structure of a given mass consisting of slender elements, the spruce structure would be stronger. Conversely, for a given strength requirement, the spruce structure would be much lighter. This explains why spruce was a very appropriate material for early airplanes, whose loading was such that frame construction was appropriate.

In a biplane wing structure as shown in the figure below, the primary loads on the spars are tension or compression.

Courtesy of Skytamer.com

So, the main mode of failure is buckling from compressive loads. Early monoplanes had wire-braced wings which also gave rise to compressive spar loads.

Both airplanes shown were made primarily from wood. This allowed the fabrication of lightweight frame structures appropriate for the low-powered engines of that time.

Now, consider a column of the same mass and length as before, made from a carbon-fiber epoxy composite:

$$\rho_{comp} = 1.6 \text{ g/cm}^3 \ (99.8 \text{ lb/ft}^3)$$

$$E_{comp} = 217.6 \text{ GPa} = 217.6 \times 10^5 \text{ N/cm}^2 \ (31.3 \times 10^6 \text{ psi})$$

$$d_{comp} = 0.719 \text{ cm} (0.2831 \text{ in}); \quad I_{comp} = 0.0223 \text{ cm}^4 \ (0.000535 \text{ in}^4)$$

This gives the following buckling load for the carbon-fiber epoxy column:

$$P_{comp} = \frac{\pi^2 \times 217.6 \times 10^5 \times 2.23 \times 10^{-2}}{91.4^2} = 573.3 \text{ N} (128.9 \text{ lb})$$

Although this is nearly twice the buckling value for the spruce column, this is not as great a difference as that between the spruce and aluminum columns. The lesson from this is that the loading on certain types of structures would not necessarily justify the expense and fabrication complexity of composite material.

Now, in fairness to the aluminum example, if the cross-section was tubular instead of a solid square, the performance would be much better (as shown in a further example). Also, it is clear that most modern airplanes are made from aluminum, as stated earlier. The appropriateness of this material has to do with loading. Wood is difficult to beat for lightly loaded structures. However, as the loads go up, the required wooden members would have to increase in cross-sectional area and weight until the failure mode is crushing and not buckling. At that point, other materials, in members sized to the loads, might be more appropriate.

Introduction

Upon returning to the spruce column example, note that its compressive stress at buckling is:

$$(\sigma_{compress})_{spruce} = P_{spruce}/A = 335.5/1.613 = 208.0\,\text{N/cm}^2\,(301.7\,\text{psi})$$

This is well below the compressive stress failure of $\sigma_{compress} = 2585.5\,\text{N/cm}^2\,(3750\,\text{psi})$.

Now consider a spruce column designed to fail in Euler buckling at the stress for compressive failure. The dimension $(d_{failure})_{spruce}$ for this is found from:

$$I = \frac{PL^2}{\pi^2 E} = \frac{\sigma d^2 L^2}{\pi^2 E} = \frac{d^4}{12} \rightarrow d = \left[\frac{12\sigma L^2}{\pi^2 E}\right]^{1/2} = (d_{failure})_{spruce} = 4.4774\,\text{cm}\,(1.763\,\text{in})$$

So, the failure load is:

$$P_{failure} = (\sigma_{compress})_{spruce}(d_{failure})^2_{spruce} = 2585.5 \times 4.4774^2 = 51{,}831.8\,\text{N}\,(11{,}652.3\,\text{lb})$$

Consider what an astonishing capacity this is for a piece of wood (spruce) that is easily picked up. In fact, the mass of this column is:

$$m_{spruce} = 0.513 \times 91.4 \times 4.4774^2 = 939.97\,\text{g}\,(2.072\,\text{lb})$$

Now, consider a square-sectioned aluminum column designed to buckle at that load:

$$I = \frac{PL^2}{\pi^2 E} = \frac{d^4}{12} \rightarrow d = \left[\frac{12PL^2}{\pi^2 E}\right]^{1/4} = (d_{buckling})_{al} = \left[\frac{12 \times 51831.8 \times 91.4^2}{\pi^2 \times 73.0 \times 10^5}\right]^{1/4} = 2.914\,\text{cm}\,(1.147\,\text{in})$$

Therefore, the mass of the aluminum column is:

$$m_{al} = 2.70 \times 91.4 \times 2.914^2 = 2095.5\,\text{g}\,(4.620\,\text{lb})$$

It is seen that the square-section aluminum column is 2.23 times heavier than the spruce square-section column at that load.

The ultimate compressive stress of the aluminum is generally taken to be that for the ultimate tensile stress which, for the popular aeronautical alloy of 6061-T6, is $45{,}000\,\text{psi}\,(31{,}026\,\text{N/cm}^2)$. Therefore, the compressive failure load is given by:

$$(P_{compress\,failure})_{al} = (\sigma_{compress})_{al}(d_{buckling})^2_{al} = 263{,}454.8\,\text{N}\,(59{,}227\,\text{lb})$$

This is 5.1 times higher than the buckling load, which shows that the example aluminum column's failure mode is to buckle at the load that would cause compressive failure for the spruce column.

From this exercise, one would conclude that wood is far superior to aluminum for frame structures. However, this doesn't account for the fact that aluminum may be readily formed into different cross-sections other than square, namely tubular or "hat" sections that produce given area moments of inertia, I, at lower mass values. In that case, the aluminum structure may be competitive to the wood structure.

For example, consider an aluminum tube of $1.5\,\text{in}\,(3.81\,\text{cm})$ outer diameter. If the thin-wall assumption is made, then the area moment of inertia is given by:

$$I_{tube} \approx \pi R^3 t$$

where R is the tube radius and t is the wall thickness. In order to match the column buckling load of the spruce example, one obtains that:

$$I = \frac{PL^2}{\pi^2 E} = \pi R^3 t \to t = \frac{PL^2}{\pi^3 E R^3} = \frac{51831.8 \times 91.4^2}{\pi^3 \times 73.0 \times 10^5 \times 1.905^3} = 0.277 \text{ cm} (0.109 \text{ in})$$

This is a relatively thin wall and produces a cross-sectional area approximated by the following equation:

$$A_{\text{tube}} \approx 2\pi R t = 2 \times \pi \times 1.905 \times 0.277 = 3.316 \text{ cm}^2$$

So, the mass of the column is:

$$m_{\text{al tube}} = \rho_{\text{al}} A_{\text{tube}} L \approx 2.70 \times 3.316 \times 91.4 = 818.3 \text{ g} (1.804 \text{ lb})$$

this is now 87% the weight of the spruce example.

The compressive stress of the tube is:

$$\sigma_{\text{tube}} = P/A_{\text{tube}} = 51831.8/3.316 = 15630.8 \text{ N/cm}^2$$

which is half that for compressive failure. The tube diameter could be further increased, with corresponding thinner walls and lighter weight. However, another type of buckling called "shell buckling" could occur in the thin walls, so there is a limit to this.

The choice of the material for a structure also depends on the loading and the cross-sectional space available. If, for example, the column is constrained to the dimensions of the compressed spruce example, but the imposed load is larger than what spruce can take, then the material must simply be changed. Recall that the spruce column with a 4.4774 cm square cross-section could carry a maximum load of 51831.8 N. However, an aluminum column with the same cross-section can carry a maximum load of 288,837 N, which is 5.6 times greater. The weight of the aluminum column is also 5.3 times greater but, as was seen, re-configuring the cross-section can reduce this number. Nonetheless, the higher stresses experienced by high-performance aircraft require materials that can accommodate these stresses, even at a weight penalty. This is why high-performance airplanes generally have higher wing loadings. Using wood for an F-14 fighter would be ridiculous.

The figure below illustrates that when the available volume for a structure is limited, such as within a wing, and the properties of wood do not provide sufficient strength even if the whole space is filled, then another material and structural configuration is required.

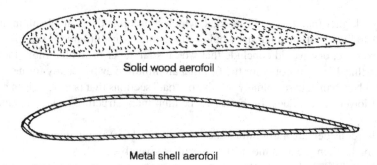

Solid wood aerofoil

Metal shell aerofoil

The lesson from this is that above a certain stress, the appropriate material to sustain the load will change from spruce to aluminum (or composites).

Finally, consider a square cross-section column made from a carbon-fiber/epoxy composite designed to sustain a buckling load of 51831.8 N. One obtains that:

Introduction

$$d = \left[\frac{12PL^2}{\pi^2 E}\right]^{1/4} = (d_{\text{buckling}})_{\text{comp}} = \left[\frac{12 \times 51831.8 \times 91.4^2}{\pi^2 \times 217.6 \times 10^5}\right]^{1/4} = 2.218\,\text{cm}\,(0.873\,\text{in})$$

which gives a mass of:

$$m_{\text{comp}} = \rho_{\text{comp}}(d_{\text{buckling}})^2_{\text{comp}} L = 1.6 \times 2.218^2 \times 91.4 = 719.4\,\text{g}\,(1.59\,\text{lb})$$

Amongst the square cross-section examples, this is the lightest case at 77% the weight of the spruce column and 34% the weight of the aluminum column. The comparison with the aluminum tube is closer, being 88% the weight of the tube. This shows that a composite structure may not be that competitive with a well-designed aluminum structure. This is illustrated by the case in the next section.

Finally, up to this point, only square cross-section spruce and composite columns have been considered. One may well ask if tubular spruce and composite columns would offer the same advantages as the tubular aluminum example. First of all, wood doesn't readily lend itself to tubular fabrication (though closed box sections are possible). Also, there would be concern about internal mold or rot within the tube. However, carbon-fiber/epoxy composite tubes are becoming increasingly available. The fabrication techniques are still more demanding than those for metal tubes, but such composite construction has proven crucial for the success of very lightweight human-powered aircraft. Also, composites in other shapes (such as sheets) are increasingly being incorporated in commercial and military aircraft. As stated before, though, it has to be used in a way that makes sense for the strength and stiffness required and the sought-after weight reduction.

ILLUSTRATIVE EXAMPLES OF WOOD, METAL AND COMPOSITE AIRPLANES

A comparison of the primarily aluminum (some fairings, etc. are composite) Piaggio "Avanti" business airplane with the all-composite Beech "Starship" shows how a well-designed primarily aluminum structure can be more than competitive with an all-composite structure.

Piaggio Avanti

Beech Starship

The Avanti out-performed the Starship in every way, as shown by the following comparison table:

	Piaggio Avanti	Beech Starship
Takeoff weight (lb)	10,810	14,400
Specific range (nautical mile per lb of fuel)	0.82	0.55
Cruise speed	391 knots	315 knots
Maximum lift coefficient	1.38	0.89
Landing distance over 50 ft obstacle	2300 ft	2700 ft

Besides the structural differences, the Avanti is probably more aerodynamically efficient than the Starship. As the chapter on canard airplanes shows, such a configuration is inherently less efficient than a tail-aft design (unless the canard airplane has a stability augmentation system). The Avanti is what is known as a three-surface configuration, with both canard surfaces and an aft tail. Such a design can be very efficient if the incidence angles on the three surfaces are optimized.

Next, an example of an airplane with an inventive wooden wing structure is the 1918 Fokker D.VIII, which was a German fighter airplane from World War I.

Notice that, unlike its contemporaries, there are few struts (only to attach the wing to the fuselage) and no bracing wires. This is because of an innovative structural design based on the concept of using a thick airfoil. The credit for this is often given to Reinhold Platz, the head designer at the Fokker Factory. The thick airfoil allowed for the incorporation of deep spars of sufficient strength and stiffness to resist the bending loads. A photo supplied by Koloman Mayrhofer of Craftlab in Austria, where accurate reproductions of the Fokker D.VIII are being made, shows this internal wing structure.

Courtesy of Koloman Mayrhofer

Introduction

This is then covered, both sides, with a 1.2-mm thick plywood sheet which provides a stiff and strong "torsion box" to resist twisting loads. This approach was in complete contrast to the thin-airfoil structural-design practices of other contemporary airplane builders.

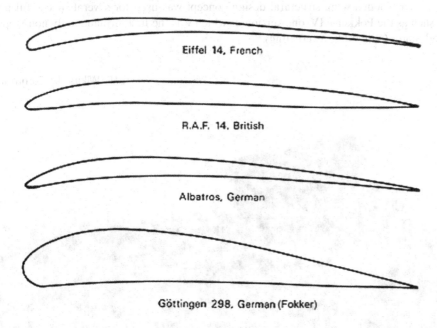

The reason for this had to do with airfoil performance from wind-tunnel tests. Most wind tunnels of that time were only capable of low Reynolds-number flow. In fact, the role of Reynolds number was generally not well understood. Thin, cambered airfoils generally perform much better than thick airfoils at low Reynolds numbers.

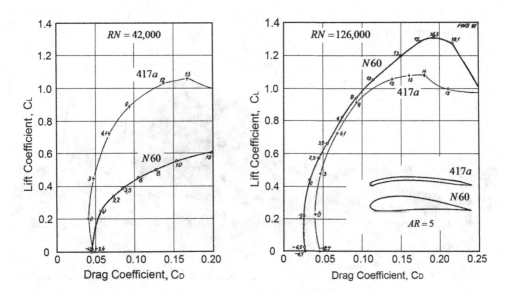

Based on these data, designers used thin cambered airfoils. Such airfoils have a small cross-sectional area and depth within which to incorporate structure. Therefore, external bracing was required to provide sufficient strength and stiffness. However, airplanes at that time were actually operating at much higher Reynolds numbers, for which a thicker airfoil is superior as seen above. It has been stated by John D. Anderson in "The Airplane, A History of Its Technology" (American Institute of Aeronautics and Astronautics, 2002, page 146) that airfoil tests in the large Göttingen wind tunnel revealed this, and that this information was possibly shared with the Fokker Company.

This Fokker wooden-wing structural-design concept was used for several successful post-war aircraft, such as the Fokker F.IV, one version of which was the first airplane to fly non-stop across the United States (with mid-air refueling).

USAF (Wikipedia Commons)

Another Fokker product was the F.VII "Tri-Motor", which was used for commercial transportation as well as exploration.

USAF (Wikipedia Commons)

However, on 31 March 1931, a Fokker Tri-Motor crashed in Kansas while on a routine flight.

Introduction

Kansa State Historical Society

This was newsworthy because a famous college football coach from the University of Notre Dame, Knute Rockne, was killed. A crash investigation concluded that the cause was deteriorated and failed glue joints. This destroyed public confidence in wooden airplanes for air transport and laid the groundwork for enthusiastic acceptance of metal airplanes, such as the Douglas DC-2 shown below.

U.S. Navy (Wikipedia Commons)

Recall the earlier observation that fabrication of a wooden (and composite) structure requires considerable care with the preparation of the components and adhesive, and careful clamping because the adequacy of the resulting joint is virtually impossible to inspect. This is not the case with a riveted metal joint, which requires less skill and can be easily inspected. Again, this gave confidence in metal airplanes (and still does).

Now, no discussion of wooden airplanes would be complete without mentioning the Hughes H-4 "Hercules" (aka, "The Spruce Goose"). This monster had a 98-m (322 ft) wingspan and is the largest wooden airplane ever made. The purpose of this aircraft was to provide trans-Atlantic transportation during World War II, safe from the privations of submarine warfare. Wood was chosen (birch, but no spruce!) because of the possible wartime shortage of aluminum. In the event, only one airplane was completed after the war, and it made only one flight (on 2 November 1947). This was at a low altitude and covered a distance of approximately one mile (1.6 km). A reporter on board noticed a loud banging noise from the tail area, and this portion of the structure was subsequently reinforced. However, the airplane never flew again, and it is not known if the wooden structure would have held up under extended flight conditions. It may well be that this was simply too large for that type of material.

It is worth noting that this airplane had a semi-monocoque construction, much like the "Mosquito" mentioned earlier. The skin panels were glued-together laminations formed on shaped molds under heat and pressure (the "Duramold Process"). This technique was later used for making wooden speedboats.

Introduction

Wood, metal, and composites are not the only materials from which airplanes have been made. An interesting example is the Goodyear "Inflatoplane" from the early 1960s.

Courtesy of the University of Akron

This was intended to be a compact package that could be delivered to downed pilots trapped behind enemy lines. The pilot then used the engine to power a compressor that would inflate the airplane. Afterwards, the pilot would take off and fly back to base (theoretically). The skin was rubberized fabric, and the shape was provided by hundreds of internal woven "yarns" constraining the inflated fabric into wing, tail and fuselage shapes. The internal pressure was 7 psi (48.3 kPa) which gave adequate structural rigidity and strength for normal operations. The airframe empty weight was very reasonable at 315 lb (143 kg) for a 23-ft (7.01 m) wingspan. However, fear of puncture and deflation kept this idea from becoming popular.

EXAMPLES WING STRUCTURES

During the 1990s, as part of a project for designing a high-altitude remotely piloted microwave-powered airplane (the SHARP program), several example wing structures were built and tested in the wind-tunnel laboratory at the University of Toronto Institute for Aerospace Studies. The construction methods mainly followed those for model airplanes, though some innovations were introduced. Also, the example airfoil (Sokolov) was a relatively thin, highly cambered section suitable for low Reynolds-number operation. Therefore, the internal structural design was crucial for giving adequate stiffness and strength with no external bracing. The examples tested are illustrated below:

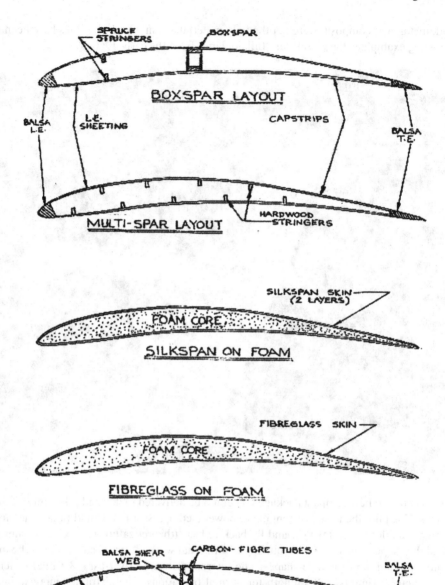

These were tested for different properties, as well as ultimate strength. The first graph shows the weights of the samples.

ASIDE

Note the interesting optical illusion that the first two columns appear divergent, but are actually parallel. These figures are shown in this chapter as they were presented in the original report.

END ASIDE

Introduction

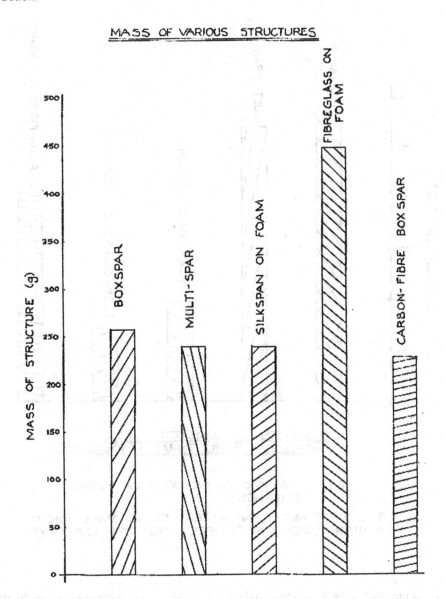

All the wings had similar weights except for the "fiberglass-on-foam" sample, which was nearly twice as heavy as the others. This type of construction was typical for the Rutan homebuilt airplanes. This can give a very molded shape with compound curves, but not a particularly light one.

Observe that EI is a measure of the wing's bending stiffness, and the next graph gives the specific stiffness, EI/m, (bending stiffness per weight) for the sample wings. It is seen that the all-wood "boxspar" and "multi-spar" are best, and the "fiberglass-on-foam" sample is the worse.

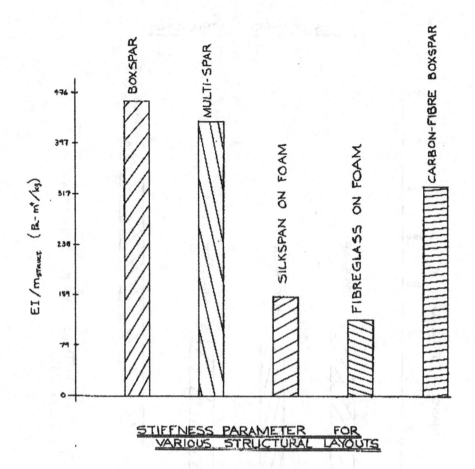

EI TERM IS GOOD INDICATOR OF WING STIFFNESS.

THE STIFFNESS PARAMETER EI/m WILL SHOW WHICH LAYOUT IS STIFFEST FOR THE LEAST WT.

Next, GJ is a measure of the wing's torsional stiffness, and the following graph gives the specific stiffness, GJ/m, (torsional stiffness per weight) for the sample wings. The "silkspan-on-foam" sample is surprisingly best, even better than the "fiberglass-on-foam" wing (which is worse).

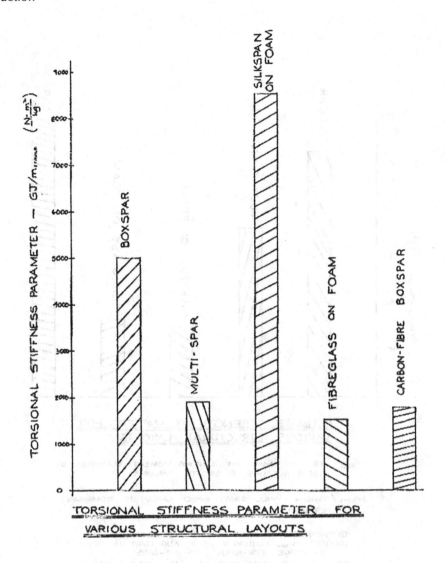

TORSIONAL STIFFNESS PARAMETER FOR VARIOUS STRUCTURAL LAYOUTS

ASIDE

Silkspan is a type of thick paper made from randomly laid natural fibers. It has been popular for model-airplane covering.

END ASIDE

The final graph gives the specific failure values due to bending loads, M_{fail}/m. It is seen that the "carbon-fiber-boxspar" wing is the best, with the all-wood "boxspar" a close second. The "silkspan-on-foam" wing is worst, with the "fiberglass-on-foam" wing being only slightly better.

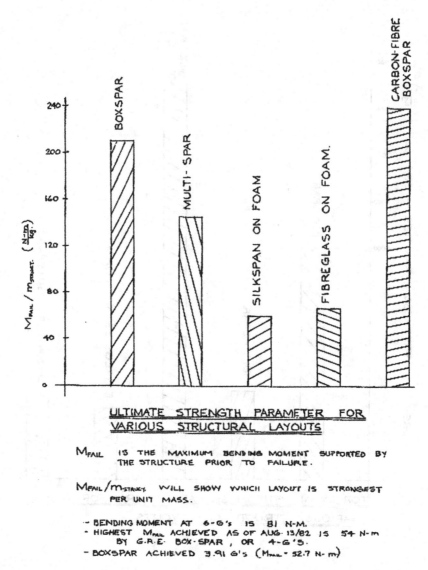

This exercise, of course, doesn't cover all possible types of wing construction. For example, one might conclude that the all-wood "boxspar" design is the best. However, "carbon-fiber-on foam" wasn't tested. Also, there are other structural configurations involving shear webs and torsion boxes that might better utilize the material properties. Nonetheless, this exercise does give insights into the appropriateness of different materials and how they may be used.

Finally, below is a drawing of a wing section used for a successful model ornithopter.

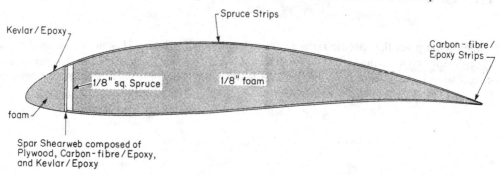

Introduction

The main structural element is the leading-edge spar. This has a plywood shear web, faced with unidirectional carbon-fiber sheet oriented span-wise to resist bending loads. The twisting loads are resisted by a "torsion box" formed by the shear web and the Kevlar skin, over a foam core. The ribs are simply thin sheets of foam capped with wood strips, and the trailing edge is formed by strips of unidirectional carbon fiber. This gave a very light but strong wing, able to resist the cyclic bending loads added to the mean flight loads, which are particular to an ornithopter.

The lessons learned from the model ornithopter were incorporated into the design of a full-scale ornithopter (www.ornithopter.ca or www.ornithopter.net).

The wing construction and materials used were the same as for the model ornithopter, the difference being that the plywood shear web was covered with Kevlar and capped with laminated carbon-fiber strips. As for the rest of the aircraft, the horizontal tail (stabilator) was constructed in much the same way as the wing, the rear fuselage was fabricated from aluminum tubing, the center fuselage (below the wing) is steel tubing, and the front fuselage is titanium tubing. Also, the main undercarriage is carbon fiber and the vertical tail is wood. This is an example of the appropriate material and structural design being selected in accordance with the calculated loads. This structural philosophy is, in fact, the main teaching from this chapter.

2 Aerodynamic Review

AIRFOILS

The figure below shows a typical airfoil section, and some of the standard parameters for describing the airfoil are illustrated:

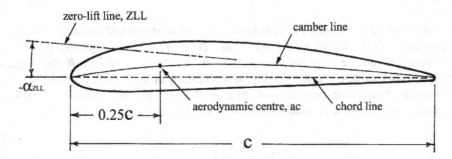

The length, c, is called the chord, and it is the distance from the leading edge to the trailing edge. The chord line is the basic reference line for the airfoil, in that the coordinates defining the airfoil's shape are typically given as percentages of the chord, above and below the chord line, at given distances along the chord line. Also, the angle-of-attack of the airfoil is defined as the angle between the chord line and the free-stream velocity.

Next, note the camber line. It touches the chord line at the leading and trailing-edge points and is midway between the airfoil's upper- and lower-curve. The airfoil shown is said to have a positive camber, in that the camber line is above the chord line all along its length.

Another parameter is the aerodynamic center, ac. This is defined to be the reference point about which the airfoil's pitching moment, M_{ac}, is given (leading-edge up is positive). From thin- theory, M_{ac} is constant with the angle of attack. This is close to being true for actual airfoils with none-separated flow airfoils, which makes this definition very useful for complete airplane analyses (as shown later).

The last parameter illustrated is the zero-lift line, ZLL. When this line is parallel with the free-stream velocity vector, the airfoil produces zero lift. The angle between the chord line and the zero-lift line is defined as the zero-lift angle, α_{ZLL}. This has a negative value when the zero-lift line is above the chord line, as shown in the figure. An interesting finding from subsonic airfoil theory is that the angle of the zero-lift line is nearly midway between the chord line and the tangent to the camber line at the trailing edge.

Pressure and viscous forces act on the airfoil's surface, as illustrated below:

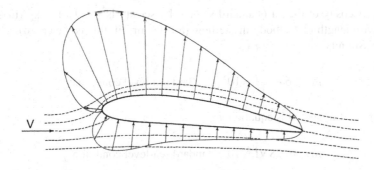

The pressure vectors are normal (perpendicular) to the airfoil's surface, and the viscous stresses are tangent to the surface, as shown below:

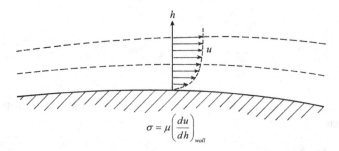

$$\sigma = \mu \left(\frac{du}{dh}\right)_{wall}$$

The local flow velocity diminishes to zero upon approach to the airfoil's surface (which explains why a dirty car can't be cleaned by driving it fast). Therefore, one has a velocity profile, and the slope of this on the surface gives the viscous shear stress, σ. For most aircraft, the height of the velocity profile is very thin compared with the dimensions of the airfoil, and this is referred to as the "boundary layer". The shear stresses integrate along the length of the airfoil to give a skin-friction coefficient, C_{fric}, which provides the airfoil's friction drag:

$$D_{fric} = \frac{\rho V^2}{2} C_{fric} S_{wet}$$

where S_{wet} is the "wetted area" (a notion from naval architecture), which is the surface area (top and bottom of the airfoil) and ρ is the atmospheric density (standard sea-level value is 1.225 kg/m³, 0.002378 slugs/ft³).

There are considered to be two types of boundary layers: laminar and turbulent. For most aircraft, a laminar boundary layer is unstable and difficult to maintain, so the turbulent boundary layer is most often encountered. In this case, the skin-friction coefficient is given by:

$$C_{fric} = \frac{0.455}{(\log_{10} RN)^{2.58}}$$

The RN term is the Reynolds number, which is a non-dimensional aerodynamic scale effect. Physically, this is considered to be the ratio of inertial forces in the flow relative to viscous forces, as may be seen from the equation for RN:

$$RN = \frac{\rho VL}{\mu}$$

where μ is the viscosity of the air (standard sea-level viscosity is 1.789×10^{-5} kg/(m – sec)) and L is the stream-wise length of the body in the flow (for the airfoil, this is the chord). Therefore, one obtains that for SI units:

$$RN = 6.85 \, VL \times 10^4 \text{ (standard sea-level conditions)}$$

For imperial units (ft-lb-sec), this equation is:

$$RN = 6.28 \, VL \times 10^3 \text{ (standard sea-level conditions)}$$

Aerodynamic Review

Now, a very useful concept is the center-of-pressure, *cp*. This is considered to be the balance point of all the integrated surface pressures (including viscous shear), as illustrated in the figure below:

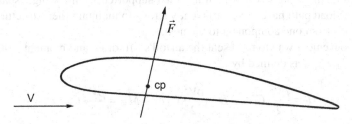

It is important to realize that this is only a mathematical definition and not an actual concentrated surface force. For example, consider the center-of-pressure for a body of revolution shown below. In this case, the *cp* is actually off the body's surface.

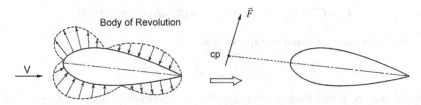

ASIDE

Observe that the body-of-revolution has a very strong upsetting moment. This has important consequences for the stability effect of airplane fuselages, as shown later, as well as for the stability of airships. This is easily demonstrated by dropping an elongated air-filled balloon with its long axis in the downward direction. The balloon will turn sideways within a drop distance of its own length.

END ASIDE

In the early days of experimental aerodynamics, results were often presented in terms of *cp* and force vectors. However, it was eventually found most convenient to use a different representation, as shown below:

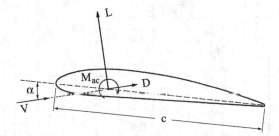

This is the familiar "wind-axis" system of lift, L, drag, D, and moment about the aerodynamic center, M_{ac}. Also, note that lift is perpendicular to the free-stream velocity, and drag is parallel to this. The fact that these are oriented in this way to the "wind" gives the system its name.

It bears repeating, especially for structural considerations, that lift, drag and moment are only mathematical concepts. The actual loading on an aerodynamic body consists of distributed pressure and viscous shear forces. On fabric-covered wings, the fabric on the upper surface often bows out in cruising equilibrium flight (as if the wings were slightly pressurized). For a traditional fabric/rib/spar structure, these aerodynamic pressures, and shear stress on the fabric integrate to running loads transmitted to the wing's ribs, which are then supported by the wing's spars. All of these components of the load path have to be sufficiently strong to maintain their structural integrity and transmit their loads from one component to the next.

Now, a very convenient way to represent the airfoil's lift, drag, and moment values is with their coefficients, C_l, C_d, C_{mac}, as defined by:

$$L = \frac{\rho V^2}{2} C_l c, \quad D = \frac{\rho V^2}{2} C_d c, \quad M_{ac} = \frac{\rho V^2}{2} C_{mac} c^2$$

Observe that for this "two-dimensional" airfoil, lower-case subscripts are used for the coefficients. Also, for attached flow and the relatively low angles-of-attack associated with this, the lift coefficient may be expressed as:

$$C_l = C_{l0} + C_{l\alpha} \alpha, \quad \text{where } C_{l\alpha} \equiv dC_l/d\alpha \approx 2\pi \quad (1/\text{rad})$$

Also, the drag coefficient may often be expressed as:

$$C_d = (C_d)_{C_l=0} + A C_l + B C_l^2$$

where A and B may be found from the curve-fitting of experimental data. The figure below, from "Summary of Low-Speed Airfoil Data, Vol. 3", C. A. Lyon, et al., SoarTech Publications, 1997 (courtesy of Professor Michael Selig of the University of Illinois), shows results for a popular example airfoil tested at different Reynolds numbers,

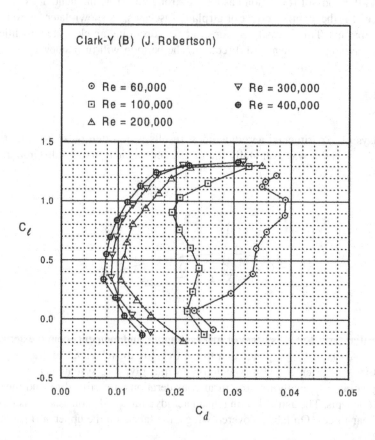

Aerodynamic Review

It is seen that for the higher RN values, a polynomial representation is reasonable. However, that is not so for low RN values. In this case, for studying the characteristics of aircraft flying in a low RN regime, the curve fit may be localized to a segment of the experimental data. Therefore, it has to be understood that the validity of the "polar" equation above is limited to the C_l range for which the curve fit was performed. This will have a consequence on the range of validity of the subsequent equations that depend on this, such as the drag polar for the entire aircraft.

Regarding the coefficient of pitching moment C_{mac}, this may be considered to be approximately constant for attached flow. If experimental values are not available, a method for estimating this will be presented further in this chapter.

Another axis system for representing the integrated forces on the airfoil is shown below:

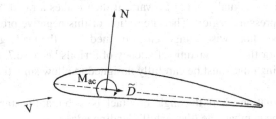

This is known as the "body-fixed" axis system, consisting of a normal force N, perpendicular to the chord line, and an axial drag $\tilde{D}$, parallel to the chord line. The moment about the aerodynamic center M_{ac}, remains the same as for the wind-axes system. It can be seen for small angles and an efficient airfoil (high values of L/D) that $N \approx L$.

The axial drag force, $\tilde{D}$, is composed of a component from friction drag and the axial camber force, combined in the term D_a. Also, for an airfoil with a properly rounded leading-edge, there is a forward force called "leading-edge suction", T_s, as illustrated below. As will be seen, this has important consequences for the airfoil's overall efficiency.

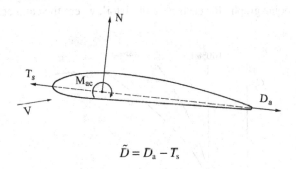

$$\tilde{D} = D_a - T_s$$

ASIDE

In the construction of wings, considerable care must be given to the accurate fabrication of the leading-edge geometry.

END ASIDE

Leading-edge suction is caused by the low-pressure region created by the airflow around the leading edge of the airfoil, as illustrated:

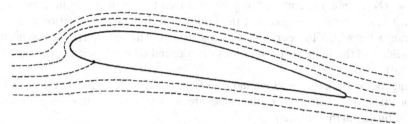

In this case, one sees that the stagnation point, where the streamline velocity goes to zero, is on the forward underside of the airfoil. It is interesting to see that ahead of the stagnation point, near the airfoil's surface, the flow is actually going forward. It then makes a rapid "U-turn" at the leading edge, setting up a low-pressure region. The integration of this negative pressure then gives a forward-acting force, whose chord-wise component is defined to be the leading-edge suction term, T_s. This is very important for the aerodynamic efficiency of airfoils because T_s opposes the axial drag term, D_a; thus, the leading edge must be carefully designed to allow smooth flow at all operational angles of attack.

Now, consider the situation where the angle-of-attack is such that the stagnation point is exactly at the leading edge of the camber line (the airfoil's leading edge):

The airfoil, in this case, is at the "ideal angle-of-attack" because this is the situation where the airfoil's drag is usually at a minimum value. Observe that the leading-edge suction is zero for this one specific angle and is thus irrelevant in the net-drag calculation. This leads to an interesting comparison between this "thick" airfoil and a very thin cambered airfoil, which typically has very little capability for leading-edge suction because of the sharpness of its leading edge. The plots below show the typical drag polar graphs for each type of airfoil, where the camber lines are considered to be identical:

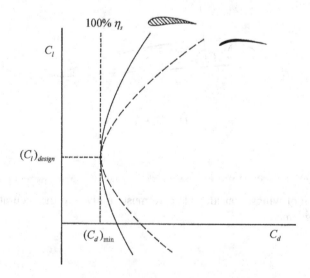

Aerodynamic Review

It is seen that at the ideal angle-of-attack, the thin airfoil can have nearly the same minimum drag as the thick airfoil. This condition is called the "design point" for both airfoils, and the purpose of the camber is to provide the required lift coefficient, $(C_l)_{design}$, at this minimum-drag condition. However, airfoils are also required to operate at "off-design" conditions throughout an airplane's flight regime, and it is seen that the thicker airfoil out-performs the thinner one. This is because, relative to the thin airfoil, it can draw upon its higher leading-edge suction efficiency η_s, to reduce the drag increase at lift coefficients away from the design value, $(C_l)_{design}$.

It is interesting to see, from the example above, that if η_s is 100% no camber would be required for efficient airfoil operation. This is not the case for any actual airfoil, where the highest values of η_s are typically around 90% (depending on RN). However, historically, a non-cambered airfoil was used by the Boeing Company for the B-17 bomber, and a large transport aircraft called the "Stratoliner", shown below from "Paul Matt Scale Airplane Drawings, Vol. 1" (aviation-heritage.com).

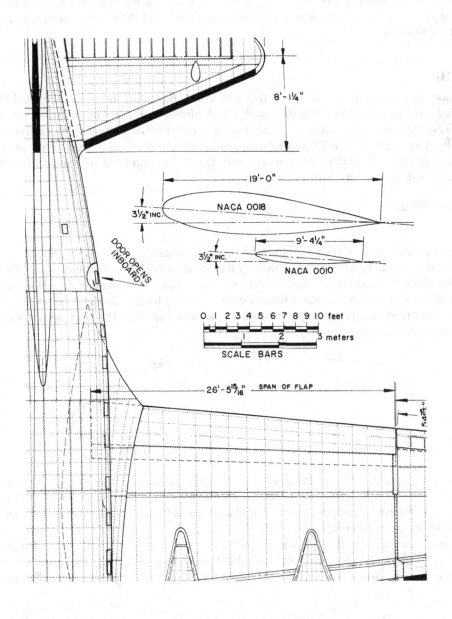

The bomber was successfully used in World War II, but it should be noted that its bomb load was considerably less than that for the similarly sized British "Lancaster" bomber that incorporated cambered airfoils.

Again, what has been presented is the information that airfoils operate at less than 100% of their theoretical leading-edge suction, and camber is incorporated in order to achieve the minimum drag at the design lift coefficient (the ideal angle-of-attack). As seen above, this fact is particularly important for thin airfoils. Also, this explanation is for airfoils with attached flow, not those airfoils that operate in separated-flow conditions.

It is also interesting that at *very low* Reynolds numbers, thin cambered airfoils can out-perform thicker airfoils. This is because the boundary layer at low RN's tends to be laminar, which is more susceptible to flow separation in adverse pressure gradients, such as encountered on the aft upper surface of the "thick" airfoil. However, thin airfoils with their relatively sharp leading edges produce a phenomenon called "leading-edge separation bubble", which is a localized flow separation on the forward upper surface of the airfoil. This starts as laminar flow, then transitions to turbulent flow and reattaches to the upper surface as a turbulent boundary layer. A turbulent boundary layer is much more resistant to flow separation, and the overall performance of the airfoil is improved.

ASIDE

Historically, very early airplanes used thin cambered airfoils. This is because the small low-speed wind tunnels of the time showed that such airfoils were superior to thicker airfoils. However, this was very misleading, and it wasn't until faster and larger wind tunnels were built, which could achieve RN's closer to that of the full-sized aircraft, that the superiority of thicker airfoils at those RN's was demonstrated. This had an important consequence for wing structural design, as was discussed in Chapter 1.

END ASIDE

Thin airfoils are also important for supersonic aircraft, though for different reasons. In this case, the wave drag has to be minimized by making the airfoil as thin as is structurally possible. Also, the design lift coefficient is very small, so that any required camber is small or zero. However, such an airfoil is a very poor performer at lower speeds, as during loiter or landing and take-off, where the required lift coefficients are higher. Therefore, camber is introduced by mechanically morphing its shape, as shown below:

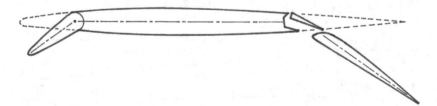

If the camber is adjusted to give the ideal angle-of-attack for that required lift coefficient, then the behavior of this reconfigured airfoil approximates that for a thick subsonic airfoil at that lift coefficient. Note that even high-speed subsonic airplanes benefit from this camber-changing capability, as seen on any airliner coming in for a landing.

In most cases, one would seek wind-tunnel data for a candidate airfoil. Also, if that is not available, sophisticated programs, such as XFOIL may be used to find the airfoil's behavior. However, there are also relatively simple methods for estimating the aerodynamic characteristics

Aerodynamic Review

of conventional-looking airfoils with fully-attached flow. For example, a semi-graphical method for finding the $C_{m_{ac}}$ and α_{ZLL} of an airfoil will be demonstrated. This was described in a 1932 book called "Applied Wing Theory" by Elliott G. Reid (McGraw-Hill). Certainly as stated, more sophisticated computer-based methods are now available; but this simple procedure has sufficient accuracy for most airfoils, and it also has the merit of illustrating the strong effect of the aft portion of the camber line on determining the moment and zero-lift line orientation.

Consider the NACA 4412 airfoil illustrated below:

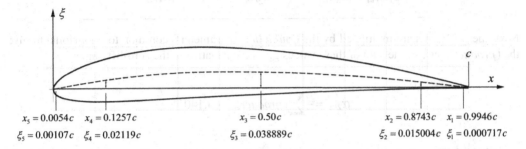

$x_5 = 0.0054c$ $x_4 = 0.1257c$ $x_3 = 0.50c$ $x_2 = 0.8743c$ $x_1 = 0.9946c$
$\xi_5 = 0.00107c$ $\xi_4 = 0.02119c$ $\xi_3 = 0.038889c$ $\xi_2 = 0.015004c$ $\xi_1 = 0.000717c$

The Reid method is to draw the airfoil to a large scale, mark the x_i locations and measure the ξ_i heights of the camber line from the chord. However, because this airfoil has a camber line defined by the following two equations (which is common to all NACA four-digit airfoils), the ξ_i values may be calculated,

$$\xi = m\frac{x}{p^2}\left(2p - \frac{x}{c}\right), \quad 0 \le \frac{x}{c} \le p, \quad \xi = m\frac{(c-x)}{(1-p^2)}\left(1 + \frac{x}{c} - 2p\right), \quad p \le \frac{x}{c} \le 1$$

where:
- m = the maximum camber as a decimal fraction of the chord.
- p = the decimal fraction of the location of the maximum camber along the chord.
- c = the chord length.

For the NACA 4412, $m = 0.04$ and $p = 0.40$.

ASIDE

The code for the NACA four-digit airfoils is that the first digit is $m \times 100$ and the second digit is $p \times 10$. The last two digits are the maximum thickness of the airfoil in per cent of the chord.

END ASIDE

For this airfoil, the camber-line equations become:

$$\frac{\xi}{c} = 0.25x\,(0.8 - x), \quad 0 \le \frac{x}{c} \le 0.4, \quad \frac{\xi}{c} = \frac{0.04}{0.36}(1-x)(0.2x), \quad 0.4 \le \frac{x}{c} \le 1.0$$

These give the ξ_i values illustrated above. Observe that the "1" subscripted values are near the trailing edge, and the "5" subscripted values are near the leading edge. This is a consequence of the fact that this methodology is actually a semi-graphical solution of the classical thin-airfoil equations.

Now, the ξ_i ordinates are used to construct the following table:

Index	x_i/c	ξ_i/c	(multiplier)$_i$	(product)$_i$ = (multiplier)$_i \times \xi_i/c$
1	0.9946	0.000717	1252.0	0.8977
2	0.8743	0.015004	109.0	1.6354
3	0.5	0.038889	32.6	1.2678
4	0.1257	0.02119	15.68	0.3323
5	0.0054	0.00107	5.978	0.0064

Next, the ξ_i/c terms are multiplied by the (multiplier)$_i$ parameters (common to all airfoils) to give the (product)$_i$ terms. The sum of these values gives the negative of the zero-lift angle:

$$\alpha_{ZLL} = -\sum_1^5 (product)_i = -4.140°$$

From NACA TR 824 ("Summary of Airfoil Data" by I. H. Abbott, A. E. von Doenhoff and L. S. Stivers), the zero-lift angle from experiment at $RN = 9.0 \times 10^6$ is $(\alpha_{ZLL})_{exp} \approx -3.9°$. This is a difference of 6% from this theoretical methodology.

Observe that the relative magnitudes of the (multipliers)$_i$ parameters show the importance of the camber line's aft geometry (x_1, x_2 and x_3 dominate the results). Also, the comparison with experiment, for this particular airfoil is encouraging.

Next, the value of C_{mac} may be found from the following equations:

$$C_{mac} = -0.027414(\alpha_{mo} - \alpha_{ZLL}), \quad \text{where } \alpha_{mo} = 62.63(\xi_1/c - \xi_2/c) \text{ (degrees)}$$

For this airfoil, $\alpha_{mo} = 62.63(0.000717 - 0.015004) = -0.8948°$, so that:

$$C_{mac} = -0.027414(-0.8948 + 4.140) = -0.089$$

This compares with $(C_{mac})_{exp} \approx -0.09$ from NACA TR 824. Again, the comparison is very good for this airfoil.

Note the importance of the two α terms: for airfoils with a positive camber along their length, such as for this example, both α_{mo} and α_{ZLL} are negative. However, the two angles are given opposite signs in the C_{mac} equation. Because the magnitude of α_{ZLL} is greater than α_{mo}, though, this gives a negative (leading-edge down) C_{mac}, which is typical for most positively cambered airfoils.

For some applications (described later), it is useful to have an airfoil with a positive C_{mac}. Clearly, as shown above, the way to do this is by modifying the aft portion of the camber line. Such a camber line is shown below:

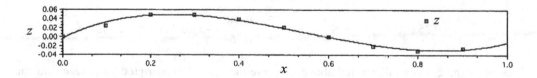

Data from "PSU-1 Reflex Airfoil"

$$z = 0.47619x - 1.2698x^2 + 0.79365x^3$$

Aerodynamic Review

From classical thin-theory (which, as mentioned before, the previous semi-graphical method is also based on), the following values were obtained:

$$\alpha_{ZLL} = 3.414 \text{deg}, \quad C_{mac} = 0.1715$$

However, such reflexed airfoils are generally poor performers, compared with typical non-reflexed airfoils, because the ideal angle-of-attack occurs at a small lift coefficient. If one has a high-speed aircraft, though, for which $(C_l)_{design}$ is small, then such an airfoil offers the opportunity for a tail-less design (as explained in Chapter 4).

A simple method also exists for estimating an airfoil's profile drag if the airfoil's camber line approximates a circular arc (the reflexed airfoil would not be eligible). From the author's publication: "Drag of Wings with Cambered Airfoils and Partial Leading-Edge Suction", Journal of Aircraft, October 1983, profile drag may be expressed as:

$$C_d = (C_d)_{fric} + (1 - \eta_{le}) 2\pi \alpha^2$$

where η_{le} is the leading-edge suction efficiency. And noting that:

$$C_l = C_{l0} + C_{l\alpha} \alpha = C_{l\alpha}(-\alpha_{ZLL} + \alpha) \approx 2\pi(-\alpha_{ZLL} + \alpha)$$

one may solve for α and substitute this into the C_d equation to give:

$$C_d \approx (C_d)_{fric} + (1 - \eta_{le})(2\pi \alpha_{ZLL}^2 + 2\alpha_{ZLL} C_l + C_l^2/(2\pi))$$

Recall that the profile-drag equation has been expressed as:

$$C_d = (C_d)_{C_l=0} + A C_l + B C_l^2$$

Therefore, one has that:

$$(C_d)_{C_l=0} \approx (C_d)_{fric} + (1 - \eta_{le}) 2\pi \alpha_{ZLL}^2, \quad A \approx 2(1 - \eta_{le}) \alpha_{ZLL}, \quad B \approx (1 - \eta_{le})/(2\pi)$$

For airfoils with well-shaped leading edges at high RN, η_{le} may be greater than 0.9. However, at lower RN (such as 10^5), η_{le} may be closer to 0.7. For sharp leading edges, often encountered with airfoils designed for very low RN, η_{le} is zero.

Finally, the friction-drag coefficient $(C_d)_{fric}$, is a function of the skin-friction drag coefficient, C_{fric}, as augmented by the thickness, t, of the airfoil. This relationship is given by the following equation from Sighard Hoerner ("Fluid Dynamic Drag", Page 6-6, Hoerner Fluid Dynamics, 1965):

$$(C_d)_{fric} = 2 C_{fric} [1 + 2(t/c) + 60(t/c)^4]$$

Note, again, that this assumes attached flow. Also, the C_{fric} equation previously presented is for a turbulent boundary layer.

WINGS

Consider the swept straight-tapered wing shown below:

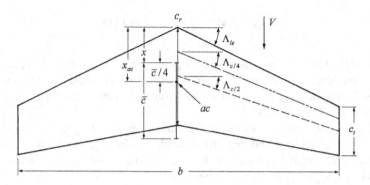

The standard parameters are illustrated, just as was done for the airfoil. These are:

- b = the "flattened-out" span (the dihedral angle is conceptually flattened).
- S = the "flattened-out" projected area.
- c_r = the wing's chord line at the root (the center).
- c_t = the wing's chord line at the tips.

The "Aspect Ratio" is defined as $Aspect\ Ratio \equiv AR = b^2/S$.

The "mean aerodynamic chord", $\bar{c}$, for a straight-tapered wing, as shown, is given by:

$$\bar{c} = \frac{2}{3}c_r\left(\frac{1+\lambda+\lambda^2}{1+\lambda}\right), \quad \text{where } \lambda \equiv \frac{c_t}{c_r} \text{ is the "taper ratio"}$$

The "aerodynamic center" for a straight-tapered wing is given by:

$$x_{ac} = \chi + \frac{\bar{c}}{4}, \quad \text{where } \chi = \left(\frac{1+2\lambda}{12}\right)c_r AR \tan\Lambda_{le}$$

The wing's aerodynamic forces and moment (note upper-case subscripts) are given by:

$$L_w = (C_L)_w\frac{\rho V^2}{2}S, \quad D_w = (C_D)_w\frac{\rho V^2}{2}S, \quad (M_{ac})_w = (C_{Mac})_w\frac{\rho V^2}{2}\bar{c}S$$

where the lift coefficient is assumed to be given by:

$$(C_L)_w = (C_{L0})_w + (C_{L\alpha})_w \alpha_w$$

and the lift-curve slope is given by the Lowry and Polhamus equation (NACA TN 3911, 1957)

$$(C_{L\alpha})_w = \frac{2\pi AR}{2+\left[\frac{AR^2}{\kappa^2}\left(1+\tan^2\Lambda_{c/2}\right)+4\right]^{1/2}} \quad (1/\text{rad})$$

where,

$$\kappa \equiv \frac{\text{the airfoil's actual lift-curve slope}}{2\pi} = \frac{(C_{l\alpha})_{\text{airfoil}}}{2\pi}$$

$\Lambda_{1/2} \equiv$ *the sweepback of the half-chord line (see figure)*

Aerodynamic Review

This equation is well-suited for a wide range of aspect ratios (infinity down to ≈ 3).

The wing's angle of attack, α_w, is the angle of the chord line at the root relative to the free-stream velocity vector.

The term, $(C_{L0})_w$, is the wing's lift coefficient when $\alpha_w = 0$, and this may be found from:

$$(C_{L0})_w = (C_{L\alpha})_w \left[-(\alpha_{ZLL})_r - J\tau \right]$$

Where $(\alpha_{ZLL})_r$ is the zero-lift line angle for the root airfoil and τ is the aerodynamic linear spanwise twist angle at the wing-tip chords relative to the root chord (τ is positive for wash-in, which means that the tip chord line is at a higher angle of attack than the root chord line) and J is given by the figure below, from NACA TR 572 by R. F. Anderson:

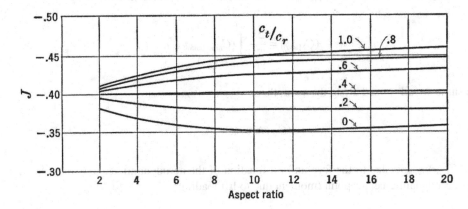

Next, the wing's drag is given by:

$$(C_D)_w = [(C_D)_{C_L=0}]_w + A_w (C_L)_w + B_w (C_L)_w^2 + \frac{(C_L)_w^2 (1+\delta)}{\pi\, AR}$$

where A_w and B_w are the profile-drag curve-fit terms for the airfoil. If the airfoil varies along the span (in both RN and shape), then these will have to be approximated by choosing average values.

Observe that $[(C_D)_{C_L=0}]_w$ is the wing's drag coefficient when its lift is equal to zero.

The last term in the equation is the "Induced Drag", which is due to the finite aspect ratio of the wing. The δ term is the Glauert correction factor for non-elliptical wings. As seen from the figure below (adapted from Sighard Hoerner, "Fluid-Dynamic Drag", Self Published, 1965, page 7–5), it is a function of AR and λ.

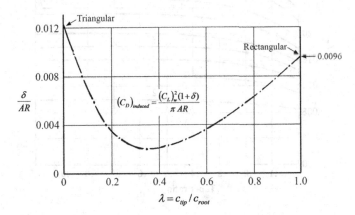

Observe that if $\lambda \approx 0.4$, the straight-tapered wing performs nearly as well as the ideal elliptical wing.

The remaining terms come from the airfoil's characteristics, as described earlier. It should be noted that this equation is only rigorously true for a rectangular un-twisted wing. However, it is a very reasonable approximation for the planform shapes and twists normally encountered in aircraft designs.

The next term to be estimated is the pitching-moment coefficient about the aerodynamic center, $(C_{Mac})_w$. This is given by:

$$(C_{Mac})_w = (C_{Mac})_0 + (C_{Mac})_{loading}$$

The $(C_{Mac})_0$ term is due to the $(C_{mac})_{airfoil}$ of the airfoils incorporated into the wing:

$$(C_{Mac})_0 = \frac{2b}{S^2} \int_0^{b/2} (C_{mac})_{airfoil} \, c^2 \, dy$$

If the same airfoil is used along the span, then:

$$(C_{Mac})_0 = (C_{mac})_{airfoil}$$

Next, $(C_{Mac})_{loading}$ is due to the fore-and-aft location of the aerodynamic center line relative to the wing's aerodynamic-center point (moment due to lift loading):

$$(C_{Mac})_{loading} = -\frac{G\tau}{57.3}(\bar{C}_{l\alpha})_w \, AR \tan \Lambda_{c/4}$$

The factor G is a function of AR and λ, as shown in the graph below from NACA TR 572. Also, note that τ is the wing tip's aerodynamic linear twist angle in *degrees* and $(\bar{C}_{l\alpha})_w$ is the average of the airfoils' lift-curve slopes, $C_{l\alpha}(y)$, along the wing (usually $\approx 2\pi$/rad).

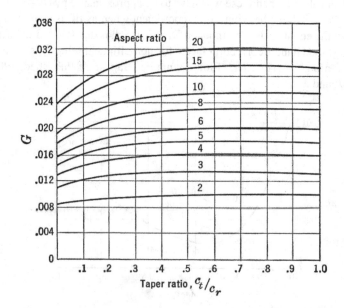

Aerodynamic Review

As an example, this methodology is applied to the wing illustrated below:

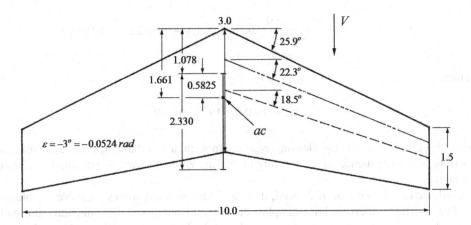

This is swept, straight-tapered and has a linear symmetrical twist along both semi-spans. Also, the airfoil is symmetric $((\alpha_{ZLL})_{airfoil} = 0, (C_{m_{ac}})_{airfoil} = 0)$. The geometric properties are:

$$b=10.0, \quad S=22.5, \quad AR=4.444, \quad \lambda=0.5, \quad \tau=-3° \text{ (washout)}, \quad (\bar{C}_{l\alpha})_w = 2\pi$$

From the equations presented earlier, it is obtained that:

$$\bar{c} = \frac{2 \times 3.0}{3}\left(\frac{1+0.5+0.25}{1+0.5}\right) = 2.33, \quad \chi = \left(\frac{1+2 \times 0.5}{12}\right) \times 3.0 \times 4.444 \times \tan 25.9° = 1.078$$

Therefore:

$$x_{ac} = 1.078 + 2.33/4 = 1.661$$

The lift-curve slope is given by:

$$(C_{L\alpha})_w = \frac{2\pi \times 4.444}{2+[(4.444)^2 \times (1+\tan^2 18.5°)+4]^{1/2}} = 3.935/\text{rad}$$

And the zero-angle lift coefficient is found from:

$$(C_{L_0})_w = 3.935[0 - (-0.41) \times (-0.0524)] = -0.0845$$

Thus, the lift-coefficient equation becomes:

$$(C_L)_w = -0.0845 + 3.935\alpha_w \quad (\alpha_w \text{ in rad})$$

$$(C_L)_w = -0.0845 + 0.0687\alpha_w \quad (\alpha_w \text{ in deg.})$$

Notice that the lift is negative at $\alpha_w = 0$ because of the washout twist, τ.

As the airfoils are symmetric, $(C_{Mac})_0 = 0$. However, the lift-loading contribution is:

$$(C_{Mac})_{loading} = -\frac{0.017 \times (-3.0)}{57.3}(6.2832) \times 4.444 \times \tan 22.3° = 0.01019$$

Therefore:

$$(C_{Mac})_w = 0 + 0.01019 = 0.01019$$

It is seen that this gives a positive (leading-edge up) moment about the wing's aerodynamic center. This will have a consequence on the stability of "flying wing" designs, as explained in Chapter 4 in this book.

It should be mentioned that, if desired, there are other methodologies available to obtain airfoil and finite-wing properties. For example, one may seek out airfoil experimental data, such as that posted on the internet by Prof. Michael Selig of the University of Illinois (the polar graphs of the Clark-Y-PT presented earlier come from that source). Also, an excellent program for calculating airfoil properties is "XFOIL" provided by Prof. Mark Drela of MIT, also available on the internet.

For finite wings, numerical methods are well described by Prof. John Anderson (University of Maryland) in Chapter 5 of "Fundamentals of Aerodynamics" (McGraw-Hill, 1984 and later editions). These include the popular vortex-lattice method. However, if the wing being analyzed does not differ much from the linearly-swept, linearly-twisted and straight-tapered type described here, then the previously described methodologies do a very reasonable job of characterizing the wing's aerodynamic properties. Otherwise, a good approximate methodology for swept and twisted wings with multiple tapers and varying airfoil characteristics is presented in the Chapter 4.

BODIES

The dominant body on most airplanes (except for flying wing designs) is the fuselage, and two representative shapes are shown below:

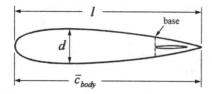
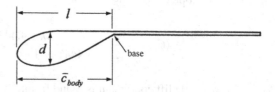

The lift of the body is dependent on several factors, such as cross-sectional shape, the locations of the attached wing and tail, etc. However, a very approximate equation (for a round or oval cross-section) is:

$$(C_L)_{body} \approx [2(k_2 - k_1) S_{base}/S_b] \alpha_b \equiv (C_{L\alpha})_{body} \alpha_b$$

where S_{base} is the body's cross-sectional area at the point where the leading edge of the stabilizer intersects the body. Also, S_b is the body's reference area and, for airplanes, it is usually chosen as the body's maximum cross-sectional area.

Aerodynamic Review

As an aside, though, for *airships* this is usually chosen to be the 2/3 power of the molded volume:

$$S_b = (Vol_{body})^{2/3}$$

Also, note that α_b is the angle of attack of the body's reference line (for a body of revolution, this would be the center line). It is assumed (or the reference line is so chosen) that the body's lift is zero when its angle of attack is zero.

The terms k_1 and k_2 are referred to as the "apparent-mass coefficients" and the expression $(k_2 - k_1)$ may be found from the following graph where l is the total body length and d is the maximum diameter of a circular cross-section body.

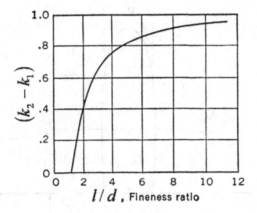

Otherwise, the equivalent diameter for other shapes would be:

$$d_{equiv} = (4 S_{max}/\pi)^{1/2}, \quad \text{where } S_{max} \text{ is the body's maximum cross-sectional area}$$

It is important to note that $(C_{L\alpha})_{body}$ is also dependent on the body's cross-sectional shape. According to S. F. Hoerner and H. V. Borst ("Fluid Dynamic Lift", 1975, chapter 19, figures 10 and 11), if the shape is rectangular instead of oval, then the calculated $(C_L)_{body}$ should be at least doubled. Therefore, this result is definitely an approximate value. Fortunately, it is usually a very small value compared with the wing lift. Incidentally, observe that for the "pod and boom" fuselage illustrated above, the base area is essentially zero. Hence, the body lift for this configuration is also essentially zero.

The pitching-moment coefficient of the body may be estimated from:

$$(C_M)_{body} = (C_{M\alpha})_{body} \alpha_b \approx 2(k_2 - k_1) Vol_{body}/(S_b \bar{c}_b) \alpha_b$$

Again, this is a very approximate equation. However, this is generally a more important term, in the context of analyzing the whole aircraft, than is the body lift. Therefore, it is worth presenting some alternative methods. Also, it is important to observe that the body's pitching moment, for all of these methods, is assumed to be a pure couple. That is, the moment doesn't have a specific point of application.

A semi-empirical method was presented by Gilruth and White, in NACA TR-711, and developed further by Perkins and Hage (p. 229). The equation is presented as:

$$(C_{M\alpha})_{body} = \frac{180}{\pi} \times \frac{K_b w_b^2 l_b}{S_b \bar{c}_b}, \text{ per radian}$$

where w_b and l_b are the maximum width and length of the body, respectively, and K_b is given by:

$$K_b = 0.2931\zeta^3 - 0.1025\zeta^2 + 0.0339\zeta + 0.0012$$

$\zeta \equiv$ (length from the body nose to the 0.25% point of the root chord)/(total body length)

Another, more complex, method is known as the Multhopp fuselage model (described in **Appendix A**). The numerical implementation of this is presented in the USAF DATCOM, Section 4.2.2.1, (February 1972 revision), in which the fuselage of a conventional airplane configuration is divided into 14 segments as shown in the figure below:

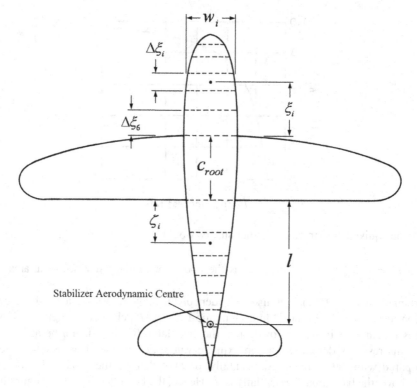

The corresponding equation for calculating $(C_{M\alpha})_{body}$ is:

$$(C_{M\alpha})_{body} = \frac{\pi}{2 S_b \bar{c}_b} \int_0^{l_b} w^2 \left(\frac{\partial \varepsilon_u}{\partial \alpha} + 1 \right) dx$$

$$\approx \frac{\pi}{2 S_b \bar{c}_b} \sum_1^5 w_i^2 \left[\left(\frac{\partial \varepsilon_u}{\partial \alpha} \right)_i + 1 \right] \Delta \xi_i + \frac{\pi}{2 S_b \bar{c}_b} w_6^2 \left[\left(\frac{\partial \bar{\varepsilon}_u}{\partial \alpha} \right)_6 + 1 \right] \Delta \xi_6$$

$$+ \frac{\pi}{2 S_b \bar{c}_b} \sum_7^{14} w_i^2 \left[\left(\frac{\partial \varepsilon_u}{\partial \alpha} \right)_i + 1 \right] \Delta \xi_i$$

l_b is the total length of the body, w_i are the side-to-side widths of the body segments (taken at the middle of the segment for segments 1 through 5), and $\Delta \xi_i$ are the incremental axial lengths of the body segments. Also, ε_u is the up-wash at the middle of each body segment, and its variation with angle of attack is $\partial \varepsilon_u / \partial \alpha$.

Aerodynamic Review

For the first *five* segments, the up-wash rate is given by the plot from Perkins and Hage's figure 5–15:

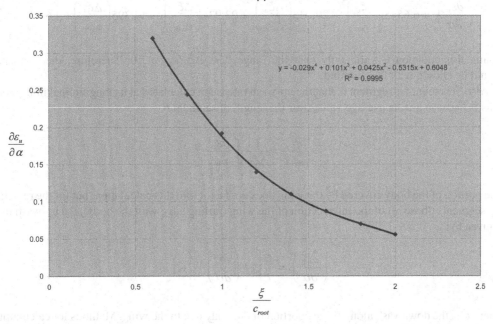

which is curve-fitted to give:

$$\left(\frac{\partial \varepsilon_u}{\partial \alpha}\right)_i = -0.029\left(\frac{\xi_i}{c_{root}}\right)^4 + 0.101\left(\frac{\xi_i}{c_{root}}\right)^3 + 0.0425\left(\frac{\xi_i}{c_{root}}\right)^2 - 0.5315\left(\frac{\xi_i}{c_{root}}\right) + 0.6048$$

This equation is valid only in the range shown $0.6 \leq \xi/c_{root} \leq 2.0$. Otherwise, the plot may be extrapolated, bearing in mind that it asymptotically converges to zero as $\xi/c_{root} \to \infty$.

For segment 6, the plot of the up-wash rate is shown below:

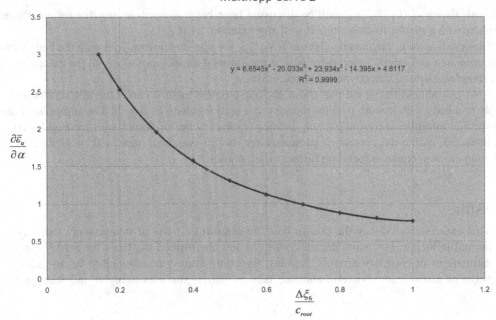

which is curve-fitted to give:

$$\left(\frac{\partial \bar{\varepsilon}_u}{\partial \alpha}\right)_6 = 6.6545\left(\frac{\Delta \xi_6}{c_{root}}\right)^4 - 20.033\left(\frac{\Delta \xi_6}{c_{root}}\right)^3 + 23.934\left(\frac{\Delta \xi_6}{c_{root}}\right)^2 - 14.395\left(\frac{\Delta \xi_6}{c_{root}}\right) + 4.6117$$

Again, this equation is valid only over the range $0.3 \leq \Delta\xi_6/c_{root} \leq 1.0$. Therefore, the segments should be chosen with that in mind.

Also, it is seen, for segment 6, that the up-wash rate is not calculated at its longitudinal mid-point. This is because it is an average value over the length of segment 6.

$$\frac{\partial \bar{\varepsilon}_u}{\partial \alpha} = \frac{1}{\Delta \xi_6} \int_0^{\Delta \varepsilon_6} \frac{\partial \varepsilon_u}{\partial \alpha} d\xi$$

The portion of the body covered by the wing does not factor into this calculation, but for the remaining segments (those aft of the intersection of the wing trailing edge with the body), the up-wash rate is given by:

$$\left(\frac{\partial \varepsilon_u}{\partial \alpha}\right)_i = \frac{\varsigma_i}{l}\left[1 - \left(\frac{\partial \varepsilon}{\partial \alpha}\right)_{tail}\right] - 1$$

where ε is the downwash along the aft portion of the body due to the wing. Methods for calculating this will be presented later.

It is important to note that this method assumes a wing-body lift-curve slope, $(C_{L\alpha})_{wing\text{-}body}$, of 4.50/rad. In order to use this method for other values of $(C_{L\alpha})_{wing\text{-}body}$, the calculated values of $(\partial \varepsilon_u/\partial \alpha + 1)$ for segments 1 through 6 should be multiplied by a correction factor:

$$\left(\frac{\partial \varepsilon_u}{\partial \alpha} + 1\right)_{corrected} = \left(\frac{\partial \varepsilon_u}{\partial \alpha} + 1\right) \times \frac{(C_{L\alpha})_{wing\text{-}body}}{4.50}$$

The calculation of $(C_{L\alpha})_{wing\text{-}body}$ will be discussed later, but it is worth noting that for an efficient airplane with a slender fuselage, this is well approximated by $(C_{L\alpha})_w$.

Observe that these calculated values of $(C_{M\alpha})_{body}$ are non-dimensionalized with the body's own reference area and length. Later, these will be transformed for incorporation into the calculation of the *complete* airplane's aerodynamic characteristics.

Also, one should recognize that these pitching-moment methods are generally understood to apply to a body with smooth oval (or round) cross-sections along its length. The author has found no specific information for the pitching moment characteristics of bodies with rectangular cross-sections. So, unlike the correction parameter for the lift equation, none will be applied to the pitching-moment equation (pending further information!).

ASIDE

This exercise introduces the notion that the aircraft consists of components that have mutual influence. For many decades, this concept has formed the basis for aerodynamic estimations of complete aircraft. Namely, the configuration is assumed to be an assemblage of distinct aerodynamic entities, such as lifting surfaces (wing(s) and stabilizer) held together with a body. The individual aerodynamic characteristics of these components are modified by correction factors to account for mutual interference. For an aircraft design that has clearly identifiable components, this approach has proven to be very useful. However, a "blended" design with no clear distinction between the wing and body would prove problematic for this approach. In this case, a numerical fluid-mechanics analysis is most appropriate. For that matter, any aircraft configuration would benefit from this. The intent of this book, though, (as stated in the preface) is to present straight-forward methodologies that allow ready and reasonably accurate estimations of a design's aerodynamic characteristics.

END ASIDE

Now, the drag of the body is considered. This may be estimated by an equation of the form:

$$(C_D)_{body} \approx (C_{Do})_{body} + (C_{L\alpha})_{body}(\alpha)_b^2$$

The term $(C_{Do})_{body}$ consists of skin-friction drag, form drag, and excrescence drag:

1. The skin-friction drag, $(C_D)_{fric}$, is due to the "wetted" surface area of the body, in the same way as the friction drag of an airfoil was obtained previously. An equation from Hoerner's "Fluid Dynamic Drag" (pages 6–17, 1965 edition) gives the friction drag for a streamlined body of revolution as:

$$(C_D)_{fric} = \frac{(C_{fric})_{body}(S_{wet})_{body}}{S_b}[1 + 1.50(w_b/l_b)^{1.5} + 7.0(w_b/l_b)^3]$$

As before, w_b is the maximum body width (or diameter in this case) and l_b is the body length. For a body with a rectangular cross-section, one should use this equation with the lesser value of either the maximum depth of the side view or the maximum width of the top view. Also, the friction drag coefficient may be calculated from the turbulent boundary-layer equation as given before, with the Reynolds number based on the body length.

2. The form drag is due to an unbalanced fore-and-aft pressure distribution on the body. In most cases, this is due to flow separation on the aft portion of the body. For most well-designed aircraft, this is minimized. However, for a blunt-ended aircraft like a blimp, the drag of the "boat tail" must be accounted for. This is best obtained from wind-tunnel tests.

3. The excrescence drag is caused by miscellaneous protrusions, such as the engine cylinder heads, cooling scoops, screw heads, panel gaps, etc. Again, these drag values are best obtained from wind-tunnel tests, though some estimates may be made as demonstrated later.

If experimental values of $(C_D)_{body}$ cannot be readily obtained, one may refer to published values for similar body shapes. Except for ultra-streamlined shapes, as for the fuselages of competition sailplanes, a typical value is $(C_D)_{body} \approx 0.15$ (with reference to the body's maximum cross-sectional area). Some example values from wind-tunnel tests by Diehl (NACA TR 230) are shown below:

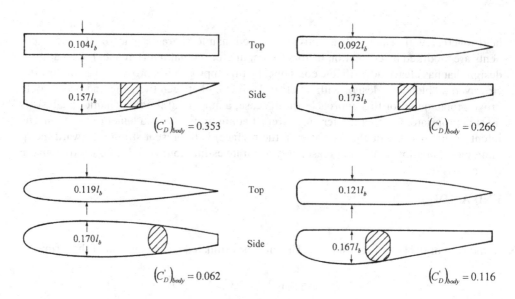

UNDERCARRIAGE

A major contributor to the aircraft's total drag is the extended undercarriage. In general, one may consider this to be the sum of the wheel and its supporting structure. Typical values for wheel drag were obtained from Herrenstein and Bierman (NACA TR 522) and are presented below in the coefficient form. Observe that the reference area for these coefficients is the wheel diameter times the maximum width. Also, this data was obtained at a Reynolds number of approximately 1.6×10^6 (based on wheel diameter). However, these coefficients would be a reasonable approximation for lower RN's. It should also be observed that the illustration below is not drawn to scale.

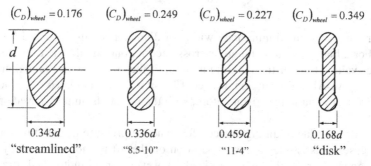

From the same reference, the drag of the complete undercarriage unit (wheel plus support) was measured, and the results for two cases are presented in the figure below:

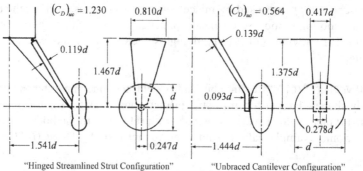

Aerodynamic Review

The Reynolds number is the same, and the reference area for the coefficients is the *single* wheel reference area defined before. Also, even though the left half of the undercarriage is illustrated, the drag coefficient is for the complete unit (both sides). For the first case, which is the "hinged streamlined strut" with the "8.5–10" wheel, one may draw upon the drag coefficient for the isolated wheel to state that the drag coefficient for both wheels is:

$$(C_D)_{\text{both wheels}} = (C_D)_{\text{left wheel}} + (C_D)_{\text{right wheel}} = 0.249 + 0.249 = 0.498$$

When this value is subtracted from the total drag coefficient of the undercarriage, one obtains that the drag coefficient of the supporting structure is 0.732 (in this exercise, the mutual interferences between the wheels and the supporting structure are ignored). Therefore, the supporting structure for the first case contributes 1.47 as much drag as the wheels alone. Upon performing the same exercise for the second case, a supporting structure drag is obtained, which is 0.60 of the wheels alone. Also, the overall drag is less than half that of the first case. This was a great improvement at that time (1935) in undercarriage design for small aircraft, made possible by the ability to form and temper sprung aluminum beams. Today, with carbon-fiber materials, this type of low-drag cantilevered "leg" is ubiquitous for small aircraft with non-retractable undercarriage.

The cross-sectional shape of the second undercarriage leg was not described in the report, but from the photos of the aircraft using such a design, it was probably rectangular with rounded edges. Small remotely-piloted aircraft often simply use a bent spring-steel wire for a leg. In that instance, its drag contribution may be estimated from the well-documented drag of a circular cylinder in crossflow. At Reynolds numbers below 1.6×10^5 (based on wire diameter), a drag coefficient of 1.2 may be chosen (again, based on wire diameter times unit length as the reference area). Above this Reynolds number (called the "critical" Reynolds number), the drag coefficient is approximately 0.3.

As an exercise, consider such an aircraft with two wheels that have a diameter of 68.8 mm and a maximum width of 24.5 mm. The geometry of these wheels is such that one may draw upon the previous data to say that $(C_D)_{\text{wheel}} \approx 0.25$. Also, the supporting structure consists of two wire legs of 4 mm diameter and a total length of 36 mm. The RN is such that the drag coefficients of the legs are sub-critical. Therefore, the total drag coefficient may be estimated from:

$$(C_D)_{\text{uc}} = 2(C_D)_{\text{wheel}} + \frac{S_{\text{legs}}}{S_{\text{wheel}}} (C_D)_{\text{legs}} = 2(C_D)_{\text{wheel}} + \frac{\text{wire dia.} \times \text{total length}}{\text{wheel dia.} \times \text{wheel width}} \times (C_D)_{\text{cylinder}} =$$

$$2 \times 0.25 + \frac{4.0 \times 36.0}{68.8 \times 24.5} \times 1.2 = (C_D)_{\text{uc}} = 0.5 + 0.103 = \underline{\underline{(C_D)_{\text{uc}} = 0.603}}$$

It is seen from this exercise that the undercarriage legs contribute 20% to the total drag. Also, this value is comparable to that for case 2 above.

At the time that the tests in TR 522 were performed (1934), almost all airplanes were "tail draggers", in that the undercarriage consisted of the two-wheeled "main gear" and a small tail wheel (or skid). The airplane thus had a nose-up attitude on the ground, and take-off consisted of accelerating in this configuration until such speed was reached that the tail would lift into a flight attitude. However, many modern airplanes now use a "tricycle" undercarriage, where the main gear is aft of the airplane's mass center, and the nose is supported by a "nose gear". The configuration of this typically consists of a single (or double) wheel and a single supporting strut. Again, the estimation of the nose gear's contribution to the total undercarriage drag coefficient may follow the exercise above, with the addition of the nose wheel's drag coefficient and that of the supporting structure.

It is important to note, though, that all drag coefficients for the undercarriage assembly should be converted to the reference area of a single main wheel, $S_{\text{main wheel}}$.

$$(C_D)_{\text{nose gear}} = (C_D)_{\text{nose wheel}} \times \frac{S_{\text{nose wheel}}}{S_{\text{main wheel}}} + (C_D)_{\text{nose leg}} \times \frac{S_{\text{nose leg}}}{S_{\text{main wheel}}}$$

Later, the contribution of the *complete* undercarriage to the aircraft's total drag coefficient will require converting this to a value based on the aircraft's reference area (usually the wing area). It should also be observed that the undercarriage drag generally produces a nose-down pitching moment about the airplane's mass center. This fact will be incorporated into the expressions for the complete aircraft's pitching-moment coefficient.

WING-BODY COMBINATION

At this point, it is important to recall that the aerodynamic characteristics of the complete aircraft involve components with mutual interferences. In this case, the influence of the body on the wing's efficiency will be considered. The drawing below illustrates how the wing's span-wise lift distribution is modified by the presence of a body:

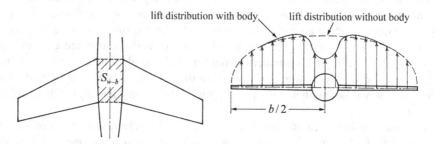

When the wing conceptually passes through the body, there is a deficit in its lift distribution. The amount is a function of the fraction of the wing's planform that is "covered" by the body, S_{w-b}/S_w. From the analysis of airship fin efficiencies, the author has derived an equation that has proven to be very useful:

$$\eta_w = 1 - 1.4(S_{w-b}/S_w) + 0.4(S_{w-b}/S_w)^2$$

The lift of the isolated wing is then multiplied by the "efficiency factor" to obtain its lift in the presence of the body. This may be expressed by the lift-slope coefficient:

$$(C_{L\alpha})_w = (C^*_{L\alpha})_w \eta_w$$

where $(C^*_{L\alpha})_w$ is the value for the isolated wing, as presented earlier.

Other characteristics of the wing are likewise modified, such as drag and moment. For example, the body's effect on the wing's induced drag may be approximately accounted for by using the wing's modified lift for this calculation. As for the friction drag, the wing area covered by the body is clearly subtracted from the wing's wetted area for the purpose of skin-friction calculations. For a typically slender fuselage, these changes would be small.

The change in the wing's pitching moment might be more of an issue if the fuselage isn't particularly slender. The middle of a wing is usually the "sweet spot", where its sectional behavior is closest to the two-dimensional ideal. Therefore, the change in $(C_{Mac})_{\text{wing}}$ may be significant. References

Aerodynamic Review

such as "Aerodynamics of the Airplane" by Schlichting and Truckenbrodt (McGraw Hill, 1979) and the USAF Stability and Control DATCOM address this in detail, treating the wing and body as a whole. For the purposes of this book, however, it is reasonable to modify the wing's pitching-moment coefficient in the same way as the lift-curve slope:

$$(C_{\text{Mac}})_w = (C^*_{\text{Mac}})_w \eta_w$$

It should be noted that the wing's effective aerodynamic center is also slightly shifted by the presence of the body, in that the defined planform is no longer exactly a straight-tapered wing. The portion of the wing covered by the body (as shown in the figure above) is un-swept and un-tapered. However, considering the other approximations in the analysis (including the methods for the body), it is reasonable to assume that the aerodynamic center is unchanged if S_{w-b} is less than 10% of the wing area. This is particularly true if the wing is un-swept.

Although the wing-body interaction has been discussed, this also applies to any lifting surface covered by the body, such as the stabilizer on a tail-aft airplane or the canard wings on a tail-forward design.

COMPLETE-AIRCRAFT AERODYNAMICS (FOR A TAIL-AFT MONOPLANE)

As a demonstration of the conceptual piecemeal aerodynamic assembly of a complete aircraft, estimates of the lift, drag and moment characteristics of a "conventional" tail-aft airplane will be analytically described. Such an airplane configuration is represented by the following drawing:

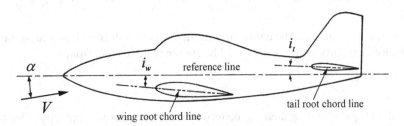

A reference line is shown (sometimes called the "loft line"). This is usually the body's reference line, as described earlier. Relative to this line, the root chords of the wing and stabilizer have built-in incidence angles, i_w and i_s (these incidences are measured relative to the root chords as extended into the middle of the body, with no change to the sweep or taper). As shown in the drawing, the angle between the reference line and the airplane's velocity vector is defined to be the airplane's angle of attack, α.

> **ASIDE**
>
> In the drawing above, a tail configuration called "cruciform" is illustrated. This is common for most airplanes and consists of a horizontal surface (stabilizer) and vertical surface (fin). In light of that, the subscript pertaining to the parameters for the stabilizer could be "s" instead of "t". However, some airplanes use a "V-Tail", as shown. Such a configuration combines the functions of both stabilizer and fin, and may be simply denoted as the "tail". The lift-curve slope of a V-Tail is obtained from:
>
> $$(C^*_{L\alpha})_t = (C^*_{L\alpha})_{\text{flattened-tail}} \cos^2 \Gamma_t, \quad \text{where } \Gamma_t = \text{Tail Dihedral Angle}$$

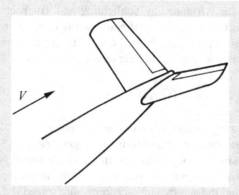

END ASIDE

LIFT

The total lift of the aircraft may be expressed as:

$$L = L_{\text{wing}} + L_{\text{tail}} + L_{\text{body}} + L_{\text{prop}} \rightarrow$$

$$L = \rho V^2 / 2 [(C_L)_w S_w + (C_L)_t S_t + (C_L)_{\text{body}} S_b + (C_L)_{\text{prop}} S_p] \rightarrow$$

$$L = \frac{\rho V^2}{2} S C_L \quad \text{where } C_L = (C_L)_w \frac{S_w}{S} + (C_L)_t \frac{S_t}{S} + (C_L)_{\text{body}} \frac{S_b}{S} + (C_L)_{\text{prop}} \frac{S_p}{S}$$

where S is the reference area for the entire aircraft. In most cases, this is chosen to be the wing area S_w. However, for generality in this case, S will be chosen to be distinct from S_w.

ASIDE

Earlier in this chapter, the wing area was defined for a straight-tapered swept wing. This is a clear definition for an isolated wing. However, for a wing covered by a body, a previous figure showed that the conceptual wing planform differs, in that the covered portion is non-tapered and un-swept. This, plus the swept and tapered uncovered portions, should be considered to be the "actual" planform shape. However, if the sweep and taper are very moderate and the body is sufficiently slender, the geometrical characteristics of the isolated wing may be considered to adequately represent the covered wing. That is, S_w is chosen to be the "flattened" (no dihedral angle) planform area of the isolated wing. Likewise, the previously described calculations for the mean aerodynamic chord and aerodynamic center of the isolated wing are considered to apply to the covered wing.

END ASIDE

Now, from previous work, one has that the wing's lift coefficient is given by:

$$(C_L)_w = [(C_{Lo}^*)_w + (C_{L\alpha}^*)_w \alpha_w] \eta_w$$

As defined before, the * coefficients are those for the isolated wing; and η_w is the wing-body efficiency factor.

The wing's angle of attack, α_w, is given by $\alpha_w = \alpha + i_w$, so the previous equation becomes:

$$(C_L)_w = [(C_{Lo}^*)_w + (C_{L\alpha}^*)_w \alpha + (C_{L\alpha}^*)_w i_w] \eta_w$$

Aerodynamic Review

Likewise, the tail's contribution to the lift is given by:

$$(C_L)_t = [(C_{Lo}^*)_t + (C_{L\alpha}^*)_t \alpha + (C_{L\alpha}^*)_t (i_t - \varepsilon_t)]\eta_t$$

For the purposes of obtaining $(C_{L0}^*)_t$ and $(C_{L\alpha}^*)_t$, the previously described wing-calculation methods are applied. Again, because a portion of the tail is generally covered by the body, a tail-body efficiency factor η_t is calculated, and the isolated-tail coefficients are multiplied by this.

Another parameter that appears in the equation is the downwash ε_t. This is a downward *angular* deflection of the flow caused by the wing and is in direct proportion to the wing's lift coefficient.

Therefore, it may be represented as:

$$\varepsilon_t = (\varepsilon/C_L)_w (C_L)_w$$

The tail lift equation thus becomes:

$$(C_L)_t = [(C_{Lo}^*)_t + (C_{L\alpha}^*)_t i_t]\eta_t - (C_{L\alpha}^*)_t(\varepsilon/C_L)_w[(C_{Lo}^*)_w + (C_{L\alpha}^*)_w i_w]\eta_w \eta_t$$
$$+ (C_{L\alpha}^*)_t \eta_t [1 - (\varepsilon/C_L)_w (C_{L\alpha}^*)_w \eta_w]\alpha$$

ASIDE

The tail usually has an untwisted planform, so that,

$$(C_{Lo}^*)_t = -(C_{L\alpha}^*)_t (\alpha_{ZLL})_t$$

END ASIDE

There are various methods for estimating the downwash term, each with different levels of sophistication. The simplest expression is:

$$\left(\frac{\varepsilon}{C_L}\right)_w = \frac{2}{\pi (AR)_w}$$

A more exact (but empirical) method is offered in the USAF Stability and Control DATCOM, where:

$$\left(\frac{\varepsilon}{C_L}\right)_w = \frac{(\partial \varepsilon / \partial \alpha)_w}{(C_{L\alpha})_w}, \quad \left(\frac{\partial \varepsilon}{\partial \alpha}\right)_w = 4.44[K_A K_\lambda K_H (\cos \Lambda_{c/4})_w^{1/2}]^{1.19}$$

The first "K" term is due to the wing's aspect ratio and is given by:

$$K_A = \frac{1}{(AR)_w} - \frac{1}{1 + (AR)_w^{1.7}}$$

The second "K" term is a function of the wing's taper ratio:

$$K_\lambda = \frac{10 - 3\lambda}{7}$$

The third "K" term is a function of the tail's location relative to the wing:

$$K_H = \frac{1 - |h_H/b_w|}{(2 l_H/b_w)^{1/3}}$$

$l_H \equiv$ the longitudinal distance between the aerodynamic centers of the wing and the tail, measured along a line parallel to the root chord of the wing (the "projected line").
$h_H \equiv$ the vertical distance from the projected line (measured normal to the line) to the tail's aerodynamic center. Observe that its absolute value is used in the equation for K_H.

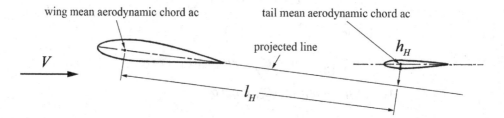

The body's lift contribution has already been presented, so what now remains is the lift from the propeller. Analysis by Herbert Ribner (NACA WR-L-217) showed that when a spinning propeller is pitched or yawed, a side force is produced (normal to the rotational axis) that is proportional to the pitch or yaw angle. In fact, after a very complex derivation was performed, Ribner noted that the resulting force was very close to that produced by an untwisted and uncambered wing whose planform shape is that of the spinning propeller's sideways shadow, as illustrated.

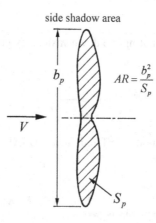

Therefore, one may write that:

$$(C_L)_{prop} = (C_{L\alpha})_{prop} \alpha \quad \text{where} \quad (C_{L\alpha})_{prop} \approx \frac{2\pi}{[1 + 2/(AR)_p]}$$

Aerodynamic Review

Ribner recommends an alternate "effective" aspect ratio of 8.0, which was appropriate for the type of full-scale military propellers of the time (1943).

Observe that the propeller's reference area for this exercise is the previously mentioned side-shadow area.

ASIDE

As a graphic example of the way in which a spinning propeller can act like a lifting surface, R. J. Hoffman, in a self-published booklet for model-airplane builders ("Model Aeronautics Made Painless", 1955), described a rubber-band powered model whose tail was replaced with a propeller. It was stable when the propeller was powered, but unstable when the power was expended (tumbling to the ground). A conceptual sketch of the model is shown below:

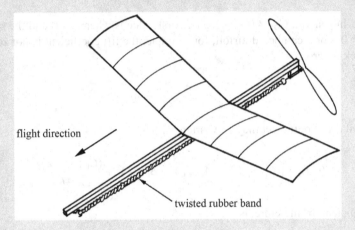

END ASIDE

Now, all the terms required for the total lift-coefficient expression have been obtained. Upon substitution and rearrangement, the equation takes on the following form:

$$C_L = C_{L0} + C_{L\alpha}\alpha$$

This describes the un-stalled lifting behavior of the entire airplane in terms of the reference line angle-of-attack, α.

DRAG

The total drag of the airplane may be expressed as:

$$D = D_{wing} + D_{tail} + D_{body} + D_{uc} \rightarrow$$

$$D = \rho V^2/2[(C_D)_w S_w + (C_D)_t S_t + (C_D)_{body} S_b + (C_D)_{uc} S_{uc}] \rightarrow$$

$$D = \frac{\rho V^2 S}{2} C_D \quad \text{where} \quad C_D = (C_D)_w \frac{S_w}{S} + (C_D)_t \frac{S_t}{S} + (C_D)_{body} \frac{S_b}{S} + (C_D)_{uc} \frac{S_{uc}}{S}$$

Recall that the wing's drag-coefficient is given by:

$$(C_D)_w = [(C_D)_{C_L=0}]_w + A_w (C_L)_w + B_w (C_L)_w^2 + \frac{(C_L)_w^2 (1+\delta_w)}{\pi AR_w}$$

Where A_w and B_w are the profile-drag terms, explained earlier, which provide a curve-fit equation to the portion of the airfoil's C_d vs. C_l experimental data in the range of interest.

Also, the wing's lift coefficient was expressed as:

$$(C_L)_w = [(C_{Lo}^*)_w + (C_{L\alpha}^*)_w \alpha + (C_{L\alpha}^*)_w i_w] \eta_w = (C_{Lo}^*)_w \eta_w + (C_{L\alpha}^*)_w \eta_w \alpha + (C_{L\alpha}^*)_w \eta_w i_w$$

These may be combined to give the wing's drag coefficient in terms of α:

$$(C_D)_w = (C_{D0})_w + (C_{D1})_w \alpha + (C_{D2})_w \alpha^2$$

ASIDE

Observe that, in general, $C_{D0} \ne (C_D)_{C_L=0}$. The first value is when $\alpha = 0$ and the second value is when $C_L = 0$. For a cambered airfoil, for example, the lift coefficient is not zero when its angle of attack is zero.

END ASIDE

Likewise, the tail drag coefficient may be written as:

$$(C_D)_t = [(C_D)_{C_L=0}]_t + A_t (C_L)_t + B_t (C_L)_t^2 + \frac{(C_L)_t^2 (1+\delta_t)}{\pi AR_t}$$

And its lift coefficient, from before, is:

$$(C_L)_t = [(C_{Lo}^*)_t + (C_{L\alpha}^*)_t i_t] \eta_t - (C_{L\alpha}^*)_t (\varepsilon/C_L)_w [(C_{Lo}^*)_w + (C_{L\alpha}^*)_w i_w] \eta_w \eta_t$$
$$+ (C_{L\alpha}^*)_t \eta_t [1 - (\varepsilon/C_L)_w (C_{L\alpha}^*)_w \eta_w] \alpha$$

Therefore, upon substitution into the $(C_D)_t$ equation, one may obtain an expression of the form:

$$(C_D)_t = (C_{D0})_t + (C_{D1})_t \alpha + (C_{D2})_t \alpha^2$$

ASIDE

A more correct representation would take into account the fact that the tail's lift vector gives a drag component in the V direction because L_t is perpendicular to the V_t direction.

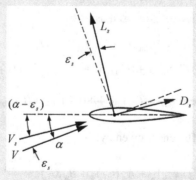

Aerodynamic Review

$$\delta D_t = L_t \varepsilon_t \rightarrow \delta(C_D)_t = (C_L)_t \varepsilon_t$$

From previous equations, one may then obtain an expression of the form:

$$\delta(C_D)_t = (C_{D0})_\delta + (C_{D1})_\delta \alpha + (C_{D2})_\delta \alpha^2$$

This may be directly added to the equation for $(C_D)_t$. However, for lightly loaded tails (which is typical of a properly trimmed and balanced airplane), $\delta(C_D)_t \approx 0$.

END ASIDE

All of these contributions may now be added to give an expression for the total drag of the airplane:

$$C_D = C_{D0} + C_{D1}\alpha + C_{D2}\alpha^2$$

In principle, one could perform the algebra to obtain closed-form solutions for these coefficients, as well as those for the lift equation. However, it is actually easier to program all of these contributions and calculate C_L and C_D while stepping through a sequence of α values. Then, one may curve fit these results to obtain C_{L0}, C_{D0}, C_{D1}, etc. Upon considering the numerous assumptions in this analysis, no accuracy is lost by such an approach.

PITCHING MOMENT (WITH NON-EXTENDED UNDERCARRIAGE)

It is assumed that the α values are sufficiently small (below stall) that the force vectors perpendicular to the reference line is adequately represented by the lift vectors. Also, it is assumed that the vertical distances from the mass center, cm, of the airplane's components are relatively small, so that their drag values do not significantly contribute to the total pitching moment about the mass center. However, the thrust vector, whose magnitude is equal to the drag in level flight, is explicitly included in order to account for cases where this vector has a significant vertical displacement from the mass center:

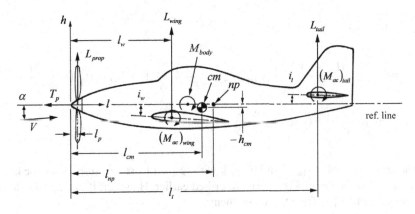

$$M_{cm} = L_{prop}(l_{cm} - l_p) + L_{wing}(l_{cm} - l_w) + L_{tail}(l_{cm} - l_t) + T_p h_{cm}$$
$$+ (M_{ac})_{wing} + (M_{ac})_{tail} + M_{body}$$

> **ASIDE**
>
> Observe that pitching-moment effect of the body's lift, L_{body}, is not included in this equation because that effect is implicitly accounted for by M_{body}.
>
> **END ASIDE**

In coefficient form, the pitching-moment equation becomes:

$$(C_M)_{cm} = (C_L)_{prop}\frac{(l_{cm}-l_p)}{\bar{c}}\frac{S_p}{S} + (C_L)_w\frac{(l_{cm}-l_w)}{\bar{c}}\frac{S_w}{S} + \eta_q(C_L)_t\frac{(l_{cm}-l_t)}{\bar{c}}\frac{S_t}{S} - C_T\frac{h_{cm}}{\bar{c}}\frac{\tilde{S}}{S}$$

$$+ (C_{Mac})_w\frac{S_w\bar{c}_w}{S\bar{c}} + \eta_q(C_{Mac})_t\frac{S_t\bar{c}_t}{S\bar{c}} + (C_M)_{body}\frac{S_b\bar{c}_b}{S\bar{c}} \quad \text{where}$$

$$\eta_q = \frac{\text{dynamic pressure at tail}}{\text{free-stream dynamic pressure}} = \frac{q_t}{q} \quad \text{(this is usually} \approx 1\text{)}$$

In this formulation, the thrust line is drawn to be coincident with the reference line. If this is not the case, then the term h_{cm} may be replaced with $(h_{cm}-h_p)$, where h_p is the vertical distance of the thrust-line in accordance with the body-fixed $l-h$ coordinate system defined in the figure.

Also in the figure, the thrust line is drawn parallel to the reference line. If, instead, it has a certain small angle ($\leq 10°$), then the value of h_p should be measured at the mass-center location.

The moment equation may be used for calculating the airplane's equilibrium trim state, where the moment about the mass center is zero. Also, the airplane's static longitudinal stability may be assessed from the derivative of this equation with respect to α.

First, recall that:

$$(C_L)_w = [(C^*_{L0})_w + (C^*_{L\alpha})_w\alpha + (C^*_{L\alpha})_w i_w]\eta_w$$

$$(C_L)_t = [(C^*_{L0})_t + (C^*_{L\alpha})_t\alpha + (C^*_{L\alpha})_t(i_t - \varepsilon_t)]\eta_t$$

Therefore, one has that:

$$\frac{d(C_M)_{cm}}{d\alpha} = (C_{L\alpha})_{prop}\frac{(l_{cm}-l_p)}{\bar{c}}\frac{S_p}{S} + (C^*_{L\alpha})_w\eta_w\frac{(l_{cm}-l_w)}{\bar{c}}\frac{S_w}{S} + (C^*_{L\alpha})_t\eta_t\eta_q\left(1-\frac{\partial\varepsilon}{\partial\alpha}\right)\frac{(l_{cm}-l_t)}{\bar{c}}\frac{S_t}{S}$$

$$- C_{T\alpha}\frac{h_{cm}}{\bar{c}}\frac{\tilde{S}}{S} + (C_{M\alpha})_{body}\frac{c_b}{\bar{c}}\frac{S_b}{S}$$

Note that the derivatives of $(C_{Mac})_w$ and $(C_{Mac})_t$ with respect to α are zero. Also, the derivative of the downwash, $\partial\varepsilon/\partial\alpha$, is the $(\partial\varepsilon/\partial\alpha)_w$ term described earlier. However, if one has the downwash at the tail expressed as $(\varepsilon/C_L)_w$ from the wing, then:

$$\frac{\partial\varepsilon}{\partial\alpha} = \frac{\partial}{\partial\alpha}\left[\left(\frac{\varepsilon}{C_L}\right)_w(C_L)_w\right] = \left(\frac{\varepsilon}{C_L}\right)_w(C_{L\alpha})_w\eta_w$$

Aerodynamic Review

For "neutral stability", one has $d(C_M)_{cm}/d\alpha = 0$. The value of l_{cm} that gives this is defined to be the "neutral point" location, l_{np}. Therefore, from the previous equation for $d(C_M)_{cm}/d\alpha$, one obtains:

$$\frac{l_{np}}{\bar{c}}\left[(C_{L\alpha})_{prop}\frac{S_p}{S}+(C^*_{L\alpha})_w\eta_w\frac{S_w}{S}+(C^*_{L\alpha})_t\eta_q\eta_t\left(1-\frac{\partial\varepsilon}{\partial\alpha}\right)\frac{S_t}{S}\right]$$

$$=(C_{L\alpha})_{prop}\frac{l_p}{\bar{c}}\frac{S_p}{S}+(C^*_{L\alpha})_w\eta_w\frac{l_w}{\bar{c}}\frac{S_w}{S}+(C^*_{L\alpha})_t\eta_q\eta_t\left(1-\frac{\partial\varepsilon}{\partial\alpha}\right)\frac{l_t}{\bar{c}}\frac{S_t}{S}+C_{T\alpha}\frac{h_{cm}}{\bar{c}}\frac{\tilde{S}}{S}$$

$$-(C_{M\alpha})_{body}\frac{c_b}{\bar{c}}\frac{S_b}{S}$$

The airplane's total lift-curve slope, ignoring body lift, is given by:

$$\hat{C}_{L\alpha}=(C_{L\alpha})_{prop}\frac{S_p}{S}+(C^*_{L\alpha})_w\eta_w\frac{S_w}{S}+(C^*_{L\alpha})_t\eta_q\eta_t\left(1-\frac{\partial\varepsilon}{\partial\alpha}\right)\frac{S_t}{S}$$

Therefore, the solution for the neutral point becomes:

$$l_{np}=\frac{\bar{c}}{\hat{C}_{L\alpha}}\left[(C_{L\alpha})_{prop}\frac{l_p}{\bar{c}}\frac{S_p}{S}+(C^*_{L\alpha})_w\eta_w\frac{l_w}{\bar{c}}\frac{S_w}{S}+(C^*_{L\alpha})_t\eta_w\eta_t\left(1-\frac{\partial\varepsilon}{\partial\alpha}\right)\frac{l_t}{\bar{c}}\frac{S_t}{S}\right]$$

$$+\frac{\bar{c}}{\hat{C}_{L\alpha}}\left[C_{T\alpha}\frac{h_{cm}}{\bar{c}}\frac{\tilde{S}}{S}-(C_{M\alpha})_{body}\frac{c_b}{\bar{c}}\frac{S_b}{S}\right]$$

All of the terms and their estimations have been defined previously, except for the propeller-thrust derivative $C_{T\alpha}$. Adapting the development in Perkins and Hage (Airplane Performance Stability and Control, Wiley, 1949, page 233), the thrust coefficient is defined as:

$$C_T=\frac{2T_p}{\rho V^2\tilde{S}}$$

where $\tilde{S}=(propeller\ diameter)^2$

Next, one has that the thrust times the velocity is equal to the propeller's output power:

$$T_p\times V=P\eta_p\rightarrow T_p=P\eta_p/V$$

where η_p is the propeller's propulsive efficiency and P is the input power from the motor.

These equations combine to give:

$$C_T=\frac{2P\eta_p}{\rho V^3\tilde{S}}$$

Now, for equilibrium flight, one has that:

$$L=\frac{\rho V^2}{2}C_L S=Weight\equiv W\rightarrow V=\left(\frac{2W}{\rho C_L S}\right)^{1/2}$$

This combines with the previous equations to give:

$$C_T = \frac{2\rho^{1/2} C_L^{3/2} P \eta_p}{(2W/S)^{3/2} \tilde{S}} \to C_T = K_p \eta_p C_L^{3/2}, \quad \text{where } K_p \eta_p \equiv \frac{2\rho^{1/2} P \eta_p}{(2W/S)^{3/2} \tilde{S}}$$

Therefore, one may obtain:

$$C_{T\alpha} = \frac{dC_T}{d\alpha} = \frac{dC_T}{dC_L}\frac{dC_L}{d\alpha} = \frac{3}{2} K_p \eta_p C_L^{1/2} C_{L\alpha}$$

ASIDE

It is seen from the above equation that $C_{T\alpha} \geq 0$. It is interesting to understand the physical meaning of this. First of all, if the airplane's angle-of-attack is increased, the lift coefficient likewise increases. Therefore, for equilibrium flight where lift equals weight, the velocity must decrease. However, the propulsive power remains the same. Since this equals thrust times velocity, the thrust must increase. Hence, one sees how thrust increases with the angle of attack.

END ASIDE

Now, define the simplified term:

$$C_{M\alpha} \equiv d(C_M)_{cm}/d\alpha$$

This is the parameter that determines static longitudinal stability. In particular:

$C_{M\alpha} < 0$ represents positive static longitudinal stability.
$C_{M\alpha} = 0$ represents neutral static longitudinal stability,
$C_{M\alpha} > 0$ represents negative static longitudinal stability (statically unstable).

From the previous equations, $C_{M\alpha}$ may be written as:

$$C_{M\alpha} = \hat{C}_{L\alpha} \frac{l_{cm}}{\bar{c}} - \left[(C_{L\alpha})_{prop} \frac{l_p}{\bar{c}} \frac{S_p}{S} + (C_{L\alpha}^*)_w \eta_w \frac{l_w}{\bar{c}} \frac{S_w}{S} + (C_{L\alpha}^*)_t \eta_q \eta_t \left(1 - \frac{\partial \varepsilon}{\partial \alpha}\right) \frac{l_t}{\bar{c}} \frac{S_t}{S} \right]$$
$$- C_{T\alpha} \frac{h_{cm}}{\bar{c}} \frac{\tilde{S}}{S} + (C_{M\alpha})_{body} \frac{c_b}{\bar{c}} \frac{S_b}{S}$$

Next, with the substitution of the equation for the neutral point, this becomes:

$$C_{M\alpha} = \hat{C}_{L\alpha} \left(\frac{l_{cm}}{\bar{c}} - \frac{l_{np}}{\bar{c}} \right) = \hat{C}_{L\alpha} \frac{(l_{cm} - l_{np})}{\bar{c}}$$

The following parameter, defined by:

$$(l_{np} - l_{cm})/\bar{c} \equiv \hat{l}_{np} - \hat{l}_{cm}$$

Aerodynamic Review

is called the "static margin" and is commonly accepted as a measure of static longitudinal stability. In general, an acceptable range is:

$$0.05 \leq \text{Static Margin} \leq 0.15$$

Note the important result that the mass center must be forward of the neutral point for static longitudinal stability. Therefore, in order for a trimmed state to be achieved, the airplane must be configured to give a constant positive pitching moment about the neutral point.

First of all, it can be seen that the equation for $(C_M)_{cm}$ may be now written as:

$$(C_M)_{cm} = C_{M0} + C_{M\alpha}\alpha_{total}$$

where

$$\alpha_{total} = \alpha_{ZLL} + \alpha, \quad \alpha_{ZLL} = C_{L0}/C_{L\alpha}$$

$$C_{M0} = \left(\frac{l_{cm} - l_w}{\bar{c}}\right)\frac{S_w}{S}\eta_w[(C_{L0}^*)_w + (C_{L\alpha}^*)_w i_w]$$

$$+ \left(\frac{l_{cm} - l_t}{\bar{c}}\right)\frac{S_t}{S}\eta_t\eta_q[(C_{L0}^*)_t + (C_{L\alpha}^*)_t i_t - (C_{L\alpha}^*)_t \varepsilon_0]$$

$$- C_{T0}\frac{h_{cm}}{\bar{c}}\frac{\tilde{S}}{S} + (C_{Mac}^*)_w \eta_w \frac{S_w}{S}\frac{\bar{c}_w}{\bar{c}} + (C_{Mac}^*)_t \eta_t \eta_q \frac{S_t}{S}\frac{\bar{c}_t}{\bar{c}}$$

$C_{M\alpha}$ is given previously.

> **ASIDE**
>
> Observe that because the propulsion thrust line is assumed to be parallel to the reference line, $(C_{L0})_p = 0$.
>
> **END ASIDE**

Now, recall that:

$$\varepsilon_0 = \left(\frac{\varepsilon}{C_L}\right)_w [(C_{L0}^*)_w + (C_{L\alpha}^*)_w i_w]\eta_w$$

When $l_{cm} = l_{np}$, then $C_{M\alpha} = 0$ and the equation above gives the constant moment about the neutral point:

$$(C_{M0})_{np} = \left(\frac{l_{np} - l_w}{\bar{c}}\right)\frac{S_w}{S}\eta_w[(C_{L0}^*)_w + (C_{L\alpha}^*)_w i_w]$$

$$+ \left(\frac{l_{np} - l_t}{\bar{c}}\right)\frac{S_t}{S}\eta_t\eta_q[(C_{L0}^*)_t + (C_{L\alpha}^*)_t i_t - (C_{L\alpha}^*)_t \varepsilon_0]$$

$$- C_{T0}\frac{h_{cm}}{\bar{c}}\frac{\tilde{S}}{S} + (C_{Mac}^*)_w \eta_w \frac{S_w}{S}\frac{\bar{c}_w}{\bar{c}} + (C_{Mac}^*)_t \eta_t \eta_q \frac{S_t}{S}\frac{\bar{c}_t}{\bar{c}}$$

Therefore, it is seen that the neutral point fulfills the same role, for the entire airplane, as the aerodynamic center does for a wing. Namely, it is the point where the pitching moment is constant with the angle of attack. Further, it can be seen that for positive static stability, where the mass center is forward of the neutral point, the pitching moment must be positive for an equilibrium trim state to be achieved:

$$(C_M)_{np} > 0$$

PITCHING MOMENT (WITH EXTENDED UNDERCARRIAGE)

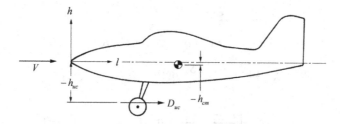

The extended undercarriage gives a pitching moment about the mass center, which is given by,

$$M_{uc} = D_{uc}(h_{uc} - h_{cm})$$

Therefore, the contribution to the airplane's total pitching moment about the mass center is:

$$\Delta(C_M)_{uc} = (C_D)_{uc} \frac{(h_{uc} - h_{cm})}{\bar{c}} \frac{S_{uc}}{S}$$

It is seen that this is not a function of α, so the presence of the undercarriage does not change the previously derived equations for the neutral point and $C_{M\alpha}$. Therefore, this term need only be added to the equation for the pitching moment about the neutral point:

$$(C_{M0})_{np} = \left(\frac{l_{np} - l_w}{\bar{c}}\right) \frac{S_w}{S} \eta_w [(C_{L0}^*)_w + (C_{L\alpha}^*)_w i_w]$$

$$+ \left(\frac{l_{np} - l_t}{\bar{c}}\right) \frac{S_t}{S} \eta_t \eta_q [(C_{L0}^*)_t + (C_{L\alpha}^*)_t i_t - (C_{L\alpha}^*)_t \varepsilon_0]$$

$$- C_{T0} \frac{h_{cm}}{\bar{c}} \frac{\tilde{S}}{S} + (C_{Mac}^*)_w \eta_w \frac{S_w}{S} \frac{c_w}{\bar{c}} + (C_{Mac}^*)_t \eta_t \eta_q \frac{S_t}{S} \frac{c_t}{\bar{c}} + (C_D)_{uc} \frac{(h_{uc} - h_{cm})}{\bar{c}} \frac{S_{uc}}{S}$$

NUMERICAL EXAMPLE

These equations are applied to the large model airplane shown below, with the following properties:

Aerodynamic Review

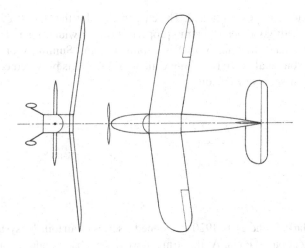

Tail-Aft Monoplane

WING

$$b_w = 2.0\,\text{m}, \quad S_w = 0.556\,\text{m}^2, \quad (\Lambda_{le})_w = (\Lambda_{c/4})_w = (\Lambda_{c/2})_w = 8°, \quad (\tau)_w = 0, \quad i_w = 3°$$

From this, one may calculate that $(AR)_w = 7.2$. Also, although this has shaped wing tips, the chord is constant over most of the span. Therefore, it is reasonable to choose $(\lambda)_w \approx 1.0$, which gives $\bar{c}_w = (c_r)_w = 0.3\,\text{m}$.

Now, observe the top view of the airplane. If the wing's leading edges are extended to the centerline of the fuselage, this defines a virtual apex point at an axial distance of 0.343 m (in the l, h reference frame shown above). The distance of the wing's aerodynamic center from that apex point is obtained from the equation given in the "Wings" section of this chapter:

$$(x_{ac})_w = \frac{(\bar{c})_w}{4} + \left(\frac{1+2\lambda}{12}\right)(c_r)_w (AR)_w \tan(\Lambda_{le})_w = \frac{0.3}{4} + \frac{3}{12} \times 0.3 \times 7.2 \tan(8°) = 0.151$$

Upon adding this to the apex distance, one obtains:

$$(l_{ac})_w \equiv l_w = 0.494\,\text{m}$$

ASIDE

This is a very conventional configuration, which is a good starting point for this book. Things get less conventional in chapters 4 and 5.

END ASIDE

Next, the wing's lift-curve slope is obtained. The chosen airfoil is the "Clark-Y", which is a popular easy-to-build section with good aerodynamic properties over a wide range of Reynolds numbers. The aerodynamic data (and drawing below) is obtained from "Summary of Low-Speed Airfoil Data, Vol. 3", C. A. Lyon, et al., SoarTech Publications, 1997, and is provided courtesy of Professor Michael Selig of the University of Illinois.

Clark–Y

ASIDE

Col. Virginious Clark, in the early 1920s, designed a series of airfoils by systematically varying their geometrical parameters. At that time most airfoils had "under-camber", which was a concave curve on the underside. This under-camber complicated fabrication, and Col. Clark was curious to see if "flat-bottomed" airfoils could give good performance. It is said that his approach was to replace the under-curve on certain Göttingen sections with a straight line. Each candidate airfoil was then evaluated with wind-tunnel testing. Among these, the "Y" design stood out with the best overall performance and it found widespread use on numerous airplanes up to recent times.

END ASIDE

The lift-coefficient data for different values of RN is given below:

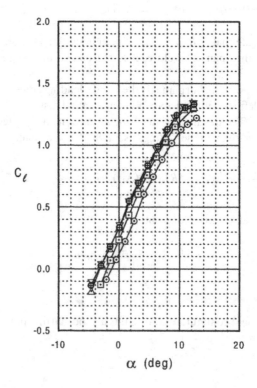

Aerodynamic Review

In chapter 7, it is calculated that this model flies at $V \approx 14.5 \, \text{m/sec}$, which gives an $RN = 3.0 \times 10^5$. So, these wind-tunnel experiments (inverted triangle data points) give that $(C_{l\alpha})_{\text{exp}} \approx 6.10/\text{rad}$. Thus $\kappa = 6.10/2\pi = 0.97$ and the *isolated* wing's lift-curve slope is:

$$(C_{L\alpha}^*)_w = \frac{2\pi \, AR}{2 + \left[\dfrac{AR^2}{\kappa^2}(1 + \tan^2 \Lambda_{c/2}) + 4\right]^{1/2}} = \frac{2\pi \times 7.2}{2 + \left[\dfrac{(7.2)^2}{(0.97)^2}(1 + \tan^2(8^0)) + 4\right]^{1/2}}$$

$$= \frac{45.239}{2 + [55.10 \times 1.020 + 4]^{1/2}} = \frac{45.239}{2 + 7.758} \rightarrow \underline{\underline{(C_{L\alpha}^*)_w = 4.636 \ (1/\text{rad})}}$$

Also, the wing is untwisted, $\tau = 0$, and the airfoil's zero-lift angle of attack, α_{ZLL}, is $-3.4°(-0.0593 \, \text{rad})$. Therefore, the zero-angle lift coefficient of the isolated wing is:

$$(C_{L0}^*)_w = -(C_{L\alpha}^*)_w (\alpha_{\text{ZLL}}) = 4.636 \times 0.0593 = 0.275$$

Next, with reference to the Clark-Y drag polar plots shown previously, in the "Airfoils" section, it is seen that a reasonable polynomial fit may be obtained between $-0.109 \leq C_l \leq 9.17$:

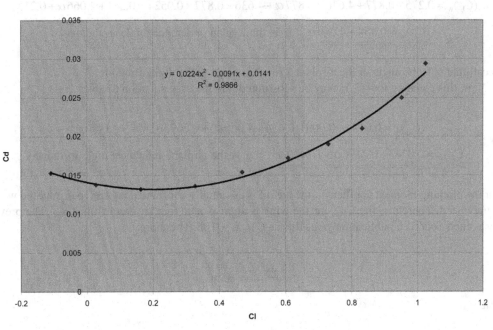

Clark-Y Cd vs. Cl Polar

From this, one obtains:

$$[(C_d)_{C_l=0}]_w = 0.0141, \quad A_w = -0.0091, \quad B_w = 0.0224$$

Recall that the induced drag is given by:

$$(C_{Dw})_{\text{induced}} = \frac{(C_L)_w^2 (1 + \delta_w)}{\pi \, AR_w}$$

From the graph in the "Wings" section $\delta_w \approx 0.065$, so that $(C_{Dw})_{induced} = 0.0471(C_L)_w^2$. All together then, the wing's drag is given by:

$$(C_D)_w = 0.0141 - 0.0091(C_L)_w + 0.0695(C_L)_w^2$$

where, from before, $(C_L)_w$ is given by:

$$(C_L)_w = (C_{Lo}^*)_w \eta_w + (C_{L\alpha}^*)_w \eta_w \alpha + (C_{L\alpha}^*)_w \eta_w i_w$$

For this example, $S_{w-b} = 0.05\,\text{m}^2$. So, the wing-body efficiency factor is obtained, as described before, from:

$$\eta_w = 1 - 1.4\left(\frac{S_{w-b}}{S_w}\right) + 0.4\left(\frac{S_{w-b}}{S_w}\right)^2 = 1 - 1.4 \times \frac{0.05}{0.556} + 0.4\left(\frac{0.05}{0.556}\right)^2 = 0.877$$

Thus, one obtains:

$$(C_L)_w = 0.275 \times 0.877 + 4.636 \times 0.877\alpha + 4.636 \times 0.877 \times 0.0524 = 0.241 + 4.066\alpha + 0.213$$

$$\underline{\underline{(C_L)_w = 0.454 + 4.066\alpha}}, \quad \alpha \text{ is the airplane reference angle in radians}$$

Recall that α is the angle of the airplane's reference line to the flight direction.

Now, this equation for $(C_L)_w$ may be substituted into the drag equation to give:

$$(C_D)_w = 0.0141 - 0.0091(0.454 + 4.066\alpha) + 0.0695(0.454 + 4.066\alpha)^2$$

$$\underline{\underline{(C_D)_w = 0.0243 + 0.2196\alpha + 1.1490\alpha^2}}, \quad \alpha \text{ is the airplane reference angle in radians}$$

For the pitching-moment coefficient, the airfoil is essentially constant along the span, and the wing is untwisted. Therefore, the C_{Mac} for the wing is approximate that for the airfoil, C_{mac}. The previously cited SoarTech publication gives this as $C_{mac} \approx -0.08$. Therefore:

$$\underline{\underline{(C_{Mac}^*)_w \approx -0.08}}$$

TAIL

$$b_t = 0.75\,\text{m}, \quad S_t = 0.14\,\text{m}^2, \quad (\Lambda_{le})_t = (\Lambda_{c/4})_t = (\Lambda_{c/2})_t = 0°, \quad (\tau)_t = 0, \quad i_t = 0°$$

From this, one may calculate that $(AR)_t = 4.02$. Also, although this has shaped tips like the wing, the chord is likewise constant over most of the span. Therefore, it is reasonable to choose $(\lambda)_w \approx 1.0$, which gives $\bar{c}_t = (c_r)_t = 0.2\,\text{m}$. Also, because the tail is unswept, $(x_{ac})_t = \bar{c}_t/4 = 0.5\,\text{m}$; therefore, upon adding this to the l distance to the tail's leading edge, one obtains:

$$(l_{ac})_t \equiv l_t = 1.3\,\text{m}$$

Aerodynamic Review

Next, the tail's lift-curve slope is obtained. The chosen airfoil is the SD8020, designed by Prof. Michael Selig, which is a symmetrical section well suited to a wide range of Reynolds numbers.

Its aerodynamic properties are obtained from "Summary of Low-Speed Airfoil Data, Vol. 1", Michael Selig, et al. SoarTech Publications, 1995.

Because the wing's RN, based on $\bar{c}_w$, is 3.0×10^5, the tail's RN based on $\bar{c}_t$ is 2.0×10^5. Thus, the wind-tunnel experiments give that $(C_{l\alpha})_{exp} \approx 7.40$/rad (greater than 2π!).

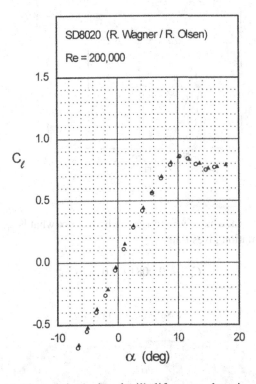

Therefore, $\kappa = 7.40 / 2\pi = 1.18$, and the *isolated* tail's lift-curve slope is:

$$(C_{L\alpha}^*)_t = \frac{2\pi AR}{2 + \left[\frac{AR^2}{\kappa^2}(1 + \tan^2 \Lambda_{c/2}) + 4\right]^{1/2}} = \frac{2\pi \times 4.02}{2 + \left[\frac{(4.02)^2}{(1.18)^2}(1 + \tan^2(0°)) + 4\right]^{1/2}}$$

$$= \frac{25.258}{2 + [11.61 \times 1 + 4]^{1/2}} = \frac{25.258}{2 + 3.951} \rightarrow \underline{(C_{L\alpha}^*)_t = 4.244 \text{ (1/rad)}}$$

Also, the tail is untwisted, $\tau = 0$, and the airfoil's zero-lift angle of attack, α_{ZLL}, is $0°$. Therefore, the zero-angle lift coefficient of the isolated tail is:

$$(C_{L0}^*)_t = -(C_{L\alpha}^*)_t (\alpha_{ZLL}) = 0$$

Next, the drag-polar plots are shown below, from which it is seen that a reasonable polynomial fit for the data at $RN = 2.0 \times 10^5$ may be obtained between $-0.659 \leq C_l \leq 0.574$.

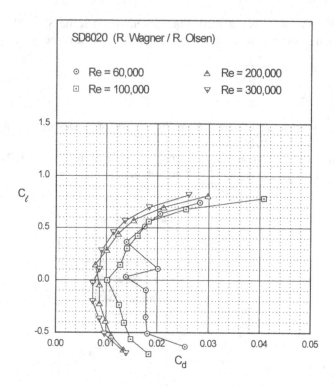

It is seen that the graph is somewhat asymmetrical, which is not what is expected for a symmetrical airfoil. A simple transformation gives that:

$$(C_d)_t = 0.0081 + 0.0161(C_l)_t^2$$

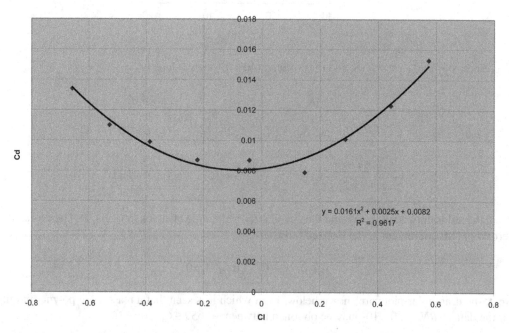

Aerodynamic Review

Therefore, the coefficients are:

$$[(C_d)_{C_l=0}]_t = 0.0081, \quad A_t = 0, \quad B_t = 0.0161$$

ASIDE

From the "Airfoils" section, an approximate equation for B is given as:

$$B \approx 2\pi(1 - \eta_{le})$$

where η_{le} is the leading-edge suction. For this case, the leading-edge suction efficiency is:

$$(\eta_{le})_t \approx 0.997 \rightarrow 99.7\%$$

This level of efficiency can only be attained with very precise construction.

END ASIDE

Recall that the induced drag is given by:

$$(C_{Dt})_{\text{induced}} = \frac{(C_L)_t^2 (1 + \delta_t)}{\pi AR_t}$$

From the graph in the "Wings" section, $\delta_t \approx 0.030$, so that $(C_{Dt})_{\text{induced}} = 0.0816(C_L)_t^2$. All together then, the tail's drag is given by:

$$(C_D)_t = 0.0081 + 0.0977(C_L)_t^2$$

where, from before, $(C_L)_t$ is given by:

$$(C_L)_t = [(C_{L0}^*)_t + (C_{L\alpha}^*)_t \alpha + (C_{L\alpha}^*)_t (i_t - \varepsilon_t)] \eta_t$$

For this example, $S_{t-b} = 0.004\,\text{m}^2$. So, the tail-body efficiency factor is obtained, as described before, from:

$$\eta_t = 1 - 1.4\left(\frac{S_{t-b}}{S_t}\right) + 0.4\left(\frac{S_{t-b}}{S_t}\right)^2 = 1 - 1.4 \times \frac{0.004}{0.140} + 0.4\left(\frac{0.004}{0.140}\right)^2 = 0.960$$

Also, the downwash term has to be calculated, where, from before:

$$\varepsilon_t = (\varepsilon/C_L)_w (C_L)_w$$

The simpler method presented earlier in the "Wings" section gives that:

$$\left(\frac{\varepsilon}{C_L}\right)_w = \frac{2}{\pi(AR)_w} = \frac{2}{\pi \times 7.2} = 0.0884 \text{ (in radians)}$$

The more-complete method, from the USAF Stability and Control DATCOM, was also presented earlier, which gives, with $l_H = 0.813\,\text{m}$ and $|h_H| = 0.093\,\text{m}$ that:

$$K_H = \frac{1-|h_H/b_w|}{(2l_H/b_w)^{1/3}} = \frac{1-|0.093/2.0|}{(2\times 0.813/2.0)^{1/3}} = \frac{0.953}{0.933} = 1.021$$

$$K_\lambda = \frac{10-3\lambda}{7} = \frac{10-3\times 1}{7} = 1.0,$$

$$K_A = \frac{1}{(AR)_w} - \frac{1}{1+(AR)_w^{1.7}} = \frac{1}{7.2} - \frac{1}{1+(7.2)^{1.7}} = 0.139 - 0.034 = 0.105$$

So,

$$\left(\frac{\partial \varepsilon}{\partial \alpha}\right)_w = 4.44[K_A K_\lambda K_H (\cos\Lambda_{c/4})_w^{1/2}]^{1.19} = 4.44[0.105\times 1.0\times 1.021(\cos 8°)^{1/2}]^{1.19} = 0.310$$

Thus,

$$\left(\frac{\varepsilon}{C_L}\right)_w = \frac{(\partial\varepsilon/\partial\alpha)_w}{(C_{L\alpha})_w} = \frac{0.310}{4.066} = 0.076 \text{ (in radians)}$$

That is 14% smaller than the simpler estimate, but this is the value chosen for the subsequent calculations.

Upon noting that $(C_{L0}^*)_t = 0$ and $i_t = 0$, the tail lift equation becomes:

$$(C_L)_t = [(C_{L0}^*)_t + (C_{L\alpha}^*)_t i_t]\eta_t - (C_{L\alpha}^*)_t(\varepsilon/C_L)_w[(C_{L0}^*)_w + (C_{L\alpha}^*)_w i_w]\eta_w \eta_t$$

$$+ (C_{L\alpha}^*)_t \eta_t [1-(\varepsilon/C_L)_w(C_{L\alpha}^*)_w \eta_w]\alpha$$

$$(C_L)_t = [0+4.244\times 0]\times 0.96 - 4.244\times 0.076\times[0.275+4.636\times 0.0524]\times 0.877\times 0.960$$

$$+4.244\times 0.960\times[1-0.076\times 4.636\times 0.877]\alpha$$

$$\underline{(C_L)_t = -0.141 + 2.815\alpha}, \quad \alpha \text{ is the airplane reference angle in radians}$$

It is seen that the tail's lift-curve slope, $(C_{L\alpha})_t$ is 66% of that for the isolated tail, $(C_{L\alpha}^*)_t$. This is a consequence of the strong downwash effect: $\alpha_t = [1-(\partial\varepsilon/\partial\alpha)_w]\alpha$.

Now, this equation for $(C_L)_t$ may be substituted into the drag equation to give:

$$(C_D)_t = 0.0081 + 0.0977(C_L)_t^2 = 0.0081 + 0.0977(-0.141+2.815\alpha)^2$$

$$(C_D)_t = 0.0081 + 0.0977(0.0199 - 0.794\alpha + 7.924\alpha^2)$$

$$\underline{(C_D)_t = 0.010 - 0.0776\alpha + 0.774\alpha^2}, \quad \alpha \text{ is the airplane reference angle in radians}$$

Aerodynamic Review

ASIDE

Although the example tail airfoil is a smooth symmetrical section, quite often a simpler flat-plate section is used, even on some full-scale aircraft:

In this case, the lift-curve slope is chosen to be 2π:

$$(C^*_{L\alpha})_t = 2\pi, \quad (C^*_{Lo})_t = 0$$

The zero-lift (and zero-angle, in this case) drag coefficient may be obtained from the equations previously given in the "Airfoils" section:

$$(C_d)_{C_l=0} \approx (C_d)_{\text{fric}} + (1-\eta_{\text{le}})2\pi\alpha^2_{\text{ZLL}} = (C_d)_{\text{fric}} \approx 2C_{\text{fric}} \approx \frac{2 \times 0.455}{(\log_{10} RN)^{2.58}}$$

For $RN = 2.0 \times 10^5$, $(C_{d0})_t = 0.0123$. Further, from the "Airfoils" section, one has:

$$B \approx (1-\eta_{\text{le}})/(2\pi)$$

For a thin airfoil like this, even if the leading edge is nicely smoothed and rounded, the lead-edge suction efficiency is rarely above 0.40. So,

$$B \approx 0.6/(2\pi) = 0.0955$$

The drag-coefficient equation for this airfoil is thus:

$$(C_d)_t^{\text{flat plate}} \approx 0.0123 + 0.0955(C_l)_t^2$$

Compare this with the corresponding equation for the S8020 airfoil:

$$(C_d)_t^{\text{S8020}} = 0.0081 + 0.0161(C_l)_t^2$$

It is seen that the Selig section is a much better performer. It is clear from a comparison of the $(C_{d0})_t$ values that a portion of the Selig section must have a laminar boundary layer. This gives a lower drag than the wholly-turbulent boundary layer of the flat-plate section. Further, the higher leading-edge suction efficiency of the Selig section gives a B value that is much lower than that for the flat-plate section. However, note that these results came from tests of a perfectly finished wind-tunnel model. Such performance might not be expected from the typical construction of an airplane.

END ASIDE

Finally, because the tail's airfoil is symmetric and the tail is untwisted, one obtains that:

$$\underline{\underline{(C^*_{\text{Mac}})_t = 0.0}}$$

BODY

The body's lift contribution is usually very small for most "conventional" configurations. However, as an approximation, the previously discussed equation will be used:

$$(C_L)_{body} \approx [2(k_2 - k_1) S_{base}/S_b] \alpha_b \equiv (C_{L\alpha})_{body} \alpha_b$$

$$S_{base} = 0.053 \times 0.053 = 0.0028\, m^2, \quad S_b = 0.25 \times 0.167 = 0.0417\, m^2,$$

$$d_{equiv} = [4(height_{max} \times width_{max})/\pi]^{1/2} = (4 \times 0.25 \times 0.167/\pi)^{1/2} = 0.230$$

$$\text{Fineness Ratio} \approx l/d_{equiv} = \frac{1.4}{0.230} = 6.09 \rightarrow (k_2 - k_1) \approx 0.86$$

Also, because the body cross-section is rectangular, the value from the above equation is doubled:

$$(C_L)_{body} \approx 2 \times (2 \times 0.86 \times 0.0028/0.0417)\alpha = \underline{\underline{(C_L)_{body} \approx 0.231\alpha}}$$

Again, α is the airplane reference angle in radians.

It is assumed that the body's drag does not vary much with α, within the airplane's normal range of α values. Therefore:

$$(C_D)_{body} \approx (C_{D0})_{body}$$

From the values for typical fuselages, presented in the "Bodies" section, a representative value of 0.27 is chosen. So,

$$\underline{\underline{(C_D)_{body} \approx 0.27}}$$

Finally, the pitching-moment slope of the body is estimated. From the previous "Bodies" section, the first method uses the following equation:

$$(C_{M\alpha})_{body}^{Method\,1} \approx \frac{2(k_2 - k_1) Vol_{body}}{S_b \bar{c}_b}$$

where the body's volume is $0.0352\, m^3$ and $\bar{c}_b = 1.4\, m$, the body's overall length. So,

$$(C_{M\alpha})_{body}^{Method\,1} \approx \frac{2 \times 0.86 \times 0.0352}{0.0417 \times 1.4} = 1.04, \text{ per radian}$$

The second method is from Gilruth and White, which, as described before, gives that:

$$(C_{M\alpha})_{body}^{Method\,2} = \frac{180}{\pi} \times \frac{K_b w_b^2 l_b}{S_b \bar{c}_b}$$

where l_b and w_b are the maximum length and width of the body (1.4 m and 0.167 m, respectively), and K_b is given by:

$$K_b = 0.2931\zeta^3 - 0.1025\zeta^2 + 0.0339\zeta + 0.0012$$

ζ = (length from the nose of the body to 25% of the root chord)/(total body length)

Aerodynamic Review

For this example, $\zeta = 0.425/1.40 = 0.304 \rightarrow K_b = 0.0103$. So,

$$(C_{M\alpha})_{body}^{Method\ 2} \approx \frac{180}{\pi} \times \frac{0.0103 \times (0.167)^2 \times 1.4}{0.0417 \times 1.4} = 0.394, \text{ per radian}$$

This is a much lower value than that from the previous method, which is puzzling when considering the effects of the wing's up-wash on the fore body, and its downwash on the aft body. It is now interesting to compare these results with that from the Multhopp method (also described in the "Bodies" section). The required inputs are shown below:

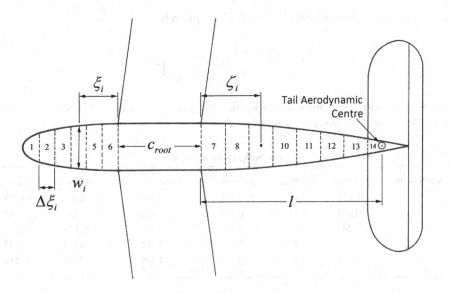

Fore body

Segment no.	ξ_i (m)	w_i (m)	$\Delta \xi_i$ (m)
1	0.321	0.078	0.0583
2	0.263	0.117	0.0583
3	0.204	0.140	0.0583
4	0.146	0.153	0.0583
5	0.088	0.162	0.0583
6	0.029	0.166	0.0583

Aft body

Segment no.	ζ_i (m)	w_i (m)	$\Delta \xi_i$ (m)
7	0.043	0.165	0.0863
8	0.130	0.162	0.0863
9	0.216	0.151	0.0863
10	0.303	0.135	0.0863
11	0.389	0.116	0.0863
12	0.475	0.094	0.0863
13	0.562	0.068	0.0863
14	0.680	0.028	0.150

As given earlier, the up-wash angle for the first five segments is given by:

$$\left(\frac{\partial \varepsilon_u}{\partial \alpha}\right)_i = -0.029\left(\frac{\xi_i}{c_{\text{root}}}\right)^4 + 0.101\left(\frac{\xi_i}{c_{\text{root}}}\right)^3 + 0.0425\left(\frac{\xi_i}{c_{\text{root}}}\right)^2 - 0.5315\left(\frac{\xi_i}{c_{\text{root}}}\right) + 0.6048$$

The averaged up-wash angle for segment 6 is given by:

$$\left(\frac{\partial \overline{\varepsilon}_u}{\partial \alpha}\right)_6 = 6.6545\left(\frac{\Delta\xi_6}{c_{\text{root}}}\right)^4 - 20.033\left(\frac{\Delta\xi_6}{c_{\text{root}}}\right)^3 + 23.934\left(\frac{\Delta\xi_6}{c_{\text{root}}}\right)^2 - 14.395\left(\frac{\Delta\xi_6}{c_{\text{root}}}\right) + 4.6117$$

And the up-wash angle for segments 7 through 14 are given by:

$$\left(\frac{\partial \varepsilon_u}{\partial \alpha}\right)_i = \frac{\varsigma_i}{l}\left[1 - \left(\frac{\partial \varepsilon}{\partial \alpha}\right)_{\text{tail}}\right] - 1$$

where $l = 0.657\,\text{m}$ and, from previous work, $(\partial \varepsilon / \partial \alpha)_{\text{tail}} = 0.310$ and $c_{\text{root}} = 0.30\,\text{m}$. Therefore, the tables for calculation may be set up as follows:

Fore-body, up to segment 6

Segment no.	ξ_i (m)	ξ_i/c_{root}	$(\partial \varepsilon_u/\partial \alpha)_i$	$(\partial \varepsilon_u/\partial \alpha)_i + 1$	w_i^2 (m²)	$\Delta\xi_i$ (m)
1	0.321	1.070	0.171	1.171	0.0061	0.0583
2	0.263	0.876	0.223	1.223	0.0137	0.0583
3	0.204	0.681	0.288	1.288	0.0196	0.0583
4	0.146	0.487	0.366	1.366	0.0235	0.0583
5	0.088	0.292	0.456	1.456	0.0262	0.0583

Fore-body, segment 6

Segment no.	ξ_6 (m)	ξ_6/c_{root}	$(\partial \overline{\varepsilon}_u/\partial \alpha)_6$	$(\partial \overline{\varepsilon}_u/\partial \alpha)_6 + 1$	w_6^2 (m²)	$\Delta\xi_6$ (m)
6	0.0293	0.098	3.41	4.41	0.0276	0.0583

Aft-body, from segment 7 to 14

$$\left(\frac{\partial \varepsilon_u}{\partial \alpha}\right)_i = \frac{\varsigma_i}{l}\left[1 - \left(\frac{\partial \varepsilon}{\partial \alpha}\right)_{\text{tail}}\right] - 1 = \frac{\varsigma_i}{0.657}(1 - 0.310) - 1 = 1.050\varsigma_i - 1$$

Segment no.	ζ_i (m)	$(\partial \varepsilon_u/\partial \alpha)_i$	$(\partial \varepsilon_u/\partial \alpha)_i + 1$	w_i^2 (m²)	$\Delta\xi_i$ (m)
7	0.086	−0.910	0.090	0.0272	0.0863
8	0.130	−0.864	0.137	0.0262	0.0863
9	0.216	−0.773	0.227	0.0228	0.0863
10	0.303	−0.682	0.318	0.0182	0.0863
11	0.389	−0.592	0.408	0.0134	0.0863
12	0.475	−0.501	0.499	0.0088	0.0863
13	0.562	−0.410	0.590	0.0046	0.0863
14	0.680	−0.286	0.714	0.0008	0.0863

Aerodynamic Review

Therefore, the moment contribution from the fuselage, as given in the "Bodies" section, is:

$$(C_{Mac})_{body}^{Method\ 3} \approx \frac{\pi}{2 S_b \bar{c}_b} \sum_{1}^{5} w_i^2 \left[\left(\frac{\partial \varepsilon_u}{\partial \alpha}\right)_i + 1\right]\Delta\xi_i + \frac{\pi}{2 S_b \bar{c}_b} w_6^2 \left[\left(\frac{\partial \bar{\varepsilon}_u}{\partial \alpha}\right)_6 + 1\right]\Delta\xi_6$$

$$+ \frac{\pi}{2 S_b \bar{c}_b} \sum_{7}^{14} w_i^2 \left[\left(\frac{\partial \varepsilon_u}{\partial \alpha}\right)_i + 1\right]\Delta\xi_i$$

$$= \frac{\pi}{2 \times 0.0417 \times 1.4}(0.00696 + 0.00710 + 0.00260) = 0.448$$

This value is comparable to that from Method 2 and much smaller than that from Method 1. As a check on the robustness of this method, it is informative to redo the fore-body calculation with different segment spacing,

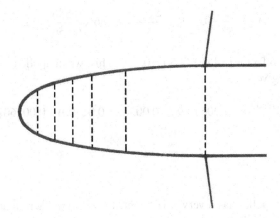

Bear in mind the statement in the "Bodies" section about the regions of validity of the up-wash equations. This spacing fulfills that criteria much more than the previous spacing.

Segment no.	ξ_i (m)	w_i (m)	$\Delta\xi_i$ (m)
1	0.332	0.061	0.0337
2	0.298	0.096	0.0337
3	0.266	0.115	0.0337
4	0.231	0.130	0.0337
5	0.182	0.145	0.0643
6	0.075	0.164	0.150

From this and the above equations, one obtains:

Segment no.	ξ_i (m)	ξ_i/c_{root}	$(\partial\varepsilon_u/\partial\alpha)_i$	$(\partial\varepsilon_u/\partial\alpha)_i + 1$	w_i^2 (m^2)	$\Delta\xi_i$ (m)
1	0.332	1.107	0.162	1.162	0.0037	0.0337
2	0.298	0.994	0.189	1.189	0.0092	0.0337
3	0.266	0.888	0.219	1.219	0.0132	0.0337
4	0.231	0.770	0.257	1.257	0.0169	0.0337
5	0.182	0.607	0.316	1.316	0.0210	0.0643

Segment no.	ζ_6 (m)	ζ_6/c_{root}	$(\partial\bar{\varepsilon}_u/\partial\alpha)_6$	$(\partial\bar{\varepsilon}_6/\partial\alpha)_6 + 1$	w_6^2 (m²)	$\Delta\xi_6$ (m)
6	0.075	0.250	2.222	3.222	0.0269	0.150

which gives the following results:

$$(C_{Mac})_{body}^{Method\ 3} \approx \frac{\pi}{2 \times 0.0417 \times 1.4}(0.00355 + 0.01300 + 0.00260) = 0.515$$

This is 15% larger than the result for the previous segment spacing. It must be said that the author has never come across any publication that gives recommended ideal segment spacing, but it is encouraging that the method is robust enough to give comparable results for two very-different spacing schemes.

Finally, the "Bodies" section stated that the coefficients for the Multhopp method are based on $(C_{L\alpha})_{wing-body} = 4.5/rad$. For example:

$$(C_{L\alpha})_{wing-body} = (C_{L\alpha})_{wing}\frac{S_w}{S} + (C_{L\alpha})_{body}\frac{S_{body}}{S} = 4.066 \times \frac{0.556}{0.556} + 0.231 \times \frac{0.0417}{0.556} = 4.083$$

Therefore, the correction factor is $4.083/4.50 = 0.907$. This, when applied to the first six segments of the Method 3 result, gives:

$$(C_{M\alpha})_{body}^{Corrected\ Method\ 3} \approx 26.906\,[0.907(0.00355 + 0.01300) + 0.00260] = 0.474 \rightarrow$$

$$(C_{M\alpha})_{body} \approx 0.47$$

Propeller

For this example, the propeller has a very small contribution to the aerodynamics of the complete airplane, as will be seen in subsequent calculations:

$$S_p \approx 0.006\,m^2, \quad (C_{L\alpha})_p \approx \frac{2\pi}{[1 + 2/8.0]} \rightarrow (C_{L\alpha})_p \approx 5.0/rad$$

Also, from the "Longitudinal Numerical Example" section of Chapter 6, one obtains:

$$C_{T\alpha} = 3.231$$

Of course, the drag of the propeller is not a factor unless the engine has stopped.

Undercarriage

As stated earlier, in the Undercarriage section, the reference area is the diameter times the maximum cross-sectional width of *one* wheel:

$$S_{uc} \approx 0.093 \times 0.040 = 0.0037\,m^2$$

Also, the design of the undercarriage is close to that of the "unbraced cantilever configuration"; so, the drag coefficient for this is chosen:

$$(C_D)_{uc} \approx 0.564$$

Aerodynamic Review

The lift of the undercarriage is not considered to be significant.

COMPLETE AIRPLANE LIFT AND DRAG COEFFICIENTS

From the "Complete-Aircraft Aerodynamics" section, one has:

$$C_L = (C_L)_w \frac{S_w}{S} + (C_L)_t \frac{S_t}{S} + (C_L)_b \frac{S_b}{S} + (C_L)_p \frac{S_p}{S} \rightarrow$$

$$C_L = (0.454 + 4.066\alpha)\frac{0.556}{0.556} + (-0.141 + 2.815\alpha)\frac{0.140}{0.556} + (0.231\alpha)\frac{0.0417}{0.556} + (5.0\alpha)\frac{0.006}{0.556}$$

$$= 0.454 + 4.066\alpha - 0.036 + 0.709\alpha + 0.017\alpha + 0.05\alpha$$

$$\underline{\underline{C_L = 0.418 + 4.842\alpha,}} \quad \alpha \text{ is the airplane reference angle in radians}$$

$$C_D = (C_D)_w \frac{S_w}{S} + (C_D)_t \frac{S_t}{S} + (C_D)_b \frac{S_b}{S} + (C_D)_{uc} \frac{S_{uc}}{S} \rightarrow$$

$$C_D = (0.0243 + 0.2196\alpha + 1.1490\alpha^2)\frac{0.556}{0.556} + (0.010 - 0.0776\alpha + 0.774\alpha^2)\frac{0.140}{0.556}$$

$$+ 0.27 \times \frac{0.0417}{0.556} + 0.564 \times \frac{0.0037}{0.556}$$

$$= 0.0243 + 0.2196\alpha + 1.1490\alpha^2 + 0.0025 - 0.0195\alpha + 0.1949\alpha^2 + 0.0203 + 0.0038$$

$$\underline{\underline{C_D = 0.0509 + 0.2001\alpha + 1.3439\alpha^2,}} \quad \alpha \text{ is the airplane reference angle in radians}$$

The airplane polar plot will be calculated, but first, it should be noted that the propeller lift is usually not included. Therefore, the C_L equation is determined as if the airplane is a glider:

$$\underline{\underline{(C_L)_{\text{Glider}} \equiv (C_L') = 0.418 + 4.788\alpha}}$$

The resulting plot is shown below, where the minimum α value is $-10°$ and the maximum α value is $15°$; and these are incremented by $1.0°$ indicated by the data points.

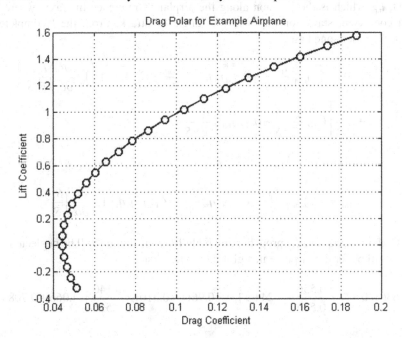

One could derive the drag-polar equation from the preceding equations, but it is simpler to perform a curve-fitting process on the data points. A perfect polynomial curve fit is provided by:

$$C_D = 0.0437 - 0.0072(C_L') + 0.0586(C_L')^2$$

A valuable measure of an airplane's aerodynamic efficiency is its maximum lift/drag ratio. The equation gives a means for calculating this. First, divide by C_L':

$$\frac{C_D}{C_L'} = \frac{0.0437}{C_L'} - 0.0072 + 0.0586 C_L'$$

Then, find the minimum of this by taking the derivative and setting this equal to zero:

$$\frac{d(C_D/C_L')}{dC_L'} = \frac{-0.0437}{(C_L')^2} + 0.0586 = 0 \rightarrow C_L' = \left(\frac{0.0437}{0.0586}\right)^{1/2} \rightarrow (C_L')_{\text{MaxL/D}} = 0.864$$

From this, and the previous equation, one may now calculate the maximum L/D value:

$$\left(\frac{C_L'}{C_D}\right)_{\text{Max}} = \left(\frac{C_D}{C_L'}\right)_{\text{Min}}^{-1} = \left[\frac{0.0437}{0.864} - 0.0072 + 0.0586 \times 0.864\right]^{-1} = (0.0940)^{-1} \rightarrow$$

$$(C_L'/C_D)_{\text{Max}} = 10.64$$

Further, the α value for which this occurs may be found from the C_L' equation:

$$(C_L')_{\text{MaxL/D}} = 0.418 + 4.788(\alpha)_{\text{MaxL/D}} \rightarrow (\alpha)_{\text{MaxL/D}} = (0.864 - 0.418)/4.788 \rightarrow$$

$$(\alpha)_{\text{MaxL/D}} = 0.0932 \text{ rad} = 5.34°$$

COMPLETE AIRPLANE MOMENT CHARACTERISTICS

In order to characterize the airplane's pitching-moment behavior, the first step is to determine the neutral point, l_{np}, which is the location along the airplane's reference-line axis where the pitching-moment coefficient stays constant with the angle of attack. From the "Complete-Airplane Aerodynamics" section, one has:

$$l_{np} = \frac{\bar{c}}{\hat{C}_{L\alpha}}\left[(C_{L\alpha})_{\text{prop}} \frac{l_p}{\bar{c}} \frac{S_p}{S} + (C_{L\alpha}^*)_w \eta_w \frac{l_w}{\bar{c}} \frac{S_w}{S} + (C_{L\alpha}^*)_t \eta_q \eta_t \left(1 - \frac{\partial \varepsilon}{\partial \alpha}\right) \frac{l_t}{\bar{c}} \frac{S_t}{S}\right]$$

$$+ \frac{\bar{c}}{\hat{C}_{L\alpha}}\left[C_{T\alpha} \frac{h_{cm}}{\bar{c}} \frac{\tilde{S}}{S} - (C_{M\alpha})_{\text{body}} \frac{c_b}{\bar{c}} \frac{S_b}{S}\right]$$

where,

$$\hat{C}_{L\alpha} = (C_{L\alpha})_{\text{prop}} \frac{S_p}{S} + (C_{L\alpha}^*)_w \eta_w \frac{S_w}{S} + (C_{L\alpha}^*)_t \eta_q \eta_t \left(1 - \frac{\partial \varepsilon}{\partial \alpha}\right) \frac{S_t}{S}$$

For this particular calculation, the propeller and body-lift contribution will be neglected. Therefore, the result, l_{np}', will be for the airplane in a gliding mode. First,

$$\hat{C}_{L\alpha}' = 4.636 \times 0.877 \times \frac{0.556}{0.556} + 4.244 \times 1.0 \times 0.96(1 - 0.310) \times \frac{0.140}{0.556} = 4.066 + 0.708 = 4.774$$

Aerodynamic Review

Next,

$$l'_{np} = \frac{\bar{c}}{\hat{C}'_{L\alpha}} \left[(C^*_{L\alpha})_w \eta_w \frac{l_w}{\bar{c}} \frac{S_w}{S} + (C^*_{L\alpha})_t \eta_q \eta_t \left(1 - \frac{\partial \varepsilon}{\partial \alpha}\right) \frac{l_t}{\bar{c}} \frac{S_t}{S} - (C_{M\alpha})_{body} \frac{c_b}{\bar{c}} \frac{S_b}{S} \right]$$

$$= \frac{0.3}{4.774} \left[4.636 \times 0.877 \times \frac{0.494}{0.3} \times \frac{0.556}{0.556} + 4.244 \times 1.0 \times 0.96 \times (1 - 0.310) \times \frac{1.30}{0.3} \times \frac{0.140}{0.556} \right]$$

$$- \frac{0.3}{4.774} \left(0.470 \times \frac{1.40}{0.3} \times \frac{0.0417}{0.556} \right)$$

$$= 0.0628(6.695 + 3.067) - 0.0628 \times 0.165 = 0.6131 - 0.0104 \rightarrow$$

$$l'_{np} = 0.603 \, \text{m}$$

As expected for this configuration, l'_{np} is aft of the wing's aerodynamic center, $(l_{ac})_w = 0.494$ m. Also, it is interesting to note the relatively small effect that the body has on the neutral-point location (after all those $(C_{M\alpha})_{body}$ calculations!).

The next step is to calculate the airplane's pitching moment about its neutral point, $(C_{M0})_{np}$. As presented before in the "Complete-Airplane Aerodynamics" section, the equation is, upon neglecting the propeller-thrust term:

$$(C'_{M0})_{np} = \left(\frac{l'_{np} - l_w}{\bar{c}}\right) \frac{S_w}{S} \eta_w \left[(C^*_{L0})_w + (C^*_{L\alpha})_w i_w\right]$$

$$+ \left(\frac{l'_{np} - l_t}{\bar{c}}\right) \frac{S_t}{S} \eta_t \eta_q \left[(C^*_{L0})_t + (C^*_{L\alpha})_t i_t - (C^*_{L\alpha})_t \varepsilon_0\right]$$

$$+ (C^*_{Mac})_w \eta_w \frac{S_w}{S} \frac{\bar{c}_w}{\bar{c}} + (C^*_{Mac})_t \eta_t \eta_q \frac{S_t}{S} \frac{\bar{c}_t}{\bar{c}}$$

Recall that,

$$\varepsilon_0 = \left(\frac{\varepsilon}{C_L}\right)_w [(C^*_{L0})_w + (C^*_{L\alpha})_w i_w] \eta_w = 0.076 \times (0.275 + 4.636 \times 0.05236) \times 0.877 = 0.0345$$

Therefore, one obtains:

$$(C'_{M0})_{np} = \left(\frac{0.603 - 0.494}{0.3}\right) \times \frac{0.556}{0.556} \times 0.877 \times (0.275 + 4.636 \times 0.05236)$$

$$+ \left(\frac{0.603 - 1.3}{0.3}\right) \times \frac{0.140}{0.556} \times 0.960 \times 1.0 \times (0.0 + 4.244 \times 0.0 - 4.244 \times 0.0345)$$

$$+ (-0.08) \times 0.877 \times \frac{0.556}{0.556} \times \frac{0.3}{0.3} + 0.0 \times 0.960 \times 1.0 \times \frac{0.140}{0.556} \times \frac{0.20}{0.30}$$

$$= 0.1650 + 0.0822 - 0.0702 - 0.0 \rightarrow (C'_{M0})_{np} = 0.1770$$

$$C'_{M\alpha} = \hat{C}'_{L\alpha} (l_{cm} - l'_{np})/\bar{c} = 4.774(0.497 - 0.603)/0.30 \rightarrow C'_{M\alpha} = -1.6868$$

$(C'_{M0})_{np}$ is positive, and $C'_{M\alpha}$ is negative, so a trimmed and statically stable flight condition may be obtained with the mass center ahead of the neutral point. This will be described in Chapter 7.

ASIDE

Observe that, for this example, the excrescence drag has not been explicitly included. As described in the "Bodies" section, this is caused by protrusions in the airflow, such as scoops, gaps, fabrication irregularities, etc. There are also interference-drag effects, caused by the junctions of the wings and tail to the body as well as the stabilizer and fin junctions to each other. It has been assumed that the conservative choice of body drag coefficient, $(C_D)_{body} \approx 0.27$, accounts for this. Otherwise, one can't go too far wrong to add 5 or 10% to the best estimated zero-lift drag coefficient for the clean configuration.

END ASIDE

A FINAL COMMENT

Professor Walter Vincenti, at Stanford University, would emphasize that it made no sense to calculate a number to the umpteenth decimal place if the analysis being used is approximate. This is likewise the case here. It would be amazing if the coefficients of the lift equation matched reality to the second decimal place, and it would be positively miraculous if the coefficients of the drag equation matched to the third decimal place. These discrepancies are due both to the approximations in the analyses and the inevitable imperfections of an as-built aircraft. This should be borne in mind.

3 Propeller Analysis

SIMPLE BLADE-ELEMENT ANALYSIS

A method of propeller analysis was developed by Stefan Drzewiecki in 1885 (summarized in "Theórie Général de l'Hélice Propulsive", published by Gauthier-Villars, Paris, 1920), where he assumed that the propeller is a twisted wing and that each segment follows a helical path along a relative-velocity vector, generating elemental lift and drag forces.

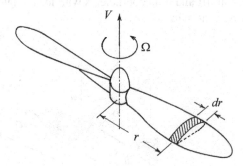

This is called "simple blade-element theory", and it assumes that the flow velocity through the propeller disc ("through-flow velocity") is the same as the flight speed. If the propeller is lightly loaded and is operating at a high advance ratio ($J \equiv V/(2R \times RPS)$), this is not a bad assumption, and the simple blade-element theory gives reasonable results.

> **ASIDE**
>
> Since aeronautical propellers were not a major consideration in 1885, Drzewiecki developed his analysis for ship propellers. These are typically wide-bladed, so the segment must be a portion of an annulus as shown.
>
>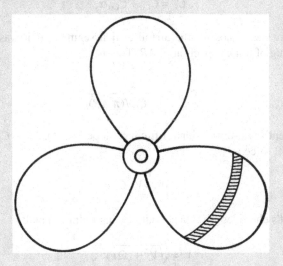

However, for thin-bladed aeronautical propellers, this is not usually a necessary refinement and one may assume a straight cut as shown above. Note, however, that optimum propellers for very high altitude airplanes can have wider blades than those for lower altitudes, in which case the blade's cross-section will have to follow the arc of the annulus.

END ASIDE

Other assumptions for the simple blade-element theory are that each segment operates aerodynamically independent of the others (like "strip theory" in aeroelasticity) and that there is no significant radial flow. Since the effects of shed vorticity are not fundamentally accounted for in this theory, one assumes that the segmental lift and drag coefficients with local angle-of-attack are those for a wing with a certain finite aspect ratio (usually 6).

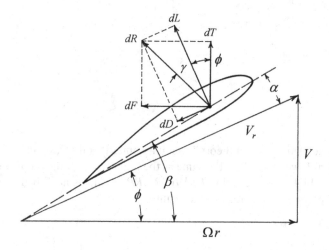

Therefore, with reference to Chapter 2, the segment's lift coefficient is given by:

$$C_L = C_{L0} + C_{L\alpha}\alpha$$

where C_{L0} and $C_{L\alpha}$ are the values for the airfoil at that segment, if it was incorporated into an untwisted elliptical wing of finite aspect ratio, AR. That is:

$$C_{L\alpha} = \frac{C_{l\alpha}}{1 + C_{l\alpha}/(\pi\, AR)}$$

where $C_{l\alpha}$ is the segment's two-dimensional lift-curve slope ($\approx 2\pi$). Also, $C_{L0} = -C_{L\alpha}\alpha_{ZLL}$
So, the segment's lift force is:

$$dL = 0.5\rho V_r^2\, C_L\, c\, dr$$

where c is the segment's chord length (blade width at that radius, r) and:

$$V_r = \sqrt{V^2 + (\Omega r)^2}\ .$$

Propeller Analysis

Also, the segment's drag coefficient is given by,

$$C_D = (C_D)_{C_L=0} + A C_L + B C_L^2 + \frac{C_L^2}{\pi \, AR}$$

So, an angle γ may be defined as:

$$\gamma = \tan^{-1}(C_D/C_L)$$

where C_D/C_L is the inverse of the airfoil's lift/drag ratio at that particular angle of attack, α.

Thus, the segment's resultant force is given by:

$$dR = dL/\cos\gamma$$

From this, the segment's contribution to the propeller's thrust is:

$$dT = dR \cos(\phi + \gamma)$$

Further, since:

$$V_r = V/\sin\phi$$

one obtains:

$$dT = \frac{0.5 \rho V^2 C_L \, c \, dr \cos(\phi + \gamma)}{\sin^2 \phi \cos\gamma}$$

It is common practice to define a parameter K as:

$$K \equiv \frac{C_L \, c}{\sin^2 \phi \cos\gamma}$$

So that a segmental thrust factor, T_c, is given as:

$$T_c = K \cos(\phi + \gamma)$$

Therefore, one has that the segment's thrust force is:

$$dT = 0.5 \rho V^2 T_c \, dr$$

If N is defined as the number of blades and R is the propeller's radius, then the total thrust is given by:

$$T = 0.5 \rho V^2 N \int_0^R T_c \, dr$$

Further, the segment's contribution to the propeller's torque is:

$$dQ = r \, dF = r \, dR \sin(\phi + \gamma)$$

Upon defining a segmental torque factor by:

$$Q_c = K r \sin(\phi + \gamma)$$

It is obtained that:

$$dQ = 0.5\rho V^2 Q_c\, dr \rightarrow Q = 0.5\rho V^2 N \int_0^R Q_c\, dr$$

The power absorbed by the propeller (or "torque power") is given by:

$$P_{in} = Q\Omega = Q(2\pi \times RPS) = Q\,(120\pi \times RPM)$$

and the delivered power (or "thrust power") is $P_{out} = TV$, so the propulsive efficiency is:

$$\eta = P_{out}/P_{in}$$

A major limitation to the simple blade-element analysis is the assumption that the through-flow velocity is equal to the flight speed. As subsequent analysis shows, this is not generally so.

ACTUATOR-DISK MOMENTUM THEORY

In 1889 the English researcher, Robert Edmond Froude, formulated the fundamental concept of the "ideal" propeller in a fluid stream. In the figure illustrated below, the propeller is modelled by an "actuator disk" that imparts a uniform pressure jump across its surface. Also, this disk is encompassed by a fluid streamtube that has no radial swirl. Further, there is no pulsing from discrete blades.

It is also important to note that the flow is with respect to the actuator disk. That is, the disk is actually moving through a quiescent fluid field at velocity V; but, for this analysis, it is most convenient to use a disk-fixed reference system (as if the propeller was in a wind tunnel).

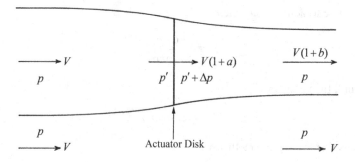

Far ahead of the disk, the flow velocity equals the flight velocity V, at a static pressure of p. Therefore, from Bernoulli's equation, the total (stagnation) pressure p_0, ahead of the disk is:

$$p_0 = p + 0.5\rho V^2$$

Likewise, when the stream-tube contracts, the flow speeds up to a velocity of $V(1+a)$ through the disk. However, the total pressure ahead of the disk is still p_0, so that:

$$p_0 = p' + 0.5\rho V^2(1+a)^2 = p + 0.5\rho V^2$$

The action of the propeller is to add energy to the flow, which manifests itself as a jump in static pressure across the disk (recall that pressure has the units of energy per unit volume). Therefore, across the disk, the total pressure is:

$$p_0' = p' + \Delta p + 0.5\rho V^2(1+a)^2 = p_0 + \Delta p$$

Propeller Analysis

Also, at a far distance downstream, the static pressure reverts to the atmospheric pressure p, though the flow velocity within the streamtube has accelerated to $V(1+b)$. Therefore,

$$p'_0 = p + 0.5\rho V^2 (1+b)^2$$

Now, since $\Delta p = p'_0 - p_0$, one has that:

$$\Delta p = [p + 0.5\rho V^2(1+b)^2] - [p + 0.5\rho V^2] = \rho V^2 b(1+0.5b)$$

Choose A to be the area of the disk, so that the propeller's thrust is given by:

$$T = A\Delta p = A\rho V^2 b(1+0.5b)$$

Also, from momentum theory, the thrust is equal to the total change of axial momentum per unit time:

$$T = \text{mass per unit time} \times \text{total increase in streamtube velocity}$$

$$T = A\rho V(1+a) \times bV = A\rho V^2(1+a)b$$

If this is equated to the previous thrust expression, one obtains:

$$T = A\rho V^2 b(1+0.5b) = A\rho V^2(1+a)b \rightarrow a = 0.5b$$

So, according to the Froude actuator-disk theory, the through-flow velocity is half of the distance downstream velocity.

ASIDE

In real fluids, the viscosity would dissipate the downstream jet. However, the veracity of this analysis can be qualitatively felt (literally) by feeling the flow ahead, immediately behind, and further behind a floor fan.

END ASIDE

There are other interesting results from this actuator-disk analysis. The power imparted to the fluid stream by the disk is given by:

$$P_{in} = \text{mass per unit time} \times \text{increase in kinetic energy per unit mass}$$

$$P_{in} = A\rho V(1+a) \times \frac{[V^2(1+b)^2 - V^2]}{2} = A\rho b V^3 (1+a)^2$$

Recall that the output power is the thrust times the flight speed, $P_{out} = TV$, so that the efficiency is given by:

$$\eta = \frac{P_{out}}{P_{in}} = \frac{TV}{A\rho b V^3 (1+a)^2} = \frac{A\rho V^3 (1+a)b}{A\rho b V^3 (1+a)^2} = \frac{1}{1+a}$$

This is, of course, an ideal efficiency that does not account for losses due to viscosity, swirl, multiple blades (and their mutual interference), etc. However, it is a useful concept by which a propeller's actual efficiency may be compared. Also, this provides guidance for what factors increase a propeller's efficiency. For example, return to the thrust expression given previously:

$$T = A\rho V^2 (1+a) b$$

With the introduction of $\eta = 1/(1+a)$ and $a = 0.5b$, this equation becomes:

$$T = \frac{2A\rho V^2 (1-\eta)}{\eta^2}$$

Define a thrust coefficient:

$$C'_T \equiv \frac{T}{0.5\rho V^2 A}$$

Then one obtains a relationship between the ideal efficiency and the thrust coefficient:

$$\frac{\eta^2}{1-\eta} = \frac{4}{C'_T}$$

One may calculate η for different values of C'_T. For example, if $C'_T = 1$, then:

$$\eta^2 = 4(\eta - 1) \rightarrow \eta^2 + 4\eta - 4 = 0 \rightarrow \eta = 0.828$$

Likewise, if $C'_T = 10$, then:

$$\eta^2 = 0.4(\eta - 1) \rightarrow \eta^2 + 0.4\eta - 0.4 = 0 \rightarrow \eta = 0.463$$

This shows that the ideal efficiency decreases as the thrust coefficient increases. One may look again at the definition of C'_T to observe what design or operational factors affect efficiency:

1. For a given flight speed, actuator area and density, efficiency decreases with thrust; that is, the higher the propeller is loaded, the less efficient it becomes.
2. For a given thrust, actuator area and density, the efficiency increases with flight speed (increasing advance ratio for an actual spinning propeller).
3. For a given thrust, actuator area and flight speed, the efficiency increases with fluid density.
4. For a given thrust, flight speed and density, the efficiency increases with actuator disk area.

The last statement is particularly valuable because it illustrates that a large-diameter slow-rotating propeller is a more efficient than a smaller, highly-rotating propeller. This often makes it worth incorporating a drive reduction unit, because the increase in propulsive efficiency frequently outweighs the mechanical losses from the unit.

EXTENSIONS TO THE SIMPLE BLADE-ELEMENT ANALYSIS

As stated previously, an upgrade to the simple blade-element analysis is to incorporate the actuator-disk theory to account for a through-flow velocity that differs from flight speed. In this case, one may choose two-dimensional aerodynamic values for every segment's airfoil because the actuator-disk theory gives the effective average "downwash" velocity that modifies the local angles of attack. Recall that the simple blade-element analysis had approximated this effect by assuming a finite aspect ratio in order to calculate each segment's C_L vs. α and C_D vs. C_L expressions. Now, with the incorporation of the actuator-disk through-flow velocity, the aerodynamic values at each segment are those for a wing with an infinite aspect ratio (the "profile" values).

Propeller Analysis

The solution is iterative, and the first step is to calculate the propeller's thrust without the through-flow velocity. Then, upon knowing this, the flight velocity and the disk area, one may calculate a through-flow velocity and thus a revised thrust (which will usually be lower). The iteration is continued until satisfactory convergence is achieved.

Observe that this has assumed a constant through-flow velocity across the propeller's disk. However, in reality, the through-flow velocity is rarely uniform across the disk, and refinement is to perform the calculations for annuli formed by the rotational sweep of the propeller segments. One may draw upon a variation of the actuator-disk derivation, applied to the annuli, to obtain:

$$dT = \text{mass per unit time} \times \text{total increase in annular-streamtube velocity}$$

$$dT = 2\pi r\, dr\, V(1+a)\rho \times 2aV = 4\pi \rho V^2 (1+a) a\, r\, dr$$

Therefore, the same solution iteration is performed as described above, but only for each annulus. However, momentum theory offers yet another refinement. Up to this point, the rotational swirl imparted to the flow has been neglected. In truth, the swirl ahead of the propeller is negligible. However, the flow is given a significant swirl upon passing through the propeller disk. A simple experiment with a floor fan and a thread taped to a slender stick would clearly illustrate this.

The same reasoning that was applied to the axial-momentum theory can likewise be used to obtain the relationship between the torque of an annular segment and the rotational momentum imparted to that segment, Accounting for the fact that the swirl tangential velocity is doubled aft of the disk:

$$dQ = \text{mass per unit time} \times \text{increase in swirl tangential velocity} \times \text{radius}$$

$$dQ = 2\pi r\, dr\, V(1+a)\rho \times 2w \times r$$

It is seen that w is the variable chosen to represent the swirl tangential velocity at the specific radius on the disk. Also, if a "rotational interference factor" a', is introduced as:

$$a' = \frac{w}{\Omega r} = \frac{w}{2\pi r \times RPS}$$

then dQ may be expressed as:

$$dQ = 4\pi \rho\, V(1+a)(2\pi r \times RPS) a' r^2 dr$$

In light of these factors, a revised blade-element diagram may be drawn:

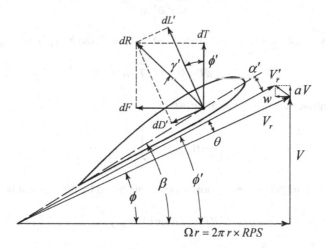

From this, one has that:

$$dL' = 0.5\rho(V_r')^2 C_l\, c\, dr$$

where C_l is the two-dimensional lift coefficient of the segment's airfoil:

$$C_l = C_{lo} + C_{l\alpha}\alpha'$$

Likewise, the segment's profile drag is given by:

$$C_d = (C_d)_{C_l=0} + AC_l + BC_l^2$$

Therefore, γ' is given by:

$$\gamma' = \tan^{-1}(C_d/C_l)$$

And the resultant force on the segment is:

$$dR = 0.5\rho(V_r')^2 C_l\, c\, dr / \cos\gamma'$$

Thus, the element's thrust is given by:

$$dT = dR\cos(\phi' + \gamma')$$

Further, since one has that:

$$V_r' = V(1+a)/\sin\phi'$$

then the segmental thrust is:

$$dT = \frac{0.5\rho V^2 (1+a)^2 C_l\, c\, dr \cos(\phi' + \gamma')}{\sin^2\phi' \cos\gamma'}$$

Now, let the parameter K become:

$$K \equiv \frac{C_l\, c\, (1+a)^2}{\sin^2\phi' \cos\gamma'}$$

Also, the segment's thrust factor becomes redefined as $T_c \equiv K\cos(\phi' + \gamma')$, which then gives that:

$$dT = 0.5\rho V^2 T_c\, dr$$

As before, this integrates to give the total thrust:

$$T = 0.5\rho V^2 N \int_0^R T_c\, dr$$

Next, the torque about the propeller rotational axis contributed by the element is given by:

$$dQ = r\, dR \sin(\phi' + \gamma')$$

Propeller Analysis

Upon defining a segment torque factor as $Q_c \equiv K r \sin(\phi' + \gamma')$, then one has that:

$$dQ = 0.5\rho V^2 Q_c \, dr$$

Thus, the complete torque value is obtained from:

$$Q = 0.5\rho V^2 N \int_0^R Q_c \, dr$$

The propulsive efficiency equation remains the same as that presented for the simple blade-element analysis:

$$\eta = \frac{P_{out}}{P_{in}} = \frac{TV}{Q\Omega} = \frac{TV}{Q 2\pi \times RPS} = \frac{TV}{120\pi Q \times RPM}$$

As stated before, the strength of this extended blade-element analysis is that the two-dimensional aerodynamic characteristics of each element may be used (two-dimensional lift and drag coefficients with respect to the local angles of attack). However, the limitations to this theory are that attached flow is assumed, and there is negligible radial velocity. Subject to this, the analysis gives very good comparisons with experimental data, as seen in the "Numerical Example" section.

METHOD OF CALCULATION

The solution follows the methodology described by William H. Miller and Edward Harpoothian (Report 3190, Curtiss Aeroplane and Motor Company, 1928), which depends upon finding a solution for the "downwash" angle θ. The first step in achieving this is to equate the expressions for dT from the momentum theory and the blade-element theory (accounting for the number of blades):

$$dT = 4\pi \rho V^2 (1+a) a r \, dr = \frac{0.5 \rho V^2 N (1+a)^2 C_l \, c \, dr \cos(\phi' + \gamma')}{\sin^2 \phi' \cos \gamma'} \rightarrow$$

$$\frac{a}{(1+a)} = \frac{N C_l c \cos(\phi' + \gamma')}{8\pi r \sin^2 \phi' \cos \gamma'}$$

A new parameter is defined, which is the ratio of the circumference of the annulus to the total blade width:

$$S \equiv \frac{2\pi r}{Nc}$$

Then one obtains that:

$$\frac{a}{1+a} = \frac{C_l \cos(\phi' + \gamma')}{4 S \sin^2 \phi' \cos \gamma'} \tag{3.1}$$

A similar exercise may be performed by equating the momentum and blade-element (BE) expressions for torque:

$$(dQ)_{momentun} = 4\pi \rho V (1+a) \Omega r a' r^2 dr$$

$$(dQ)_{BE} = dR r \sin(\phi' + \gamma') = 0.5 N \rho (V_r')^2 C_l c r \sin(\phi' + \gamma') dr / \cos \gamma'$$

Then $(dQ)_{\text{momentum}} = (dQ)_{\text{BE}}$ gives,

$$4\pi\Omega a' r^2 = \frac{0.5 N V C_1 c \sin(\phi'+\gamma')(1+a)}{\sin^2\phi'\cos\gamma'} \rightarrow$$

$$\frac{a'}{(1+a)} = \frac{N V c}{2\pi r^2 \Omega} \frac{C_1 \sin(\phi'+\gamma')}{4\sin^2\phi'\cos\gamma'} \rightarrow \frac{Nc}{2\pi r} \frac{V}{\Omega r} \frac{C_1 \sin(\phi'+\gamma')}{4\sin^2\phi'\cos\gamma'} \rightarrow$$

$$\frac{a'}{1+a} = \frac{C_1 \tan\phi \sin(\phi'+\gamma')}{4S\sin^2\phi'\cos\gamma'} \tag{3.2}$$

From equations (3.1) and (3.2), one may obtain:

$$a' = a\tan\phi\tan(\phi'+\gamma') \tag{3.3}$$

Also, from the illustration of the blade element, it is seen that:

$$\tan\phi = \frac{V}{\Omega r} = \frac{V}{\Omega r}\frac{(1-a')}{(1-a')} = \frac{(1-a')}{(1+a)}\frac{V(1+a)}{(\Omega r - \Omega r a')} = \frac{(1-a')}{(1+a)}\frac{V(1+a)}{(\Omega r - w)} = \frac{(1-a')}{(1+a)}\tan\phi'$$

Therefore, a solution for a' is:

$$a' = 1 - (1+a)\frac{\tan\phi}{\tan\phi'} \tag{3.4}$$

Upon equating equations (3.3) and (3.4), the following expression is found:

$$a\tan\phi\tan(\phi'+\gamma') = (1+a) - a - (1+a)\tan\phi/\tan\phi' \rightarrow$$

$$a + a\tan\phi\tan(\phi'+\gamma') = (1+a) - (1+a)\tan\phi/\tan\phi' \rightarrow$$

$$a + a\tan\phi\tan(\phi'+\gamma') = (1+a)\tan\phi'/\tan\phi' - (1+a)\tan\phi/\tan\phi' \rightarrow$$

$$a[1+\tan\phi\tan(\phi'+\gamma')] = (1+a)\cot\phi'(\tan\phi' - \tan\phi) \rightarrow$$

$$\frac{a}{(1+a)} = \frac{\cot\phi'(\tan\phi' - \tan\phi)}{1+\tan\phi\tan(\phi'+\gamma')} \tag{3.5}$$

Equations (3.1) and (3.5) may now be equated to give:

$$\frac{S}{C_1} = \frac{\cos(\phi'+\gamma') + \tan\phi\sin(\phi'+\gamma')}{4\sin\phi'\cos\phi'(\tan\phi' - \tan\phi)\cos\gamma'} = \frac{1 - \tan\gamma'\tan(\phi'-\phi)}{4\sin\phi'\tan(\phi'-\phi)}$$

And since $\theta = \phi' - \phi$, this may be written as:

$$\frac{S}{C_1} = \frac{1 - \tan\gamma'\tan\theta}{4\sin(\phi+\theta)\tan\theta} \tag{3.6}$$

Propeller Analysis

It is seen that this is an implicit equation for θ, and an iterative solution is required. An example solution will illustrate this. First, though, one more factor is required, which is obtained from equation (3.5):

$$1 + a = \frac{\tan\phi'[1 + \tan\phi\tan(\phi' + \gamma')]}{\tan\phi[1 + \tan\phi'\tan(\phi' + \gamma')]} \tag{3.7}$$

NUMERICAL EXAMPLE

A numerical example will be chosen from "Aircraft Propeller Design" by Fred E. Weick, McGraw-Hill, 1930. This particular propeller is one of many different designs that were tested in the Stanford University wind tunnel in the 1920s.

r (m)	c (m)	β (deg)	$(C_l)_{C_l=0}$	$C_{l\alpha}$ (1/deg)	t (thickness in m)	A	B
0.0686	0.0686	56.1	0.062	0.11	0.0254	−0.006	0.05
0.1372	0.0719	36.6	0.465	0.11	0.0186	−0.046	0.05
0.2057	0.0762	26.4	0.635	0.11	0.0133	−.063	0.05
0.2743	0.0719	20.4	0.564	0.11	0.0096	−0.056	0.05
0.3429	0.0604	16.6	0.456	0.11	0.0065	−0.045	0.05
0.4115	0.0412	13.9	0.374	0.11	0.0037	−0.037	0.05

Further, the diameter is 0.9144 m, the flight speed is $V = 17.88$ m/sec, and the rotational speed is 30 $RPS \rightarrow \Omega = 188.5$ rad/sec. Also, recall that A is the coefficient in the profile-drag equation for C_l and B is the coefficient in the profile-drag equation for C_l^2.

Note that only six segments are given, for which the values in the table are taken at the midpoints of the segments. This may seem like crude spacing, but it will be shown that this is actually sufficient. Now, consider the fifth segment:

$$\phi = \tan^{-1}\left(\frac{V}{r\Omega}\right) = 15.46°, \quad S = \frac{2\pi r}{Nc} = 17.835$$

Also, $\alpha' = \beta - \phi - \theta = (16.6° - 15.46°) - \theta \rightarrow \alpha' = 1.14° - \theta$.

At the time Weick's book was written, computerized solutions were not possible. So, he described a hand-solution iterative method where an initial guess was made for θ, from which an initial value of α' was calculated. Then, with this α', C_l was obtained. Also, he observed that for equation (3.6), $\tan\gamma'\tan\theta \ll 1$. Therefore, the equation simplifies to:

$$\frac{S}{C_l} = \frac{1}{4\sin(\phi+\theta)\tan\theta} \tag{3.8}$$

Upon the substitution of C_l, S and ϕ, one may solve for a revised value of θ from charts in his book. The iteration is then continued until the θ value converges.

However, with modern programming capabilities, a solution may be obtained in a very straightforward manner. First, equation (3.8) is rewritten as:

$$\frac{S}{C_l} - \frac{1}{4\sin(\phi+\theta)\tan\theta} = \delta$$

Then, the δ value is calculated for numerous small steps of θ over a large range. When δ changes sign, interpolation is used to find the θ value for which $\delta = 0$. In this numerical example, $\theta = 1.243° \rightarrow \alpha' = -0.10°$.

Even easier, as seen in the Matlab program presented later, one may simply use the **fsolve** operation to obtain θ.

The γ' term is important for subsequent calculations, so this is obtained by first determining C_d:

$$C_d = (C_d)_{C_l=0} + AC_l + BC_l^2$$

As stated before, $(C_d)_{C_l=0}$ is the profile drag when the lift coefficient equals zero. If aerodynamic data for this airfoil at the appropriate RN is not readily found, the $(C_d)_{C_l=0}$ value may be adequately estimated in the manner described in the "Airfoils" section of Chapter 2.

$$(C_d)_{C_l=0} = (C_d)_{\text{fric}} + (1-\eta_{le})2\pi\alpha_{ZLL}^2 = 2C_{\text{fric}}[1+2(t/c)+60(t/c)^4]+2\pi(1-\eta_{le})\alpha_{ZLL}^2$$

where α_{ZLL} is found from:

$$\alpha_{ZLL} = -C_{lo}/C_{l\alpha} = -0.456/0.11 = -4.15° = -0.072 \text{ rad}$$

and η_{le} is found from:

$$B = (1-\eta_{le})/(2\pi) = 0.05 \rightarrow (1-\eta_{le}) = 0.05 \times 2\pi = 0.314 \rightarrow \eta_{le} = 0.686$$

Therefore,

$$(C_d)_{C_l=0} = 2C_{\text{fric}}[1+2(t/c)+60(t/c)^4]+2\pi(1-0.686)(-0.072)^2 \rightarrow$$

$$(C_d)_{C_l=0} = 2.447 C_{\text{fric}} + 0.010$$

where,

$$C_{\text{fric}} = \frac{0.455}{(\log_{10} RN)^{2.58}}, \quad RN \approx 6.85 V_r\, c \times 10^4 \text{ (standard sea-level conditions)}$$

Now,

$$V_r = [V^2 + (\Omega r)^2]^{1/2} = [17.88^2 + 64.64^2]^{1/2} = 67.07 \text{ m/sec} \rightarrow RN = 2.775 \times 10^5 \rightarrow$$

$$C_{\text{fric}} = 0.0058 \rightarrow (C_d)_{C_l=0} = 0.0242$$

Note that the A term in the profile-drag equation was obtained from:

$$A = 2(1-\eta_{le})\alpha_{ZLL} = 2 \times 0.314 \times (-0.072) = -0.0452$$

Propeller Analysis

Next, the C_l value was calculated in the course of the program (in order to evaluate the S/C_l term) and was found to be $C_l = 0.445$. Of course, this may also be confirmed by using the $\alpha' = -0.10$ value to calculate:

$$C_l = C_{lo} + C_{l\alpha}\alpha' = 0.456 + 0.11 \times (-0.10) = 0.445 \quad \text{(note: } \alpha' \text{ in degrees)}$$

Further,

$$(C_d)_{C_l=0} + AC_l + BC_l^2 = 0.0242 - 0.045 \times 0.445 + 0.05 \times (0.445)^2 = 0.0141$$

Therefore, γ' is now found:

$$\gamma' = \tan^{-1}(C_d/C_l) = \tan^{-1}(0.0141/0.445) = 1.81°$$

> **ASIDE**
>
> The value of $\tan\gamma' \tan\theta$ may be calculated and is found to be 0.0007, which confirms Weick's statement that this is negligible compared with unity.
>
> **END ASIDE**

Now, the value of ϕ' is found from:

$$\phi' = \phi + \theta = 15.46° + 1.24° = 16.70°$$

Therefore, equation (3.7) may now be used to find $(1+a)$:

$$1 + a = \frac{\tan 16.70°[1 + \tan 15.46° \tan(16.70° + 1.81°)]}{\tan 15.46°[1 + \tan 16.70° \tan(16.70° + 1.81°)]} = \frac{0.300[1 + 0.277 \times 0.335]}{0.277[1 + 0.300 \times 0.335]}$$

$$(1+a) = \frac{0.3278}{0.3048} = 1.075$$

This gives all the values required to obtain K:

$$K \equiv \frac{C_l c\, (1+a)^2}{\sin^2\phi' \cos\gamma'} = \frac{0.445 \times 0.0604 \times (1.075)^2}{\sin^2 16.7° \cos 1.81°} = \frac{0.03106}{0.08254} = 0.3763$$

Finally, the segment's thrust factor, T_c is found from:

$$T_c = K \cos(\phi' + \gamma') = 0.3763 \cos(16.70° + 1.81°) = 0.3568$$

Likewise, the torque factor is given by:

$$Q_c = K r \sin(\phi' + \gamma') = 0.3763 \times 0.3429 \sin(16.70° + 1.81°) = 0.0410$$

Upon programming this analysis, the T_c and Q_c values for all six segments are found to be:

r(m)	T_c(m)	Q_c(m²)
0.0686	0.0023	0.0004
0.1372	0.0811	0.0091
0.2057	0.2211	0.0248
0.2743	0.3222	0.0366
0.3429	0.3579	0.0412
0.4115	0.3080	0.0360

Observe that the T_c and Q_c values for the fifth segment differ slightly from those obtained by the previous hand calculation. This is because of the much-reduced round-off error from the program.

The final step is the integration to obtain:

$$T = 0.5\rho V^2 N \int_0^R T_c \, dr, \quad Q = 0.5\rho V^2 N \int_0^R Q_c \, dr$$

For discrete-element integration, this is rather crude spacing. Modern computer programming would allow many more segments in order to converge on an accurate solution. However, each segment would require specification of that airfoil's aerodynamic properties. Since the airfoils generally differ along the radius of a propeller, it could be a time-consuming task to obtain the properties of each. Therefore, a different solution method will be used, based on that described by Weick. Although this method was motivated by the limitations of hand calculation, it is interesting that a computerized version of this is still most suitable for the solution.

The first step is to use the above values to obtain curve-fitted equations, the plots of which are shown:

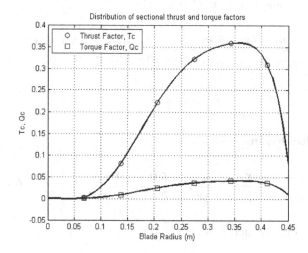

It is found that a seventh-order polynomial suffices to give excellent curve fits. Then, for sea-level conditions, $\rho = 1.225\,\text{kg/m}^2$, and two blades, $N = 2$, the numerical integration of these polynomials results in the following thrust and torque values:

$$T = 0.5\rho V^2 N \int_0^R T_c \, dr = 34.402\,N, \quad Q = 0.5\rho V^2 N \int_0^R Q_c \, dr = 3.940\,\text{N-m}$$

Propeller Analysis

Further, the input power and propulsive efficiency are calculated to be:

$$P_{in} = 742.65\,W \rightarrow \eta = 0.828$$

From wind-tunnel experiments on this propeller, the following values were obtained:

$$T_{exp} = 34.563\,N, \quad (P_{in})_{exp} = 800.14\,W, \quad \eta_{exp} = 0.771$$

These are reasonable comparisons, which speaks well for the veracity of the analysis.

Recall that fully-attached flow has been assumed. This has to be confirmed for every solution. In this case, a plot of α' vs. r is shown below:

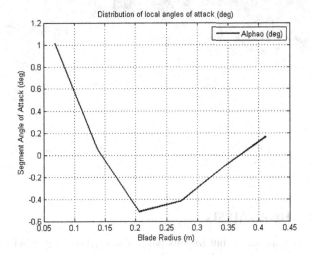

Since the highest value of α' is $\approx 1.0°$, it may be safely assumed that all segments are operating well below stall.

ASIDE

Weick also calculated the performance of the example propeller using simple blade-element analysis. What he obtained is:

$$T = 33.006\,N, \quad P_{in} = 710.94\,W, \quad \eta = 0.830$$

These values are not quite as close to the experimental values as those from the extended analysis. However, they are surprisingly "ballpark" accurate. That is, the simple blade-element analysis should generally give good initial estimates, as long as the fundamental assumptions of the theory are fulfilled.

A photo from "The New Art of Flying" by Waldemar Kaempffert (Dodd, Mead, and Company, 1911) shows a Wright Brothers' propeller that was designed by using a version of the simple blade-element theory (performing the calculations on the backs of scrap wallpaper!). Although Wilber calculated an η of 0.66, modern wind-tunnel tests on replica propellers show η values closer to 0.8. One need only look at the elegant shape of their design to not be surprised by this.

END ASIDE

APPLICATION OF THE ANALYSIS

There are various ways to proceed, but for this particular methodology it will be assumed that an aircraft and engine have been identified and that it is now required to design or select an appropriate propeller.

First of all, it must be decided which flight condition the propeller must be designed for. This could be maximum speed, maximum climb rate, cruising level flight, and maximum C_L/C_D, etc. In this case, the propeller will be selected to operate efficiently at a given level-flight cruise condition, identified by a specific value of lift coefficient, $(C_L)_{design}$. The steps are now as follows:

1. With the design lift coefficient, the flight speed may be calculated from:

$$V = \left(\frac{2W}{\rho C_L S} \right)^{1/2}$$

   where W is the aircraft's weight.
2. The propeller's required thrust is now found from:

$$T = D = 0.5 C_D \rho V^2 S$$

   And, from the "Complete Aircraft Aerodynamics" section of Chapter 2, one has:

$$C_D = C_{D0} + C_{D1}\alpha + C_{D2}\alpha^2$$

Propeller Analysis

where,

$$\alpha = \frac{C_L - C_{L0}}{C_{L\alpha}}$$

3. The required power for flight is now simply obtained from:

$$P_{out} = D \times V$$

4. At this point, one should look at the power and torque curves for the candidate motor (or motor-gearbox unit)

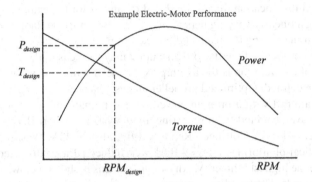

A direct-current electric motor has been chosen for this exposition, but this methodology would also work for the power and torque curves of an internal combustion engine.
Now, because $\eta = P_{out}/P_{in}$, the required design power is larger than the required output power:

$$P_{in} = P_{design} = P_{out}/\eta$$

Upon having confidence that an efficient propeller will result from these calculations, an initial estimate of $\eta_{estimated} = 0.8$ may be made. Then the resulting value of P_{design} should be compared with the motor's power curve. If this value is above the motor's maximum power, then clearly, the aircraft is not capable of sustained flight. If properly selected, though, the design power P_{design}, will fall comfortably upon the left-hand side of the curve, as shown above. This gives a power margin for climbing and maximum speed.

ASIDE
One could also match power on the right side of the curve. However, the low torque available and corresponding high *RPM* would give a very inefficient arrangement.

END ASIDE

5. After identifying the P_{design} point on the power curve, one may now find the *RPM* value and the corresponding torque that will be supplied to the propeller. Therefore, the task is to design a propeller to a specified thrust and torque, given an *RPM* value.
6. The next parameter that needs to be determined is the propeller's diameter. Generally speaking, a large diameter is desired to maximize the propulsive efficiency, as described

in the "Actuator-Disk Momentum Theory" section. However, for a given *RPM*, too large a diameter may result in "toothpick" thin blades with low sectional aerodynamic efficiency. Also, propeller diameter may be limited by ground clearance considerations. At any rate, an initial value should be chosen based on typical sizes for similar aircraft. Also, if the diameter is particularly restricted, propellers with more than two blades may be required.

7. With the flight speed and the *RPM*, the ϕ value for every segment may be found:

$$\phi = \tan^{-1}\left(\frac{V}{\Omega r}\right) = \tan^{-1}\left(\frac{V}{2\pi r\, RPS}\right) = \tan^{-1}\left(\frac{30V}{\pi r\, RPM}\right)$$

To this, an additional angle is added to give the pitch angle β. It suffices, for the beginning of this iterative process, to calculate ϕ at the 0.75 Blade Radius location and add an angle of 2° to fix β at this location (recall the θ and α' angles for the example propeller). Then upon assuming a linearly-varying helical twist, with 90° at Blade Radius = 0, the β distribution is determined along the blade. Of course, if one is trying to assess the suitability of a manufactured propeller, then the β distribution may be measured.

8. The next part of the iteration is the assumption of the chord distribution. For this, a blade shape has to be defined. Optimized propeller analysis has become popular in recent years for low-speed aircraft with minimum power, such as human and solar-powered airplanes. Such aircraft have benefited from revisiting an analysis by Albert Betz ("Airscrews with Minimum Energy Loss", Göttingen Reports, 1919, also NACA, Washington DC, 1922). Just as an elliptical-planform wing gives the lowest induced drag, a distorted elliptical blade shape can give the highest efficiency. An example of this is shown below, which is a 44 cm long propeller blade developed for a microwave-powered airplane ("SHARP": Stationary High-Altitude Relay Platform): by Roland Lorenz ("An Experimental Investigation of Low-Speed Single-and Dual-Rotating Propellers", MASc. Thesis, University of Toronto Institute for Aerospace Studies, 1987).

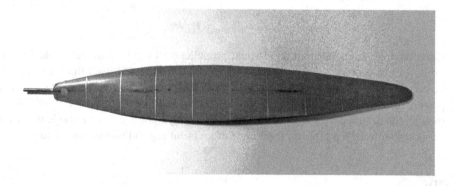

However, just as good-flying airplanes exist with straight-tapered and even rectangular wings, one needn't exactly match a Betz blade shape to have a satisfactory propeller. Instead, a reasonable elliptic curve may be assumed. This can be expressed in equation form with a scaling parameter in order to readily change the chord widths during the iteration process. Again, though if a manufactured propeller is being analyzed, the chord distribution is simply measured.

9. At this point, the initial performance values of thrust and torque are obtained, as well as the efficiency, and if these do not match the desired values, the iterative process is continued with different pitch and chord-width distributions. Note that the new η values must be used to adjust P_{design}. With computerized methods, it doesn't take long to identify a design space with the two scaling parameters for $\beta(r)$ and $c(r)$.

PROPELLER AIRFOILS

The airfoil shapes along the length of a propeller blade can change considerably, and not all of these have wind-tunnel data available. Therefore, sophisticated prediction programs (such as XFOIL) may be used to obtain this information. However, for the type of simple propellers typically made for model airplanes, the methods described in the "Airfoils" section of Chapter 2 often suffice. Indeed, these were used for the numerical-example propeller.

It is worth revisiting this methodology, as applied to model-airplane propellers that are often flat-bottomed with fairly-sharp leading edges:

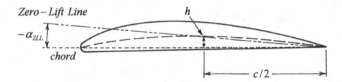

First of all, if the camber line approximates a circular arc, then the zero-lift line is approximately identified as passing through two points: the trailing edge and the camber line at the half-chord location. Therefore, the angle of the zero-lift line to the chord is:

$$\alpha_{ZLL} \approx -\tan^{-1}\left(\frac{h}{c/2}\right) \rightarrow C_{lo} \approx -C_{l\alpha}\,\alpha_{ZLL}$$

Further, if the leading edge is sharp enough, the chord line coincides with the flat bottom, which makes measuring the pitch angle β much easier.

Based on what was discussed about leading-edge suction in the "Airfoils" section, it would seem that a considerable penalty is incurred by using a sharp leading edge. However, as the numerical example showed, the airfoils for an efficient propeller may operate at surprisingly low angles of attack (even slightly negative for some segments). This means that these airfoils may be close to their "ideal angles of attack", for which leading-edge suction efficiency is not a major factor. Also, at the low RN's of model-airplane propeller airfoils, a fairly sharp leading edge has some aerodynamic advantages because of a phenomenon known as the "leading-edge separation bubble". With the leading edge designed right, this may enhance the airfoil's lift characteristics.

For small-scale model-airplane propellers, the leading-edge suction efficiency may be assumed to be zero. Of course, for larger-scale propellers, the factors that are important for a wing's airfoil also apply to those for the propeller's airfoils.

MATLAB PROGRAM

A computer program is presented for the calculation of a candidate propeller's thrust, torque, power absorbed, P_{in}, and propulsive efficiency. This uses the extended blade-element analysis described previously, and the values in the program are for the numerical example. Also, metric units are used. However, these may be easily converted to the Imperial (ft, lb) system by changing the units for the propeller's dimensions, atmospheric density and the equation for the Reynolds number (the imperial equation for RN is given in Chapter 2.

```
clc
clear all
%Date: 23 February, 2014
%Propeller Analysis using The Weick Method
%Stanford Propeller
ro=1.225; %Atmospheric density in kg/m^3
R=0.4572; %Propeller radius (m)
```

```
N=2; %Number of blades
RPM=1800.0;
RPS=RPM/60;
V=17.88; %Flight speed (m/sec)
%Stations along the blade (m)
r(1)=0.0686;
r(2)=0.1372;
r(3)=0.2057;
r(4)=0.2743;
r(5)=0.3429;
r(6)=0.4115;
%Chord widths along the blade (m)
c(1)=0.0686;
c(2)=0.0719;
c(3)=0.0762;
c(4)=0.0719;
c(5)=0.0604;
c(6)=0.0412;
%Lift coefficient of section when alphao equals zero
Clo(1)=0.062;
Clo(2)=0.465;
Clo(3)=0.635;
Clo(4)=0.564;
Clo(5)=0.456;
Clo(6)=0.374;
%2D Lift-curve slope of sections (per degree)
Clalpha(1)=0.11;
Clalpha(2)=0.11;
Clalpha(3)=0.11;
Clalpha(4)=0.11;
Clalpha(5)=0.11;
Clalpha(6)=0.11;
%Geometrical pitch angle along the blade (deg)
beta(1)=56.1;
beta(2)=36.6;
beta(3)=26.4;
beta(4)=20.4;
beta(5)=16.6;
beta(6)=13.9;
%Segment thickness (m)
t(1)=0.0254;
t(2)=0.0186;
t(3)=0.0133;
t(4)=0.0096;
t(5)=0.0065;
t(6)=0.0037;
%Profile drag variation with lift coefficient
A(1)=-0.006;
A(2)=-0.046;
A(3)=-0.063;
A(4)=-0.056;
A(5)=-0.045;
A(6)=-0.037;

%Profile drag variation with lift-coefficient squared
B(1)=0.05;
```

Propeller Analysis

```
B(2)=0.05;
B(3)=0.05;
B(4)=0.05;
B(5)=0.05;
B(6)=0.05;

for I=[1:1:6]
vt=2*pi*r(I)*RPS; %Section tangential velocity
phirad=atan(V/vt);
phideg=phirad*180/pi;
Sfactor=2*pi*r(I)/(N*c(I));
Alphafactor=beta(I)-phideg;
Cl0=Clo(I);
ClAlpha=Clalpha(I);

%Use of MATLAB non-linear equation solver
x0=[1]; % initial guess for theta value
options=optimset('Display','iter'); % history of the iteration
[X]=fsolve('prop1',x0,options,Sfactor,Cl0,ClAlpha,Alphafactor,phirad);

Theta=X(1); % solution for theta (deg)
Alpha0(I)=Alphafactor-Theta; %Calculation of segment angle of attack
CL=Clo(I)+Clalpha(I)*Alpha0(I); %Calculation of segment lift coefficient
Vrel=sqrt(V^2+vt^2); %Approximate relative velocity
RN=6.71*Vrel*c(I)*10^4; %Segment Reynolds number in metric system
Cf=0.455/((log10(RN))^2.58);
alphaZLL=(Clo(I)/Clalpha(I))/57.3;
etals=1-2*pi*B(I);
%Profile drag coeff. at zero lift coeffient
Cdo=2*Cf*(1+2.0*(t(I)/c(I))+60.0*(t(I)/c(I))^4)+2*pi*(1-
etals)*alphaZLL^2;
CD=Cdo+A(I)*CL+B(I)*CL^2;
gammarad=atan(CD/CL);
gammadeg=gammarad*180/pi;
phiodeg=phideg+Theta;
phiorad=phiodeg*pi/180;
Afactor=tan(phiorad)/tan(phirad)*(1+tan(phirad*tan(phiorad+gamma
rad)))/...
   (1+tan(phiorad*tan(phiorad+gammarad)));
K=CL*c(I)*Afactor^2/((sin(phiorad))^2*cos(gammarad));
Tc(I)=K*cos(phiorad+gammarad);
Qc(I)=K*r(I)*sin(phiorad+gammarad);
end

%Curve fitting of Tc and Qc:
Order-7;
radius=[0.0,r,R];
deltathrust=[0.0,Tc,0.0];
PTc=polyfit(radius,deltathrust,Order);
rp=0:0.01:R;
tp=polyval(PTc,rp);
deltatorque=[0.0,Qc,0.0];
PQc=polyfit(radius,deltatorque,Order);
rp=0:0.01:R;
qp=polyval(PQc,rp);
%Plotting of results
```

```
plot(r,Tc,'bo')
grid on
xlabel('Blade Radius (m)')
ylabel('Tc, Qc')
title('Distribution of sectional thrust and torque factors')
hold on
plot(r,Qc,'rs')
hold on
legend('Thrust Factor, Tc','Torque Factor, Qc','Location','NW')
plot(rp,tp,'linewidth',2)
hold on
plot(rp,qp,'linewidth',2)

%Numerical integration for Thrust and Torque
ThrustCoeff=integral(@(xp)PTc(1)*xp.^7+PTc(2)*xp.^6+PTc(3)*xp.^5+PT
c(4)*xp.^4+PTc(5)*xp.^3+PTc(6)*xp.^2+PTc(7)*xp,0,R);
TorqueCoeff=integral(@(xp)PQc(1)*xp.^7+PQc(2)*xp.^6+PQc(3)*xp.^5+PQ
c(4)*xp.^4+PQc(5)*xp.^3+PQc(6)*xp.^2+PQc(7)*xp,0,R);

%Results
Thrust=0.5*ro*V^2*N*ThrustCoeff
Torque=0.5*ro*V^2*N*TorqueCoeff
Efficiency=Thrust*V/(Torque*2*pi*RPS)
PowerAbsorbedWatts=Torque*2*pi*RPS

figure
JA=[1:1:6];
plot(r(JA),Alpha0(JA),'linewidth',2)
grid on
xlabel('Blade Radius (m)')
ylabel('Segment Angle of Attack (deg)')
title('Distribution of local angles of attack (deg)')
legend('Alphao (deg)','Location','B')
```

Note that the following subroutine stands as a separate program within the same folder of programs.

```
function F=prop1(X,Sfactor,Cl0,ClAlpha,Alphafactor,phirad)

thetadeg=X(1);

F=Sfactor/(Cl0+ClAlpha*(Alphafactor-thetadeg))-...
    1/(4*sin(phirad+thetadeg*pi/180)*tan(thetadeg*pi/180));
```

4 Flying Wings (or Tailless Airplanes)

"FLYING PLANKS"

Flying-Wing type airplanes can take on a variety of configurations. The most basic configuration is the so-called "Flying Plank", which consists of an un-swept wing (usually rectangular) with no twist and a reflexed airfoil. An example of this is the 1922 Simplex-Arnoux racer shown below (provided by www.passionpourlaviation.fr):

Courtesy of www.passionpourlaviation.fr

This innovative airplane flew successfully and was designed by Rene Arnoux, who is credited with inventing the Flying-Plank concept. By the way, the beer-keg-looking thing on the top is a Lamblin radiator used for engine cooling. It was a common feature for airplanes of that period, though it was drag-producing and, in this case, obstructive to forward vision.

A Flying Plank has to use a reflexed airfoil to achieve inherent stable trim:

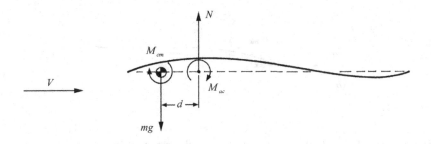

DOI: 10.1201/9781315228167-4

For example, the moment about the mass center is given by:

$$M_{cm} = M_{ac} - Nd, \quad \text{where } N \text{ is the aerodynamic force normal to the chord}$$

For the trim condition, $M_{cm} = 0$, therefore one has that:

$$M_{ac} = Nd \rightarrow d = M_{ac}/N$$

In coefficient form this becomes:

$$d = \frac{0.5\rho V^2 C_{Mac} \bar{c} S}{0.5\rho V^2 C_N S} = \frac{C_{Mac} \bar{c}}{C_N} \rightarrow \frac{d}{\bar{c}} = \frac{C_{Mac}}{C_N}$$

In typical equilibrium flight C_N is positive, so d is a positive value only if C_{Mac} is positive. This is the case if the airfoil is sufficiently reflexed, such as for an example in Chapter 2.

A positive value of d means that, for the trim condition, the mass center is forward of the aerodynamic center. This also means that static stability is achieved, as demonstrated below:

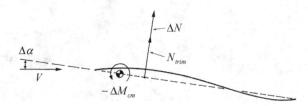

For a positive pitch perturbation, $\Delta\alpha$, the normal force, N, is increased by an amount ΔN. This causes a negative (nose down) pitching moment about the mass center, $-\Delta M_{cm}$, that drives the wing back to its trim condition. The opposite occurs for a negative pitch perturbation:

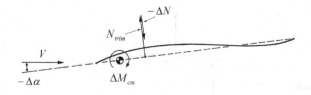

Therefore, the reflexed airfoil allows a design that can fly in a stable trim condition.

In order to demonstrate this inherent stability, a model Flying-Plank glider was built by the author for his Aircraft Design course. Note the degree of reflex required for the airfoil operating at such a low Reynolds Number:

Flying Wings

"Flying-Plank" Glider

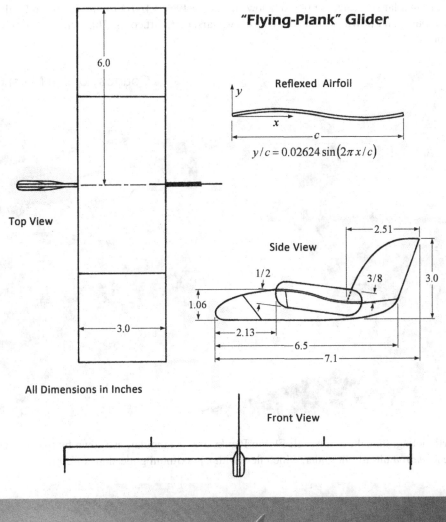

$$y/c = 0.02624 \sin(2\pi x/c)$$

All Dimensions in Inches

On a somewhat larger scale, the photo below shows the 1957 *XM-1D* Flying Plank glider that was designed, built, and flown by Jim Marske (www.marskeaircraft.com). This achieved a maximum glide slope of 28:1.

It's worth noting, though, that not all Flying Planks are rectangular, as shown by the elegant 1974 Marske *Monarch* ultra-light glider, which flew with a maximum glide slope of 20:1.

The main performance limitation of Flying Planks is that the reflexed airfoil is aerodynamically inferior to positively-cambered (hence unstable) airfoils. This is because the ideal angle-of-attack

Flying Wings

(minimum-drag condition) occurs at a small value of lift coefficient, as mentioned in the "Airfoils" section of Chapter 2. Therefore, Flying-Plank airplanes with inherent stability (no stability augmentation systems [SAS]) will generally not be as aerodynamically efficient as a tail-aft configuration. However, with some geometrical modifications, the Flying Plank can morph into a much more efficient design.

SWEPT FLYING WINGS

A more efficient Flying Wing is obtained by adding sweep and span-wise twist (washout). The idea is to provide a positive pitching moment from the twist and not necessarily from a reflex on the airfoil, thus allowing more efficient airfoils to be used. This design approach was used by the Horten brothers of Germany to develop an extraordinary series of Flying Wings. Shown below is the Horten brothers' 1942 H-IV glider:

The way in which span-wise twist (washout) can provide inherent stable trim is demonstrated by the following example. For simplicity, a symmetrical airfoil is assumed.

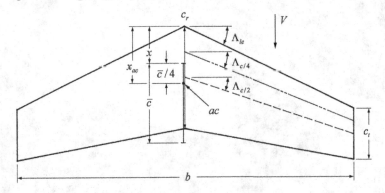

The trim state may be found by setting the pitching moment about the mass center, M_{cm}, equal to zero:

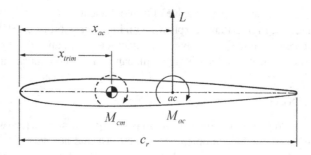

$$M_{cm} = M_{ac} - L(x_{ac} - x_{trim}) = 0 \rightarrow x_{trim} = x_{ac} - M_{ac}/L$$

In coefficient form this becomes:

$$x_{trim} = x_{ac} - C_{Mac}\bar{c}/C_L$$

This shows that for the mass center to be ahead of the aerodynamic center, $(C_{Mac})_{wing}$ must be positive (as was the case for the Flying Plank).

Now, for the numerical example in the "Wings" section of Chapter 2, one has that:

$$x_{ac} = 1.661, \quad \bar{c} = 2.33, \quad (C_{Mac})_w = 0.01019$$

So, for an example lift coefficient of $(C_L)_{trim} = 0.5$, $x_{trim} = 1.614$. Because this is ahead of the aerodynamic center, this wing has inherent (but slight) static stability at that lift coefficient.

The notion of using sweep, twist, and reflex to produce a stable flight vehicle originated with the Zanonia seed, which breaks off and glides some distance from its tree.

Scott Zona, Wikipedia Commons

Flying Wings

This was a source of inspiration to several aviation pioneers. In particular, it leads to the 1914 Dunne D8 biplane Flying Wing (s?) shown below:

This was one of a series, which was characterized by their extraordinary stability. Note how sweep and twist are incorporated into the design.

A considerately more modern example of a successful swept-wing design is the 2009 Marske *Pioneer 3*.

In this case, the wings are swept forward, which offers certain aerodynamic advantages, such as suppression of tip stall. This sophisticated design achieves a remarkable maximum lift/drag ratio of 45:1. A new version of this, the Pioneer 4 with a laminar airfoil, promises to significantly exceed this value.

PARAGLIDERS

A Flying Wing may also be stabilized by its mass-center location. The photo by Rafa Tecchio (cropped by the author) shows a paraglider, which is a gliding, steerable form of parachute. The airfoil is not necessarily reflexed, and the stability is provided by the extremely low position of the pilot's weight.

Rafa Tecchio, Wikipedia Commons

The diagram below shows how this works. In the trim position, the net moment about the mass center, M_{trim}, is zero. When the glider has a positive pitch perturbation, the lift and drag increase by Δ amounts. If the Lift/Drag ratio is high enough, these give rise to a ΔM about the mass center that acts to restore the pitch angle to the trim condition. Such aircraft have proven to be very stable, but it is important to not get into a negative-lift situation.

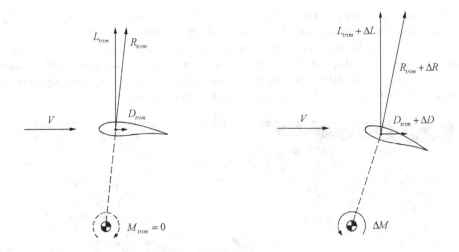

ROGALLO-TYPE HANG GLIDERS

Most flying-wing hang gliders use a combination of stabilizing features in their designs. For example, the most popular hang gliders use sweep and washout, along with a low mass-center position. This flexible-wing configuration was originated by NACA researcher Francis Rogallo in 1948. It was under consideration as a form of parachute for re-entry vehicles, but it found its true application as a recreational hang glider. The photo below, by Clement Bucco-Lechat, shows the basic simple design.

This particular design has battens (the flexible strips inserted into the wing in approximately the chord-wise direction), which can take a reflexed shape under aerodynamic loads.

The basic Rogallo-Wing configuration underwent an evolution to higher aspect ratios as shown in the photo by Rob (https://www.flickr.com/people/49503214348@N01) from Cambridge, MA.

The author has cropped this and added labels for the major components of a typical hang glider. The two men running along the ground are assisting in the training of the pilot.

This structure aero-elastically deforms to produce a stable and aerodynamically efficient shape. Elements controlling this shape are the wing battens and rigging lines (note the degree of washout). Control is provided by the pilot's shifting mass center relative to the wing's aerodynamic center. This means for control dates back to the earliest days of hang gliding.

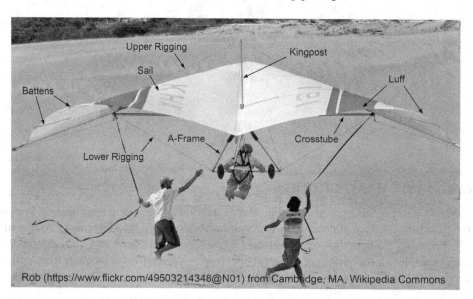

SPAN-LOADER FLYING WINGS

Returning to the Flying Plank concept, the Helios high-altitude solar-powered remotely-piloted aircraft, designed by AeroVironment Inc., is an extreme example of this configuration. The airfoil's mild reflex can be seen in the photo by NASA:

Additionally, the stability is augmented by the low mass center-locations of the streamlined housings ("pods") for the heavy batteries and avionics. These pods are positioned along the span of the airplane in order to achieve what is called a "span-loaded wing". The idea is that the distributed aerodynamic loading along the wing's span is approximately balanced by the distributed weight of the wing and the pods. Doing this greatly reduces the structural bending moments and shear forces and allows a much lighter structure to be used. Another NASA picture of the Helios clearly elucidates this:

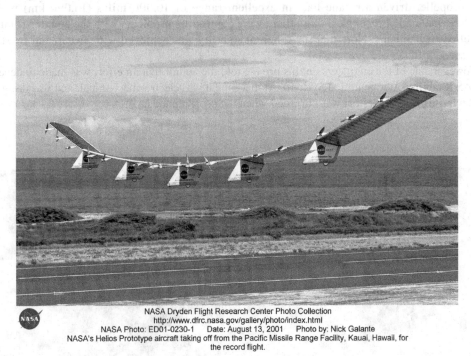

NASA Dryden Flight Research Center Photo Collection
http://www.dfrc.nasa.gov/gallery/photo/index.html
NASA Photo: ED01-0230-1   Date: August 13, 2001   Photo by: Nick Galante
NASA's Helios Prototype aircraft taking off from the Pacific Missile Range Facility, Kauai, Hawaii, for the record flight.

The span-loading concept had been considered previously for other Flying-Wing designs, such as for a proposed Boeing aircraft named the "Distributed Load Freighter":

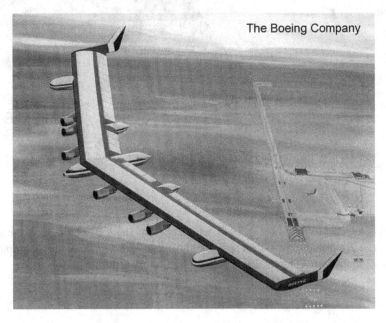

This feature was also cited for the famous Northrop B-35 Flying-Wing bomber shown below (though it was not as "pure" a span loader as the Helios). Another advantage stated was its small turn radius, in that the airplane could be highly banked (tilted laterally) and made to turn tightly while "standing on a wing tip". The Northrop bombers were the most successful large Flying-Wing aircraft until the B-2 bomber, and were the result of a long development process involving models and small-scale proof-of-concept piloted experimental designs. This propeller-driven airplane had an excellent range of 10,000 miles (16,090 km) with a 10,000 lb payload (4,535 kg), though there were continuing problems with the counter-rotating propeller mechanisms. Note that the inherent stability was achieved with sweep and twist, as discussed earlier.

However, one of the engineers on this project told the author that an effort was made to develop an SAS (Stability-Augmentation System). At that time it was electro-mechanical, and was designed to be wind-driven. The effort was curtailed when the project ended.

The B-35 was converted to jet power in 1947, which became the YB-49. Because the aft-mounted propellers on the B-35 had given directional stability (as described in the "Complete-Aircraft Aerodynamics" section in Chapter 2), the B-49 had to incorporate four vertical tails as seen below:

Courtesy of Gerald Balzer

The range was reduced to 4,900 miles (7,865 km) because of the less efficient jet engines of that time. Also, the speed was limited to 400 mph (642 km/h) because of the thick airfoil, which caused locally sonic flow above that speed.

Courtesy of Gerald Balzer

The thick wing had previously been cited as an advantage for offering a large internal volume within which the crew, engines, payload, etc. could be stored. However, the speed limitation, among other things, caused the cancellation of the project in 1949.

A CANADIAN FLYING-WING GLIDER

There are many other Flying-Wing configurations that are variations on the basic concepts discussed. The Canadian NRC Tailless Glider from 1946 is based on design ideas from Geoffrey Hill, a British pioneer of flying-wing research, and this configuration incorporates a straight center section with an efficient airfoil in this most efficient portion of the wing. The inherent stability is then provided by the swept and twisted outer portions of the wing. However, the amount of twist required gives an inefficient span loading and compromises the overall aerodynamic efficiency. The following RCAF photograph, along with a further description of the project, is found at: http://silverhawkauthor.com/canadian-warplanes-5-the-post-war-piston-era-nrc-tailless-glider_875.html.

A similar configuration was used for the powered Armstrong-Whitworth AW-52, as shown in the general-arrangement drawing (courtesy of aerofred.com):

Flying Wings

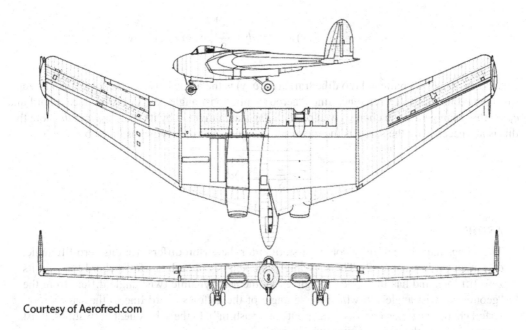

Courtesy of Aerofred.com

These two examples of aircraft differ from the swept straight-tapered linear-twisted configurations previously discussed. Calculations for the parameters: $x_{ac}, \bar{c}, C_L$ and C_{Mac} are more involved for such compound planforms. A reference discussing these, as well as other aspects of flying-wing designs, is "Tailless Aircraft in Theory and Practice" by Karl Nickel and Michael Wohlfahrt, published by Elsevier, 1994.

AN APPROXIMATE METHOD FOR ESTIMATING THE AERODYNAMIC CHARACTERISTICS OF WINGS WITH VARIABLE TWIST, TAPER AND SWEEP

An example of an arbitrary wing configuration is given below. This illustration is an adaptation of a rather lima-bean shaped planform found in the previously-stated reference:

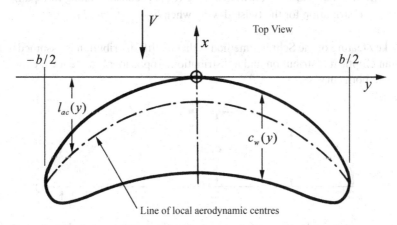

An approximate, but useful, methodology may be obtained by drawing upon research by Oster Schrenk ("A Simple Approximation Method for Obtaining the Spanwise Lift Distribution", NACA Technical Memorandum 948, 1940). The first step is to calculate the wing's zero-lift line angle relative to the aircraft's reference line (usually the wing's root-chord line for a "flying wing"). This is assumed, by Schrenk, to be an average along the span given by:

$$\bar{\delta}_{ZLL} = \frac{2}{c_{avg} b} \int_0^{b/2} \delta(y) \times c_w(y) dy = \frac{2}{S} \int_0^{b/2} \delta(y) \times c_w(y) dy$$

where S is the wing's flattened (no dihedral) area, $\delta(y)$ is the angle between the local airfoil's zero-lift line and the aircraft's reference line ("aerodynamic twist angle"), $c_w(y)$ is the local chord and c_{avg} is the mean geometric chord, given by the wing's area divided by its span, $c_{avg} = S/b$. Note that this is *not* necessarily the same as the mean aerodynamic chord, $\bar{c}$, which is given by:

$$\bar{c} = \frac{2}{S} \int_0^{b/2} c_w^2(y) dy$$

ASIDE

Two things must be carefully noted. First, $\delta(y)$ for the airfoil differs from its zero-lift angle, α_{ZLL} (defined in Chapter 2). The α_{ZLL} angle is taken between the airfoil's chord line and its zero-lift line, and has the opposite sign. Also, the aerodynamic twist angle differs from the "geometric twist angle", ε, which is the angle of the airfoil's chord line to the wing's root-airfoil chord line (referred to as "wash-out" or "wash-in"). In the subsequent derivations, twist always refers to the aerodynamic twist angles.

If $\varepsilon(y)$ is the geometric twist-angle distribution and i_w is the incidence angle of the root-airfoil's chord line relative to the aircraft's reference line, then one sees that:

$$\delta(y) = i_w + \varepsilon(y) - \alpha_{ZLL}(y)$$

END ASIDE

Now, Schrenk assumed that the lift distribution along the wing, $l(y)$, is composed of two parts:

$\hat{l}_{basic}(y)$, the lift distribution for an untwisted wing ($\delta(y)$ is constant along the span),
$\hat{l}_{twist}(y)$, the lift distribution for the twisted wing when $\bar{\delta}_{ZLL} = 0$, (*net lift = 0*)

For $\hat{l}_{basic}(y)$, a key feature of the Schrenk method is that the lift distribution is assumed to be an average between an elliptical distribution and a distribution proportional to the local chord, with equal areas, S, beneath both curves.

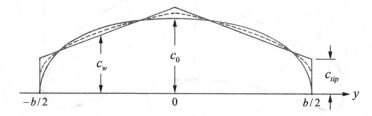

This intermediate curve is derived in Chapter 5, and the resulting non-dimensional distribution function is:

$$fn(y) = \frac{1}{2}\left[\frac{c_w(y)}{c_0} + \left\{ 1 - \left(\frac{y}{b/2}\right)^2 \right\}^{1/2} \right]$$

Flying Wings

Upon noting that the total lift force is given by:

$$L_{basic} = 2\int_0^{b/2} \hat{l}_{basic}(y)\,dy = q\,C_{L\alpha}\,\bar{\alpha}_{basic}\,2c_0\int_0^{b/2} fn(y)\,dy = q\,C_{L\alpha}\bar{\alpha}_{basic}\,S = qC_L S$$

One sees that:

$$\hat{l}_{basic}(y) = q\,C_{L\alpha}\,\bar{\alpha}_{basic}\,c_0\,fn(y) = qC_L c_0\,fn(y)$$

where q is the dynamic pressure, $q = 0.5\rho V^2$, $C_{L\alpha}$ is the wing's lift-curve slope, and $\bar{\alpha}_{basic} = \bar{\delta}_{ZLL} + \alpha$, where α is the untwisted wing's angle-of-attack. Note that $\bar{\alpha}_{basic}$ is a *constant* value along the span.

For $l_{twist}(y)$, Schrenk assumes that this is a situation where the integrated distributed lift equals zero, and the local lift-curve slope is *half* of that for a two-dimensional section with no downwash effects, $(C_{l\alpha})_{2D}$. Therefore, the distributed lift is given by:

$$\hat{l}_{twist}(y) = 0.5q(C_{l\alpha})_{2D}\Delta\delta(y)c_w(y)$$

where $\Delta\delta(y)$ is the difference between the local airfoil's zero-lift line angle and the twisted wing's zero-lift line angle, relative to the reference line, as shown in the picture below:

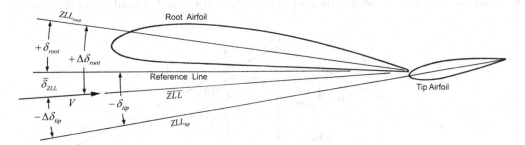

$$\Delta\delta(y) = \delta(y) - \bar{\delta}_{ZLL}$$

ASIDE

The reason behind Schrenk's assumptions for the previous equation, the 0.5 factor and the two-dimensional lift coefficient, is not clearly explained in his report. However, the justification appears to be the close correlation with results from a much more sophisticated analysis. Also, in "Airplane Flight Dynamics and Automatic Flight Controls" by Jan Roskam (1979, Roskam Aviation and Engineering Corporation), Roskam states that for this twisted wing situation: "The factor π should theoretically be 2π (theoretical section lift-curve slope); however, it has been found that leaving off the 2 tends to account better for three dimensional (induction) effects".

END ASIDE

These two lift-distribution equations combine to give the total lift distribution:

$$\hat{l}(y) = \hat{l}_{basic}(y) + \hat{l}_{twist}(y) = qC_{L\alpha}\bar{\alpha}_{basic}\,c_0\,fn(y) + 0.5q\,(C_{l\alpha})_{2D}\,\Delta\delta(y)c_w(y) \rightarrow$$

$$\hat{l}(y) = q[C_{l\alpha}(y)]_{basic}\,c_w(y)\bar{\alpha}_{basic} + 0.5q[C_{l\alpha}(y)]_{twist}\,c_w(y)\Delta\delta$$

A useful parameter for the *untwisted* wing is the "basic lift distribution", defined as:

$$\text{Basic Lift Distribution} \equiv BLD = [C_{l\alpha}(y)]_{\text{basic}} \bar{\alpha}_{\text{basic}} c_w(y)/c_{\text{avg}} = [C_l(y)]_{\text{basic}} c_w(y)/c_{\text{avg}}$$

From the previous equation, this is given by:

$$BLD = C_{L\alpha} \bar{\alpha}_{\text{basic}} fn(y) c_0/c_{\text{avg}} = C_L fn(y) c_0/c_{\text{avg}}$$

The corresponding parameter for the *twisted wing*, at $\bar{\alpha}_{\text{basic}} = 0$, is the "twisted lift distribution", defined as:

$$\text{Twisted Lift Distribution} \equiv TLD = 0.5[C_{l\alpha}]_{\text{twist}} \Delta\delta(y) c_w(y)/c_{\text{avg}}$$

From the previous equation, this is given by:

$$TLD = 0.5(C_{l\alpha})_{\text{2D}} \Delta\delta(y) \times c_w(y)/c_{\text{avg}}$$

EXAMPLE 1, STRAIGHT-TAPERED LINEAR-TWISTED WING

Consider the previous example of a straight-tapered wing with a linear twist and symmetrical airfoil:

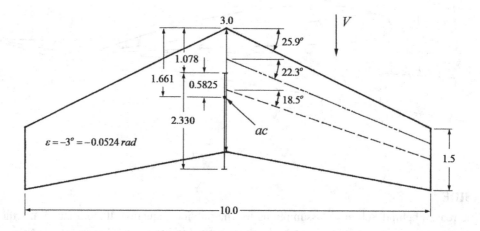

First of all, the average chord, c_{avg}, is given by:

$$c_{\text{avg}} = S/b = 22.5/10 = 2.25$$

Notice how this value differs from the mean aerodynamic chord, $\bar{c} = 2.33$. The definition of $\bar{c}$ is biased toward the more aerodynamically efficient center of the wing.

Next, c_0 for the equivalent elliptical planform is given by:

$$c_0 = \frac{4S}{\pi b} = \frac{4 \times 22.5}{\pi \times 10.0} = 2.865$$

The chord-distribution function, $c_w(y)$, is:

$$c_w(y) = 3.0 - (3.0 - 1.5)/5 = 3.0 - 0.30 y, \quad 0 \leq y \leq 5$$

So $fn(y)$ is given by:

$$fn(y) = \frac{1}{2}\left[\frac{(3.0-0.30y)}{2.865} + \left\{1-\left(\frac{y}{5}\right)^2\right\}^{1/2}\right]$$

And thus, the basic lift distribution is:

$$BLD = C_L\, fn(y)\frac{c_0}{c_{avg}} = 0.6367\left[\frac{(3.0-0.30y)}{2.865} + \left\{1-\left(\frac{y}{5}\right)^2\right\}^{1/2}\right]C_L$$

For different values of lift coefficient, this plots out as follows:

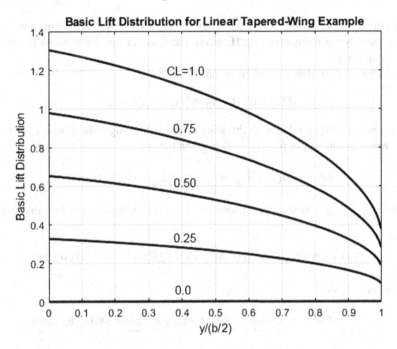

Next, the wing's geometric zero-lift angle will be found:

$$\bar{\delta}_{ZLL} = \frac{1}{c_{avg}}\frac{2}{b}\int_0^{b/2}\delta(y)\times c_w(y)\,dy$$

Recall that $\delta(y)$ is the local angle between the airfoil's zero-lift line and the wing's reference line. In this case, the reference line is the root chord line. Therefore, since this wing has a linear washout twist of $-3°\,(-0.0524\,\mathrm{rad})$, and symmetrical airfoils (each airfoil's zero-lift line is coincident with its chord line), one has:

$$\delta(y) = -0.0524\, y/(b/2) = -0.0524\, y/5 = -0.01048\, y,\ \mathrm{radians}$$

So, the equation becomes:

$$\bar{\delta}_{ZLL} = \frac{1}{2.25}\times\frac{2}{10}\int_0^5 -0.01048\, y(3.0-0.30y)\,dy$$

$$= -0.08889\int_0^5 (0.031446 y - 0.003144\, y^2)\,dy$$

$$\bar{\delta}_{ZLL} - 0.08889 \times \left| \frac{0.03144\, y^2}{2} - \frac{0.003144\, y^3}{3} \right|_0^5 = -0.08889 \times (0.3930 - 0.1310)$$

$$\bar{\delta}_{ZLL} = -0.02329 \text{ radians } (-1.334°)$$

From the same example in Chapter 2, the methodology based on the charts in NACA TR 572 gives:

$$(\bar{\delta}_{ZLL})_w = -(C_{L0})_w / (C_{L\alpha})_w = -0.0845/3.935 = -0.021474 \text{ radians, } (-1.2304°)$$

This is undoubtedly more correct since this analysis is biased toward the inner part of the wing, which is more aerodynamically efficient. However, the results from the Schrenk methodology will be used for consistency.

Now, for the twisted lift distribution equation:

$$TLD \equiv 0.5 (C_{l\alpha})_{2D} \Delta\delta(y) \times c_w(y)/c_{avg}$$

the twist distribution, $\Delta\delta(y)$, relative to the wing's zero-lift configuration (where, in this case, the root airfoil's zero-lift line angle, δ_{root}, is zero) is given by:

$$\Delta\delta(y) = \delta(y) - \bar{\delta}_{ZLL} = -0.010472\, y + 0.02329, \text{ (rad)}$$

Also, the value of $(C_{l\alpha})_{2D}$ may be approximated by $2\pi = 6.2832/\text{rad}$. Therefore, the resulting equation is:

$$TLD = 0.5 \times 6.2832(-0.010472\, y + 0.02329)(3.0 - 0.30\, y)/2.25$$

This is plotted as follows:

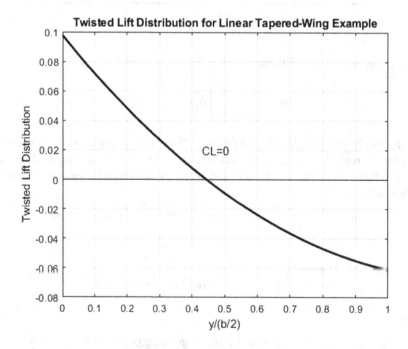

The sum of the basic lift distribution and twisted lift distribution gives the total lift distribution:

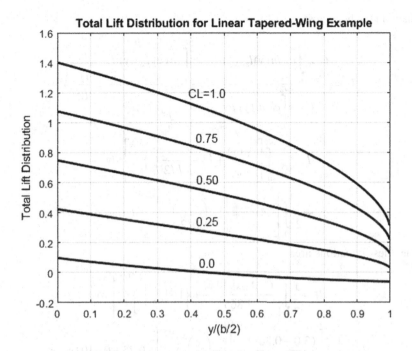

Finally, the equation for obtaining the total spanwise loading is given by:

$$\hat{l}(y) = q\, c_{\text{avg}} \times (\textit{Total Lift Distribution})$$

Certainly, there are far more sophisticated methodologies for obtaining spanwise wing loading, as described in Appendix B. However, under the right conditions (moderate to high aspect ratios, small sweep angles, and attached flow), the Schrenk approximation is surprisingly accurate and persists as a useful tool for preliminary estimates as well as a "reality check" on more sophisticated calculations.

Now, these Schrenk results will be evaluated for further aerodynamic calculations. First of all, the aerodynamic center is given by the following equation:

$$\overline{l}_{\text{ac}} = \frac{\int_0^{b/2} C_{l\alpha}(y)\, c_w(y)\, l_{\text{ac}}(y)\, dy}{\int_0^{b/2} C_{l\alpha}(y)\, c_w(y)\, dy}$$

where $C_{l\alpha}(y)\, c_w(y)$ is obtained from the basic lift distribution.

From before, one has that:

$$\hat{l}_{\text{basic}}(y) = q\, C_{L\alpha}\, \overline{\alpha}_{\text{basic}}\, c_0\, \textit{fn}(y)$$

This can be set equal to the equation describing the distributed lift in terms of the local lift-curve slope coefficients, $C_{l\alpha}(y)$:

$$\hat{l}_{\text{basic}}(y) = q\, C_{l\alpha}(y)\, \overline{\alpha}_{\text{basic}}\, c_w(y)$$

When these are set equal to one another, one obtains an equation for $C_{l\alpha}(y)\, c_w(y)$:

$$C_{l\alpha}(y)\, c_w(y) = C_{L\alpha}\, c_0\, \textit{fn}(y)$$

The l_{ac} equation now becomes:

$$\bar{l}_{ac} = \frac{C_{L\alpha}\int_0^{b/2} c_0\, fn(y) l_{ac}(y)\,dy}{C_{L\alpha}\int_0^{b/2} c_0\, fn(y)\,dy} = \frac{\int_0^{b/2} fn(y)\, l_{ac}(y)\,dy}{\int_0^{b/2} fn(y)\,dy} \equiv \frac{I_2}{I_1} \rightarrow$$

$$I_1 = \frac{1}{2}\int_0^{b/2}\left[\frac{c_w(y)}{c_0} + \left\{1-\left(\frac{y}{b/2}\right)^2\right\}^{1/2}\right]dy$$

$$I_2 = \frac{1}{2}\int_0^{b/2}\left[\frac{c_w(y)}{c_0} + \left\{1-\left(\frac{y}{b/2}\right)^2\right\}^{1/2}\right] l_{ac}(y)\,dy$$

For example, these equations are:

$$I_1 = \frac{1}{2}\int_0^5\left[\frac{(3.0-0.30y)}{2.865} + \left\{1-\left(\frac{y}{5}\right)^2\right\}^{1/2}\right]dy$$

$$I_2 = \frac{1}{2}\int_0^5\left[\frac{(3.0-0.30y)}{2.865} + \left\{1-\left(\frac{y}{5}\right)^2\right\}^{1/2}\right](0.75+0.4101y)\,dy$$

Numerical integration of these functions gives:

$$I_1 = 3.9270, \quad I_2 = 6.4436, \quad \rightarrow \bar{l}_{ac} = 1.6409$$

This value for l_{ac} is 1.21%, which is less than that calculated in "Wings" section of Chapter 2 for the numerical example. This isn't much, but it should be possible to do better.

Therefore, an extension to the Schrenk model is proposed based on an illustration adapted from the previously referenced Nickel and Wohlfahrt book:

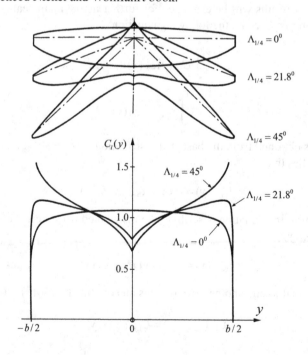

Flying Wings

It is seen that as the wing is swept back, the outer portions become increasingly loaded. This is a significant issue for swept-back wings, in that they are susceptible to tip stall. This is usually alleviated with washout and, with swept-back flying wings, significant washout is a feature.

A modification to the Schrenk model that can account for outer-panel loading is to introduce an additional term to the elliptical distribution function:

$$fn'(y) = \left\{1 - \left(\frac{y}{b/2}\right)^2\right\}^{1/2} + \eta \frac{\Delta l_{ac}(y)}{c'_0}$$

where η is a constant of proportionality on the effect of sweep and $l_{ac}(y)$ is the stream-wise distance from the wing's apex to its span-wise distributed aerodynamic centers, as shown in an earlier figure.

$$\Delta l_{ac} = l_{ac}(y) - (l_{ac})_{root}$$

The value of c'_0 is found by equating the area under the $c'_0 \, fn'(y)$ curve to the wing's planform area:

$$2c'_0 \int_0^{b/2} fn'(y)\,dy = S \rightarrow 2c'_0 \int_0^{b/2} \left\{1 - \left(\frac{y}{b/2}\right)^2\right\}^{1/2} dy + 2\eta \int_0^{b/2} \Delta l_{ac}(y)\,dy = S \rightarrow$$

$$\frac{\pi b c'_0}{4} + 2\eta \int_0^{b/2} \Delta l_{ac}(y)\,dy = S \rightarrow \underline{c'_0 = \frac{4}{\pi b}\left(S - 2\eta \int_0^{b/2} \Delta l_{ac}(y)\,dy\right)}$$

From this, the new distribution function describing the shape of the wing loading is:

$$fn(y) = \frac{1}{2}\left[\frac{c_w(y)}{c'_0} + \left\{1 - \left(\frac{y}{b/2}\right)^2\right\}^{1/2} + \eta \frac{\Delta l_{ac}(y)}{c'_0}\right]$$

As before, one has that $\overline{l}_{ac}$ is given by:

$$\overline{l}_{ac} = \frac{\int_0^{b/2} fn(y)\, l_{ac}(y)\,dy}{\int_0^{b/2} fn(y)\,dy} \equiv \frac{I_2}{I_1}$$

where, in this case:

$$I_1 = \frac{1}{2}\int_0^{b/2}\left[\frac{c_w(y)}{c'_0} + \left\{1 - \left(\frac{y}{b/2}\right)^2\right\}^{1/2} + \eta \frac{\Delta l_{ac}(y)}{c'_0}\right] dy$$

$$I_2 = \frac{1}{2}\int_0^{b/2}\left[\frac{c_w(y)}{c'_0} + \left\{1 - \left(\frac{y}{b/2}\right)^2\right\}^{1/2} + \eta \frac{\Delta l_{ac}(y)}{c'_0}\right] l_{ac}(y)\,dy$$

Note that for the example straight-tapered wing, one has that:

$$\Delta l_{ac}(y) = y \tan \Lambda_{ac} = y \tan(22.3°) = 0.4101\,y$$

Now, η is chosen to give a $\overline{l}_{ac}$ close to that from the Chapter 2 example. The value obtained is $\eta = 0.19$. Therefore, c'_0 is found to be:

$$c'_0 = \frac{4}{10\pi}\left(22.5 - 2 \times 0.19 \int_0^5 0.4101\,y\,dy\right) = 0.1273\left(22.5 - 0.19 \times 0.8202\,\frac{25}{2}\right) \rightarrow \underline{c'_0 = 2.616}$$

So, the integrals are:

$$I_1 = \frac{1}{2} \int_0^5 \left[\frac{(3.0-0.30y)}{2.616} + \left\{1-\left(\frac{y}{5}\right)^2\right\}^{1/2} + 0.0298y \right] dy$$

$$I_2 = \frac{1}{2} \int_0^5 \left[\frac{(3.0-0.30y)}{2.616} + \left\{1-\left(\frac{y}{5}\right)^2\right\}^{1/2} + 0.0298y \right](0.75+0.4101y) dy$$

Numerical integration of these functions gives:

$$I_1 = 4.2992, \quad I_2 = 7.1468, \quad \to \bar{l}_{ac} = \underline{1.662}$$

This value for $\bar{l}_{ac}$ is essentially identical to that calculated in the "Wings" section of Chapter 2 for the numerical example, as was planned for the selection of η.

It is interesting to look at the lift distributions given by this revised analytical model. The basic lift distribution is shown below:

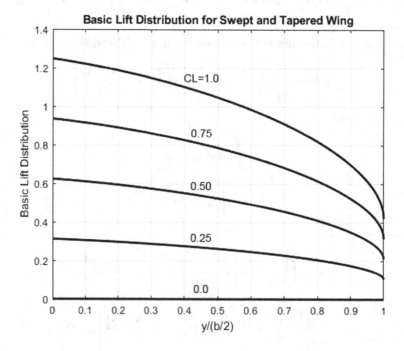

It is seen that how the lift distribution is now more biased toward the wing tip. For direct comparison with the Nickel and Wohlfahrt illustration, the lift-coefficient distribution is obtained from the previous equations:

$$BLD = [C_{l\alpha}(y)]_{basic} c_w(y)/c_{avg} \times \bar{\alpha}_{basic} = [C_l(y)]_{basic} c_w(y)/c_{avg} \to$$

$$[C_l(y)]_{basic} = \frac{BLD}{c_w(y)/c_{avg}}$$

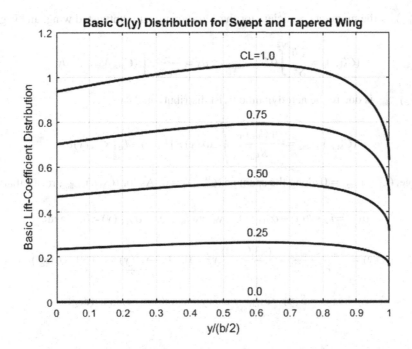

The increase in $[C_l(y)]_{basic}$ toward the wing tip is clear. The increase is even sharper with the Nickel and Wohlfahrt example because of that particular wing's rapidly decreasing tip chords.

Regarding the twisted lift distribution, this is the same as before; and, when added to the basic lift distribution, gives the total lift distribution shown below:

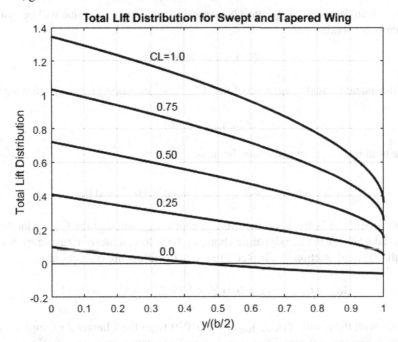

Next, as described in Chapter 2, the pitching moment about the aerodynamic center is composed of two parts:

$$C'_{Mac} = (C'_{Mac})_0 + (C'_{Mac})_{loading}$$

where $(C_{Mac})_0$ is the moment coefficient for the aerodynamically untwisted wing, and is given by:

$$(C'_{Mac})_0 = \frac{2b}{S^2}\int_0^{b/2}(C_{mac})_{airfoil}\,c_w^2\,dy = \frac{2}{S\,c_{avg}}\int_0^{b/2}(C_{mac})_{airfoil}\,c_w^2\,dy$$

Also, $(C_{Mac})_{loading}$ is due to the aerodynamic twist distribution, $\Delta\delta(y)$:

$$(C'_{Mac})_{loading} = -\frac{(C_{l\alpha})_{2D}}{S\,c_{avg}}\int_0^{b/2}\Delta\delta(y)\times[l_{ac}(y)-\bar{l}_{ac}]\times c_w(y)\,dy$$

For example, $(C'_{mac})_{airfoil} = 0$ along the span; so $(C'_{Mac})_0 = 0$. As for $(C'_{Mac})_{loading}$, recall that:

$$\delta(y) = i_w + \varepsilon(y) - \alpha_{ZLL}(y), \quad \text{where } i_w = 0, \quad \alpha_{ZLL}(y) = 0, \quad \text{and}$$

$$\varepsilon(y) = \frac{(\varepsilon_{tip}-\varepsilon_{root})}{b/2}y = \frac{(-3.0-0)}{5}y = -0.60\,y \to \delta(y) = -0.60\,y,\,(\text{deg.})$$

$$\Delta\delta(y) = \delta(y) - \bar{\delta}_{ZLL} \to \Delta\delta(y) = -0.60\,y + 1.334,\,(\text{deg.})$$

Also,

$$l_{ac}(y) - \bar{l}_{ac} = 0.25\,c_{root} + \Delta l_{ac}(y) - \bar{l}_{ac} = 0.25\times 3.0 + 0.4101\,y - 1.662 \to$$

$$l_{ac}(y) - \bar{l}_{ac} = -0.9120 + 0.4101\,y$$

Because $\Delta\delta(y)$ is in degrees, $(C_{l\alpha})_{2D}$ must be expressed per degree. Its value will be assumed to be 2π/radian; thus one obtains that:

$$(C_{l\alpha})_{2D} = 0.10966/\text{deg}.$$

From these inputs, numerical integration of the $(C'_{Mac})_{loading}$ equation gives the following result:

$$(C'_{Mac})_{loading} = 0.0120$$

and thus, the total pitching moment about the ac is:

$$C'_{Mac} = (C'_{Mac})_0 + (C'_{Mac})_{loading} = 0.0 + 0.0120 = 0.0120$$

Note that this coefficient is based on the average chord, c_{avg}, whereas the C_{Mac} value from NACA TR 572 is based on the mean aerodynamic chord, $\bar{c}$. Therefore, a meaningful comparison with the example in the "Wings" section of Chapter 2 requires the following correction:

$$C_{Mac} = C'_{Mac}\,c_{avg}/\bar{c} = 0.0120\times 2.25/2.33 \to \underline{C_{Mac} = 0.0116}$$

This compares with the result of $(C_{Mac})_{loading} = 0.01019$ from the Chapter 2 example, and is 13.8% larger. This is not an insignificant difference, although the values are relatively close.

Flying Wings

A likely reason for this difference may be found in the illustration below, adapted from Schrenk's report. This shows a comparison between Schrenk's prediction of twisted-wing lift and that obtained by a more exact analysis from Hans Multhopp. Schrenk doesn't provide the reference, but it is probably covered in "Methods for calculating the lift distribution of wings (subsonic lifting-surface theory)", Aeronautical Research Council (British) R & M 2884, January 1950.

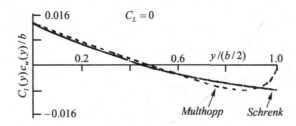

Schrenk's example wing has a straight-tapered planform with $c_{tip}/c_{root} = 0.5$, $-3°$ aerodynamic twists (washout) and $AR = 6.67$. The two non-dimensional lift-distribution curves track surprisingly well, up to $y/(b/2) \approx 0.8$. Beyond that they begin to diverge, especially near the tip. Now, for the purpose of contributing positive (nose-up) moments about the aerodynamic center, this portion of the swept-wing has the longest moment arms. So, the fact that the magnitude of this negative outboard lift goes to zero for the Multhopp analysis, and does not for the Schrenk approximation, is probably why C_{Mac} is somewhat over-predicted.

In any case, barring further comparisons with results from more sophisticated analyses applied to the example wing, the $C_{Mac} = 0.0116$ value will be chosen as consistent with this approximate analytical model.

ASIDE

In all fairness to Oster Schrenk, his method was probably developed only for the purposes of quickly predicting wing loads. In truth, this extension by the author to provide initial estimates for the aerodynamic characteristics of wings with arbitrary swept-back planforms and twist distributions seems to work surprisingly well (judging by comparison with the numerical example). A further development of the twist-distribution function would be a promising direction, as described in Appendix B.

END ASIDE

Next, from a derivation performed earlier in this chapter, it's possible to calculate the static margin for this example as a function of the equilibrium lift coefficient:

$$\text{Static Margin} = \frac{d}{\bar{c}} = \frac{C_{Mac}}{C_N} \approx \frac{C_{Mac}}{C_L} \rightarrow \frac{0.0116}{C_L}$$

A plot of this function is shown below:

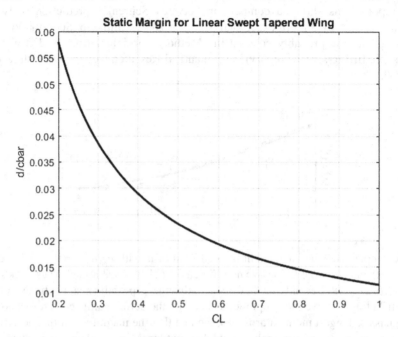

For a tailed airplane, the generally-stated criterion is that the static margin should fall between 0.05 and 0.15 (5% to 15% of the mean aerodynamic chord, $\bar{c}$). For a tailless aircraft, however, note that pitch control is dependent on control surfaces built into the wing, rather than elevator control on a dedicated stabilizing surface at a distance from the mass center. In a way, the pitch control on a tailless aircraft can be envisioned as being dependent on the ability to change the wing's in-flight C_{Mac} value. To quantify this, the previous equation can be extended to express the control power for changing the wing's lift coefficient:

$$\Delta C_L = \frac{\Delta C_{Mac}}{(d/\bar{c})}$$

For a given aircraft design, it is seen that the smaller the static margin, the more sensitive is the pitch control. Of course, this also causes a reduction in static stability, so there is a fine balance between the goals of controllability and stability. Jim Marske, whose remarkable designs have been described earlier in this chapter, informed the author that he typically flies his aircraft at a 5% static margin. For the numerical example, however, it is seen that the static margins are very small for the useful range of lift coefficients. For example, if $C_L = 0.6$, then $d/\bar{c} = 0.0193 \rightarrow 1.93\%$. Clearly, this design would require an additional twist to meet that criterion.

ASIDE

There has been some discussion and controversy about the afore-mentioned static-margin criterion. Some aircraft designers and pilots make a strong case that aircraft with acceptable trim and handling can function with static margins much less and much greater than the 5–15% range. What really decides this is a dynamic-stability and control-response analysis for that particular aircraft.

END ASIDE

Flying Wings

EXAMPLE 2, DOUBLE-SWEPT AND DOUBLE-TAPERED WING

Now, as mentioned before, the purpose of this Schrenk-derived analysis is to provide a methodology for estimating the aerodynamic characteristics of wings with planforms and twist distributions that are other than straight-tapered linearly-twisted configurations. Such an example wing is illustrated below:

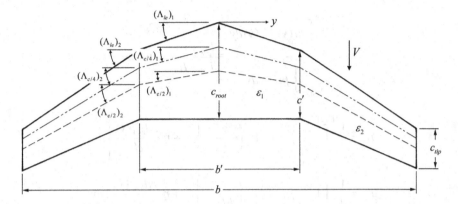

> **ASIDE**
>
> The cognoscenti of aviation history will recognize this planform as similar to that for the Armstrong-Whitworth AW52 experimental flying wing, which was illustrated previously.
>
> **END ASIDE**

This planform is composed of two distinct swept & tapered panels, where the line of aerodynamic centers is assumed to be at the quarter-chord locations. The innermost panel is designated as number 1, and the outer panel is number 2. Thus, the distribution function is:

$$fn(y) = \frac{1}{2}\left[\frac{c_w(y)}{c_0'} + \left\{1 - \left(\frac{y}{b/2}\right)^2\right\}^{1/2} + \eta\frac{\Delta l_{ac}(y)}{c_0'}\right] = fn_1(y) + fn_2(y), \quad \text{where}$$

$$fn_1(y) = \frac{1}{2}\left[\frac{c_{w1}(y)}{c_0'} + \left\{1 - \left(\frac{y}{b/2}\right)^2\right\}^{1/2} + \eta\frac{\Delta l_{ac1}(y)}{c_0'}\right], \quad 0 \le y \le b'/2$$

$$fn_2(y) = \frac{1}{2}\left[\frac{c_{w2}(y)}{c_0'} + \left\{1 - \left(\frac{y}{b/2}\right)^2\right\}^{1/2} + \eta\frac{\Delta l_{ac2}(y)}{c_0'}\right], \quad b'/2 \le y \le b/2$$

And c_0' is given by:

$$c_0' = \frac{4}{\pi b}\left(S - 2\eta\int_0^{b'/2}\Delta l_{ac1}(y)dy - 2\eta\int_{b'/2}^{b/2}\Delta l_{ac2}(y)dy\right)$$

Also, $c_{w1}(y)$ and $\Delta l_{ac1}(y)$ are given by:

$$c_{w1}(y) = c_{root} + \frac{(c' - c_{root})}{b'/2}y, \quad \Delta l_{ac1}(y) = y\tan(\Lambda_{c/4})_1, \quad 0 \le y \le b'/2$$

Likewise, $c_{w2}(y)$ and $\Delta l_{ac2}(y)$ are given by:

$$c_{w2}(y) = c' + \frac{(c_{tip} - c')}{(b - b')/2}(y - b'/2), \quad \Delta l_{ac2}(y) = \frac{b'}{2}\tan(\Lambda_{c/4})_1 + (y - b'/2)\tan(\Lambda_{c/4})_2 \quad b'/2 \leq y \leq b/2$$

Now, the wing's aerodynamic center is given by:

$$\bar{l}_{ac} = \frac{\int_0^{b/2} fn(y)\, l_{ac}(y)\, dy}{\int_0^{b/2} fn(y)\, dy} \equiv \frac{I_{2a} + I_{2b}}{I_{1a} + I_{1b}}, \quad \text{where}$$

$$I_{1a} = \frac{1}{2}\int_0^{b'/2}\left[\frac{c_{w1}(y)}{c'_0} + \left\{1 - \left(\frac{y}{b/2}\right)^2\right\}^{1/2} + \eta\,\frac{\Delta l_{ac1}(y)}{c'_0}\right] dy$$

$$I_{1b} = \frac{1}{2}\int_{b'/2}^{b/2}\left[\frac{c_{w2}(y)}{c'_0} + \left\{1 - \left(\frac{y}{b/2}\right)^2\right\}^{1/2} + \eta\,\frac{\Delta l_{ac2}(y)}{c'_0}\right] dy$$

$$I_{2a} = \frac{1}{2}\int_0^{b'/2}\left[\frac{c_{w1}(y)}{c'_0} + \left\{1 - \left(\frac{y}{b/2}\right)^2\right\}^{1/2} + \eta\,\frac{\Delta l_{ac1}(y)}{c'_0}\right] l_{ac1}(y)\, dy$$

$$I_{2b} = \frac{1}{2}\int_{b'/2}^{b/2}\left[\frac{c_{w2}(y)}{c'_0} + \left\{1 - \left(\frac{y}{b/2}\right)^2\right\}^{1/2} + \eta\,\frac{\Delta l_{ac2}(y)}{c'_0}\right] l_{ac2}(y)\, dy$$

where,

$$l_{ac1} = 0.25\, c_{root} + \Delta l_{ac1}(y), \quad l_{ac2} = 0.25\, c_{root} + \Delta l_{ac2}(y)$$

These equations will be used for the numerical example shown below:

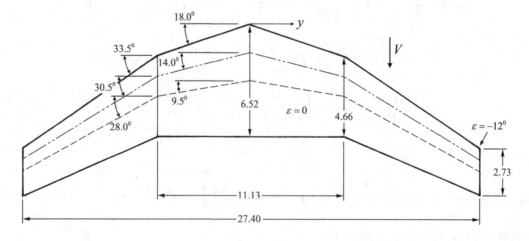

The panel areas are $S_1 = 31.11$ and $S_2 = 30.06$, so that the total area is:

$$S = 2 \times (31.11 + 30.06) = 122.34 \rightarrow c_{avg} = S/b = 122.34/27.4 = 4.465$$

Flying Wings

Also, $\eta = 0.19$ as chosen for the Example 1 single-swept wing.

From the previous equations, one obtains:

$$c_{w1}(y) = 6.52 + \frac{(4.66 - 6.52)}{11.13/2} y = 6.52 - 0.334 y, \quad \Delta l_{ac1} = y \tan(14°) = 0.249 y$$

$$c_{w2}(y) = 4.66 + \frac{(2.73 - 4.66)}{(27.40 - 11.13)/2}(y - 5.565) = 4.66 - 0.237(y - 5.565),$$

$$\Delta l_{ac2}(y) = 5.565 \tan(14°) + (y - 5.565) \tan(30.5°) = 1.388 + 0.589 (y - 5.565)$$

And c_0' is obtained from:

$$c_0' = \frac{4}{\pi b}(S - I_3 - I_4), \quad \text{where } I_3 \equiv 2\eta \int_0^{b'/2} \Delta l_{ac1}(y) dy = \eta \frac{(b')^2}{4} \tan(\Lambda_{c/4})_1, \quad \text{and}$$

$$I_4 \equiv 2\eta \int_{b'/2}^{b/2} \Delta l_{ac2}(y) dy = \eta \frac{b'(b - b')}{2}[\tan(\Lambda_{c/4})_1 - \tan(\Lambda_{c/4})_2] + \eta \tan(\Lambda_{c/4})_2 \frac{[b^2 - (b')^2]}{4}$$

So,

$$I_3 = 0.19 \times \frac{11.13^2}{4} \tan(14°) = 1.4671$$

$$I_4 = 0.19 \times 11.13 \frac{(27.40 - 11.13)}{2}[\tan(14°) - \tan(30.5°)] + 0.19 \times \tan(30.5°) \frac{(27.40^2 - 11.13^2)}{4} \rightarrow$$

$$I_4 = -5.8442 + 17.5399 = 11.6959$$

which gives that:

$$c_0' = \frac{4}{27.4 \pi}(122.34 - 1.4671 - 11.6959) = \underline{\underline{c_0' = 5.0733}}$$

Next, the integrals for obtaining the distance from the wing's forward apex to the aerodynamic center, $\bar{l}_{ac}$, were evaluated with numerical integration, and the results are:

$$I_{1a} = 5.8421, \quad I_{2a} = 13.4470, \quad I_{1b} = 6.2148, \quad I_{2b} = 32.1926 \rightarrow$$

$$\bar{l}_{ac} = \frac{I_{2a} + I_{2b}}{I_{1a} + I_{1b}} = \frac{13.4470 + 32.1926}{5.8421 + 6.2148} \rightarrow \underline{\underline{\bar{l}_{ac} = 3.7853}}$$

ASIDE

Upon using a methodology described in the USAF DATCOM, page 2.2.2-5, April 1978 Revision, one obtains that $(\bar{l}_{ac})_{DATCOM} = 3.7122$, which differs only by 1.9% from the previous value.

END ASIDE

Next, the C'_{Mac} will be calculated. Recall that this is composed of two parts: $(C'_{\text{Mac}})_0$ from the airfoils' distributed $C_{\text{mac}}(y)$, and the twist loading of the wing, $(C'_{\text{Mac}})_{\text{loading}}$. For this example, the inner portion of the wing is untwisted and is assumed to have an NACA 2412 airfoil, as shown by the following drawing (plotted from www.airfoiltools.com/plotter/index):

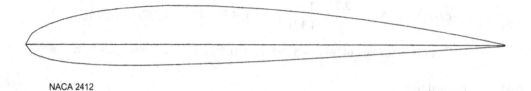

NACA 2412

The outer portion has a matching NACA 2412 airfoil at the crank, which then linearly morphs to a NACA 0009 symmetrical airfoil at the tip, illustrated from the same reference:

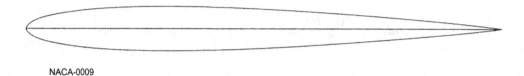

NACA-0009

So, $(C'_{\text{Mac}})_0$ is given by:

$$(C'_{\text{Mac}})_0 = \frac{2}{S c_{\text{avg}}} \int_0^{b/2} C_{\text{mac}}(y) c_w^2(y) dy \rightarrow$$

$$(C'_{\text{Mac}})_0 = \frac{2}{S c_{\text{avg}}} \left[\int_0^{b'/2} C_{\text{mac}1}(y) c_{w1}^2(y) dy + \int_{b'/2}^{b/2} C_{\text{mac}2}(y) c_{w2}^2(y) dy \right]$$

where,

$$C_{\text{mac}1}(y) = (C_{\text{mac}})_{\text{root}} + 2\left[\frac{(C_{\text{mac}})_{\text{crank}} - (C_{\text{mac}})_{\text{root}}}{b'}\right] y, \quad 0 \leq y \leq b'/2$$

$$C_{\text{mac}2}(y) = (C_{\text{mac}})_{\text{crank}} + 2\left[\frac{(C_{\text{mac}})_{\text{tip}} - (C_{\text{mac}})_{\text{crank}}}{b - b'}\right] (y - b'/2), \quad b'/2 \leq y \leq b/2$$

ASIDE

Observe that the "crank" of the wing is that location where the sweep angle changes from $(\Lambda_{c/4})_1$ to $(\Lambda_{c/4})_2$.

END ASIDE

Data on the NACA 2412 and NACA 0009 airfoils are found on pages 394 and 390 in the NACA Technical Report 824 by Ira Abbott, Albert E. von Doenhoff, and Louis S. Stivers, Jr., 1945. In particular, $(C_{\text{mac}})_{2412} \approx -0.05$ and $(C_{\text{mac}})_{\text{tip}} = 0.0$. Therefore, the above equations become:

$$C_{\text{mac}1}(y) = -0.05$$

$$C_{mac2}(y) = -0.05 + 2\left[\frac{0.0+0.05}{16.27}\right](y-5.565) = -0.084204 + 0.006146\,y$$

Upon substituting into the previous equations and, with numerical integration, one obtains:

$$(C'_{Mac})_0 = \frac{2}{122.34 \times 4.465}(-8.7750 - 3.3233) = -0.0443$$

For $(C_{Mac})_{loading}$, note that the geometrical twist, $\varepsilon(y)$, is zero for the inner portion, and varies linearly from $\varepsilon_{crank} = 0$ to $\varepsilon_{tip} = -12.0°$ along the outer portion:

$$\varepsilon_1(y) = 0.0, \quad 0 \leq y \leq b'/2,$$

$$\varepsilon_2(y) = \varepsilon_{crank} + \frac{2\,(\varepsilon_{tip} - \varepsilon_{crank})}{(b-b')}(y - b'/2), \quad b'/2 \leq y \leq b/2 \rightarrow$$

$$\varepsilon_2(y) = 0 + \frac{2\,(-12.0 - 0)}{16.27}(y - 5.565) = 8.2089 - 1.4751\,y, \text{ (deg.)}$$

Next, the zero-lift angle of the NACA 2412 airfoil is $\alpha_{ZLL} = -2.1°$ (from NACA 824), and varies linearly in the outer portions from $(\alpha_{ZLL})_{crank} = -2.1°$ to $(\alpha_{ZLL})_{tip} = 0°$:

$$\alpha_{ZLL1}(y) = -2.1, \quad 0 \leq y \leq b'/2,$$

$$\alpha_{ZLL2}(y) = (\alpha_{ZLL})_{crank} + \frac{2\,[(\alpha_{ZLL})_{tip} - (\alpha_{ZLL})_{crank}]}{(b-b')}(y - b'/2), \quad b'/2 \leq y \leq b/2 \rightarrow$$

$$\alpha_{ZLL2}(y) = -2.1 + 0.2581(y - 5.565) = -3.5366 + 0.2581\,y, \text{ (deg.)}$$

If the chord line of the inner portion is chosen to be the wing's reference line ($i_w = 0.0$), then the aerodynamic twist distributions are:

$$\delta_1(y) = i_w + \varepsilon_1(y) - \alpha_{ZLL1}(y) = 0 + 0 - (-2.1) = 2.1$$

$$\delta_2(y) = i_w + \varepsilon_2(y) - \alpha_{ZLL2}(y) = 11.7455 - 1.7332\,y, \quad (\delta \text{ in deg.})$$

From this, one may calculate the angle of the wing's zero-lift line, $\bar{\delta}_{ZLL}$ relative to the reference line (the root chord line, in this case):

$$\bar{\delta}_{ZLL} = \frac{2}{S}\left[\int_0^{b'/2} \delta_1(y) \times c_{w1}(y)\,dy + \int_{b'/2}^{b/2} \delta_2(y) \times c_{w2}(y)\,dy\right]$$

From numerical integration, one obtains the following result for this example:

$$\bar{\delta}_{ZLL} = \frac{2}{122.34}(65.3275 - 130.3430) \rightarrow \underline{\bar{\delta}_{ZLL} = -1.0629°}$$

Thus, the $\Delta\delta(y)$ distributions are given by:

$$\Delta\delta_1(y) = \delta_1(y) - \bar{\delta}_{ZLL} = 2.1 - (-1.0629) \rightarrow \underline{\Delta\delta_1(y) = 3.1629°}$$

$$\Delta\delta_2(y) = \delta_2(y) - \bar{\delta}_{ZLL} = 11.7455 - 1.7332\,y - (-1.0629) \rightarrow$$

$$\underline{\Delta\delta_2(y) = 12.8084 - 1.7332\,y, \text{ (deg.)}}$$

This, plus previous results, now allows the calculation of $(C'_{Mac})_{loading}$ from:

$$(C'_{Mac})_{loading} = -\frac{(C_{l\alpha})_{2D}}{S c_{avg}} \left[\int_0^{b'/2} \Delta\delta_1(y) \times [l_{ac1}(y) - \bar{l}_{ac}] \times c_{w1}(y) dy \right.$$

$$\left. + \int_{b'/2}^{b/2} \Delta\delta_2(y) \times [l_{ac2}(y) - \bar{l}_{ac}] \times c_{w2}(y) dy \right] \rightarrow$$

$$(C'_{Mac})_{loading} = -\frac{0.10966}{122.34 \times 4.465}(-147.5939 - 243.7763) = 0.0786$$

Therefore,

$$C'_{Mac} = (C'_{Mac})_0 + (C'_{Mac})_{loading} \rightarrow C'_{Mac} = 0.0343$$

Upon converting this to the $\bar{c}$ reference length:

$$\bar{c} = \frac{2}{S}\left[\int_0^{b'/2} c_{w1}^2(y)dy + \int_{b'/2}^{b} c_{w2}^2(y)dy\right] \rightarrow \frac{2}{122.34}(175.5001 + 113.5925) = 4.7261$$

$$C_{Mac} = C'_{Mac} c_{avg}/\bar{c} = 0.0343 \times 4.464/4.7261 \rightarrow C_{Mac} = 0.0324$$

The static margin for this example is obtained from:

$$\text{Static Margin} = \frac{d}{\bar{c}} = \frac{C_{Mac}}{C_N} \approx \frac{C_{Mac}}{C_L} \rightarrow \frac{0.0324}{C_L}$$

which plots out as shown:

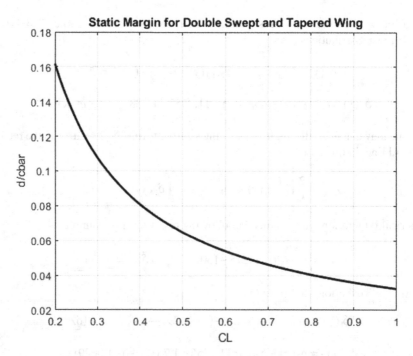

In this case, the desired static margin of 0.05 (5%) occurs within a range of useful lift coefficients. However, in order to achieve this, the outer portions of the wing required a geometric twist of −12.0° (and an aerodynamic twist of −14.1°). This might seem extreme, resulting in a small range of angles-of-attack where the flow is non-separated everywhere on the wing. However, recall that for an un-twisted swept wing, the span-wise lift coefficients sharply increase toward the wing tips. This was seen for the previous example, and the same effect is even more pronounced for this example:

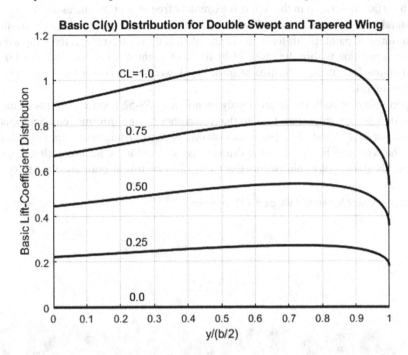

However, when the twist is added, the total span-wise lift-coefficient distribution is given by the following plot:

In this case, the lift coefficients, $C_l(y)$, actually decrease toward the wing tip. At a Reynolds Number of 3.0×10^6, NACA TR 824 gives the lift-coefficient range for the airfoils' attached flow:

NACA 2412: $-1.0 \leq C_l \leq 1.5$, Smooth Surface; $-1.0 \leq C_l \leq 1.2$, "Std. Roughness"
NACA 0009: $-1.3 \leq C_l \leq 1.3$, Smooth Surface; $-0.9 \leq C_l \leq 0.9$, "Std. Roughness"

The smooth surface described in the report is extremely free of protuberances and flaws, even small ones. The standard roughness, on the other hand, is that which might occur due to dust, small grit, or a coat of flat-finished paint. If this numerical example had a smooth surface, the wing would experience no flow separation within the range of its lift coefficient, C_L, between 0.0 and 1.0. However, for a slightly imperfect surface, the area near the crank would experience local flow separation at a lift coefficient of 1.0.

Now, upon referring back to the previously-mentioned AW-52 upon which this example's planform is based, it is very important to note that the author has no information about this aircraft's actual airfoils or wing twist. The previously-shown general arrangement drawing clearly shows washout at the wing tip. However, if the drawing is assumed to be accurate, this is far less than −12 degrees. The photo below also shows that the airfoils on this aircraft are close to symmetrical.

Imperial War Museum, Wikipedia Commons

If, in fact, they were truly symmetrical all along the span, then the amount of washout required for virtually the same static margin is 5.2 degrees. That is:

$$\varepsilon_{root} = \varepsilon_{crank} = 0, \quad \varepsilon_{tip} = -5.2°$$

This value is much closer to that shown on the general arrangement drawing, which makes a strong case for the Armstrong-Whitworth AW-52 actually having symmetrical airfoils.

Another important design feature is the lift-curve slope of the wing, $C_{L\alpha}$. Section 4.1.3.2-7 of the USAF DATCOM states that the Lowry and Polhamus equation, as given in "Wings" section of Chapter 2, may be used if $\Lambda_{c/2}$ is replaced by $(\Lambda_{c/2})_{eff}$

$$C_{L\alpha} = \frac{2\pi AR}{2 + \left\{ \frac{AR^2}{\kappa^2}[1+\tan^2(\Lambda_{c/2})_{eff}] + 4 \right\}^{1/2}} \quad (1/\text{rad})$$

where,

$$(\Lambda_{c/2})_{eff} = \cos^{-1}\left[\frac{2}{S} \sum_{1}^{n} (\cos \Lambda_{c/2})_i \, S_i \right]$$

Flying Wings

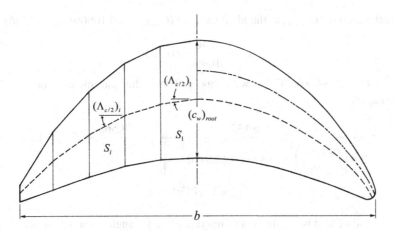

ASIDE

The illustrated planform is based on the 1938 12 m-span "Parabel" glider from the Horten brothers. This innovative design was never flown because winter storage caused un-fixable warpage. Therefore, it was burned. However, several large radio-controlled models have been built and successfully flown by hobbyists, proving the veracity of the design.

END ASIDE

For the double-swept numerical example,

$$(\Lambda_{c/2})_1 = 9.5°, \quad S_1 = 31.11, \quad (\Lambda_{c/2})_2 = 28°, \quad S_2 = 30.06$$

So, one obtains:

$$(\Lambda_{c/2})_{eff} = \cos^{-1} \frac{2}{122.34}[31.11 \times \cos(9.5°) + 30.06 \times \cos(28°)] \rightarrow$$

$$(\Lambda_{c/2})_{eff} = \cos^{-1}(0.9355) \rightarrow \underline{(\Lambda_{c/2})_{eff} = 20.69°}$$

Also, from NACA TR 824, the two-dimensional lift-curve slopes of the airfoils (at $RN = 3 \times 10^6$) are:

$$(C_{l\alpha})_{2412} = 0.10313/\deg, \quad (C_{l\alpha})_{0009} = 0.10625/\deg$$

These values are close enough that an average may be taken as representative of the wing:

$$(C_{l\alpha})_{avg} = 0.10469$$

ASIDE

A more precise weighted average can be calculated, accounting for the morphing of the airfoil from 2412 to 0009 on the outer panel. However, that seems excessive considering all of the approximations in this analysis.

END ASIDE

The term κ is the ratio of $(C_{l\alpha})_{\text{exp}}$ to the ideal value of $(C_{l\alpha})_{\text{ideal}} = 0.10966/\text{deg}$ ($2\pi/\text{rad}$):

$$\kappa = \frac{0.10469}{0.10966} = 0.955$$

Also, $AR = (27.4)^2/122.34 = 6.137$, so with this, and the other parameters, the wing's lift-curve slope may be calculated:

$$C_{L\alpha} = \frac{2\pi \times 6.137}{2 + \left\{\left(\frac{6.137}{0.96}\right)^2 \times [1 + \tan^2(20.69°)] + 4\right\}^{1/2}} = \frac{38.560}{9.154} \rightarrow C_{L\alpha} = 4.211/\text{rad}$$

$$C_{L\alpha} = 0.07350/\text{deg}$$

With the root-airfoil's chord being the wing's reference line for angle-of-attack measurement ($i_w = 0$), the lift coefficient for $\alpha = 0$ is:

$$C_{L0} = C_{L\alpha}\,\bar{\delta}_{\text{ZLL}} = 0.07350 \times (-1.0629) \rightarrow C_{L0} = -0.07812$$

So, the lift equation for this example is:

$$C_L = -0.07812 + 0.07350\alpha, \quad (\alpha \text{ in deg.})$$

Now, for the drag behavior of a flying wing, the span-wise twist can have a significantly larger effect than that for the typically lightly-twisted wings on tailed airplanes. For instance, consider the section lift-coefficient distribution for the zero-lift ($C_L = 0$) condition. From previous work, this is given by:

$$(C_l)_{\text{twisted}} = \frac{\textit{Twisted Lift Distribution}}{c_w(y)/c_{\text{avg}}} = TLD \times \frac{c_{\text{avg}}}{c_w(y)}$$

For the double-swept numerical example, this has the graphical form shown:

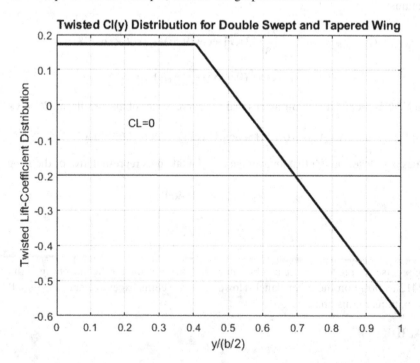

Note that, in reality, this plot would look significantly different. For example, there would be no abrupt change in values at the crank, as well as at the wing tip. However, this is what is consistent with the extended Schrenk approximation, and the analysis will continue accordingly.

Now, as described in the "Airfoils" section of Chapter 2, each airfoil experiences a profile drag coefficient that is a function of their lift coefficient, which can be generally represented by the following polynomial equation:

$$C_d = (C_d)_{C_l=0} + A C_l + B C_l^2$$

So, the zero-lift drag of the twisted wing is given by:

$$(C_D)_{C_L=0} = \frac{2}{S} \int_0^{b/2} (C_d)_{\text{twisted}} \, c_w(y) \, dy = \frac{2}{S} \int_0^{b/2} [(C_d)_{C_l=0} + A\,(C_l)_{\text{twisted}} + B(C_l)^2_{\text{twisted}}] \, c_w(y) \, dy$$

At a Reynolds Number of 3×10^6, data from NACA TR 824 is curve-fitted to give the following profile-drag polynomials:

$$(C_d)_{2412} = 0.00645 - 0.00225 C_l + 0.00618 C_l^2; \quad (C_d)_{0009} = 0.00549 + 0.00667 C_l^2$$

The profile drag of the inner panel is constant along with its span:

$$[C_d]_1 = (C_d)_{\text{root}} = (C_d)_{2412} \rightarrow [C_d(y)_{C_l=0}]_1 = 0.00645$$

However, because the airfoil morphs linearly from the NACA 2412 to the NACA 0009 along the span of the outer panel, $(C_d)_{C_l=0}$, A, and B are also assumed to vary linearly:

$$[C_d(y)_{C_l=0}]_2 = [(C_d)_{C_l=0}]_{\text{crank}} + 2 \frac{\{[(C_d)_{C_l=0}]_{\text{tip}} - [(C_d)_{C_l=0}]_{\text{crank}}\}}{(b-b')} (y - b'/2)$$

$$A_2(y) = A_{\text{crank}} + 2 \frac{(A_{\text{tip}} - A_{\text{crank}})}{(b-b')} (y - b'/2), \quad B_2(y) = B_{\text{crank}} + 2 \frac{(B_{\text{tip}} - B_{\text{crank}})}{(b-b')} (y - b'/2)$$

So,

$$[C_d(y)_{C_l=0}]_2 = 0.00645 + 2 \times \frac{(0.00549 - 0.00645)}{16.27} (y - 5.565) = 0.00710 - 0.000118\,y$$

$$A_2(y) = -0.00225 + 2 \times \frac{(0.00225)}{16.27} (y - 5.565) = -0.00379 + 0.00028\,y$$

$$B_2(y) = 0.00618 + 2 \times \frac{(0.00667 - 0.00618)}{16.27} (y - 5.565) = 0.00585 + 0.00006\,y$$

With the previously presented equations for $c_{w1}(y)$ and $c_{w2}(y)$, one may perform a numerical integration to obtain:

$$(C_D)_{C_L=0} = \frac{2}{S} \int_0^{b/2} (C_d)_{\text{twisted}} \, c_w(y) \, dy = \frac{2}{S} \int_0^{b'/2} (C_{d1})_{\text{twisted}} \, c_{w1}(y) \, dy + \frac{2}{S} \int_{b'/2}^{b} (C_{d2})_{\text{twisted}} \, c_{w2}(y) \, dy \rightarrow$$

$$(C_D)_{C_L=0} = \frac{2}{122.34}(0.1943 + 0.2065) \rightarrow (C_D)_{C_L=0} = 0.006552$$

This is 8.9% larger than the average of $(C_{do})_{2412}$ and $(C_{do})_{0009}$, which illustrates the penalty caused by the large 12° geometric twists for this example. However, as pointed out before, the AW-52 aircraft that this example's planform was based upon probably had symmetrical airfoils throughout the span, instead of the cambered airfoil assumed for the center section. Thus, the significantly lesser twist required for an acceptable static margin range would also have given a much lower C_{D0} penalty, which certainly would have worked to offset any aerodynamic advantage provided by using a cambered airfoil in the center section.

This analysis may be extended to obtain the profile drag of the wing, $(C_D)_{\text{profile}}$, as a function of lift coefficient, C_L. In this case $(C_l)_{\text{twisted}}$ is replaced with C_l, which is the total running lift coefficient along the span. This is given by:

$$C_l(y) = [C_l(y)]_{\text{basic}} + [C_l(y)]_{\text{twist}} = \frac{\text{Basic Lift Distribution}}{c_w(y)/c_{\text{avg}}} + \frac{\text{Twisted Lift Distribution}}{c_w(y)/c_{\text{avg}}}$$

$$= \frac{\text{Total Lift Distribution}}{c_w(y)/c_{\text{avg}}} = \text{Func.}(y, C_L)$$

So that,

$$(C_D)_{\text{profile}} = \frac{2}{S}\int_0^{b/2}(C_d)_{\text{profile}}\, c_w(y)\,dy = \frac{2}{S}\int_0^{b/2}[(C_d)_{C_l=0} + A\,C_l(y) + B\,C_l^2(y)]\,c_w(y)\,dy$$

For the double-swept wing example, this equation becomes:

$$(C_D)_{\text{profile}} = \frac{2}{S}\int_0^{b'/2}\{[C_d(y)_{C_l=0}]_1 + A_1(y)C_l(y) + B_1(y)C_l^2(y)\}\,c_{w1}(y)\,dy$$

$$+ \frac{2}{S}\int_{b'/2}^{b/2}\{[C_d(y)_{C_l=0}]_2 + A_2(y)C_l(y) + B_2(y)C_l^2(y)\}\,c_{w2}(y)\,dy$$

Upon substitution of values for these terms, numerically integrating, and curve-fitting the results over a C_L range of 0.0, 0.25, 0.50, 0.75, 1.0, one obtains the following polynomial equation:

$$(C_D)_{\text{profile}} = 0.006552 - 0.002609\,C_L + 0.006144\,C_L^2$$

Observe that the $(C_D)_{\text{profile}}$ value for $C_L = 0$ is previously obtained for the zero-lift case.

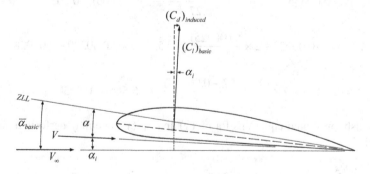

The induced drag is due to the basic-lift distribution because the twisted-lift distribution contributes no net lift. Therefore, the local lift coefficient $[C_l(y)]_{\text{basic}}$ is equal to:

$$[C_l(y)]_{\text{basic}} = (C_{l\alpha})_{\text{2D}}\,\alpha(y) \approx 2\pi\,\alpha(y)$$

Flying Wings

Further,
$$\alpha(y) = \bar{\alpha}_{basic} - \alpha_i(y)$$

where α_i is the induced angle of attack due to the finite aspect ratio of the wing. This gives that:

$$\alpha_i(y) = \bar{\alpha}_{basic} - \frac{[C_l(y)]_{basic}}{2\pi}$$

Now the induced drag of the local section is due to the rearwards tilting of the local lift vector, relative to the flight direction:

$$(C_d)_{induced} = [C_l(y)]_{basic}\, \alpha_i(y) = [C_l(y)]_{basic}\left\{\bar{\alpha}_{basic} - \frac{[C_l(y)]_{basic}}{2\pi}\right\} \rightarrow$$

$$(C_d)_{induced} = [C_l(y)]_{basic}\, \bar{\alpha}_{basic} - \frac{[C_l(y)]_{basic}^2}{2\pi}$$

Thus, the integral of this along the span gives the wing's induced drag:

$$(C_D)_{induced} = \frac{2}{S}\int_0^{b/2} (C_d)_{induced}\, c_w(y)\, dy = \frac{2\bar{\alpha}_{basic}}{S}\int_0^{b/2}[C_l(y)]_{basic}\, c_w(y)\, dy - \frac{1}{\pi S}\int_0^{b/2}[C_l(y)]_{basic}^2\, c_w(y)\, dy$$

Now,
$$C_L = C_{L\alpha}\bar{\alpha}_{basic} \rightarrow \bar{\alpha}_{basic} = C_L/C_{L\alpha} \rightarrow$$

$$(C_D)_{induced} = \frac{2C_L}{C_{L\alpha}S}\int_0^{b/2}[C_l(y)]_{basic}\, c_w(y)\, dy - \frac{1}{\pi S}\int_0^{b/2}[C_l(y)]_{basic}^2\, c_w(y)\, dy$$

Again, for the double-swept wing example, this equation becomes:

$$(C_D)_{induced} = \frac{2C_L}{C_{L\alpha}S}\int_0^{b'/2}[C_l(y)]_{basic1}\, c_{w1}(y)\, dy - \frac{1}{\pi S}\int_0^{b'/2}[C_l(y)]_{basic1}^2\, c_{w1}(y)\, dy$$

$$+ \frac{2C_L}{C_{L\alpha}S}\int_{b'/2}^{b/2}[C_l(y)]_{basic2}\, c_{w2}(y)\, dy - \frac{1}{\pi S}\int_{b'/2}^{b/2}[C_l(y)]_{basic2}^2\, c_{w2}(y)\, dy$$

And, upon substitution of values for these terms, numerically integrating, and curve-fitting the results over a C_L range of 0.0, 0.25, 0.50, 0.75, 1.0, one obtains the following polynomial equation:

$$(C_D)_{induced} = 0.07763\, C_L^2$$

It is interesting to compare this with the result from the equation for minimum induced drag:

$$(C_D)_{induced}^{min.} = \frac{C_L^2}{\pi\, AR} = \frac{C_L^2}{6.137\pi} = 0.05187\, C_L^2$$

In the previously-cited Nickle and Wohlfahrt reference, a "k-factor" is defined as:

$$k-\text{factor} \equiv (C_D)_{induced}/(C_D)_{induced}^{min.}$$

For this example, the k-factor is 1.497. This doesn't speak to the wing being particularly efficient. However, this is evidently not uncommon, since the authors state that "…k-factors from 1.1 to 1.5 (!) can easily be encountered for flying wings".

Now, the total drag coefficient may be obtained by adding the profile-drag coefficient and the induced-drag coefficient. The resulting polynomial equation is:

$$C_D = 0.006552 - 0.002609 C_L + 0.08377 C_L^2$$

From this, one may find the maximum lift/drag ratio:

$$\frac{C_D}{C_L} = \frac{0.006552}{C_L} - 0.002609 + 0.08377 C_L \rightarrow \frac{d(C_L/C_D)}{dC_L} = \frac{-0.006552}{C_L^2} + 0.08377$$

Upon setting this equal to zero, one obtains:

$$C_L^2 = \frac{0.006552}{0.08377} = 0.0.07819 \rightarrow C_L = 0.27963 \rightarrow$$

$$\left(\frac{C_D}{C_L}\right)_{min} = \frac{0.006552}{0.27963} - 0.002609 + 0.08377 \times 0.27963 = 0.04424 \rightarrow$$

$$(C_L/C_D)_{max} = 22.604$$

This is a reasonable value for a wing of this aspect ratio, which is due to the smooth-surface drag coefficients given for the assumed airfoils from NACA TR 824. If the "Standard roughness" C_d vs. C_l values are chosen (given at $RN = 5.7 \times 10^6$), which are more representative of manufacturing imperfections and operating conditions, then the curve-fit equations for these values are:

$$(C_d)_{2412} = 0.00971 - 0.00469 C_l + 0.01256 C_l^2; \quad (C_d)_{0009} = 0.00868 + 0.01389 C_l^2$$

From these, one obtains that:

$$(C_L/C_D)_{max} = 18.198 \quad \text{at} \quad C_L = 0.3352$$

This is 16.8% less than that for the smooth-surface condition, which still isn't bad for this wing. However, note that a fuselage pod, stabilizing surfaces, and inevitable excrescence-drag contributions, will lower this value of maximum lift/drag ratio even further.

As mentioned earlier, if symmetrical airfoils had been used all along the span, the required geometric twist for acceptable static margins would be reduced from $-12°$ to $-5.2°$. Such a candidate airfoil for the inner panel would be the NACA 0012, for which the profile drag curve-fit equation may be obtained from data published in "Low-Speed Aerodynamic Characteristics of NACA Aerofoil Section, including the Effects of Upper-Surface Roughness Simulating Hoar Frost", by N. Gregory and C. L. O'Reilly, NPL R&M No. 3726, January 1970:

$$(C_d)_{0012} = 0.00974 + 0.01251 C_l^2$$

From this, and the previous equations for the NACA 0009, one obtains that:

$$(C_L/C_D)_{max} = 17.105 \quad \text{at} \quad C_L = 0.328$$

The maximum lift/drag ratio is 6.0% less than that for the example incorporating the cambered NACA 2412 airfoil, and the corresponding lift coefficient is nearly identical. Although this shows a somewhat lesser performance, it's not nearly as great as one would imagine from a wing with non-cambered airfoils. This serves to confirm the previously stated notion that the profile-drag reduction with the lesser twist can offset the reduction in lift efficiency for the symmetrical airfoil.

Finally, even though wings with straight-tapered and uniformly-swept panels have been used as numerical examples, it should be clear that planforms with uniformly varying geometries may also be analyzed. Such a wing would be the "Parabel" design illustrated earlier. In this case, $\Delta l_{ac}(y)$ and $c_w(y)$ would be continuous parabolic functions from the wing's root to its tips. Likewise, the geometric twist, $\varepsilon(y)$, may be non-linear. In fact, Reimar and Walter Horten were past masters at tweaking the twist distribution to optimize efficiency. In the Nickel and Wohlfahrt reference, it is stated that "the washout distribution of the Horten H II has been: 'Additively 2° linear, 2° quadratic and 2° cubic'".

The last observation is that this analysis may also be applied to swept-forward wings, as is seen in the Marske "Pioneer" designs. In these cases, $\Delta l_{ac}(y)$ would be negative.

BIG CAVEAT

It is important to remember that, despite the calculated values in the example having numerous decimal places, this is an approximate methodology. In no way does this replace a proper vortex-lattice analysis or a properly-executed CFD analysis. A simple 1940 methodology by Oster Schrenk has been extended, perhaps well beyond Herr Schrenk's furthest imaginings, and is just as approximate as his original analysis. However, this methodology does seem to give reasonably close results by comparison with those from more exact analyses available to the author. So, it is suggested that this methodology can be used for initial-design purposes and as a "reality check" on the results from more sophisticated (but less robust) analyses.

END CAVEAT

MATLAB COMPUTER PROGRAM

A listing of the computer program used for the double-swept example is presented. This, of course, can also be used for a single-swept design. Note that neither metric nor imperial dimensions are specified. The only factor that requires knowledge of the dimensions and speed is the Reynolds Number. For the example, a Reynolds Number was assumed for the sake of presenting the methodology. However, for the purposes of analyzing a specific design, its performance will have to be estimated in order to calculate the Reynolds Number. With this knowledge, one can then obtain the corresponding profile-drag coefficients either from theory or published experimental data.

Also, skilled programmers will, no-doubt, see that there are numerous redundancies in this listing and that this can be greatly streamlined. However, this program evolved in a piece-meal fashion, with the author checking each addition. The result is a program that gave the results needed for the example, redundancies and all.

```
clc
% 2 April,2020
% Aerodynamic Properties for Linear Double-Swept Double-Tapered-Wing
Example
% Root and crank airfoil is NACA 2412
% Tip airfoil is NACA 0009
% Airfoil properties are linearly morphed between crank and tip
% airfoil characteristics are from NACA TR 824, for SMOOTH surface
% RN=3.0^10^6

% Basic Lift Distribution
clear all
croot=6.52; % root airfoil's chord
cprime=4.66; % chord at crank
ctip=2.73; % tip airfoil's chord
bprime=11.13; % span from crank-to-crank
b=27.40; % total flattened wing span
Sw1=31.11; % inner panel area
Sw2=30.06; % outer panel area
S=2*(Sw1+Sw2); % total wing area
```

```
cavg=S/b; % average chord
Cla=6.2832; % 2-D airfoil lift-curve slope
Lam1deg=14.0; % inner panel quarter-chord sweep angle, degrees
Lam2deg=30.5; % outer panel quarter-chord sweep angle, degrees
Lam1halfdeg=9.5; % inner panel half-chord sweep angle, degrees
Lam2halfdeg=28.0; % outer panel half-chord sweep angle, degrees
Lam1rad=pi*Lam1deg/180.0;
Lam2rad=pi*Lam2deg/180.0;
Lam1halfrad=pi*Lam1halfdeg/180.0;
Lam2halfrad=pi*Lam2halfdeg/180.0;
iw=0.0; % wing incidence angle relative to the reference line, deg.
epsilonroot=0.0; % geometric twist angle rel. to ref. line at root, deg
epsiloncrank=0.0; % geometric twist angle rel. to ref. line at crank, deg.
epsilontip=-12; % geometric twist angle rel. to ref. line at tip, deg.

% Properties for root airfoil
alphaZLLroot=-2.1; % zero-lift angle rel. to chord at root, deg.
Cmacroot=-0.05; % profile moment-coeff. about its aero. center
Claroot=0.10313; % 2D lift curve slope/deg
Cd0root=0.00645; % constant term in Cdprofile polynomial equation
Aroot=-0.00225; % const. for linear term in Cdprofile polynomial equation
Broot=0.00618; % const. for quadratic term in Cdprofile polynomial
equation

% Properties for crank airfoil
alphaZLLcrank=-2.1; % zero-lift angle relative to its chord, deg.
Cmaccrank=-0.05; % profile moment-coeff. about its aero. center
Clacrank=0.10313; % 2D lift-curve slope/deg.
Cd0crank=0.00645; % constant term in Cdprofile polynomial equation
Acrank=-0.00225; % const. for linear term in Cdprofile polynomial
equation
Bcrank=0.00618; % const. for quadratic term in Cdprofile polynomial
equation

% Properties for tip airfoil
alphaZLLtip=0.0; % zero-lift angle relative to its chord, deg.
Cmactip=0.0; % profile moment-coeff. about its aero. center
Clatip=0.10625; % 2D lift-curve slope/deg
Cd0tip=0.00549; % constant term in Cdprofile polynomial equation
Atip=0.0; % const. for linear term in Cdprofile polynomial equation
Btip=0.00667; % const. for quadratic term in Cdprofile polynomial
equation

% Parameters for integration
lowerlimit1=0.0;
upperlimit1=bprime/2;
lowerlimit2=bprime/2;
upperlimit2=b/2;
N=1000; % number of summation intervals

% Sweep constant of proportionality
eta=0.19;

% calculate mean aerodynamic chord, cbar
% inner portion
y=linspace(lowerlimit1,upperlimit1,N);
cw1=croot+2*((cprime-croot)/bprime)*y;
```

```
Integrand1=cw1.^2;
I1=trapz(y,Integrand1);

% outer portion
y=linspace(lowerlimit2,upperlimit2,N);
cw2=cprime+2*((ctip-cprime)/(b-bprime))*(y-bprime/2);
Integrand2=cw2.^2;
I2=trapz(y,Integrand2);

cbar=(2/S)*(I1+I2);

% Calculate c0prime
% inner portion
y=linspace(lowerlimit1,upperlimit1,N);
Deltalac1=y*tan(Lam1rad);
Integrand3=2*eta*Deltalac1;
Int3=trapz(y,Integrand3);

% outer portion
y=linspace(lowerlimit2,upperlimit2,N);
Deltalac2=(bprime/2)*tan(Lam1rad)+(y-bprime/2)*tan(Lam2rad);
Integrand4=2*eta*Deltalac2;
Int4=trapz(y,Integrand4);

c0prime=4/(pi*b)*(S-Int3-Int4);

% Test for veracity of c0prime
% inner portion
y=linspace(lowerlimit1,upperlimit1,N);
cw1=croot+2*((cprime-croot)/bprime)*y;
F2a=sqrt(1-((y/(b/2)).^2));
g1=trapz(y,F2a);
Deltalac1=y*tan(Lam1rad);
g2=trapz(y,eta*Deltalac1);
fn1=c0prime*(F2a+eta*Deltalac1/c0prime);
S1test=trapz(y,fn1);

% outer portion
y=linspace(lowerlimit2,upperlimit2,N);
cw2=cprime+2*((ctip-cprime)/(b-bprime))*(y-bprime/2);
F2b=sqrt(1-((y/(b/2)).^2));
g3=trapz(y,F2b);
Deltalac2=(bprime/2)*tan(Lam1rad)+(y-(bprime/2))*tan(Lam2rad);
g4=trapz(y,eta*Deltalac2);
fn2=c0prime*(F2b+eta*Deltalac2/c0prime);
S2test=trapz(y,fn2);
Stest=2*(S1test+S2test);

% Aerodynamic center
% inner portion
y=linspace(lowerlimit1,upperlimit1,N);
cw1=croot+2*((cprime-croot)/bprime)*y;
F1a=cw1/c0prime;
F2a=sqrt(1-((y/(b/2)).^2));
Deltalac1=y*tan(Lam1rad);
F3a=eta*Deltalac1/c0prime;
lac1=0.25*croot+Deltalac1;
```

```
Integrand5=0.5*(F1a+F2a+F3a);
Integrand6=Integrand5.*lac1;
I1a=trapz(y,Integrand5);
I2a=trapz(y,Integrand6);

% outer portion
y=linspace(lowerlimit2,upperlimit2,N);
cw2=cprime+2*((ctip-cprime)/(b-bprime))*(y-bprime/2);
F1b=cw2/c0prime;
F2b=sqrt(1-((y/(b/2)).^2));
Deltalac2=(bprime/2)*tan(Lam1rad)+(y-(bprime/2))*tan(Lam2rad);
F3b=eta*Deltalac2/c0prime;
lac2=0.25*croot+Deltalac2;
Integrand7=0.5*(F1b+F2b+F3b);
Integrand8=Integrand7.*lac2;
I1b=trapz(y,Integrand7);
I2b=trapz(y,Integrand8);

LacBar=(I2a+I2b)/(I1a+I1b);  % aerodynamic center from forward apex of wing

% Pitching-moment coefficient due to airfoils' Cmac values
% inner portion
y=linspace(lowerlimit1,upperlimit1,N);
Cmac1=Cmacroot+2*((Cmaccrank-Cmacroot)/bprime)*y;
cw1=croot+2*((cprime-croot)/bprime)*y;
Integrand9=Cmac1.*cw1.^2;
Int5=trapz(y,Integrand9);

% outer portion
y=linspace(lowerlimit2,upperlimit2,N);
Cmac2=Cmaccrank+2*((Cmactip-Cmaccrank)/(b-bprime))*(y-bprime/2);
cw2=cprime+2*((ctip-cprime)/(b-bprime))*(y-bprime/2);
Integrand10=Cmac2.*cw2.^2;
Int6=trapz(y,Integrand10);

CMac0=(2/(S*cavg))*(Int5+Int6);

% Zero-lift angle of attack for the whole wing
% inner portion
y=linspace(lowerlimit1,upperlimit1,N);
cw1=croot+2*((cprime-croot)/bprime)*y;
epsilon1=epsilonroot+2*((epsiloncrank-epsilonroot)/bprime)*y;
alphaZLL1=alphaZLLroot+(2*(alphaZLLcrank-alphaZLLroot)/bprime)*y;
delta1=iw+epsilon1-alphaZLL1;
Integrand11=delta1.*cw1;
Int7=trapz(y,Integrand11);

% outer portion
y=linspace(lowerlimit2,upperlimit2,N);
cw2=cprime+2*((ctip-cprime)/(b-bprime))*(y-bprime/2);
epsilon2=epsiloncrank+2*((epsilontip-epsiloncrank)/
 (b-bprime))*(y-bprime/2);
alphaZLL2=alphaZLLcrank+2*((alphaZLLtip-alphaZLLcrank)/(b-bprime))*...
     (y-bprime/2);
delta2=iw+epsilon2-alphaZLL2;
Integrand12=delta2.*cw2;
Int8=trapz(y,Integrand12);

deltaBarZLL=(2/S)*(Int7+Int8);
```

```
% Pitching-moment coefficient from loading
% inner portion
y=linspace(lowerlimit1,upperlimit1,N);
cw1=croot+2*((cprime-croot)/bprime)*y;
epsilon1=epsilonroot+2*((epsiloncrank-epsilonroot)/bprime)*y;
alphaZlL1=alphaZLLroot+2*((alphaZLLcrank-alphaZLLroot)/bprime)*y;
delta1=iw+epsilon1-alphaZLL1;
Ddelta1=delta1-deltaBarZLL;
Ddelta1Rad=Ddelta1*pi/180;
distlac1=croot/4+Deltalac1-LacBar;
Integrand13=Ddelta1Rad.*distlac1.*cw1;
Int9=trapz(y,Integrand13);

% outer portion
y=linspace(lowerlimit2,upperlimit2,N);
cw2=cprime+2*((ctip-cprime)/(b-bprime))*(y-bprime/2);
epsilon2=epsiloncrank+2*((epsilontip-epsiloncrank)/...
(b-bprime))*(y-bprime/2);
alphaZLL2=alphaZLLcrank+2*((alphaZLLtip-alphaZLLcrank)/(b-bprime))*...
    (y-bprime);
delta2=iw+epsilon2-alphaZLL2;
Ddelta2=delta2-deltaBarZLL;
Ddelta2Rad=Ddelta2*pi/180;
distlac2=croot/4+Deltalac2-LacBar;
Integrand14=Ddelta2Rad.*distlac2.*cw2;
Int10=trapz(y,Integrand14);

CMacLoad=-(2*pi/(S*cavg))*(Int9+Int10);

CMacprime=CMac0+CMacLoad;
CMac=CMacprime*cavg/cbar;

% Calculation of wing's lift-curve slope
Lamhalfeff=acos((2*Sw1*cos(Lam1halfrad)+2*Sw2*cos(Lam2halfrad))/S);
Lamhalfeffdeg=Lamhalfeff*180/pi;
Claavgdeg=(Claroot+Clatip)/2; % average 2D lift-curve slope of wing
sections
Claavgrad=Claavgdeg*180/pi;
k=Claavgrad/(2*pi);
AR=b^2/S;
Num=2*pi*AR;
Denom=2+sqrt(((AR/k)^2)*(1+(tan(Lamhalfeff))^2)+4);
CLa=Num/Denom;
CLadeg=CLa*pi/180;
CL0=CLadeg*deltaBarZLL;

% Calculation of wing's profile-drag coeff.
% inner portion
for J=1:1:5
    CL(J)=0.25*(J-1);
    y=linspace(lowerlimit1,upperlimit1,N);
    cw1=croot+2*((cprime-croot)/bprime)*y;
    Cd01=Cd0root+2*((Cd0crank-Cd0root)/bprime)*y;
    A1=Aroot+2*((Acrank-Aroot)/bprime)*y;
    B1=Broot+2*((Bcrank-Broot)/bprime)*y;
    epsilon1=epsilonroot+2*((epsiloncrank-epsilonroot)/bprime)*y;
    alphaZLL1=alphaZLLroot+2*((alphaZLLcrank-alphaZLLroot)/bprime)*y;
    delta1=iw+epsilon1-alphaZLL1;
```

```
        Ddelta1=delta1-deltaBarZLL;
        Integrand15=Ddelta1.*cw1;
        Cladeg=Cla*pi/180;
        TLD1=0.5*Cladeg*Integrand15/cavg; % twisted lift distribution
        TCl1=TLD1./(cw1/cavg); % local twisted lift coefficient for wing at CL=0
        F2a=sqrt(1-((y/(b/2)).^2));
        Deltalac1=y*tan(Lam1rad);
        % basic lift distribution:

BLD1=0.5*(c0prime/cavg)*((cw1/c0prime)+F2a+eta*Deltalac1/c0prime)*CL(J);
        % local basic lift coefficient:
        BCl1=BLD1*cavg./cw1;
        Cltotal1=TCl1+BCl1;
        % local airfoil drag coefficient at given CL:
        Cd1=Cd01+A1.*Cltotal1+B1.*(Cltotal1).^2;
        Integrand17=cw1.*Cd1;
        Int13(J)=trapz(y,Integrand17);
end

% outer portion
for J=1:1:5
    CL(J)=0.25*(J-1);
    y=linspace(lowerlimit2,upperlimit2,N);
    cw2=cprime+2*((ctip-cprime)/(b-bprime))*(y-bprime/2);
    Cd02=Cd0crank+2*((Cd0tip-Cd0crank)/(b-bprime))*(y-bprime/2);
    A2=Acrank+2*((Atip-Acrank)/(b-bprime))*(y-bprime);
    B2=Bcrank+2*((Btip-Bcrank)/(b-bprime))*(y-bprime);
    epsilon2=epsiloncrank+2*((epsilontip-epsiloncrank)/(b-bprime))*...
        (y-bprime/2);
    alphaZLL2=alphaZLLcrank+2*((alphaZLLtip-alphaZLLcrank)/
(b-bprime))*...
        (y-bprime/2);
    delta2=iw+epsilon2-alphaZLL2;
    Ddelta2=delta2-deltaBarZLL;
    Integrand16=Ddelta2.*cw2;
    TLD2=0.5*Cladeg*Integrand16/cavg;
    % local airfoil lift coefficient for wing at CL=0:
    TCl2=TLD2./(cw2/cavg);
    F2b=sqrt(1-((y/(b/2)).^2));
    Deltalac2=(bprime/2)*tan(Lam1rad)+(y-(bprime/2))*tan(Lam2rad);
    % basic lift distribution:

BLD2=0.5*(c0prime/cavg)*((cw2/c0prime)+F2b+eta*Deltalac2/c0prime)*CL(J);
    % local basic lift coefficient:
    BCl2=BLD2*cavg./cw2;
    Cltotal2=TCl2+BCl2;
    % local airfoil drag coefficient at given CL:
    Cd2=Cd02+A2.*Cltotal2+B2.*(Cltotal2).^2;
    Integrand18=cw2.*Cd2;
    Int14(J)=trapz(y,Integrand18);
end

% summation
for J=1:1:5
    CL(J)=0.25*(J-1);
    CDprofile(J)=(2/S)*(Int13(J)+Int14(J));
```

```
end

% curve fit and plot CDprofile
figure
x=[0 0.25 0.5 0.75 1];
y=[CDprofile(1) CDprofile(2) CDprofile(3) CDprofile(4) CDprofile(5)];
p1=polyfit(x,y,2);
xp=0:0.01:1;
yp=polyval(p1,xp);
plot(x,y,'o',xp,yp,'linewidth',1.5)
xlabel('CL');ylabel('CDprofile')
title('CDprofile vs. CL Polar Curve')

% Calculation of the basic-wing induced drag coefficient
% inner portion
for J=1:1:5
    CL(J)=0.25*(J-1);
    y=linspace(lowerlimit1,upperlimit1,N);
    cw1=croot+2*((cprime-croot)/bprime)*y;
    F2a=sqrt(1-((y/(b/2)).^2));
    Deltalac1=y*tan(Lam1rad);

BLD1=0.5*(c0prime/cavg)*((cw1/c0prime)+F2a+eta*Deltalac1/c0prime)*CL(J);
    Cl1=BLD1*cavg./cw1;
    Integrand19=Cl1.*cw1;
    Integrand20=cw1.*Cl1.^2;
    Int19=trapz(y,Integrand19);
    Int20=trapz(y,Integrand20);
    CDi1(J)=(2*CL(J)/(CLa*S))*Int19-(1/(pi*S))*Int20;
end

% outer portion
for J=1:1:5
    CL(J)=0.25*(J-1);
    y=linspace(lowerlimit2,upperlimit2,N);
    cw2=cprime+2*((ctip-cprime)/(b-bprime))*(y-bprime/2);
    F2b=sqrt(1-((y/(b/2)).^2));
    Deltalac2=(bprime/2)*tan(Lam1rad)+(y-(bprime/2))*tan(Lam2rad);

BLD2=0.5*(c0prime/cavg)*((cw2/c0prime)+F2b+eta*Deltalac2/c0prime)*CL(J);
    Cl2=BLD2*cavg./cw2;
    Integrand21=Cl2.*cw2;
    Integrand22=cw2.*Cl2.^2;
    Int21=trapz(y,Integrand21);
    Int22=trapz(y,Integrand22);
    CDi2(J)=(2*CL(J)/(CLa*S))*Int21-(1/(pi*S))*Int22;
end

% summation
for J=1:1:5
    CL(J)=0.25*(J-1);
    CDi(J)=CDi1(J)+CDi2(J);
end

% induced-drag curve fit and plot
figure
```

```
x=[0 0.25 0.5 0.75 1];
y=[CDi(1) CDi(2) CDi(3) CDi(4) CDi(5)];
p2=polyfit(x,y,2);
xp=0:0.01:1;
yp=polyval(p2,xp);
plot(x,y,'o',xp,yp,'linewidth',1.5)
xlabel('CL');ylabel('CDi')
title('CDinduced vs. CL Polar Curve')

% CL-CD Polar
for J=1:1:5
    CL(J)=0.25*(J-1);
    CD(J)=CDprofile(J)+CDi(J);
end

% curve fit and plot
figure
x=[0 0.25 0.5 0.75 1];
y=[CD(1) CD(2) CD(3) CD(4) CD(5)];
p3=polyfit(x,y,2);
xp=0:0.01:1;
yp=polyval(p3,xp);
plot(x,y,'o',xp,yp,'linewidth',1.5)
xlabel('CL');ylabel('CD')
title('CD vs. CL Polar Curve')

% maximum L/D characteristics
CLmaxLoverD=sqrt(p3(3)/p3(1)); % lift coefficient for max L/D
LoverDmin=(p3(3)/CLmaxLoverD)+p3(2)+p3(1)*CLmaxLoverD; % max L/D
LoverDmax=1/LoverDmin;

% k-factor
AR=b^2/S;
kfactor=p2(1)/(1/(pi*AR));

fprintf('Aerodynamic Characteristics of Swept Double-Tapered Linearly-
Twisted Wings')
fprintf('\n')
fprintf('\n')
fprintf('S=%6.4f, Stest=%6.4f, AR=%6.4f, k-factor=%6.4f\n'...
    ,S,Stest,AR,kfactor)
fprintf('\n')
fprintf('c0prime=%6.4f, cavg=%6.4f, cbar=%6.4f,
deltaBarZLL(deg)=%6.4f\n'...
    ,c0prime,cavg,cbar,deltaBarZLL)
fprintf('\n')
fprintf('I1a=%6.4f, I2a=%6.4f, I1b=%6.4f, I2b=%6.4f, LacBar=%6.4f\n'...
    ,I1a,I2a,I1b,I2b,LacBar)
fprintf('\n')
fprintf('CMac0=%6.4f, CMacLoad=%6.4f, CMacprime=%6.4f, CMac=%6.4f\n'...
    ,CMac0,CMacLoad,CMacprime,CMac)
fprintf('\n')
fprintf('CLalpha/rad=%6.4f, CLalpha/deg=%6.5f,
CL0=%6.5f\n',CLa,CLadeg,CL0)
fprintf('\n')
fprintf('Polynomial Terms for Profile-Drag Coefficient: ')
fprintf('CDprofile0=%7.6f, CDprofile1=%7.6f, CDprofile2=%7.6f\n'...
    ,p1(3),p1(2),p1(1))
```

```
fprintf('\n')
fprintf('Polynomial Terms for Induced-Drag Coefficient: ')
fprintf('CDinduced0=%7.6f, CDinduced1=%7.6f, CDinduced2=%7.6f\n'...
    ,p2(3),p2(2),p2(1))
fprintf('\n')
fprintf('Polynomial Terms for CD vs. CL: ')
fprintf('CD0=%7.6f, CD1=%7.6f, CD2=%7.6f\n',p3(3),p3(2),p3(1))
fprintf('\n')
fprintf('Max L/D=%7.4f, CL for Max
L/D=%7.6f\n',LoverDmax,CLmaxLoverD)

figure
% Twisted-Wing Lift Distribution
% inner portion
y=linspace(lowerlimit1,upperlimit1,N);
cw1=croot+2*((cprime-croot)/bprime)*y;
epsilon1=epsilonroot+2*((epsiloncrank-epsilonroot)/bprime)*y;
alphaZLL1=alphaZLLroot+(2*(alphaZLLcrank-alphaZLLroot)/bprime)*y;
delta1=iw+epsilon1-alphaZLL1;
Ddelta1=delta1-deltaBarZLL;
DistribT1=Ddelta1.*cw1;
Cladeg=Cla*pi/180;
TLD1=0.5*Cladeg*DistribT1/cavg;
plot(2*y/b,TLD1,'k','linewidth',2)
hold on

% outer portion
y=linspace(lowerlimit2,upperlimit2,N);
cw2=cprime+2*((ctip-cprime)/(b-bprime))*(y-bprime/2);
epsilon2=epsiloncrank+2*((epsilontip-epsiloncrank)/
(b-bprime))*(y-bprime/2);
alphaZLL2=alphaZLLcrank+2*((alphaZLLtip-alphaZLLcrank)/(b-bprime))*...
    (y-bprime/2);
delta2=iw+epsilon2-alphaZLL2;
Ddelta2=delta2-deltaBarZLL;
DistribT2=Ddelta2.*cw2;
TLD2=0.5*Cladeg*DistribT2/cavg;
plot(2*y/b,TLD2,'k','linewidth',2)
grid on
xlabel('y/(b/2)')
ylabel('Twisted Lift Distribution')
title('Twisted Lift Distribution for Double Swept and Tapered Wing')

figure
% Twisted-Wing Lift-Coefficient Distribution
% inner portion
y=linspace(lowerlimit1,upperlimit1,N);
cw1=croot+2*((cprime-croot)/bprime)*y;
epsilon1=epsilonroot+2*((epsiloncrank-epsilonroot)/bprime)*y;
alphaZLL1=alphaZLLroot+(2*(alphaZLLcrank-alphaZLLroot)/bprime)*y;
delta1=iw+epsilon1-alphaZLL1;
Ddelta1=delta1-deltaBarZLL;
DistribT1=Ddelta1.*cw1;
Cladeg=Cla*pi/180;
TLD1=0.5*Cladeg*DistribT1/cavg;
TCl1=TLD1./(cw1/cavg);
plot(2*y/b,TCl1,'k','linewidth',2)
hold on
```

```matlab
% outer portion
y=linspace(lowerlimit2,upperlimit2,N);
cw2=cprime+2*((ctip-cprime)/(b-bprime))*(y-bprime/2);
epsilon2=epsiloncrank+2*((epsilontip-epsiloncrank)/...
    (b-bprime))*(y-bprime/2);
alphaZLL2=alphaZLLcrank+2*((alphaZLLtip-alphaZLLcrank)/(b-bprime))*...
    (y-bprime/2);
delta2=iw+epsilon2-alphaZLL2;
Ddelta2=delta2-deltaBarZLL;
DistribT2=Ddelta2.*cw2;
TLD2=0.5*Cladeg*DistribT2/cavg;
TCl2=TLD2./(cw2/cavg);
plot(2*y/b,TCl2,'k','linewidth',2)
grid on
xlabel('y/(b/2)')
ylabel('Twisted Lift-Coefficient Distribution')
title('Twisted Cl(y) Distribution for Double Swept and Tapered Wing')

figure
% Basic-Lift Distributions
% inner portion
for J=1:1:5
    CL(J)=0.25*(J-1);
    y=linspace(lowerlimit1,upperlimit1,N);
    cw1=croot+2*((cprime-croot)/bprime)*y;
    F2a=sqrt(1-((y/(b/2)).^2));
    Deltalac1=y*tan(Lam1rad);

BLD1=0.5*(c0prime/cavg)*((cw1/c0prime)+F2a+eta*Deltalac1/c0prime)*CL(J);
    plot(2*y/b,BLD1,'k','linewidth',2)
    hold on
end

% outer portion
for J=1:1:5
    CL(J)=0.25*(J-1);
    y=linspace(lowerlimit2,upperlimit2,N);
    cw2=cprime+2*((ctip-cprime)/(b-bprime))*(y-bprime/2);
    F2b=sqrt(1-((y/(b/2)).^2));
    Deltalac2=(bprime/2)*tan(Lam1rad)+(y-(bprime/2))*tan(Lam2rad);

BLD2=0.5*(c0prime/cavg)*((cw2/c0prime)+F2b+eta*Deltalac2/c0prime)*CL(J);
    plot(2*y/b,BLD2,'k','linewidth',2)
    hold on
    grid on
xlabel('y/(b/2)')
ylabel('Basic Lift Distribution')
title('Basic Lift Distribution for Double Swept and Tapered Wing')
end

% Total Lift Distributions
% inner portion
figure
for J=1:1:5
    CL(J)=0.25*(J-1);
    y=linspace(lowerlimit1,upperlimit1,N);
    cw1=croot+2*((cprime-croot)/bprime)*y;
```

```
        F2a=sqrt(1-((y/(b/2)).^2));
        Deltalac1=y*tan(Lam1rad);

BLD1=0.5*(c0prime/cavg)*((cw1/c0prime)+F2a+eta*Deltalac1/c0prime)*CL(J);
        TotalLD1=BLD1+TLD1;
        plot(2*y/b,TotalLD1,'k','linewidth',2)
        hold on
end

% outer portion
for J=1:1:5
        CL(J)=0.25*(J-1);
        y=linspace(lowerlimit2,upperlimit2,N);
        cw2=cprime+2*((ctip-cprime)/(b-bprime))*(y-bprime/2);
        F2b=sqrt(1-((y/(b/2)).^2));
        Deltalac2=(bprime/2)*tan(Lam1rad)+(y-(bprime/2))*tan(Lam2rad);

BLD2=0.5*(c0prime/cavg)*((cw2/c0prime)+F2b+eta*Deltalac2/c0prime)*CL(J);
        TotalLD2=BLD2+TLD2;
        plot(2*y/b,TotalLD2,'k','linewidth',2)
        hold on
        grid on
xlabel('y/(b/2)')
ylabel('Total Lift Distribution')
title('Total Lift Distribution for Double Swept and Tapered Wing')
end

% Basic Cl(y) Distributions
% inner portion
figure
for J=1:1:5
        CL(J)=0.25*(J-1);
        y=linspace(lowerlimit1,upperlimit1,N);
        cw1=croot+2*((cprime-croot)/bprime)*y;
        F2a=sqrt(1-((y/(b/2)).^2));
        Deltalac1=y*tan(Lam1rad);

BLD1=0.5*(c0prime/cavg)*((cw1/c0prime)+F2a+eta*Deltalac1/c0prime)*CL(J);
        Cl1=BLD1*cavg./cw1;
        plot(2*y/b,Cl1,'k','linewidth',2)
        hold on
end

% outer portion
for J=1:1:5
        CL(J)=0.25*(J-1);
        y=linspace(lowerlimit2,upperlimit2,N);
        cw2=cprime+2*((ctip-cprime)/(b-bprime))*(y-bprime/2);
        F2b=sqrt(1-((y/(b/2)).^2));
        Deltalac2=(bprime/2)*tan(Lam1rad)+(y-(bprime/2))*tan(Lam2rad);

BLD2=0.5*(c0prime/cavg)*((cw2/c0prime)+F2b+eta*Deltalac2/c0prime)*CL(J);
        Cl2=BLD2*cavg./cw2;
        plot(2*y/b,Cl2,'k','linewidth',2)
        hold on
        grid on
xlabel('y/(b/2)')
ylabel('Basic Lift-Coefficient Distribution')
```

```matlab
    title('Basic Cl(y) Distribution for Double Swept and Tapered Wing')
end

% Total Lift-Coefficient Distributions
% inner portion
figure
for J=1:1:5
    CL(J)=0.25*(J-1);
    y=linspace(lowerlimit1,upperlimit1,N);
    cw1=croot+2*((cprime-croot)/bprime)*y;
    F2a=sqrt(1-((y/(b/2)).^2));
    Deltalac1=y*tan(Lam1rad);
BLD1=0.5*(c0prime/cavg)*((cw1/c0prime)+F2a+eta*Deltalac1/c0prime)*CL(J);
    TotalCl1=(BLD1+TLD1)*cavg./cw1;
    plot(2*y/b,TotalCl1,'k','linewidth',2)
    hold on
end

% outer portion
for J=1:1:5
    CL(J)=0.25*(J-1);
    y=linspace(lowerlimit2,upperlimit2,N);
    cw2=cprime+2*((ctip-cprime)/(b-bprime))*(y-bprime/2);
    F2b=sqrt(1-((y/(b/2)).^2));
    Deltalac2=(bprime/2)*tan(Lam1rad)+(y-(bprime/2))*tan(Lam2rad);
BLD2=0.5*(c0prime/cavg)*((cw2/c0prime)+F2b+eta*Deltalac2/c0prime)*CL(J);
    TotalCl2=(BLD2+TLD2)*cavg./cw2;
    plot(2*y/b,TotalCl2,'k','linewidth',2)
    hold on
    grid on
xlabel('y/(b/2)')
ylabel('Total Lift-Coefficient Distribution')
title('Total Lift-Coeff. Distrib. for Double Swept and Tapered Wing')
end

% Static Margin
for I=20:1:100
    CL(I)=I/100;
    dhat(I)=CMac/CL(I);
end

% Plot Static-Margin results
figure
I=20:1:100;
plot(CL(I),dhat(I),'k','linewidth',2)
grid on
xlabel('CL')
ylabel('d/cbar')
title('Static Margin for Double Swept and Tapered Wing')
```

"DELTA" TAILLESS AIRCRAFT

The noted German aircraft designer Alexander Lippisch researched flying wings and tailless aircraft throughout his career. His interest evolved toward low aspect-ratio designs, such as the 1931 "Delta 1". This characterized a family of related aircraft configurations that are "Flying Wings" in the technical sense of the term but are better referred to as Tailless Aircraft.

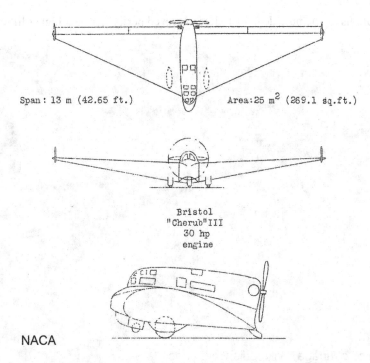

An example of a "pure" delta-winged Tailless Aircraft is the 1955 Convair F-102A "Delta Dagger" jet-propelled fighter. Note that at the high speeds flown by such aircraft as the F-102A, the lift coefficients are very low. Therefore, the induced drag is low compared with the friction and form drag, and thus the low aspect ratio is not an aerodynamic penalty in that flight regime. Also, it is seen that the delta planform gives a compact configuration with excellent structural depth.

There are variations on the pure delta that offer improved performance at higher lift coefficients:

Katsuhiko Tokunaga, Wikipedia Commons

The Swedish SAAB "Draken" from the 1950s shows a compound-delta configuration. A bound leading-edge vortex on the highly-swept inner portions enhances the flow behavior on the less-swept outer portions, thus allowing higher lift coefficients to be attained. The same idea is used on the wing geometry of the Space Shuttle (the photo is of the Enterprise during a glide test, with a tail cone attached).

NASA

A highly-swept wing at an angle of attack can generate stable bound vortices at the leading edge. These are enhanced by having a relatively sharp leading-edge radius. The drawing below illustrates this phenomenon in great detail and is obtained from Le Moigne, T., Rizzi, A., and Johansson, P.,

"CFD Simulations of a Delta Wing in High-Alpha Pitch Oscillations", AIAA Paper 01-0862, presented at the 39th Aerospace Sciences Meeting and Exhibit, 8–11 January 2001, Reno, NV.

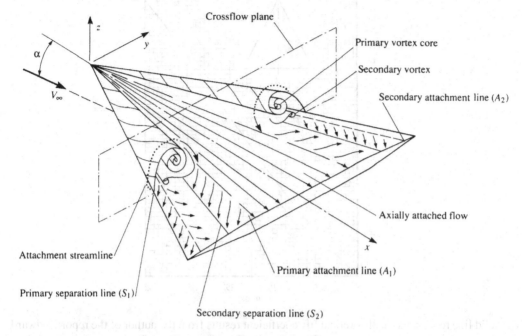

This is also seen in the water-dye visualization from ONERA, as obtained from Milton Van Dyke's book: "An Album of Fluid Motion" (The Parabolic Press, Stanford, CA, 1982):

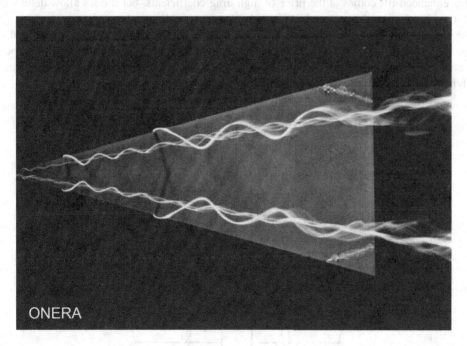

The vortices serve to energize the flow on the upper surface of the wing, allowing high lift coefficients to be attained at high angles of attack. This is seen in the following figure from NASA TN D-3767, where the dashed line is the theoretical lift coefficient without accounting for the edge vortices, and the circular points are experimental data.

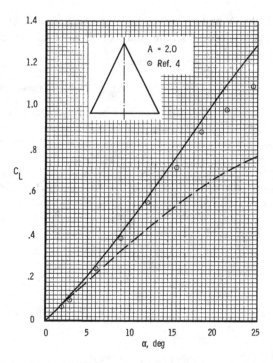

The solid line represents the theoretical lift-coefficient results from the author of the report, Edward C. Polhamus ("A Concept of the Vortex Lift of Sharp-Edge Delta Wings Based on a Leading-Edge-Suction Analogy").

This enhanced lift comes at the price of high drag coefficients, but it does allow delta-winged aircraft, with sufficient thrust, to take off and land at reasonable speeds.

The same effect is used to enhance the low-speed behavior of aircraft with compound delta wings, such as the Draken and Space Shuttle. As mentioned before, the inner highly-swept portion generates the bound vortex; but because the outer portion gives an overall higher aspect ratio, the aerodynamic efficiency of the entire combination is improved.

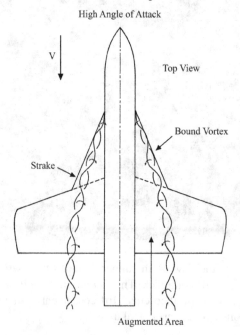

For some wing configurations, the inner highly-swept portion is called a "glove". This feature is also used on the wings of some high-speed airplanes with aft tails, such as the F-16 Fighter.

The "Concorde" supersonic airliner incorporated a "blended" compound-delta wing:

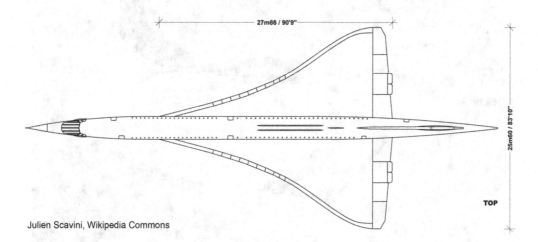
Julien Scavini, Wikipedia Commons

Extensive tests were conducted in water tunnels using tracer dyes to shape the wing for optimum vortex behavior. The picture below shows the Concorde taking off at a steep angle.

Eduard Marmet, Creative Commons

This bound-vortex lift-enhancement feature allowed the Concorde to operate from existing airports, with no need to extend the runway. Also, there was no need to incorporate flaps and slots like those for conventional airliners.

There are other notable low aspect-ratio tailless airplane concepts. One of these is the 1930s "Arup" design by Dr C. L. Snyder. He was a podiatrist and was inspired by the shape of a heel insert. Although the aircraft didn't have high aerodynamic efficiency, it was compact and strong. Raoul Hoffman, a professional aeronautical engineer, refined the design. The Arup was stable and practical, and safely performed hundreds of flights, demonstrating its remarkable flight-speed range of 37–156 km/h. However, the company failed because of poor management.

Another and extraordinary design was the Vought V-173 "Flying Flapjack" by NACA engineer Charles Zimmerman (the aircraft was sometimes called the "Zimmer Skimmer").

The idea is that the aerodynamic inefficiency of the low aspect ratio is overcome by the propellers counter-rotating against the wing's tip vortices. Also, the stall angle of low aspect-ratio wings is very high, so that this combined with the high thrust from the huge propellers (rotors, actually), gave STOL (Short Take-Off and Landing) ability. The success from the V-173 tests lead to the development of a carrier-based fighter, the Chance-Vought XF5U-1:

Transmission problems with the propeller drive delayed testing until after World War II. Although taxi trials were conducted, the project was canceled before the airplane was flown. Unfortunately, it was scrapped, but the V-173 still exists in the custody of the Smithsonian's National Air and Space Museum.

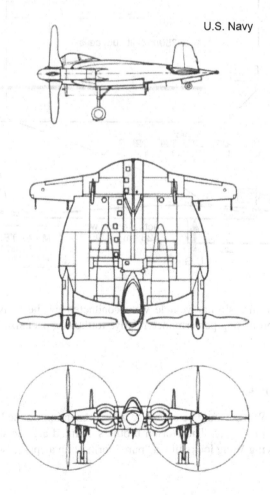

Most recently, the Zimmerman concept was revisited by the U.S. Naval Research Laboratory for Micro Air Vehicle (MAV) applications:

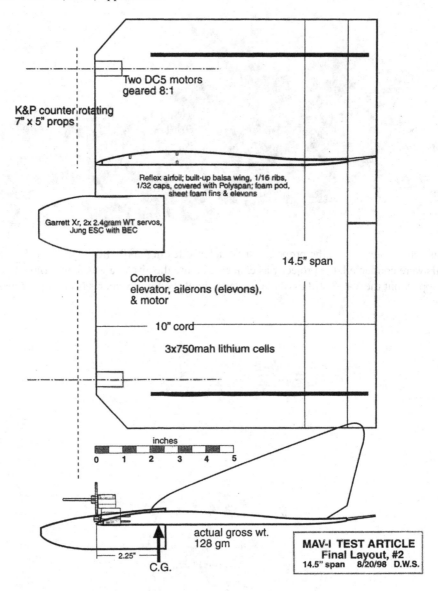

Also, wind tunnel tests on a similar vehicle were conducted by the University of Notre Dame. The results show very promising performance relative to the compact size limitations imposed on MAV's.

FINAL OBSERVATIONS

No discussion of Flying Wings would be complete without mentioning the Horten Ho-VI sailplane, which had an aspect ratio of 32.4. The Horten brothers claimed a glide slope of 45:1, but this is unlikely. Although a Flying Wing looks like a "pure" form of an airplane, with no tail, fuselage, or excrescences, the aerodynamic compromises required to give it inherent longitudinal stability (as discussed previously) make it less efficient than a tail-aft design. Nonetheless, this aircraft has a breathtaking beauty:

However, the goal of a pure flying wing with efficiencies exceeding those of tail-aft designs may now be realized with the incorporation of a computerized SAS. The picture below shows the Northrop B-2 bomber that is inherently controls-fixed unstable because it lacks twist or reflex.

However, the onboard SAS provides the stability that allows it to fly at efficient trim states. Modern SAS's may permit flying wings to finally achieve the performance that their clean geometry always seemed to promise. This would realize the dreams of Flying-Wing pioneers, such as Jack Northrop.

When the B-2 program began, it was in utmost secrecy. By this time, Jack Northrop was long retired and in poor health. However, arrangements were made to bring him into a secure conference room to see a model of the proposed airplane. His comment was: "Now I know why God has kept me alive until now".

The story isn't over, though, for Flying Wings with inherent stability. During World War II, the German aircraft manufacturer Blohm and Voss proposed a unique design, the P 208.02, as shown:

The wing incorporates an efficient airfoil and has no twist. The stability is provided by the outrigger fins, which also act like modern winglets. These utilize the tip vortices to produce leading-edge thrust, thus off-setting the drag. Model glider tests at the U.S. Naval Research Laboratory have shown high efficiency for the design, and these results have been confirmed by wind-tunnel tests at the University of Toronto.

In fairness, however, the outrigger fins appear to be morphing to a tail-aft design. This brings to mind a quote attributed to Igor Sikorsky, who stated that there is nothing wrong with a Flying Wing that an aft tail wouldn't fix.

Finally, Jack Northrop's name has been mentioned more than once in this chapter. He was the premier champion of Flying Wings in the United States, as were the Horten brothers in Germany. This interest began early, as shown by the stick-and-tissue proof-of-concept model he made while young, which is now on display at the Smithsonian National Air and Space Museum.

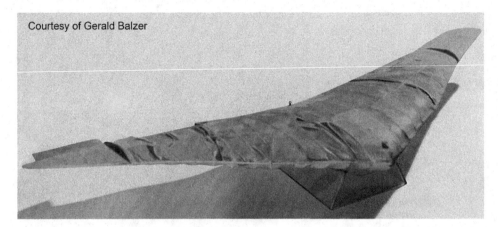

Courtesy of Gerald Balzer

Such models are often used by designers to obtain an initial evaluation of their concept. It is said that early designs for the space-shuttle configuration were tested by NASA engineers gliding balsa-wood models while standing on their chairs. It was the author's experience, while working at NASA in the 1960s, that many aerospace designers and researchers had a background in aeromodeling, which gave them the skills to build and test their ideas.

In summary, it is seen that numerous designers were enticed by the idea of a "pure" aircraft being a Flying Wing, which eliminates the complication and drag of an aft tail and extended fuselage. However, it is also shown that for inherent longitudinal stability, geometrical compromises are required that reduce the overall aerodynamic efficiency. This may not be a factor, in future, if a modern compact SAS is incorporated. The dreams and intentions of these designers would then be realized.

5 Canard Airplanes and Biplanes

CANARD AIRPLANES

Canard airplanes are configured to have their horizontal tail forward of the wing, as shown in the picture of the 1908 Wright Model A. The horizontal tail itself is referred to as the "canard". Such configurations have existed throughout the history of aviation, and designers have had various motivations for using these. In the Wrights' case, they wanted the canard to provide protection in case of a crash (a "bumper").

Wikipedia Commons

ASIDE

The term "canard" comes from the French word for "duck" because the extended forward fuselage seemed to be akin to the extended long neck of a duck in full flight.

END ASIDE

Modern canard-airplane designers claim certain performance advantages for their designs.

Courtesy of Velocity, Inc.

One claimed feature is that at high angles of attack, the canard will stall before the wing. Therefore, the nose will dip down and stall recovery will occur with minimum loss of altitude.

The following picture shows the famous Wright Flyer, which is accepted as the World's first successful airplane:

Wikipedia Commons

Its essential design features are mirrored in the modern Velocity canard airplane shown before. For example, besides incorporating a canard surface, the propeller is mounted aft as a "pusher" design. Also, twin vertical tails are mounted rearward. However, there are very different philosophies regarding longitudinal and lateral stability.

Canard Airplanes and Biplanes

In the 1980s, the Wright Flyer design was extensively assessed by aerospace researchers at the United States National Air and Space Museum and Systems Technology, Inc. This involved wind-tunnel tests on rigid and aeroelastic models. The following six figures are from Systems Technology, Inc. (STI) Paper No. 362, "Flight Control Dynamics of the Wright Flyer" by Henry R. Jex and Fred E. C. Culick, 1985, and are presented here courtesy of STI.

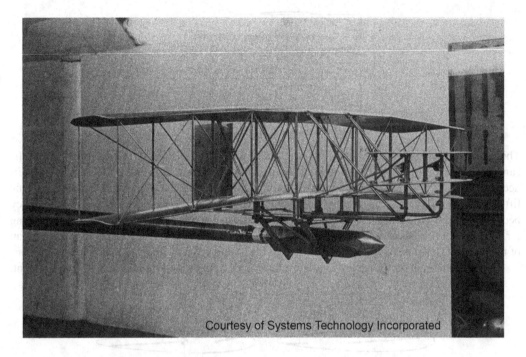

Courtesy of Systems Technology Incorporated

These wind-tunnel experiments were complemented with a theoretical vortex lattice analysis.

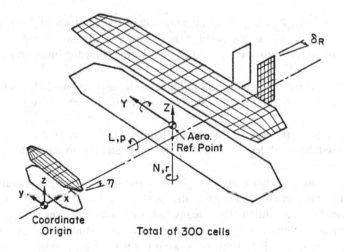

Approximation to the Wright Flyer
for Vortex Lattice Calculations

The purpose of this work was to characterize the Wright Flyer's engineering design and performance in order to allow a more accurate assessment of the historical record. Questions had existed about its aerodynamic efficiency and flight-dynamic stability and control.

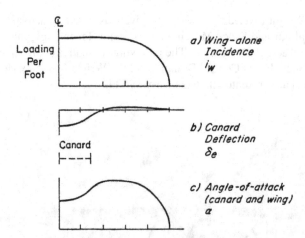

This study, as well as others, showed that the wing's lift is diminished by the downwash from the canard. The top figure shows the wing-alone lift distribution. The middle figure shows the wing's lift decrement caused by the canard; and in the bottom figure, the decrement is added to the wing-alone lift to give the modified span-wise lift distribution. This significantly reduces the wing's efficiency because the span-wise loading is far from the elliptical shape that gives the minimum induced drag for a given total lift. Observe in the middle figure how the decrement approximately follows the span of the canard. This observation will prove useful for an approximate analysis described further on.

Another limitation for the canard configuration comes from its longitudinal stability requirement.

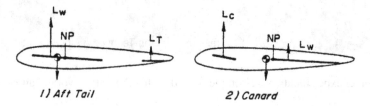

The figure on the left shows the trim arrangement (zero moment about the mass-center) for a tail-aft airplane. Furthermore, as described in the "Complete-Aircraft Aerodynamics" section of Chapter 2, the mass-center must be forward of the neutral point for inherent longitudinal static stability.

Note that for this example, the wing and tail airfoils are un-cambered ($M_{ac} = 0$ for both). Also, depending on the trim condition, the lift of the horizontal tail (shown as L_T here) may be positive or negative or zero.

At any rate, for a properly designed stable and trimmed tail-aft airplane, the lift (and lift coefficient) of the tail will be much less than that for the wing, and the airplane's total lift is primarily provided by the wing.

However, for the canard airplane on the right, the canard surface has to operate at a higher lift coefficient than the wing in order to support the mass-center being forward of the neutral point, providing a balanced trim condition. This means that if the wing operates at its optimum lift coefficient, then the canard will have to operate at a higher non-optimum lift coefficient. What is done, in practice, is a compromise where neither the wing nor the canard is operating optimally for the airplane's overall lift coefficient. The result is that the canard airplane will have higher induced drag, at a given lift coefficient, than that for the equivalent tail-aft design.

An argument made by canard airplane proponents is that both the canard and wing produce positive lift (lift in the "right" direction). That's true but, as is seen, it comes at the price of increased drag and overall less aerodynamic efficiency if inherent longitudinal static stability is sought.

A way to increase the aerodynamic efficiency of canard airplanes is to allow the mass-center to move aft of the neutral point. The wing and canard could then operate at their optimum lift

Canard Airplanes and Biplanes

coefficients. However, this would give rise to a configuration with static longitudinal instability. The Wright Flyer operated in that condition and was dynamically unstable.

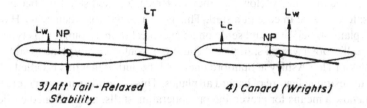

3) Aft Tail – Relaxed Stability *4) Canard (Wrights)*

The following figure shows a response to a one-degree control blip where the airplane then diverges to stall in three seconds, requiring the need for quick control action. The dynamic instability of the Wright Flyer could only be overcome by the considerable skill of the Wright brothers. Even so, the longest flight was 59 s, and the airplane was never flown again. In fact, attempts in recent years to replicate the airplane, and its flights, have not been successful.

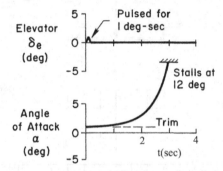

Subsequent Wright designs were less unstable, but they were never easy to fly. Designers since the Wrights have sought inherent longitudinal stability with forward mass-center locations. Also, computerized Stability Augmentation Systems (SAS) have allowed efficient "unstable" canard airplanes to fly successfully, such as the X-29 experimental airplane shown below.

Dryden Flight Research Center EC90-039-4 Photographed 1990
X-29 at an angle that highlights the forward swept wings.
NASA photo by Larry Sammons

In fairness to the Wrights, their papers make it clear that they were fully aware of the instability of their designs. In fact, on one of their airplanes, they attached iron weights to the forward skids and observed the increase in stability. However, they preferred "relaxed stability" because they considered it to be a safety feature in case of a stall. This notion came from their Kitty Hawk glider tests where, if the airplane stalled, it would settle on an even keel to the ground. This type of behavior is called a "deep stall". Otherwise, a stall would cause the front of the airplane to suddenly dip down into a dive. Professor Holt Ashley at Stanford once told the author that this is called a "hammerhead stall" and is an undesirable feature of canard airplanes. This certainly runs counter to the notion that a canard is somehow a means for preventing or moderating stalls, as mentioned earlier.

The human-powered airplanes, "Gossamer Condor" and "Gossamer Albatross" are both canard designs with the mass-centers aft of their neutral points. Therefore, both airplanes are *statically* unstable, like the Wright Flyer. However, both airplanes were readily controllable, and the Gossamer Albatross successfully flew across the English Channel in 1979.

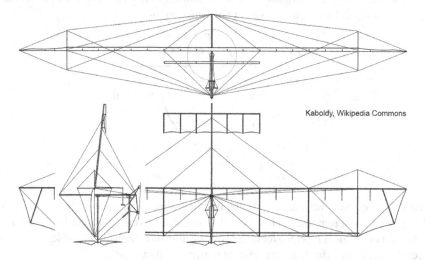

Kaboldy, Wikipedia Commons

The reason for their controllability and *dynamic* stability is that both airplanes were extraordinarily light. The physics behind this behavior is quantitatively explained in the "Longitudinal Dynamic Stability" section of Chapter 6. However, a qualitative explanation is also informative.

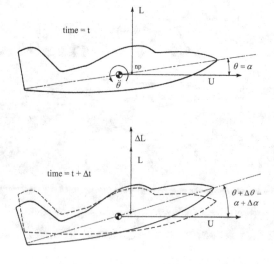

Aircraft Free to Pitch Only ("Pinned")

Canard Airplanes and Biplanes

The figure below shows an aircraft that has pitching freedom only. Namely, it is pivoted at the mass-center point. If the neutral point is forward of the lift vector, then the lift vector gives an unstable pitching moment. Thus, the static-stability criterion states that the neutral point must be aft of the mass-center for stability. This criterion has likewise been accepted by airplane designers for decades, as also giving dynamic longitudinal stability. For most airplanes, this notion is correct.

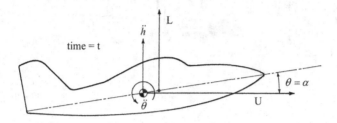

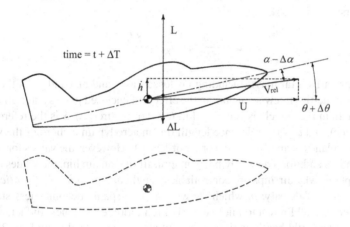

Aircraft Free to Pitch and Heave

However, if the airplane is free to both heave and pitch, as is the case in flight, then the situation differs from that for the pitching-only case. As the figure shows, the lift vector not only gives a pitching acceleration but also a heave (vertical) acceleration. At time Δt, the pitch angle has increased by $\Delta \theta$. However, the angle of attack has decreased by an amount $\Delta \alpha$. This gives a change in the lift vector ΔL, which is less, or even opposite, the value for the pinned case. Therefore, the rate of unstable divergence will be less than that of the pitch-only case.

One sees that heaving tends to stabilize an aircraft where the neutral point is forward of the mass-center. The mass for most airplanes is such that very little heaving occurs compared with pitching, which gives rise to a mode of motion called the "Short-Period Mode". Therefore, for such cases, the static-stability criterion has veracity for that mode of dynamic stability.

However, if the aircraft has a low mass, then the heaving effect becomes significant. This explains why the lightweight Gossamer Condor and Albatross, which were statically unstable, were also sufficiently dynamically stable to be controllable. This effect is also illustrated by the 1902 Wright glider, which was statically unstable in the same way as the powered airplane, but much more controllable because of being lighter.

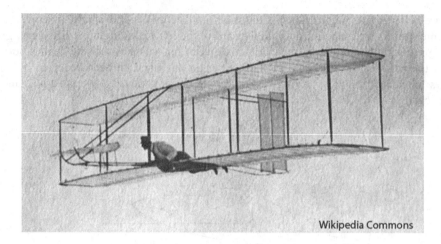

It is important to note that this observation applies to all airplane configurations and not just canard airplanes. This effect especially applies to airships, submarines and torpedoes, which are almost always statically unstable.

The "lightness" of the aircraft can be quantified by the non-dimensional mass and moment-of-inertia parameters:

$$\mu = \frac{m}{\rho S \bar{c}}, \quad i_{yy} = \frac{8 I_{yy}}{\rho S \bar{c}^3}$$

where m is the aircraft's total mass (including enclosed gases and air) and I_{yy} is the aircraft's total pitching moment of inertia (likewise including the effects of enclosed gasses and air). Further, ρ is the density of the medium the vehicle is immersed in (air, for an aircraft), S is the reference area (wing area for an airplane), and $\bar{c}$ is the reference length (mean aerodynamic chord of the wing for an airplane). For most airplanes, μ and i are of the order of 10^2–10^3. However, the values for very lightweight aircraft are much less and, for a neutrally buoyant aircraft like an airship, the values are on the order of unity. This explains why airships are controllable even though most are very *statically* unstable.

Also, note the role of density, ρ, which shows that an airplane becomes less stable as it gains altitude. This effect is well known to pilots who fly high-altitude airplanes. In fact, a U-2 pilot once told the author that he could barely maintain control at the operational altitude of 21 km and, if he lost control, he would have no choice but to ride it down to a much lower altitude before he could attempt to regain control.

Canard Airplanes and Biplanes

Wright airplanes after 1903 were somewhat better behaved because of larger canard surfaces on a longer moment arm and a more forward mass-center. However, they were still dynamically unstable and difficult to learn to fly. Other designers began to evolve configurations that had inherent longitudinal stability. A transition airplane was the 1910 Farman design, which had both a canard and an aft tail.

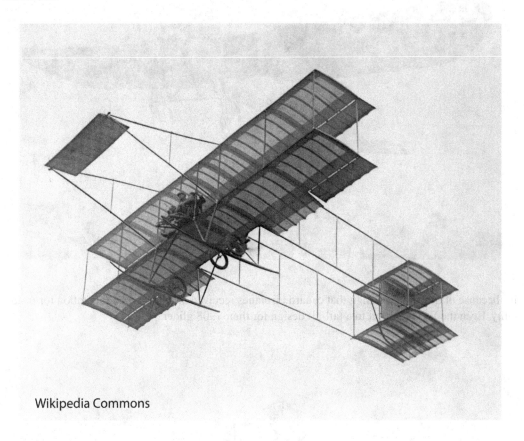

Wikipedia Commons

However, the 1909 Antoinette and 1913 Bleriot tail-aft configurations set the template for most airplane designs for the rest of the century.

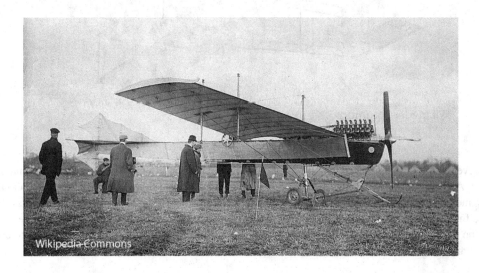

Wikipedia Commons

It's because of the Wright design that canard airplanes received an undeserved reputation for instability. Even the Wrights went to a tail-aft design for their 1908 glider.

However, as previously shown, canard airplanes can be designed to have inherent longitudinal stability. As mentioned before, the drawback is that, at moderate to high lift coefficients, the canard airplane has higher drag coefficients than the equivalent tail-aft design. However, this is less of an issue for high-speed airplanes operating at low lift coefficients.

The photo below shows the SAAB Viggen fighter airplane, where the canard surfaces are used to enhance maneuverability.

Canard Airplanes and Biplanes

As seen in the figure below, the tail-aft airplane increases its lift by first producing an opposite lift at the tail.

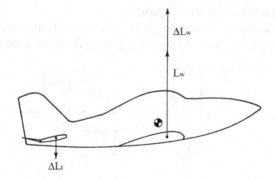

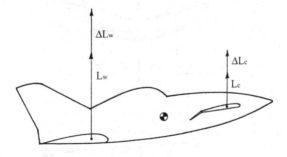

Pitch Control Comparison

This causes a nose-up pitch to the airplane, thus increasing the wing lift to a larger magnitude than the opposite value from the tail. Therefore, the airplane has a net positive lift and then climbs. By comparison, the canard surface produces the initial force in the desired direction. Therefore, the time to execute the maneuver should be less.

When the canard area approaches the wing area, the airplane becomes known as a "Tandem" configuration.

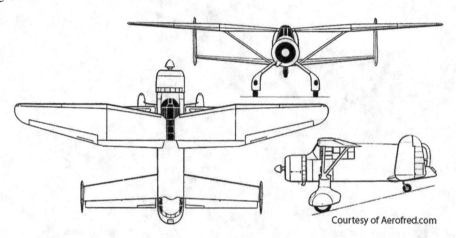
Courtesy of Aerofred.com

These are not common because they do not offer 100% of the advantages of either the tail-aft or canard design, even though they appear to be a compromise of both configurations. The picture shows a modified Westland P.12 "Wendover" airplane of World War II, which was a modification of the "Lysander" design. The Lysander's conventional aft stabilizer has been replaced by a wing in order to support the weight of the rear gun turret. This modification seemed to be a quick fix to provide the airplane with some defensive capability.

A dedicated tandem design was almost the World's first successful airplane. The drawing shows one of a series of scaled flying models ("Aerodromes") designed by Professor Samuel P. Langley in the 1890s.

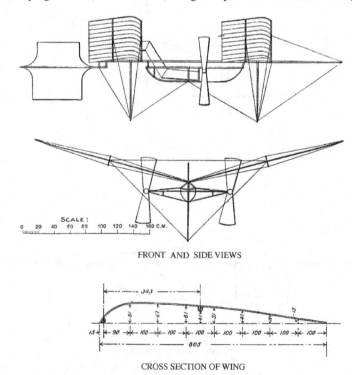

FRONT AND SIDE VIEWS

CROSS SECTION OF WING

Canard Airplanes and Biplanes

Prof. Langley was the director of the Smithsonian Institution at that time, and he thus had the resources to pursue his life-long aeronautical interest. The model shown flew successfully under steam power, thus vindicating Prof. Langley's aeronautical theories. In fact, this and another similar model (internal combustion powered) were the very first sustained engine-powered model airplane flights.

This research culminated in the piloted "Great Aerodrome" of 1903. The aircraft was launched off a houseboat in the Potomac River, as was done for the models. Two attempts were made and both ended in failure because of excessive aeroelastic deformation of the wings. Langley had used a highly cambered single-surface airfoil in wings with slender spars. Although the wings had extensive wire bracing, the basic structure was still so compliant that considerable deformation could occur under aerodynamic loads. This caused the airplane to assume an unstable configuration that could not be controlled and the Great Aerodrome crashed into the river.

By comparison with the Langley design, the Wright brothers stacked their two wings into a biplane configuration. This offers an enormous structural advantage in that the wings act as the top and bottom of a deep trussed beam. The primary loads on the wing spars are thus compression or tension, instead of bending or twisting. If the spars have sufficient buckling strength, one obtains a lightweight structure with overall considerable strength and stiffness in bending and twisting. As discussed in the "Different Materials in Column Buckling" section in Chapter 1, this is the reason why biplane structures were popular for early airplanes with their thin airfoils.

AN APPROXIMATE METHOD FOR ESTIMATING THE AERODYNAMIC CHARACTERISTICS OF CANARD AIRPLANES

This section describes a useful approximate method for estimating the aerodynamic characteristics of a canard airplane, developed by the author.

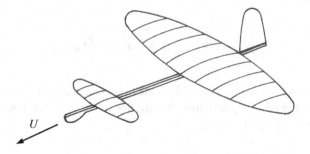

It is assumed that the major aerodynamic components are the wing and canard, so that the airplane's total lift coefficient, C_L, is given by:

$$C_L = C'_{Lw} + C_{Lc} S_c / S \tag{5.1}$$

where C'_{Lw} is the lift coefficient of the wing under the influence of the canard.

The effect of the fuselage is neglected in order to focus on how the canard and wing interact. However, the fuselage effect can be added by using the analysis presented in the "Bodies" section of Chapter 2.

The mutual interference between the canard and wing is due to:

1. Downwash from the canard on the wing,
2. Up-wash from the wing on the canard.

The (1) effect is stronger than (2), so this approximate method ignores (2).

Now the lift distribution across the wing due to the canard, $l'_w(y)$, is illustrated below. Note that the "divot" in this distribution is caused by the canard, as mentioned earlier in this chapter.

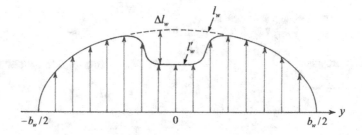

The wing's total lift, L'_w, is given by:

$$L'_w = 2 \int_0^{bw/2} l'_w \, dy = 2q \int_0^{bw/2} C'_{lw} c_w \, dy \tag{5.2}$$

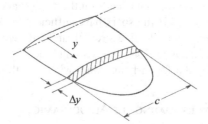

where $C'_{lw} = C_{lw} - \Delta C_{lw}$

If one assumes that the region of the lift deficit is confined to the span of the canard, b_c, then equation (5.2) becomes:

$$L'_w = 2q \int_0^{bw/2} C_{lw} c_w \, dy - 2q \int_0^{bc/2} \Delta C_{lw} c_w \, dy \equiv (L_w)_{\text{isolated}} - \Delta L_w \tag{5.3}$$

$(L_w)_{\text{isolated}}$ is the lift of the isolated wing and ΔL_w is the lift deficit due to the canard. A crude approximation to ΔC_{lw} may be obtained by assuming that the downwash from the canard is uniform across the portion of the wing that it influences:

Canard Airplanes and Biplanes

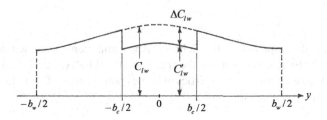

so that

$$\Delta C_{lw} = (C_{l\alpha})_w \varepsilon, \quad \text{where } \varepsilon = (\varepsilon/C_L)_c C_{Lc}$$

The downwash parameter, ε/C_L, may be found by the methods presented in the "Complete-Aircraft Aerodynamics" section in Chapter 2.

ASIDE

A wing's lift-coefficient distribution, $C_{lw}(y)$, doesn't necessarily follow the same shape as the lift distribution, $l_w(y)$. This may be seen from the equation:

$$C_{lw}(y) = \frac{l_w(y)}{q\,c_w(y)}$$

If the lift distribution, $l_w(y)$ is elliptical and the chord-wise shape, $c_w(y)$ is also elliptical, then the span-wise lift coefficient, $C_{lw}(y)$ is constant. This is, in fact, the result from Prandtl's lifting-line theory; namely that an untwisted wing with an elliptical planform has a constant C_{lw} along its span. For other wing configurations, of course, $C_{lw}(y)$ will vary.

END ASIDE

Now, the lift deficit from equation (5.3) may be expressed as

$$\Delta L_w = 2q \int_0^{bc/2} (C_{l\alpha})_w \varepsilon c_w \, dy = 2q \left(\frac{\varepsilon}{C_L}\right)_c C_{Lc} \int_0^{bc/2} (C_{l\alpha})_w c_w \, dy \tag{5.4}$$

where $(C_{l\alpha})_w$ is the span-wise distribution of the *isolated* wing's lift-curve slope. This may be estimated from the "Schrenk Approximation" ("A Simple Approximation for Obtaining The Spanwise Lift Distribution", by O. Schrenk, NACA Tech. Memorandum No. 948), which gives that an *untwisted* isolated wing's lift distribution may be expressed as:

$$l_w(y) \approx k\,q\,\hat{\alpha}\,fn(y)(c_0)_w \tag{5.5}$$

where $\hat{\alpha}$ is the wing's angle-of-attack taken to its zero-lift line, given by $\hat{\alpha} = a_w - (\alpha_{ZLL})_w$ and $(\alpha_{ZLL})_w = -(C_{L0})_w/(C_{L\alpha})_w$ (terms are defined in the "Wings" section of Chapter 2. Further, for this development, the airfoils along the span are assumed to have identical values of $(\alpha_{ZLL})_{\text{airfoil}}$, which thus means that $(\alpha_{ZLL})_w = (\alpha_{ZLL})_{\text{airfoil}}$.

ASIDE

A more complete development of the Schrenk Approximation as applied to aerodynamically-twisted wings ($\hat{\alpha}$ is variable along the span) is offered in "An Approximate Method for Estimating the Aerodynamic Characteristics of Wings with Variable Twist, Taper and Sweep" section of Chapter 4.

END ASIDE

The distribution function, $fn(y)$, is assumed to be the average between an elliptical span-wise distribution and that based on the local chord.

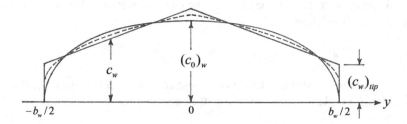

The elliptical function is given by:

$$fn_{\text{ellip}}(y) = (c_0)_w \left[1 - \left(\frac{y}{b_w/2} \right)^2 \right]^{1/2} \quad (5.6)$$

and the function based on the wing's local chord is simply:

$$fn_{\text{chord}}(y) = c_w(y) \quad (5.7)$$

Next, $(c_0)_w$ is adjusted so that the integrated areas of $fn_{\text{ellip}}(y)$ and $fn_{\text{chord}}(y)$ are equal:

$$2 \int_0^{b_w/2} fn_{\text{ellip}}(y)\,dy = \frac{\pi}{4} b_w (c_0)_w = 2 \int_0^{b_w/2} fn_{\text{chord}}(y)\,dy = S_w \rightarrow (c_0)_w = \frac{4 S_w}{\pi b_w} \quad (5.8)$$

Further, as stated before, $fn(y)$ is approximated by the average of $fn_{\text{ellip}}(y)$ and $fn_{\text{chord}}(y)$ (divided by $(c_0)_w$):

$$fn(y) = \frac{\left[fn_{\text{ellip}}(y) + fn_{\text{chord}}(y) \right]}{2(c_0)_w} \quad (5.9)$$

Upon substituting in the terms, one obtains:

$$fn(y) = \frac{1}{2} \left[\frac{c_w(y)}{(c_0)_w} + \left\{ 1 - \left(\frac{y}{b_w/2} \right)^2 \right\}^{1/2} \right] \quad (5.10)$$

Canard Airplanes and Biplanes

Although the drawing above showed a $c_w(y)$ that is trapezoidal, representing a straight-tapered wing, this methodology applies to arbitrary planforms (as long as they are not too weird!).

The next step is to find the value of "k" in equation (5.5). First of all, note that the wing's total lift-curve slope, $C_{L\alpha}$, is given by:

$$\left(C_{L\alpha}\right)_w = \frac{2}{S_w} \int_0^{bw/2} \left(C_{l\alpha}\right)_w c_w \, dy \tag{5.11}$$

Also, Equation (5.5) gives that:

$$L_w = 2 \int_0^{bw/2} l_w \, dy = 2kq\hat{\alpha}(c_0)_w \int_0^{bw/2} fn(y)\,dy \rightarrow \left(C_{L\alpha}\right)_w \hat{\alpha} q S_w = 2kq\hat{\alpha}(c_0)_w \int_0^{bw/2} fn(y)\,dy \rightarrow$$

$$\left(C_{L\alpha}\right)_w = \frac{2k\,(c_0)_w}{S_w} \int_0^{bw/2} fn(y)\,dy \tag{5.12}$$

Because $(C_{L\alpha})_w$ may be found from the method described in the "Wings" section of the Chapter 2 and, because S_w is known along with $(c_0)_w$ from equation (5.8), as well as $fn(y)$ from equation (5.10), one may solve for k.

Finally, Equation (5.5) gives that:

$$l_w(y) = \left(C_{l\alpha}\right)_w \hat{\alpha} q c_w = kq\hat{\alpha}\, fn(y)(c_0)_w \rightarrow \left(C_{l\alpha}\right)_w = k\, fn(y)(c_0)_w / c_w(y) \tag{5.13}$$

which is the span-wise distribution of lift-curve slopes for the isolated wing.

Also, equation (5.5), for the span-wise lift distribution, is useful for structural load calculations. One peculiarity is that the lift does not go to zero at the tips, like it should. However, this can be rounded off in an "eyeball" fashion if desired.

Example, Rectangular Wing

For this case, $c_w(y) = \text{constant} = c_w$. From equation (5.8) one then finds that

$$(c_0)_w = \frac{4 S_w}{\pi b_w} = \frac{4 b_w \times c_w}{\pi b_w} = \frac{4 c_w}{\pi}$$

And, from equation (5.10), one obtains:

$$fn(y) = \frac{1}{2}\left[\frac{\pi}{4} + \left\{1 - \left(\frac{y}{b_w/2}\right)^2\right\}^{1/2}\right] \tag{5.14}$$

Substituting equation (5.14) into equation (5.12) and integrating, gives the result:

$$(C_{L\alpha})_w = \frac{2k(c_0)_w}{S_w} \int_0^{b_w/2} fn(y)\,dy = \frac{2k}{b_w c_w}\left(\frac{4c_w}{\pi}\right)\int_0^{b_w/2} fn(y)\,dy \rightarrow$$

$$(C_{L\alpha})_w = \frac{8k}{\pi b_w}\int_0^{b_w/2} \frac{\pi}{8}\,dy + \frac{4k}{\pi b_w}\int_0^{b_w/2}\left\{1-\left(\frac{y}{b_w/2}\right)^2\right\}^{1/2} dy =$$

$$(C_{L\alpha})_w = \frac{k}{b_w}\left(\frac{b_w}{2}\right) + \frac{4k}{\pi b_w}\left(\frac{\pi b_w/2}{4}\right) = \frac{k}{2} + \frac{k}{2} \rightarrow (C_{L\alpha})_w = k$$

For an aspect ratio, AR, equals 4.0, the Lowry and Pohlhamus equation in the "Wings" section of Chapter 2 gives that $(C_{L\alpha})_w = 3.88$. Therefore, for this case, $k = 3.88$ and equation (5.13) gives that

$$(C_{l\alpha})_w = \frac{4.940}{2}\left[\frac{\pi}{4}+\left\{1-\left(\frac{y}{b_w/2}\right)^2\right\}^{1/2}\right] \tag{5.15}$$

This plots out as

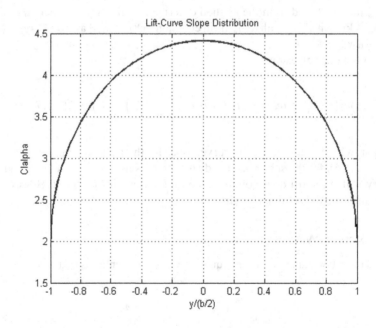

This certainly differs from the uniform distribution for an elliptical-planform wing.

Also, the lift distribution is simply given by:

$$l_w = (C_{l\alpha})_w q\hat{\alpha}c_w(y) \tag{5.16}$$

In this case c_w is constant, so the lift distribution, $l_w(y)$, has the same shape as the lift-curve slope distribution.

Canard Airplanes and Biplanes

Upon returning to Equation (5.3), one may now obtain the lift coefficient of the wing, as influenced by the canard:

$$C'_{Lw} q S_w = 2q \int_0^{b_w/2} C_{lw} c_w \, dy - 2q \int_0^{b_c/2} \Delta C_{lw} c_w \, dy \rightarrow$$

$$C'_{Lw} = \frac{2}{S_w} \int_0^{b_w/2} C_{lw} c_w \, dy - \frac{2}{S_w} \int_0^{b_c/2} \Delta C_{lw} c_w \, dy \equiv \left(C_{Lw}\right)_{isolated} - \Delta C_{Lw} \quad (5.17)$$

The isolated-wing lift coefficient may be simply obtained by the method described in the "Wings" section of Chapter 2:

$$\left(C_{Lw}\right)_{isolated} = \left(C_{L0}\right)_w + \left(C_{L\alpha}\right)_w \alpha \quad (5.18)$$

And the lift-deficit term is obtained from equation (5.4):

$$\Delta C_{Lw} q S_w = 2q \left(\frac{\varepsilon}{C_L}\right)_c C_{Lc} \int_0^{b_c/2} \left(C_{l\alpha}\right)_w c_w \, dy \rightarrow \Delta C_{Lw} = \frac{2}{S_w} \left(\frac{\varepsilon}{C_L}\right)_c C_{Lc} \int_0^{b_c/2} \left(C_{l\alpha}\right)_w c_w \, dy \quad (5.19)$$

EXAMPLE, CANARD GLIDER

The drawing below shows a balsa-wood glider that the author built as a demonstration model for his Aircraft Design class. The required characteristics of the canard, in order to determine its downwash, are:

$$b_c = 5.0'', \quad S_c = 10 \text{ inch}^2, \quad (AR)_c = 2.5, \quad h_H = 0.70'' - 7.65'' \tan(6.5°) = -0.17'', \quad \lambda_c = 1.0$$

$$l_H = 9.3 + 0.25 \times 3.0 - (2.4 + 0.25 \times 2.0) = 10.05 - 2.90 = 7.15'', \quad \Lambda_{c/4} = 0$$

The method used is that from the USAF Stability & Control DATCOM, described in the "Complete Aircraft Aerodynamics" section in Chapter 2:

$$\left(\frac{\varepsilon}{C_L}\right)_c = \frac{(\partial \varepsilon / \partial \alpha)_c}{(C_{L\alpha})_c}, \quad \left(\frac{\partial \varepsilon}{\partial \alpha}\right)_c = 4.44 \left[K_A K_\lambda K_H (\cos \Lambda_{c/4})_c^{1/2} \right]^{1.19}$$

where $K_A = \dfrac{1}{(AR)_c} - \dfrac{1}{1+(AR)_c^{1.7}} = \dfrac{1}{2.5} - \dfrac{1}{1+(2.5)^{1.7}} = 0.4 - 0.174 = 0.226$

$$K_\lambda = \frac{10 - 3\lambda_c}{7} = \frac{10 - 3}{7} = 1.0$$

$$K_H = \frac{1 - |h_H/b_c|}{(2l_H/b_c)^{1/3}} = \frac{1 - 0.17/5.0}{(2 \times 7.15/5.0)^{1/3}} = \frac{1 - 0.034}{(2.86)^{1/3}} = \frac{0.966}{1.419} = 0.681$$

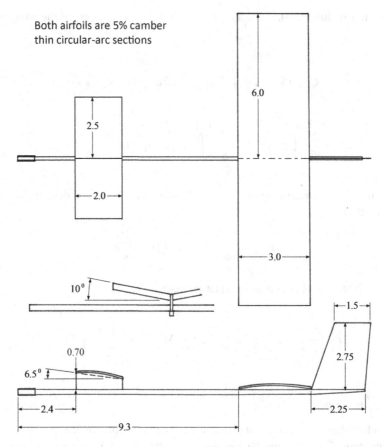

All linear dimensions are in inches

Therefore,

$$(\partial \varepsilon / \partial \alpha)_c = 4.44 \left[0.226 \times 1.0 \times 0.681 \right]^{1.19} = 0.479$$

Furthermore, from the "Wings" section of Chapter 2, the lift-curve slope of the canard is given by:

$$(C_{L\alpha})_c = \frac{2\pi \, AR_c}{2 + \left[\frac{AR_c^2}{\kappa^2}\left(1 + \tan^2 \Lambda_{c/2}\right) + 4 \right]^{1/2}} \quad (1/\text{rad})$$

The κ term is assumed to be 1.0 and $\Lambda_{c/2} = 0$, so one obtains:

$$(C_{L\alpha})_c = \frac{2\pi \times 2.5}{2 + (2.5^2 + 4)^{1/2}} = \frac{15.71}{5.20} = 3.02 \quad (1/\text{rad})$$

Thus, $(\varepsilon / C_L)_c = \dfrac{(\partial \varepsilon / \partial \alpha)_c}{(C_{L\alpha})_c} = \dfrac{0.479}{3.02} = 0.159$

Canard Airplanes and Biplanes

Now, previous work gave the distribution of $C_{l\alpha}$ over an untwisted wing of $AR = 4.0$ with identical airfoils along the span. The present example has such a wing and the values of $(C_{l\alpha})_w$ over the span region covered by the assumed wake of the canard are:

y (inches)	$(C_{l\alpha})_w$
0	4.409
0.5	4.400
1.0	4.374
1.5	4.331
2.0	4.268
2.5	4.185

The average value of $(C_{l\alpha})_w$ in this region is $\overline{(C_{l\alpha})}_w = 4.328$. This may be used with equation (5.19) to give:

$$\Delta C_{Lw} = \frac{2}{S_w}\left(\frac{\varepsilon}{C_L}\right)_c C_{Lc} \int_0^{b_c/2} (C_{l\alpha})_w c_w \, dy \approx \frac{2}{S_w}\left(\frac{\varepsilon}{C_L}\right)_c C_{Lc} \overline{(C_{l\alpha})}_w c_w \frac{b_c}{2}$$

$$\Delta C_{Lw} = \frac{2}{36} \times 0.159 \times C_{Lc} \times 4.328 \times 3.0 \times 2.5 = 0.287 C_{Lc}$$

Therefore, from equations (5.17) and (5.18), one obtains:

$$C'_{Lw} = (C_{Lw})_{isolated} - \Delta C_{Lw} = (C_{L0})_w + (C_{L\alpha})_w \alpha_w - \Delta C_{Lw}$$

As shown in the drawing of the glider, both airfoils are 5%-camber circular arcs. The methodology in the "Airfoils" section of Chapter 2 may be followed to obtain α_{ZLL}, but for such an airfoil this value is simply given by:

$$\alpha_{ZLL} = -2 \times h_{max}/c, \quad \text{where } h_{max} \text{ is the maximum height of the camber line.}$$

For this airfoil the equation becomes $\alpha_{ZLL} = -2 \times 0.05 c/c = -0.10$ rad $(-5.73 \deg)$. Therefore, $(C_{L0})_w = -(C_{L\alpha}) \alpha_{ZLL} = 0.10(C_{L\alpha})_w$.

As seen from the drawing, the characteristics of the wing are $AR_w = 4$, $\Lambda_{c/2} = 0$ and, once again, κ is chosen to be one. Therefore, as before, the following equation is used to obtain the isolated-wing's lift-curve slope:

$$(C_{L\alpha})_w = \frac{2\pi AR_w}{2+\left[\frac{AR_w^2}{\kappa^2}(1+\tan^2 \Lambda_{c/2})+4\right]^{1/2}} = \frac{2\pi \times 4.0}{2+(4.0^2+4)^{1/2}} = \frac{25.132}{6.472} = 3.88 \quad (1/\text{rad})$$

So, the wing's lift equation is:

$$C'_{Lw} = 0.388 + 3.88 \alpha_w - 0.287 C_{Lc}$$

Note that α_w is the wing's angle of attack relative to its root chord, as defined in the "Wings" section of Chapter 2. In this case, since the wing is mounted with zero incidence angle to the body of the glider, α_w is also the angle of attack of the glider, α. If the wing were mounted at an incidence angle i_w, then that would have to be accounted for as is done below for the canard.

The canard is mounted at a 6.5-degree (0.113 radian) incidence angle. So $\alpha_c = \alpha + i_c$ and the lift equation is:

$$C_{Lc} = (C_{L0})_c + (C_{L\alpha})_c (\alpha + i_c) = -(C_{L\alpha})_c (\alpha_{ZLL})_c + (C_{L\alpha})_c (\alpha + i_c) \rightarrow$$

$$C_{Lc} = 0.10 (C_{L\alpha})_c + (C_{L\alpha})_c (\alpha + i_c) \rightarrow$$

$$C_{Lc} = 0.10 \times 3.02 + 3.02 (\alpha + 0.113) = 0.302 + 0.341 + 3.02\alpha$$

$$C_{Lc} = 0.643 + 3.02\alpha$$

Upon substituting this into the wing-lift equation, one obtains:

$$C'_{Lw} = 0.388 + 3.88\alpha - 0.287(0.643 + 3.02a) = 0.388 - 0.185 + 3.88\alpha - 0.87\alpha$$

$$C'_{Lw} = 0.203 + 3.01\alpha$$

So, equation (5.1) gives the total lift equation for the glider:

$$C_L = C'_{Lw} + C_{Lc} S_c/S = 0.203 + 3.01\alpha + (0.643 + 3.02\alpha)10/36$$

$$\underline{\underline{C_L = 0.382 + 3.85\alpha}}$$

Next, the glider's drag coefficient will be found. The first step is to determine the aerodynamic characteristics of the airfoils. In lieu of wind-tunnel data, the "Airfoils" section of Chapter 2 gives an equation for estimating this value:

$$C_d \approx (C_d)_{fric} + (1 - \eta_{le})(2\pi\alpha_{ZLL}^2 + 2\alpha_{ZLL}C_l + C_l^2/(2\pi))$$

Further, because these have sharp leading edges, $\eta_{le} = 0$. Therefore, the drag equation becomes

$$C_d \approx (C_d)_{fric} + 2\pi\alpha_{ZLL}^2 + 2\alpha_{ZLL}C_l + C_l^2/(2\pi)$$

From the previous, for these 5% circular-arc sections, $\alpha_{ZLL} = -0.10$ radian. So, the equation further becomes

$$C_d \approx (C_d)_{fric} + 0.0628 - 0.20C_l + 0.1592C_l^2$$

From glide tests, the glider flight speed is approximately 8 ft/s (2.44 m/s). So, the Reynolds numbers for canard and wing are found from:

$$RN = 6.85 VL \times 10^4$$

where L for the canard is its chord of 2 inches (0.0508 m) and L for the wing is its chord of 3 inches (0.0762 m). Therefore, one has that:

$$RN_c = 8491, \quad RN_w = 12736$$

Next, the friction drag coefficient is given by $(C_d)_{fric} = 2C_{fric}$ where:

$$C_{fric} = \frac{0.455}{(\log_{10} RN)^{2.58}}$$

Therefore, $[(C_d)_{fric}]_c = 0.0267$, $[(C_d)_{fric}]_w = 0.0238$

Canard Airplanes and Biplanes

So the sectional drag coefficients for the canard and wing are:

$$(C_d)_c \approx 0.0895 - 0.20 C_{lc} + 0.1592 C_{lc}^2$$

$$(C_d)_w \approx 0.0866 - 0.20 C_{lw} + 0.1592 C_{lw}^2$$

Now, from the "Wings" section of Chapter 2, the drag coefficient for a finite untwisted wing with identical airfoils along the span is approximated by

$$C_D = (C_D)_{C_L=0} + A_w C_L + B_w C_L^2 + \frac{C_L^2(1+\delta)}{\pi AR}$$

where the wing's profile drag is given by:

$$(C_D)_{\text{profile}} = (C_D)_{C_L=0} + A_w C_L + B_w C_L^2$$

This is obtained from the integration of the airfoil's sectional drag coefficient along the span:

$$(C_D)_{\text{profile}} = \frac{2}{S} \int_0^{b/2} C_d \, c \, dy \approx \frac{2}{S} \int_0^{b/2} \left\{ (C_d)_{\text{fric}} + (1-\eta_{le}) \left[2\pi \alpha_{ZLL}^2 + 2\alpha_{ZLL} C_l + C_l^2/(2\pi) \right] \right\} c \, dy$$

For this particular example, it is seen that $(C_D)_{C_L=0}$ is given by:

$$(C_D)_{C_L=0} = \frac{2}{S} \int_0^{b/2} \left\{ (C_d)_{\text{fric}} + (1-\eta_{le}) \left[2\pi \alpha_{ZLL}^2 \right] \right\} c \, dy$$

For this case, with $(C_d)_{\text{fric}}$ and α_{ZLL} constant along the span, and $\eta_{le} = 0$, one obtains:

$$(C_D)_{C_L=0} = \frac{2}{S} \left[(C_d)_{\text{fric}} + 2\pi \alpha_{ZLL}^2 \right] \int_0^{b/2} c \, dy \rightarrow$$

$$(C_D)_{C_L=0} = (C_d)_{\text{fric}} + 2\pi \alpha_{ZLL}^2 \rightarrow \left[(C_D)_{C_L=0} \right]_c = 0.0895, \quad \left[(C_D)_{C_L=0} \right]_w = 0.0866$$

as derived before.

Next, $A C_L$ is given by:

$$A C_L = \frac{2}{S} \int_0^{b/2} (1-\eta_{le})(2\alpha_{ZLL}) C_l \, c \, dy$$

For this numerical example, one obtains:

$$A C_L = \frac{2}{S} (2\alpha_{ZLL}) \int_0^{b/2} C_l \, c \, dy = 2\alpha_{ZLL} C_L \rightarrow A_c = A_w = -0.20$$

Finally, $B C_L^2$ is given by:

$$B C_L^2 = \frac{2}{S} \int_0^{b/2} (1-\eta_{le}) \frac{C_l^2}{2\pi} c \, dy$$

where, for this numerical example, it becomes

$$BC_L^2 = \frac{1}{S\pi}\int_0^{b/2} C_l^2\, c\, dy \approx \frac{1}{2\pi}C_L^2 \to B_c = B_w = 0.1592$$

ASIDE

This methodology is based on the assumption of an untwisted elliptical planform with identical airfoils along the span, as described in the author's article: "Drag of Wings with Cambered Airfoils and Partial Leading-Edge Suction", *AIAA Journal of Aircraft*, Vol. 20, No. 10, October, 1983. When applied to experimental results for two low-AR rectangular wings, the results were very satisfactory. However, one could also use the $C_l(y)$ from the Schrenk approximation to calculate this. As $\hat{\alpha}$ is constant along the span of the wing for this particular case, these are the following relationships:

$$C_l(y) = C_{l\alpha}(y)\hat{\alpha}, \quad C_L = C_{L\alpha}\hat{\alpha}$$

Also, from the previous work for a rectangular untwisted wing,

$$C_{l\alpha}(y) = \frac{2C_{L\alpha}}{\pi}\left[\frac{\pi}{4} + \left\{1-\left(\frac{y}{b_w/2}\right)^2\right\}^{1/2}\right] \to C_l(y) = \frac{2C_L}{\pi}\left[\frac{\pi}{4} + \left\{1-\left(\frac{y}{b_w/2}\right)^2\right\}^{1/2}\right]$$

Integrating this, one obtains:

$$BC_L^2 = \frac{2}{S}\int_0^{b/2}\left(1-\eta_{le}\right)\frac{C_l^2}{2\pi}c\,dy \to BC_L^2 = 0.1624\, C_L^2 \to B_c = B_w = 0.1624$$

This is only 2% larger than the previous value. Considering that the Schrenk method is an approximation itself, the calculation will proceed with the previous value. However, note that accounting for the wing's lift deficit, due to the canard's wake, in the integration would give an even more accurate value for B_w.

END ASIDE

Now, the induced-drag term is:

$$(C_D)_{\text{induced}} = \frac{C_L^2(1+\delta)}{\pi\, AR}$$

Recall that $AR_c = 2.5$ and $AR_w = 4.0$. For a rectangular-planform wing, the graph in the "Wings" section of chapter 2 gives that $\delta_c \approx 0.024$ and $\delta_w \approx 0.038$. Therefore, one obtains:

$$(C_{Dc})_{\text{induced}} = 0.130\, C_{Lc}^2, \quad (C_{Dw})_{\text{induced}} = 0.083\, C_{Lw}^2$$

ASIDE

The δ values are for an isolated wing. This is a good assumption for the canard, but not necessarily for the wing, which has a lift-distribution deficit from the canard's wake. However δ is a relatively small value in any case, so the isolated-wing assumption is applied in this methodology.

END ASIDE

Canard Airplanes and Biplanes

Therefore, the drag-coefficient equations for the canard and wing are:

$$(C_D)_c = 0.0895 - 0.20 C_{Lc} + 0.2892 C_{Lc}^2$$
$$(C_D)_w = 0.0866 - 0.20 C_{Lw} + 0.2422 C_{Lw}^2$$

The wet area of the body stick is $(S_{wet})_{body} \approx 9.1 \text{in}^2$ ($58.7 \text{cm}^2 = 0.00587 \text{m}^2$), and the Reynolds number, upon noting the body length of 14.6 inches (0.371 m), is $RN_{body} = 6.85 \times 2.44 \times 0.371 \times 10^4 = 62009 \rightarrow (C_f)_b = 0.0080$. So, one has:

$$C_{Db} S_b = (C_f)_b (S_{wet})_b = 0.000047 \text{m}^2 = 0.0729 \text{in}^2$$

The additional drag includes the fin, riblets, etc., for which there is an approximate wetted area, S_{add}, of 20in^2 ($129.0 \text{cm}^2 = 0.0129 \text{m}^2$). Also, the average Reynolds number is $RN_{add} \approx 10{,}000$, which gives a skin-friction coefficient of $(C_f)_{add} = 0.0127 \rightarrow$

$$C_{Dadd} S_{add} = (C_f)_{add} (S_{wet})_{add} = 0.000164 \text{m}^2 = 0.254 \text{in}^2$$

So finally, the glider's total drag, using $S_w \equiv S = 0.02323 \text{m}^2 = 36 \text{inch}^2$ as the reference area, is given by:

$$C_D = C_{Dw} + C_{Dc} \frac{S_c}{S} + C_{Db} \frac{S_b}{S} + C_{Dadd} \frac{S_{add}}{S} \rightarrow$$

$$C_D = C_{Dw} + 0.278 C_{Dc} + 0.00202 + 0.00706 = C_{Dw} + 0.278 C_{Dc} + 0.00908$$

Next, the equations for C_{Dw} and C_{Dc} are expressed in terms of α:

$$C_{Dw} = 0.0866 - 0.20(0.203 + 3.01\alpha) + 0.2422(0.203 + 3.01\alpha)^2$$

$$C_{Dw} = 0.0866 - 0.0406 - 0.602\alpha + 0.2422(0.0412 + 1.222\alpha + 9.060\alpha^2)$$

$$C_{Dw} = 0.0560 - 0.3060\alpha + 2.1943\alpha^2$$

$$C_{Dc} = 0.0895 - 0.20(0.643 + 3.02\alpha) + 0.2892(0.643 + 3.02\alpha)^2$$

$$C_{Dc} = 0.0895 - 0.1286 - 0.604\alpha + 0.2892(0.4134 + 3.884\alpha + 9.120\alpha^2)$$

$$C_{Dc} = 0.0805 + 0.5193\alpha + 2.6375\alpha^2$$

Finally, the drag coefficient of the entire glider is:

$$C_D = 0.0560 - 0.3060\alpha + 2.1943\alpha^2 + (0.0805 + 0.5193\alpha + 2.6375\alpha^2)10/36$$

$$+ 0.0729/36 + 0.254/36$$

$$= 0.0560 + 0.02236 + 0.00203 + 0.0071 - 0.3060\alpha + 0.1443\alpha$$

$$+ 2.1943\alpha^2 + 0.7326\alpha^2$$

$$\underline{\underline{C_D = 0.0874 - 0.1618\alpha + 2.9269\alpha^2}}$$

With the equations for lift and drag coefficients, the glider's polar plot is obtained as discussed for the example airplane in the "Numerical Example" section of Chapter 2, as shown below. The minimum α value is $-10°$ and the maximum α value is $15°$, and the points are incremented by $1.0°$ intervals. From a polynomial curve fit to this plot, one obtains:

$$C_D = 0.1323 - 0.1929 C_L + 0.1975 C_L^2 \equiv P_0 + P_1 C_L + P_2 C_L^2$$

Now,

$$C_D/C_L = P_0/C_L + P_1 + P_2 C_L \to d(C_D/C_L)/dC_L = -P_0/C_L^2 + P_2$$

So, the lift coefficient for maximum lift/drag ratio is given by:

$$-\frac{P_0}{C_L^2} + P_2 = 0 \to (C_L)_{\text{Max L/D}} = \left(\frac{P_0}{P_2}\right)^{1/2} = \left(\frac{0.1323}{0.1975}\right)^{1/2} = 0.8185$$

And the maximum lift/drag ratio is:

$$\left(\frac{C_L}{C_D}\right)_{\text{Max}} = \left(\frac{C_D}{C_L}\right)^{-1}_{\text{Max}} = \left[\frac{P_0}{(C_L)_{\text{Max L/D}}} + P_1 + P_2 \times (C_L)_{\text{Max L/D}}\right]^{-1}$$

$$= \left[\frac{0.1323}{0.8185} - 0.1929 + 0.1975 \times 0.8185\right]^{-1} = (0.1304)^{-1} \to (C_L/C_D)_{\text{Max}} = 7.67$$

Further, the α value at which this occurs is given by:

$$(C_L)_{\text{Max L/D}} = 0.382 + 3.85(\alpha)_{\text{Max L/D}} \to (\alpha)_{\text{Max L/D}} = (0.8185 - 0.382)/3.85 \to$$

$$(\alpha)_{\text{Max L/D}} = 0.1134 \text{ rad} = 6.50°$$

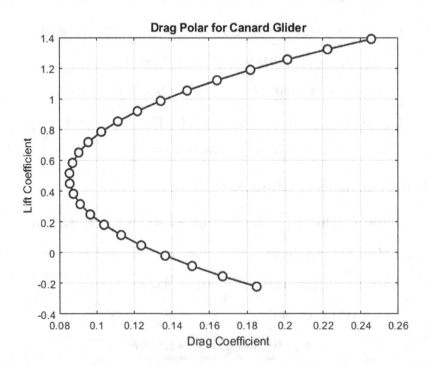

Canard Airplanes and Biplanes

Finally, the glider's neutral point and mass-center for equilibrium trimmed flight will be found. The equation for pitching moment about the mass-center (assuming body effects are negligible) is:

$$M_{cm} \approx L_c (l_{cm} - l_c) - L_w (l_w - l_{cm}) + (M_{ac})_c + (M_{ac})_w$$

The drag of the canard and fin are assumed to have a negligible effect. Now, in coefficient form, the moment equation becomes:

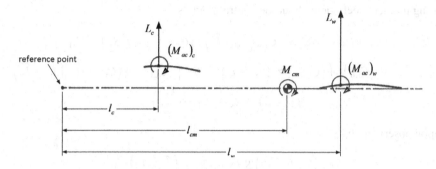

$$(C_m)_{cm} = C_{Lc}\left(\hat{l}_{cm} - \hat{l}_c\right) S_c / S + C_{Lw}\left(\hat{l}_{cm} - \hat{l}_w\right) + (C_{Mac})_c S_c c_c / (Sc) + (C_{Mac})_w$$

where $\hat{l} = l/c$ and c is the reference length (the wing chord).

The derivative of this with respect to α is:

$$(C_{M\alpha})_{cm} = (C_{L\alpha})_c \left(\hat{l}_{cm} - \hat{l}_c\right) S_c / S + (C_{L\alpha})_w \left(\hat{l}_{cm} - \hat{l}_w\right)$$

The neutral point is defined to be that cm location, $\hat{l}_{cm}$, where the moment stays constant with pitch angle. In other words, $(C_{M\alpha})_{cm} = 0$. From the previous equation, that location, $\hat{l}_{np}$, is given by:

$$(C_{L\alpha})_c \left(\hat{l}_{np} - \hat{l}_c\right) S_c / S + (C_{L\alpha})_w \left(\hat{l}_{np} - \hat{l}_w\right) = 0 \rightarrow \hat{l}_{np} = \frac{(C_{L\alpha})_c \hat{l}_c S_c / S + (C_{L\alpha})_w \hat{l}_w}{C_{L\alpha}}$$

The inputs for this example are:

$$(C_{L\alpha})_c = 3.02, \quad (C_{L\alpha})_w = 3.01, \quad \hat{l}_c = 2.9/3.0 = 0.967, \quad \hat{l}_w = 10.05/3.0 = 3.35,$$
$$S_c/S = 10/36 = 0.278, \quad C_{L\alpha} = 3.85$$

Thus, $\hat{l}_{np} = \dfrac{3.02 \times 0.967 \times 0.278 + 3.01 \times 3.35}{3.85} = \dfrac{10.895}{3.85} = \underline{\hat{l}_{np} = 2.830}$

and $l_{np} - 8.490''$ (21.56 cm = 0.0216 m).

Now, the equation for $(C_{M\alpha})_{cm}$ may be rewritten as:

$$(C_{M\alpha})_{cm} = (C_{L\alpha})_c \left[\left(\hat{l}_{cm} - \hat{l}_{np}\right) + \left(\hat{l}_{np} - \hat{l}_c\right)\right] S_c / S + (C_{L\alpha})_w \left[\left(\hat{l}_{cm} - \hat{l}_{np}\right) + \left(\hat{l}_{np} - \hat{l}_w\right)\right]$$

From before, this becomes:

$$(C_{M\alpha})_{cm} = \left(\hat{l}_{cm} - \hat{l}_{np}\right) \left[(C_{L\alpha})_c S_c / S + (C_{L\alpha})_w\right]$$

Define:
$$\text{Static Margin} \equiv \hat{x}_{np} = \left(\hat{l}_{np} - \hat{l}_{cm}\right)$$

So, the equation becomes:
$$(C_{M\alpha})_{cm} = -\hat{x}_{cm} C_{L\alpha}$$

The pitching-moment coefficient about the neutral point is:
$$(C_M)_{np} = C_{Lc}\left(\hat{l}_{np} - \hat{l}_c\right) S_c/S + C_{Lw}\left(\hat{l}_{np} - \hat{l}_w\right) + (C_{Mac})_c S_c c_c/(Sc) + (C_{Mac})_w$$
$$= \left[(C_{L0})_c + (C_{L\alpha})_c \alpha\right]\left(\hat{l}_{np} - \hat{l}_c\right) S_c/S + \left[(C_{L0})_w + (C_{L\alpha})_w \alpha\right]\left(\hat{l}_{np} - \hat{l}_w\right)$$
$$+ (C_{Mac})_c S_c c_c/(Sc) + (C_{Mac})_w$$

Again, upon observing that
$$(C_{L\alpha})_c \left(\hat{l}_{np} - \hat{l}_c\right) S_c/S + (C_{L\alpha})_w \left(\hat{l}_{np} - \hat{l}_w\right) = 0$$

this equation becomes:
$$(C_M)_{np} = (C_{L0})_c \left(\hat{l}_{np} - \hat{l}_c\right) S_c/S + (C_{L0})_w \left(\hat{l}_{np} - \hat{l}_w\right) + (C_{Mac})_c S_c c_c/(Sc) + (C_{Mac})_w$$

For the glider, the inputs are:
$$(C_{L0})_c = 0.643, \quad \hat{l}_c = 0.967, \quad (C_{L0})_w = 0.203, \quad \hat{l}_w = 3.35, \quad \hat{l}_{np} = 2.830$$

Also, for a circular-arc airfoil, $C_{mac} = \pi \alpha_{ZLL}/2$. For a 5% section, $\alpha_{ZLL} = -0.10\,\text{rad}$, so $C_{mac} = -0.1571$. Because the wing is untwisted and has identical airfoils along the span, $(C_{Mac})_c = (C_{Mac})_w = C_{mac} = -0.1571$. Therefore:

$$(C_M)_{np} = 0.643(2.830 - 0.967) \times 0.278 + 0.203(2.830 - 3.35) - 0.1571 \times 0.1852 - 0.1571$$
$$= 0.3330 - 0.1056 - 0.0291 - 0.1571 = (C_M)_{np} = 0.0412$$

Because this is a positive value, the glider is capable of trimmed flight with static longitudinal stability. The mass-center location for the different trim states is simply found from:

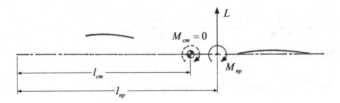

$$M_{cm} = M_{np} + L\left(l_{cm} - l_{np}\right) \rightarrow (C_M)_{cm} = (C_M)_{np} + C_L\left(\hat{l}_{cm} - \hat{l}_{np}\right) = 0 \rightarrow \hat{l}_{cm} = \hat{l}_{np} - (C_M)_{np}/C_L$$

For the glider, this becomes:
$$\hat{l}_{cm} = 2.830 - 0.0412/C_L$$

Canard Airplanes and Biplanes

Recall that the lift coefficient for maximum L/D is $(C_L)_{MaxL/D} = 0.8185$. Therefore, the mass-center location that gives this condition is $(\hat{l}_{cm})_{MaxL/D} = 2.780$. The static margin is thus:

$$\text{Static Margin} = \hat{x}_{np} = \left(\hat{l}_{np} - \hat{l}_{cm}\right) = (2.830 - 2.780) = 0.050 \to 5.0\%$$

This is at the low end of the usual acceptable range, but still sufficient to ensure stability.

If the drag of the canard and fin are included in the moment equation, it would become:

$$M_{cm} \approx L_c(l_{cm} - l_c) - L_w(l_w - l_{cm}) + (M_{ac})_c + (M_{ac})_w + D_c(h_c - h_{cm}) + D_{fin} h_{fin} \to$$

$$(C_m)_{cm} = C_{Lc}\left(\hat{l}_{cm} - \hat{l}_c\right) S_c/S + C_{Lw}\left(\hat{l}_{cm} - \hat{l}_w\right) + (C_{Mac})_c S_c c_c/(Sc) + (C_{Mac})_w$$
$$+ C_{Dc} S_c/S\left(\hat{h}_c - \hat{h}_{cm}\right) + C_{Dfin} S_{fin}/S\, \hat{h}_{fin}$$

Now, recall that for the glider

$$C_{Dc} = 0.0805 + 0.5193\alpha + 2.6375\alpha^2$$

so that

$$(C_{D\alpha})_c = d(C_{Dc})/d\alpha = 0.5195 + 5.275\alpha$$

Therefore,

$$(C_{M\alpha})_{cm} = (C_{L\alpha})_c \left(\hat{l}_{cm} - \hat{l}_c\right) S_c/S + (C_{L\alpha})_w \left(\hat{l}_{cm} - \hat{l}_w\right) + (C_{D\alpha})_c S_c/S\left(\hat{h}_c - \hat{h}_{cm}\right) \to$$

$$(C_{M\alpha})_{np} = (C_{L\alpha})_c \left(\hat{l}_{np} - \hat{l}_c\right) S_c/S + (C_{L\alpha})_w \left(\hat{l}_{np} - \hat{l}_w\right) + (C_{D\alpha})_c S_c/S\left(\hat{h}_c - \hat{h}_{cm}\right) = 0 \to$$

$$\hat{l}_{np} = \frac{(C_{L\alpha})_c \hat{l}_c S_c/S + (C_{L\alpha})_w \hat{l}_w - (C_{D\alpha})_c S_c/S\left(\hat{h}_c - \hat{h}_{cm}\right)}{C_{L\alpha}}$$

When the canard's drag is taken into account, the neutral-point location, $\hat{l}_{np}$, is now a function of α. For the example glider, assume that it is at an angle of $(\alpha)_{MaxL/D} = 0.1134 \text{ rad} = 6.50°$ and $\hat{h}_c - \hat{h}_{cm} \approx 0.25$. The neutral-point location for this condition is now

$$\hat{l}_{np} = \frac{3.02 \times 0.967 \times 0.278 + 3.01 \times 3.35 - 1.1177 \times 0.25 \times 0.278}{3.85} = \frac{10.818}{3.85} = 2.810$$

Further, $(C_M)_{np}$ becomes:

$$(C_M)_{np} = (C_{L0})_c \left(\hat{l}_{np} - \hat{l}_c\right) S_c/S + (C_{L0})_w \left(\hat{l}_{np} - \hat{l}_w\right) + (C_{Mac})_c S_c c_c/(Sc) + (C_{Mac})_w$$
$$+ C_{Dc} S_c/S\left(\hat{h}_c - \hat{h}_{cm}\right) + C_{Dfin} S_{fin}/S\, \hat{h}_{fin}$$

The fin area is 5.156 in^2 and its drag coefficient is given by:

$$C_{Dfin} = (C_f)_{fin}(S_{wet})_{fin}/S_{fin} = 0.0127 \times 10.31/5.156 = 0.0254$$

which is assumed to act a vertical distance, $h_{fin} \approx 1.2$ inches, above the mass-center.

So, for the glider at $(\alpha)_{MaxL/D}$, $(C_M)_{np}$ becomes:

$$(C_M)_{np} = 0.0412 + 0.173 \times \frac{10}{36} \times 0.25 + 0.0254 \times \frac{5.156}{36} \times 0.40 = 0.0412 + 0.012 + 0.0015$$

$$(C_M)_{np} = 0.0547$$

It is seen that the vertical fin has a small effect, but the vertical placement of the canard is significant. For this case, the mass-center location for trim is:

$$\hat{l}_{cm} = \hat{l}_{np} - (C_M)_{np}/C_L = 2.810 - 0.0547/0.8185 = 2.743$$

This is only 1.33% further forward than that value given by the previous analysis, which ignored the drag-moment effects of the canard and fin. Also, the static margin is:

$$\text{Static Margin} = 2.810 - 2.743 = 0.067 \rightarrow 6.7\%$$

There is no doubt that the drag-producing elements above the mass-center help the static stability. In fact, some large-finned configurations rely upon this. However, for most airplane designs, these effects can be ignored, as was done in the "Complete Aircraft Aerodynamics" section in Chapter 2.

Finally, the author built the canard glider in accordance with the drawing shown earlier:

Canard Airplanes and Biplanes

The mass-center location for efficient trimmed flight was:

$$l_{cm} = 8.18 \text{ inches} \rightarrow \hat{l}_{cm} = 2.723$$

This is close to the value of $\hat{l}_{cm} = 2.743$ calculated above (which includes the drag effects). In truth, considering all of the assumptions and the Reynolds number effects, the author was pleased (and surprised, to be honest) at how this worked out.

It should be noted that more sophisticated, vortex-lattice methods are now accessible for predicting the aerodynamic characteristics of canard configurations. In the author's opinion, though, this simple approximate method gives some valuable educational insights into how the components work together.

BIPLANES

As mentioned previously, the original motivation for the biplane design was to provide an adequately stiff and strong structure for the incorporation of thin airfoils. However, biplanes persisted even after this wasn't a consideration. The picture shows a Fokker D-VII fighter from World War I.

By this time the Fokker Company was using thick airfoils and wings with such stiffness and strength that external bracing was not required. In fact, the "N-struts" near the wing tips were incorporated more for confidence-giving than for any structural purpose. This was the premier fighter of the Imperial German Air Force and was specifically called out by the Treaty of Versailles for disposal. Also, the next picture shows the Fairey "Swordfish", which was a carrier-based torpedo bomber during World War II.

Royal Navy (Wikipedia Commons)

It was most noted for tracking down and disabling the German battleship "Bismarck", as well as performing notable service in the Mediterranean. In many ways, it looks even more primitive than the Fokker design, having wire bracing as well as struts. In fact, its crews affectionately referred to it as a "Stringbag". So, the question must be asked as to why such designs persisted in a time when streamlined monoplanes were possible. The answer is that biplanes can offer aerodynamic advantages for situations where the wing-spans are *fixed*. Consider the following illustration for wings of given span b:

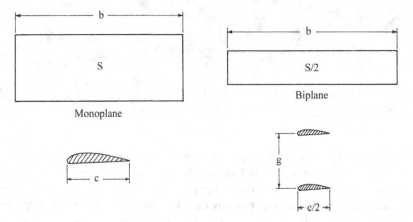

The monoplane wing of area S is conceptually split into a biplane, each wing of area $S/2$. For the monoplane, the induced drag coefficient is approximately given by:

$$\left(C_{Di}\right)_{mono} \approx \frac{C_L^2}{\pi (AR)_{mono}}$$

Canard Airplanes and Biplanes

If the gap, g, (height of one wing above the other) is sufficiently far apart, then the two biplane wings approximately act independently of one another. Therefore, total induced drag is given by:

$$(C_{Di})_{biplane} \approx (C_{Di})_{upper} \frac{S_{upper}}{S} + (C_{Di})_{lower} \frac{S_{lower}}{S} = \frac{(C_L)^2_{upper}}{\pi(AR)_{upper}} \frac{S_{upper}}{S} + \frac{(C_L)^2_{lower}}{\pi(AR)_{lower}} \frac{S_{lower}}{S}$$

For

$$S_{upper} = S_{lower} = S/2, \quad (AR)_{upper} = (AR)_{lower} = 2(AR)_{mono}$$

then

$$(C_{Di})_{biplane} \approx \frac{(C_L)^2_{upper} + (C_L)^2_{lower}}{4\pi(AR)_{mono}}$$

If

$$(C_L)_{upper} = (C_L)_{lower} = C_L$$

then

$$(C_{Di})_{biplane} \approx \frac{C_L^2}{2\pi(AR)_{mono}} = \frac{(C_{Di})_{mono}}{2}$$

This shows that, for an airplane with a fixed span and area, a biplane holds the promise of being more aerodynamically efficient than a monoplane.

The picture shows a Grumman F2F carrier-based Navy fighter of the 1930s. Storage-space limitations on carriers compelled the use of short-span airplanes, which explains why the U.S. Navy used biplanes long after the Army Air Corps adopted monoplanes.

U.S.Navy (Wikipedia Commons)

The performance characteristics were similar at that time. However, with the invention of folding wings, carrier-based monoplanes became feasible.

The notion of using more than two wings was explored by designers for early airplanes. One of the most famous is the Fokker Triplane from World War I. The intention was to enhance maneuverability by minimizing the rolling moment of inertia. In the hands of skilled pilots, like Baron Manfred von Richthofen, this could be a deadly adversary.

Adrian Pingstone, Wikipedia Commons

The most extreme multiplane example was built by Horatio Phillips in 1907. It flew briefly and held the record not only for the most wings, but also for each wing having the highest aspect ratio (152). What wasn't understood, however, is that mutual aerodynamic interference among the wings decreases the overall aerodynamic efficiency.

Canard Airplanes and Biplanes

The key interference parameter is called σ and is a function of the gap ratio, g/b, as shown in the graph for equal-span wings (from Ludwig Prandtl, *NACA Tech. Note 182*, 1924).

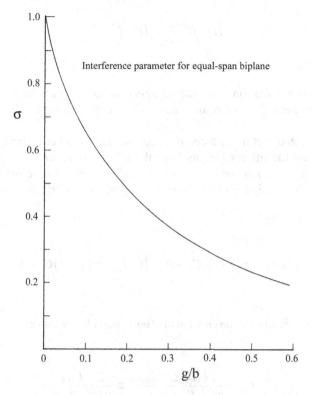

As a limiting case, consider when $g/b \to 0$. Conceptually, the two wings come together and act as one wing with half the total area of the biplane's two wings. This "single" wing has to carry the load originally carried by the two wings. Therefore, its lift coefficient is the sum of the lift coefficients (referenced to each wing's area) originally carried by the two biplane wings, and thus, its induced drag coefficient is higher than that for the two wings with a finite gap. However, as the gap increases, the mutual interference between the wings decreases. Specifically, the effect of the gap ratio on a biplane's induced-drag coefficient may be demonstrated by the following example.

Assume a special case of two identical wings with elliptic span loadings:

$$b_{upper} = b_{lower} \equiv b, \quad S_{upper} = S_{lower} \equiv S', \quad (AR)_{upper} = (AR)_{lower} \equiv AR'$$

The total induced drag coefficient of the biplane, referenced to the total wing area, $S = 2S'$ is given by

$$(C_{Di})_{biplane} = \frac{(C_L)^2_{upper}}{2\pi AR'} + \frac{\sigma (C_L)_{upper}(C_L)_{lower}}{\pi AR'} + \frac{(C_L)^2_{lower}}{2\pi AR'}$$

The previously-mentioned special case for the two wings converging to a zero-gap can be seen from these equations. For this, $\sigma = 1$ and the equation becomes:

$$(C_{Di})^{zero\,gap}_{biplane} = \frac{(C_L)^2_{upper}}{2\pi AR'} + \frac{(C_L)_{upper}(C_L)_{lower}}{\pi AR'} + \frac{(C_L)^2_{lower}}{2\pi AR'} \to$$

$$(C_{di})_{single} = \frac{\left[(C_L)_{upper} + (C_L)_{lower}\right]^2}{2\pi AR'}$$

The curve shows that for any finite gap ratio, the value of σ is less than 1.0. Therefore:

$$(C_{Di})_{biplane}^{finite\ gap} \prec (C_{Di})_{biplane}^{zero\ gap}$$

So clearly, the zero-gap situation gives the highest induced-drag coefficient, and this reduces as the gap ratio increases. This, of course, could also be seen from the above equations incorporating σ.

It should also be noted that the induced-drag coefficients from both wings are not, in general, equal. That is because the lift coefficients from the upper and lower wings are not equal, even when they are identical and mounted at the same incidence angle to the fuselage reference line. From Richard Hiscocks, *Design of Light Aircraft*, self-published, 1995, the lift coefficients may be expressed as:

$$(C_L)_{upper} = (1+A)C_L + B, \quad (C_L)_{lower} = (1-A)C_L - B$$

where C_L is the lift coefficient for the entire wing (both individual wings):

$$C_L = \frac{2}{\rho U^2} \frac{L_{upper} + L_{lower}}{S_{upper} + S_{lower}} \equiv \frac{2}{\rho U^2} \frac{L_{total}}{S}$$

ASIDE

The wing reference area for a biplane is usually taken to be the sum of the individual wing areas, even if the wings have different sizes. As defined in the "Wings" section of chapter 2, this is the "flattened" (no-dihedral) area. Also, if one wing is supported on struts above the fuselage, its area is evident. However, for a wing passing through the fuselage, the area is projected within the width of the fuselage as previously discussed. These definitions, of course, also apply to a triplane, etc.

END ASIDE

The A and B coefficients are functions of the "stagger" ratio, s/c, as shown in the figure.

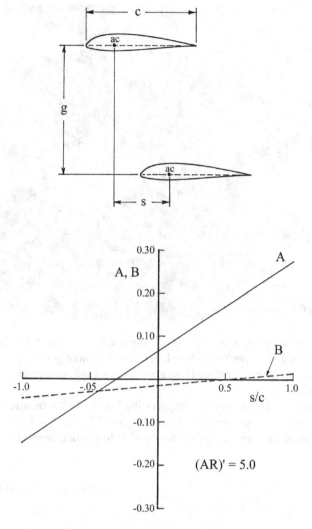

A "positive stagger" means that the upper wing is forward of the lower wing. Further, the A and B plots are for two identical wings of aspect-ratio 5.0, mounted at identical incidence angles to the airplane's reference line. For positive stagger, these plots and the previous equations give that the upper wing operates at a higher lift coefficient than the lower one. This is sometimes compensated for by mounting the upper wing at a slightly lower incidence angle than the lower wing. Such geometry is referred to as a "positive decalage angle".

It is interesting to observe that the A and B plots are not functions of the gap ratio, and thus σ are not a function of the stagger ratio. Therefore, the aerodynamic efficiency of a biplane is not a significant function of a stagger. In practice, the choice of stagger is mostly determined by pilot ergonomics, such as visibility and access. The picture below is of a Gloster "Gladiator", a biplane fighter from the late 1930s (it actually saw some action against the Italian and Soviet Air Forces early in World War II).

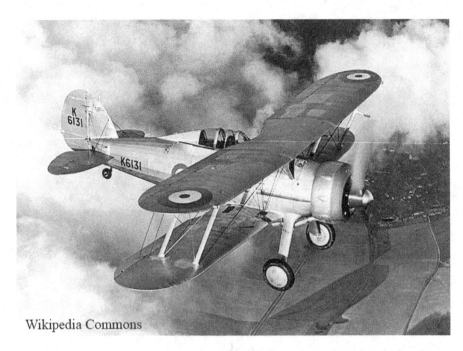

Wikipedia Commons

This airplane has positive stagger, as well as a trailing-edge cut-out. Such design features were common for biplanes, beginning in World War I, because it allowed good rear visibility. A common tactic was to approach an enemy airplane from above and behind, so rear visibility was crucial for the defending airplane.

On the other hand, the civilian Beech "Staggerwing" of the 1930s required no such combative considerations, so it was designed with a negative stagger. This gave enhanced visibility forward and up, and also allowed the structure of the lower wing to provide support for the widely-spaced main undercarriage.

U.S. Air Force (Wikipedia Commons)

Canard Airplanes and Biplanes

In the 1980s, the author and his research group were engaged in studying airplane configurations for sustained flight with beamed microwave energy from a ground station. This was sponsored by the Canadian Communications and Research Centre (CRC), and the program was called SHARP (Stationary High Altitude Relay Platform). An early, and successful, configuration is the biplane shown in the two figures below.

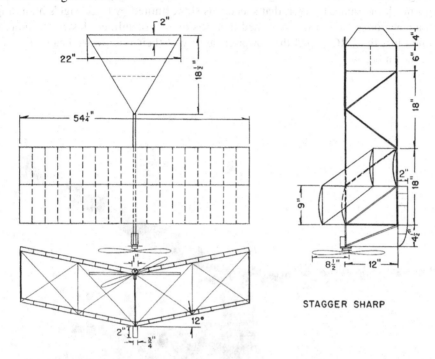

STAGGER SHARP

It had a stagger ratio of 1.0 because the underside of both wings provided a mounting surface for rectifying antennas ("rectennas") that would convert the beamed microwave energy to DC electric power. The high stagger ratio was to provide maximum exposure of the upper wing's underside to the beamed energy. This aircraft was flown at the CRC's facility at Shirleys Bay Ontario and demonstrated sustained microwave-powered flight, which was a first for a free-flying aircraft.

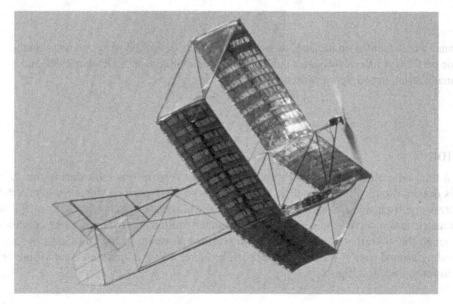

Note that the limited beam power at that time required an airplane with a very low wing loading (weight/wing area). The resulting design had a lightweight structure that incorporated thin single-surface airfoils (7.5% circular arc). This compelled the use of a biplane configuration for the reasons discussed earlier in this book. However, this design, called "Stagger SHARP", was a "single-design-point aircraft", meaning that the cruise speed, stall speed, and maximum speed were pretty much the same. Further, that speed was slow, limited by the design's high drag coefficient; so a new configuration was designed that was more streamlined, faster and more robust. At the same time, the CRC upped the power capability of their microwave beam. The resulting airplane is shown below:

Rectennas were mounted on the disk, as well as the undersides of the wing and horizontal tail. The airplane performed successfully, in a public demonstration flight, on 6 October 1988, and is now at the Canada Aviation and Space Museum.

> **ASIDE**
>
> As a final note about the SHARP program, a larger version was built during the 1990s. This design also incorporated a disk, but it was placed aft to perform the purpose of a stabilizer. At the front was a canard that only served the purpose of pitch control. Its airfoil was symmetric and it carried no aerodynamic lift under equilibrium trimmed flight conditions. Although the aircraft was intended to be microwave powered, to high altitudes, the corresponding ground station was not developed. Therefore, it flew on internal combustion, only, for lower-altitude test flights.

END ASIDE

An interesting concept is the "equivalent monoplane", which is the single wing of a given area that would produce the same induced drag as a biplane of the same total area. From Bradley Jones, "*Elements of Practical Aerodynamics*", John Wiley & Sons, 1942, the equation for the aspect ratio of the equivalent monoplane, $(AR)^*$, is

$$(AR)^* = \frac{(b^*)^2}{S} = \frac{b_1^2}{S} \frac{\mu^2(1+r)^2}{\mu^2 + 2\sigma\mu r + r^2}$$

where b_1 is the longer span of the two wings (b_2 is the span of the shorter wing) and S is the total area of the two wings. Also:

$$\mu = b_2/b_1, \quad r = S_2/S_1$$

where S_1 is the area of *Wing* 1 and S_2 is the area of *Wing* 2.

As a specific example, choose a case where $\mu = 1$, $r = 1$, $\sigma = 0.5$. The aspect ratio of the equivalent monoplane is:

$$(AR)^* = 1.333 b_1^2/S$$

Therefore,

$$(b^*)^2 = 1.333 b_1^2 \rightarrow b^* = 1.1547 b_1$$

This shows that a mere 15% increase in span (for the same area) makes the monoplane as efficient as the biplane (even ignoring the extra drag of struts, etc.).

The plot below is adapted from Sighard Hoerner, "*Fluid-Dynamic Drag*", Self Published, 1965, for a biplane with *two identical wings*. This shows again, as was discussed previously, that the aerodynamic efficiency of a biplane increases with the gap ratio. In the plot, $(AR)_{\text{effective}}$ is the aspect ratio of the monoplane that would give the same induced-drag coefficient as the biplane, whose aspect ratio is given by:

$$(AR)' = \frac{(AR)_{\text{upper}} + (AR)_{\text{lower}}}{2} = \frac{1}{2}\left(\frac{b^2_{\text{upper}}}{S_{\text{upper}}} + \frac{b^2_{\text{lower}}}{S_{\text{lower}}}\right)$$

For this case, where $b_{\text{upper}} = b_{\text{lower}} = b$, $S_{\text{upper}} = S_{\text{lower}} = S'$, one has that:

$$(AR)' = b^2/S'$$

So, $(AR)'$ is the aspect ratio of *one* of the biplane's wings. It is seen that when the gap ratio goes to infinity, $(AR)_{\text{effective}} = (AR)'$. This simply means that the biplane wings have negligible mutual interference and thus have the same induced-drag coefficient, C_{Di}, as the effective monoplane wing if both cases are carrying the same lift coefficient, C_L. At lower gap ratios, though, the mutual interference between the biplane wings means that the effective monoplane produces the same induced-drag coefficient at a lower aspect ratio than that for the biplane. For example, if the gap ratio is 0.6, then $(AR)_{\text{effective}} = 0.84(AR)'$. That is, if the biplane consists of individual wings of aspect ratio 8.0, a monoplane of aspect ratio 6.72 would produce the same induced-drag coefficient at the same lift coefficient.

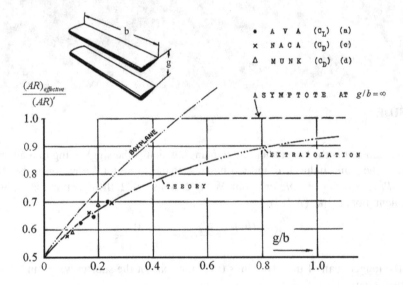

An intriguing curve from Hoerner's plot is that for a "boxwing", which is a biplane with airfoil-shaped endplates.

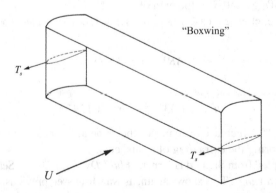

Above a gap ratio of 0.5, the effective monoplane requires a *higher* aspect ratio than that for the boxwing. This may be due to the tip plates attenuating the tip vortices while, at the same time, the airfoil shapes of the tip plates produce leading-edge suction forces off-setting their friction drag. At any rate, this is an interesting concept for an efficient biplane configuration.

In summary, this chapter has shown that biplanes can offer certain aerodynamic and structural advantages. Some designers have tried to extend these notions to multi-wing designs, such as the four-winged Armstrong-Whitworth FK-10 shown below:

The only real advantage was that the fore-and-aft center-of-pressure travel for a multi-wing design is limited so that the required area for the horizontal tail is a relatively small percentage of the total wing area.

No discussion of multi-wing aircraft would be complete without presenting the ill-fated 30 m-span, 1921 Caproni Ca-60 "Transaereo":

Although Gianni Caproni designed many successful airplanes, including several large ones, this particular example was a case of too much of everything, and it crashed during one of its tests. Imagine trying to perform an aerodynamic analysis of this, accounting for all the mutual interferences of the wings.

An Approximate Method for Estimating the Aerodynamic Characteristics of Biplanes with Wings of Equal Spans and Areas

Recall from the "Airfoils" section of Chapter 2 that an airfoil's profile drag may be estimated from:

$$C_d \approx (C_d)_{\text{fric}} + (1 - \eta_{\text{le}})\left[2\pi \alpha_{\text{ZLL}}^2 + 2\alpha_{\text{ZLL}} C_l + C_l^2/(2\pi) \right]$$

Further, it was shown in the calculations for the example canard airplane that, if an untwisted wing has the same airfoil along its span, this equation also applies to the profile drag of the finite wing:

$$\left[(C_D)_{\text{profile}} \right]_w \approx (C_D)_{\text{fric}} + (1 - \eta)\left[2\pi(\alpha_{\text{ZLL}})_w^2 + 2(\alpha_{\text{ZLL}})_w (C_L)_w + (C_L)_w^2/(2\pi) \right] \rightarrow$$

$$\left[(C_D)_{\text{profile}} \right]_w = \left[(C_D)_{C_L=0} \right]_w + A_w (C_L)_w + B_w (C_L)_w^2$$

These equations were derived in the author's article: "Drag of Wings with Cambered Airfoils and Partial Leading-Edge Suction", *AIAA Journal of Aircraft*, Vol. 20, No. 10, October 1983. The induced-drag effect was added to the profile drag to give the complete drag of the finite wing. For this biplane case, an extension to the analysis is required to account for the effect of one wing upon the other. Upon the assumption that the mutual downwash effects are *uniform* across the spans of the wings, the induced-drag terms are now:

$$(C_{Dw})_{\text{induced}} = \frac{(C_L)_w^2}{\pi(AR)_w} + (C_L)_w \varepsilon_w$$

where ε_w is the downwash angle on the wing caused by the other wing.

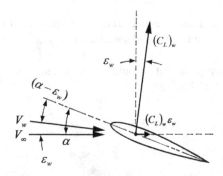

In particular, identify "w1" as the upper wing and "w2" as the lower wing so that the induced-drag coefficients for both wings, referenced to their own areas, is:

$$(C_{Dw1})_{\text{induced}} = \frac{(C_L)_{w1}^2}{\pi(AR)_{w1}} + (C_L)_{w1} (\varepsilon_{w1})_{w2}, \quad (C_{Dw2})_{\text{induced}} = \frac{(C_L)_{w2}^2}{\pi(AR)_{w2}} + (C_L)_{w2} (\varepsilon_{w2})_{w1}$$

where $(\varepsilon_{w1})_{w2}$ is the downwash angle at *wing 1* caused by *wing 2* and $(\varepsilon_{w2})_{w1}$ is the downwash angle at *wing 2* caused by *wing 1*.

Now, from Richard von Mises, "*Theory of Flight*", Dover, 1959, the induced-drag coefficient for a biplane is given by:

$$(D_i)_{\text{biplane}} = \frac{1}{q}\left(\frac{L_1^2}{b_1^2} + 2\sigma \frac{L_1 L_2}{b_1 b_2} + \frac{L_1^2}{b_2^2} \right)$$

Canard Airplanes and Biplanes

For

$$(D_i)_{\text{biplane}} = q(C_{Di})_{\text{biplane}} S, \quad \text{where } S = S_1 + S_2$$

and

$$L_1 = q(C_L)_{w1} S_1, \quad L_2 = q(C_L)_{w2} S_2$$

One obtains:

$$(C_{Di})_{\text{biplane}} = \frac{S_1}{S} \frac{(C_L)_{w1}^2}{\pi(AR)_{w1}} + \sigma \frac{2}{S} \frac{(C_L)_{w1}(C_L)_{w2} S_1 S_2}{\pi b_1 b_2} + \frac{S_2}{S} \frac{(C_L)_{w2}^2}{\pi(AR)_{w2}},$$

where S_1, b_1 are the area and span of *wing* 1 and S_2, b_2 are the area and span of *wing* 2. Also, σ is the mutual-interference parameter described earlier in this chapter.

With the assumptions of equal spans and areas, where:

$$AR' \equiv AR_1 = AR_2 = b_1^2/S_1 = b_2^2/S_2 = b^2/S$$

the equation becomes:

$$(C_{Di})_{\text{biplane}} = \frac{(C_L)_{w1}^2}{2\pi AR'} + \sigma \frac{(C_L)_{w1}(C_L)_{w2}}{\pi AR'} + \frac{(C_L)_{w2}^2}{2\pi AR'}$$

This is the same equation, as presented previously, from NACA Tech Note 182 by Ludwig Prandtl.

Now, a major assumption for this approximate method is that the contributions to the induced drag may be divided between the two wings in the following fashion:

$$S_1 (C_{Dw1})_{\text{induced}} + S_2 (C_{Dw2})_{\text{induced}} = S' \left[(C_{Dw1})_{\text{induced}} + (C_{Dw2})_{\text{induced}} \right] = S(C_{Di})_{\text{biplane}} \rightarrow$$

$$\left[(C_{Dw1})_{\text{induced}} + (C_{Dw2})_{\text{induced}} \right] = 2(C_{Di})_{\text{biplane}} \rightarrow$$

$$(C_{Dw1})_{\text{induced}} \approx \frac{(C_L)_{w1}^2}{\pi AR'} + \sigma \frac{(C_L)_{w1}(C_L)_{w2}}{\pi AR'}, \quad (C_{Dw2})_{\text{induced}} \approx \frac{(C_L)_{w2}^2}{\pi AR'} + \sigma \frac{(C_L)_{w1}(C_L)_{w2}}{\pi AR'}$$

Upon comparing these two equations with the previous ones, it is seen that the downwash-angle terms are given by:

$$\frac{(\varepsilon_{w1})_{w2}}{(C_L)_{w2}} = \frac{(\varepsilon_{w2})_{w1}}{(C_L)_{w1}} = \frac{\sigma}{\pi AR'}$$

Therefore, the lift of either wing is given by:

$$(C_L)_w = (C_{L\alpha})_w^* \left[-(\alpha_{ZLL})_w + \alpha_w + i_w - \varepsilon_w \right]$$

where $(C_{L\alpha})_w^*$ is the lift-curve slope of the isolated wing. In particular, for the two wings, one has:

$$(C_L)_{w1} = (C_{L\alpha})_{w1}^* \left[-(\alpha_{ZLL})_{w1} + \alpha_{w1} + i_{w1} - \frac{(\varepsilon_{w1})_{w2}}{(C_L)_{w2}} (C_L)_{w2} \right]$$

$$(C_L)_{w2} = (C_{L\alpha})_{w2}^* \left[-(\alpha_{ZLL})_{w2} + \alpha_{w2} + i_{w2} - \frac{(\varepsilon_{w2})_{w1}}{(C_L)_{w1}} (C_L)_{w1} \right]$$

ASIDE

The Prandtl and von Mises equations are based on a lifting-line physical model, where the lift vectors are perpendicular to the incident-velocity vectors. Therefore, this model is consistent with the inclined-lift drag model presented above.

END ASIDE

It should be noted that the wings for both the author's and von Mises' analyses assume an ideal elliptical span loading. This is one of the reasons why this is an approximate methodology. However, the examples in the Author's AIAA article showed reasonable comparisons between theory and experiment for isolated rectangular wings. A possible refinement for non-elliptical wings would be to apply the $(1+\delta)$ correction to the induced drag-coefficient equation. In any case, Elliott Reid, on page 142 of "Applied Wing Theory" (McGraw-Hill, 1932), states that "the use of the formulas for elliptic loading in a problem involving a biplane with rectangular wings of equal span leads to a smaller error than would a similar approximation in the case of monoplanes…".

EXAMPLE, "SLOW SHARP" BIPLANE

The drawing below shows a lightweight remotely-piloted biplane known as "Slow SHARP". This was a predecessor to the "Stagger SHARP" design mentioned earlier, built and flown to assess the utility of such a configuration and was not equipped with rectennas. Although this airplane had a large propeller in front, driven by a geared electric motor, the analysis will be for the propeller-less condition. Also, the peculiarity of the design will require a special approach to assessing its trim and static stability.

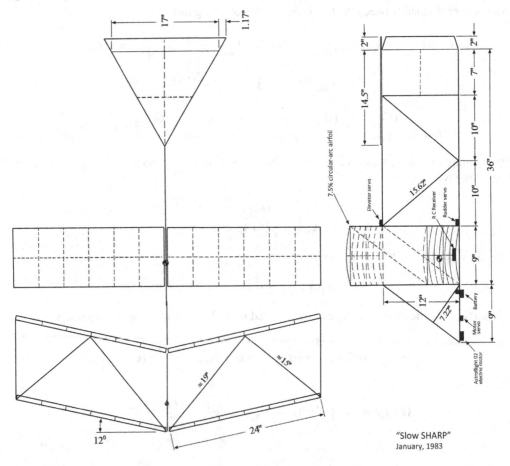

"Slow SHARP"
January, 1983

Canard Airplanes and Biplanes

Beginning with a single wing, its lift-curve slope is obtained from the Lowry and Pohlhamus equation, as described in the "Wings" section of Chapter 2:

$$C_{L\alpha} = \frac{2\pi AR}{2 + \left[\left(\frac{AR}{\kappa}\right)^2 (1+\tan^2 \Lambda_{c/2}) + 4\right]^{1/2}}$$

where $\Lambda_{c/2} = 0$, $AR = 48/9 = 5.333$, $\kappa = 1$ (assumed), so that each isolated wing's lift-curve slope is:

$$\left(C_{L\alpha}\right)^*_{w1} = \left(C_{L\alpha}\right)^*_{w2} = \frac{33.51}{2 + (28.44 + 4)^{1/2}} = 4.3543, \ (1/\text{rad})$$

The stabilizer has a triangular planform of aspect ratio $(19.34)^2/(0.5 \times 19.34 \times 16.5) = 374.04/159.6 = AR = 2.344$. Also, $\Lambda_{c/2} = 40.5°$ and $\kappa = 1$ (assumed). Therefore, the Lowry and Pohlhamus equation gives that:

$$\left(C_{L\alpha}\right)^*_s = \frac{14.73}{2 + \left[5.49(1+0.73) + 4\right]^{1/2}} = \frac{14.73}{2 + 3.674} = \left(C_{L\alpha}\right)^*_s = 2.60, \ (1/\text{rad})$$

Note that these values for $(C_{L\alpha})^*$ assume no mutual interference and each is referenced to its own area.

For the airfoils' aerodynamic characteristics, it would be ideal to have experimental results, such as was the case for the "Numerical Example" in Chapter 2. However, such information is not available for this biplane and, just as for the earlier canard example, the characteristics will have to be estimated:

The wings' airfoil is a 7.5% circular arc, for which $\alpha_{ZLL} = -0.1500 \text{ rad} = -8.59°$, so $(\alpha_{ZLL})_{w1} = (\alpha_{ZLL})_{w2} = -0.150$. Also, the incidence angle of both wings with respect to the airplane's reference line is zero, so that $i_{w1} = i_{w2} = 0$. Further, $\alpha_{w1} = \alpha_{w2} = \alpha$, the angle-of-attack of the airplane.

For this example, the two wings are of equal spans and areas, $b_1 = b_2 = 48"$, $S_1 = S_2 = 432.0 \text{ in}^2$. Further, the gap ratio is $g/b = 12/48 = 0.25 \rightarrow \sigma = 0.42$. Therefore, the mutual-downwash terms, from the previously-derived equations are:

$$\frac{(\varepsilon_{w1})_{w2}}{(C_L)_{w2}} = \frac{(\varepsilon_{w2})_{w1}}{(C_L)_{w1}} = \frac{\sigma}{\pi AR'} = \frac{0.42}{\pi \times 5.333} = 0.025$$

The wings' lift coefficient equations are thus:

$$(C_L)_{w1} = 4.354\left[0.1500 + \alpha - 0.025(C_L)_{w2}\right]$$

$$(C_L)_{w2} = 4.354\left[0.1500 + \alpha - 0.025(C_L)_{w1}\right]$$

From these, one may solve for $(C_L)_{w1}$ and $(C_L)_{w2}$ as functions of α:

$$\underline{(C_L)_{w1} = (C_L)_{w2} = 0.5890 + 3.9266\alpha}$$

It is seen that, with this methodology, both wings carry equal loads. More refined analyses and experiments show that this is not exactly true for identical wings at the same angle. Also, stagger has a significant effect, as seen in a previously presented plot. However, these limitations must be accepted for this approximate method.

Next, the stabilizer's aerodynamic contribution will be estimated. The first step will be to determine the stabilizer's aerodynamic center. A curve adapted from Hermann Schlichting and Erich Truckenbrodt, *"Aerodynamics of The Airplane"*, McGraw-Hill, 1979 is shown, from which this value will be obtained.

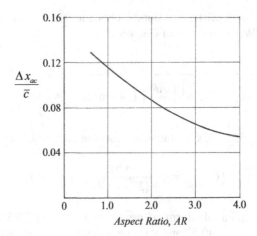

The distance from the apex (front point) to the aerodynamic center of the delta wing is given by:

$$x_{ac} = (x_{ac})_{geo} + \left(\frac{\Delta x_{ac}}{\bar{c}}\right)\bar{c}$$

where, from Schlichting and Truckenbrodt, the geometrical aerodynamic center of a delta wing is at 50% of the root chord:

$$(x_{ac})_{geo} = c_{root}/2$$

The mean aerodynamic chord of the delta wing is found by the method described in the "Wings" section of Chapter 2 and is given by $\bar{c} = 2/3 \times c_{root}$. So the plot gives that, for the stabilizer's aspect ratio of 2.344, $\Delta x_{ac}/\bar{c} \approx 0.08$. Therefore, one obtains:

$$(x_{ac})_s = \frac{16.5}{2} + 0.08 \times \frac{2}{3} \times 16.5 = 8.25 + 0.08 \times 11.0 = 9.13 \text{ inches}$$

Further, from the drawing, the horizontal distance x_s between the wing's aerodynamic centers and the stabilizer aerodynamic center is $x_s = 19.25 + 9.13 = 28.38$ inches.

The lift coefficient of the stabilizer is given by:

$$(C_L)_s = (C_{L\alpha})_s^* \left[\alpha + i_s - (\varepsilon_s)_{w1} - (\varepsilon_s)_{w2}\right]$$

where $i_s = 0$, $(\varepsilon_s)_{w1} = (\varepsilon_s/C_L)_{w1} (C_L)_{w1}$, $(\varepsilon_s)_{w2} = (\varepsilon_s/C_L)_{w2} (C_L)_{w2}$.

The downwash parameters will be calculated from the USAF DATCOM methodology described in the "Complete-Aircraft Aerodynamics" section of Chapter 2:

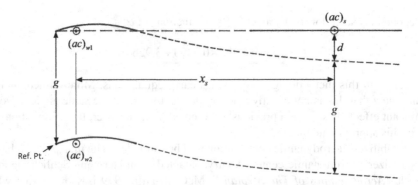

Canard Airplanes and Biplanes

$$\left(\frac{\varepsilon}{C_L}\right)_w = \frac{(\partial\varepsilon/\partial\alpha)_w}{(C_{L\alpha})_w}, \quad \left(\frac{\partial\varepsilon}{\partial\alpha}\right)_w = 4.44\left[K_A K_\lambda K_H (\cos\Lambda_{c/4})_w^{1/2}\right]^{1.19}$$

For the downwash on the stabilizer due to *wing* 1, one has that:

$$(K_A)_{w1} = \frac{1}{(AR)_{w1}} - \frac{1}{1+(AR)_{w1}^{1.7}} = \frac{1}{5.333} - \frac{1}{1+(5.333)^{1.7}} = 0.1875 - 0.0549 = 0.1326$$

Next,

$$(K_\lambda)_{w1} = \frac{10 - 3\lambda_{w1}}{7} = \frac{10-3}{7} = 1.0$$

The next term in the downwash equation is:

$$(K_H)_{w1} = \frac{1 - |(h_H)_{w1}/b_{w1}|}{\left[2(l_H)_{w1}/b_{w1}\right]^{1/3}} = \frac{1-0}{(2\times 28.38/48)^{1/3}} = \frac{1}{1.0575} = 0.9457$$

Therefore, with $(\Lambda_{c/4})_{w1} = 0$, it is derived that:

$$(\partial\varepsilon/\partial\alpha)_{w1} = 4.44(0.1326\times 1.0\times 0.9457)^{1.19} = 0.3753$$

Further, from previous work, $(C_{L\alpha})_{w1} = 3.9266$, so that

$$(\varepsilon_s/C_L)_{w1} = 0.3753/3.9266 = 0.0956 \text{ radian}, (5.48°)$$

For *wing* 2, all the terms stay the same except for K_H which, in this case, becomes:

$$(K_H)_{w2} = \frac{1 - |(h_H)_{w2}/b_{w2}|}{\left[2(l_H)_{w2}/b_{w2}\right]^{1/3}} = \frac{1-12/48}{1.0575} = \frac{1-0.25}{1.0575} = 0.7092$$

Therefore,

$$(\partial\varepsilon/\partial\alpha)_{w2} = 4.44(0.1326\times 1.0\times 0.7092)^{1.19} = 0.2665$$

Because $(C_{L\alpha})_{w2} = 3.9265$ (the same as for *wing* 1), one obtains that:

$$(\varepsilon_s/C_L)_{w2} = 0.2665/3.9266 = 0.0679 \text{ radian}, (3.89°)$$

Finally, the downwash values are given by

$$(\varepsilon_s)_{w1} = (\varepsilon_s/C_L)_{w1}(C_L)_{w1} = 0.0956\times(0.5890 + 3.9266\alpha) = 0.0563 + 0.3754\alpha$$
$$(\varepsilon_s)_{w2} = (\varepsilon_s/C_L)_{w2}(C_L)_{w2} = 0.0679\times(0.5890 + 3.9266\alpha) = 0.0400 + 0.2666\alpha$$

Therefore,

$$(C_L)_s = 2.60\left[\alpha - (0.0563 + 0.0400) - (0.3754 + 0.2666)\alpha\right] = 2.60\alpha - 0.2504 - 1.6692\alpha \rightarrow$$
$$\underline{(C_L)_s = -0.2504 + 0.9308\alpha}$$

ASIDE

Observe how much the downwash from the two wings reduces the effectiveness of the stabilizer. The lift-curve slope goes from $(C_{L\alpha})_s^* = 2.60$ to $(C_{L\alpha})_s = 0.9308$, which is a reduction of 64%. Also, the downwash effect of the upper wing is greater than that of the lower wing because the stabilizer is in the plane of the upper wing's root chord. The vertical distance between the wing root chord and the stabilizer, h_H/b, reduces the effect of the downwash and is a feature often employed in airplane design.

END ASIDE

Now, $S_{w1} = S_{w2} = 432.0 \text{ in}^2$ and $S_s = 159.6 \text{ in}^2$. Choose a reference area of $S = S_{w1} + S_{w2} = 864.0 \text{ in}^2$, so that the total lift coefficient is given by

$$C_L = (C_L)_{w1} \frac{S_{w1}}{S} + (C_L)_{w2} \frac{S_{w2}}{S} + (C_L)_s \frac{S_s}{S}$$

$$C_L = 0.5890 + 3.9266\alpha + (0.9308\alpha - 0.2504) \times 0.1847$$

$$\underline{C_L = 0.5428 + 4.0985\alpha, \quad \alpha \text{ in radians}}$$

$$\underline{C_L = 0.5428 + 0.0715\alpha, \quad \alpha \text{ in degrees}}$$

A plot of this is shown below:

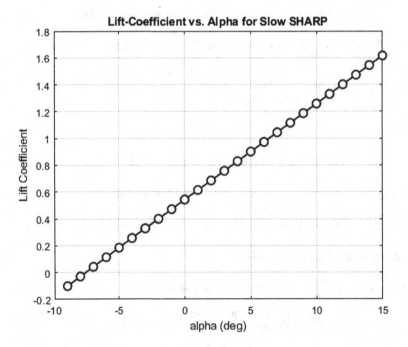

Now, for the drag coefficient, the first step is to obtain the profile drag of the circular-arc airfoil. From the "Airfoils" section in Chapter 2, one has:

$$C_d \approx (C_d)_{\text{fric}} + (1 - \eta_{\text{le}})\left[2\pi\alpha_{ZLL}^2 + 2\alpha_{ZLL}C_l + C_l^2/(2\pi)\right]$$

Canard Airplanes and Biplanes

The sharp leading edge of this airfoil gives that $\eta_{le} \approx 0$. So, the equation becomes:

$$C_d \approx (C_d)_{fric} + 2\pi\alpha_{ZLL}^2 + 2\alpha_{ZLL}C_l + C_l^2/(2\pi)$$

The cruising-flight Reynolds number for this airfoil was 76,700, so the skin-friction drag coefficient is:

$$C_{fric} = \frac{0.455}{[\log(RN)]^{2.58}} = 0.0076 \rightarrow (C_d)_{fric} = 2C_{fric} = 0.0152$$

Furthermore, the airfoil is a single surface, with exposed structure on the underside:

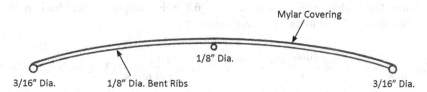

The 1/8″ diameter mid-spar and 3/16″ diameter front and rear spars are assumed to have sub-critical drag coefficients of $(C_d)_{cylinder} = 1.17$. Therefore, their contributions to the airfoil drag coefficient are given by:

$$(C_d)_{spars} \approx \frac{(C_d)_{cylinder}}{c}\Sigma(Dia.)_{spars} \approx \frac{1.17}{9}\left(2\times\frac{3}{16}+\frac{1}{8}\right) = 0.065$$

This is added to $(C_d)_{fric}$ in the profile drag-coefficient equation. Also, from before, $\alpha_{ZLL} = -0.1500$ rad, so the airfoil's profile drag coefficient is:

$$C_d = 0.0152 + 0.065 + 0.1414 - 0.300C_l + 0.1592C_l^2 = 0.2216 - 0.300C_l + 0.1592C_l^2$$

This will be assumed to be the profile drag for each wing:

$$\left(C_{Dw}\right)_{profile} = 0.2216 - 0.300(C_L)_w + 0.1592(C_L)_w^2$$

Next, the induced drag coefficient for each wing will draw upon the earlier equations:

$$\left(C_{Dw1}\right)_{induced} = \frac{(C_L)_{w1}^2}{\pi AR'} + \sigma\frac{(C_L)_{w1}(C_L)_{w2}}{\pi AR'}, \quad \left(C_{Dw2}\right)_{induced} = \frac{(C_L)_{w2}^2}{\pi AR'} + \sigma\frac{(C_L)_{w1}(C_L)_{w2}}{\pi AR'}$$

Since the C_L values for both the upper and lower wings are identical in this example, and $(AR)_{w1} = (AR)_{w2} = AR'$, the equations become:

$$\left(C_{Dw1}\right)_{induced} = \frac{(1+\sigma)(C_L)_{w1}^2}{\pi AR'}, \quad \left(C_{Dw2}\right)_{induced} = \frac{(1+\sigma)(C_L)_{w2}^2}{\pi AR'}$$

thus

$$\left(C_{Dw1}\right)_{induced} = \left(C_{Dw2}\right)_{induced}$$

For this numerical example, the gap ratio, $g/b = 12/48 = 0.25$. From the previously shown graph, this gives $\sigma \approx 0.42$. Therefore, one obtains that:

$$\left(C_{Dw1}\right)_{induced} = \left(C_{Dw2}\right)_{induced} \equiv \left(C_{Dw}\right)_{induced} = \frac{1.42}{\pi \times 5.333}(C_L)_w^2 = 0.0848(C_L)_w^2$$

Therefore, the total drag of each wing is $(C_D)_w = (C_{Dw})_{profile} + (C_{Dw})_{induced}$:

$$(C_D)_w = 0.2216 - 0.300(C_L)_w + 0.1592(C_L)_w^2 + 0.0848(C_L)_w^2 \rightarrow$$

$$\underline{(C_D)_{w1} = 0.2216 - 0.300(C_L)_{w1} + 0.2440(C_L)_{w1}^2}$$
$$\underline{(C_D)_{w2} = 0.2216 - 0.300(C_L)_{w2} + 0.2440(C_L)_{w2}^2}$$

The stabilizer has a single-surface flat-plate airfoil with an exposed structure on the underside. These components are 1/8″ diameter dowels, and it will be assumed that the primary drag-producing components are the two leading-edge dowels and the cross piece, in addition to the skin-friction drag of the covering. The leading-edge dowels are both 16.8 inches long and swept-back by 60 degrees. Therefore, their drag, relative to the stabilizer's area, is:

$$(C_{Ds})_{le\,spars} \approx \frac{(C_d)_{cylinder}(dia.)_{le\,spars} L_{le\,spars}}{S_s} \cos^3 \Lambda_{le} = \frac{1.17}{159.6} \times \frac{1}{8} \times 33.6 \cos^3 60° = 0.0039$$

The cross spar has a length of 8.5 inches, so its drag is estimated by:

$$(C_{Ds})_{cross\,spar} \approx \frac{(C_d)_{cylinder}(dia.)_{cross\,spar} L_{cross\,spar}}{S_s} = \frac{1.17}{159.6} \times \frac{1}{8} \times 8.5 = 0.0078$$

The trailing-edge spar is streamlined with the addition of the elevator surface, so its drag will be assumed to be included in the skin friction. In this case, the skin-friction drag coefficient will be chosen to be the same as for the wing. Therefore, it is obtained that:

$$(C_{Ds})_{profile} \approx (C_{Ds})_{fric} + (C_{Ds})_{le\,spars} + (C_{Ds})_{cross\,spar} = 0.0152 + 0.0039 + 0.0078 = 0.0269$$

ASIDE

This numerical example is unique in many ways. Most modern aircraft do not have as much exposed structure as this one, especially on the underside of the wings and stabilizer. For the purpose of the "Slow SHARP" there was a good reason for this, in that the design was chosen to be simple and very light. Obviously, wind-tunnel data for the profile drag of the wings and stabilizer would be preferable to the estimations made above. For example, the spar drag, in conjunction with the surface covering, may differ from that for the spar alone. Also, the cylinder drag coefficient may differ because of the surface texture of the spars. In situations like this, however, even an approximation is better than ignoring the effects altogether.

END ASIDE

As was the case for the wing, the stabilizer will be assumed to have zero leading-edge suction. Therefore, the stabilizer drag can simply be obtained by observing that, with $\eta_{le} = 0$, the lift force is perpendicular to the stabilizer's surface. Therefore the drag force, relative to the airplane's reference line, is given by

$$D_s \approx D_{profile} + L\alpha \rightarrow (C_D)_s \approx (C_{Ds})_{profile} + (C_L)_s \alpha$$
$$\underline{(C_D)_s = 0.0269 + (C_L)_s \alpha}$$

ASIDE

It should be noted that the author's profile and induced-drag methodology based on leading-edge suction considerations, described in the "Airfoils" section of Chapter 2, assumes Prandtl's equation for $C_{L\alpha}$, which is:

$$C_{L a} = \frac{2\pi \, AR}{2 + AR}$$

This equation works well for moderate to high aspect ratios, but fails for low aspect ratios. The Lowry and Pohlhamus equation, though, works well over a wide range of aspect ratios. For example, the stabilizer has an aspect ratio of 2.344 and its lift-curve slope from the Prandtl equation is 3.39, which compares to the Lowry and Pohlhamus value of 2.60. For the wing, however, its aspect ratio is 5.333 and the Prandtl value of $C_{L\alpha}$ is 4.570, which compares well to the Lowry and Pohlhamus value of 4.354.

END ASIDE

Finally, the remaining drag is due to the various struts (other than the wing spars). These are of circular cross-section so, again, a sub-critical drag coefficient of $(C_D)_{\text{cylinder}} = 1.17$ is chosen. The additional drag coefficient of these items, referenced to S, is given by:

$$(C_D)_{\text{add}} = (C_D)_{\text{wing struts}} + (C_D)_{\text{fuselage diaginals}} + (C_D)_{\text{vert. endplates}} + (C_D)_{\text{fin}}$$

where,

$$(C_D)_{\text{wing struts}} = \frac{1.17}{864.0}\left(4 \times \frac{1}{8} \times 19 + 4 \times \frac{1}{8} \times 15\right) = 0.0230$$

$$(C_D)_{\text{uselage diagonals}} = \frac{1.17}{864.0}\left(\frac{1}{8} \times 15 \cos^3 36.9° + 2 \times \frac{1}{8} \times 15.62 \cos^3 39.8°\right) = 0.0025$$

The vertical end plates and fin are covered on one side, so the skin-friction drag must be included:

$$(C_D)_{\text{vert. wing}} = \frac{1.17}{864.0}\left(6 \times \frac{1}{8} \times 12 + 3 \times \frac{1}{8} \times 15 \cos^3 36.9°\right) + 2 \times \frac{108.0}{864.0} \times 0.0152 = 0.0199$$

$$(C_D)_{\text{fin}} = \frac{1.17}{864.0}\left(\frac{1}{8} \times 12\right) + \frac{108.0}{864.0} \times 0.0152 = 0.0039$$

The additional drag coefficient is thus:

$$(C_D)_{\text{add.}} = 0.0493$$

And the total drag coefficient is:

$$C_D = (C_D)_{w1}\frac{S_1}{S} + (C_D)_{w2}\frac{S_2}{S} + (C_D)_s\frac{S_s}{S} + (C_D)_{\text{add}}$$

$$\underline{\underline{C_D = 0.5(C_D)_{w1} + 0.5(C_D)_{w2} + 0.1847(C_D)_s + 0.0493}}$$

where $(C_D)_{w1}$, $(C_D)_{w2}$ and $(C_D)_s$ are obtained from the previously derived equations.

Now, instead of expanding this to obtain C_D as a function of α, a short program is used to perform the substitution, and the following plot of C_D vs. α is obtained. Further, one may cross-plot C_D vs. C_L to obtain the polar curve. A polynomial curve-fit to this gives:

$$C_D = 0.2655 - 0.2778\, C_L + 0.2324\, C_L^2$$

which may be written as:

$$C_D = P_0 + P_1 C_L + P_2 C_L^2$$

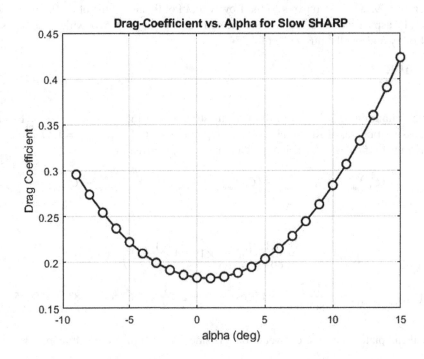

As was done for the Canard Glider example, one may find $(C_L)_{\text{Max L/D}}$ and $(L/D)_{\text{Max}}$ from

$$(C_L)_{\text{Max L/D}} = (P_0/P_2)^{1/2} = (0.2655/0.2324)^{1/2} = 1.069$$

$$(L/D)_{\text{Max}} = \left[\frac{P_0}{(C_L)_{\text{Max L/D}}} + P_1 + P_2 \times (C_L)_{\text{Max L/D}}\right]^{-1} = 4.568$$

Further, the α value at which this occurs is given by

$$(C_L)_{\text{Max L/D}} = 0.5428 + 4.0985(\alpha)_{\text{Max L/D}} \rightarrow (\alpha)_{\text{Max L/D}} = (1.069 - 0.5428)/4.0985 \rightarrow$$

$$(\alpha)_{\text{Max L/D}} = 0.1284\,\text{rad} = 7.36°$$

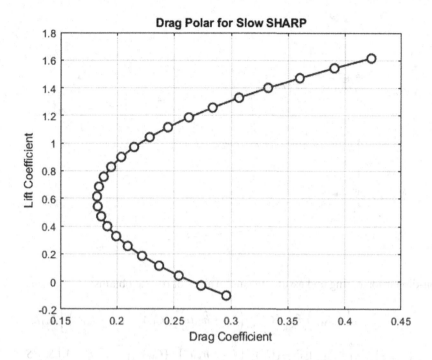

Drag Polar for Slow SHARP

ASIDE

The Slow SHARP does not have a very high aerodynamic efficiency, largely because of the amount of excrescence drag. However, this was made up for by its light construction that allowed slow flight at low power. Observe that:

$$\text{Power Required} = Drag \times Velocity = 0.5 \rho S C_D V^3$$

The power required varies linearly with drag coefficient, but varies as the cube of the velocity. If lightweight (in particular, low wing loading $\equiv S/mg$) gives low speed, then this can more than compensate for a high drag coefficient. This was the reason why very early airplanes flew with low power, despite the drag of struts and wires. Also, this is the design approach for human-powered airplanes.

END ASIDE

For the pitching-moment calculation, note that this particular example has a significant vertical dimension and the drag coefficients are high. Therefore, the moment equations about the reference point have to account for these effects:

$$M_{\text{ref}} = (M_{w1})_{\text{ref}} + (M_{w2})_{\text{ref}} + (M_s)_{\text{ref}} + (M_{\text{add}})_{\text{ref}}$$
$$(M_{w1})_{\text{ref}} = (M_{ac})_1 - L_1(l_{w1} \cos\alpha + h_{w1} \sin\alpha) + D_1(h_{w1} \cos\alpha - l_{w1} \sin\alpha)$$
$$(M_{w2})_{\text{ref}} = (M_{ac})_2 - L_2(l_{w2} \cos\alpha + h_{w2} \sin\alpha) + D_2(h_{w2} \cos\alpha - l_{w2} \sin\alpha)$$
$$(M_s)_{\text{ref}} = -L_s(l_s \cos\alpha + h_s \sin\alpha) + D_s(h_s \cos\alpha - l_s \sin\alpha)$$
$$(M_{\text{add}})_{\text{ref}} \approx D_{\text{excres}} h_{\text{excres}}$$

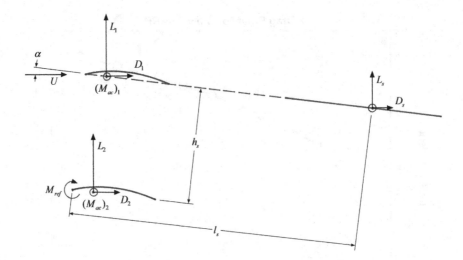

Upon non-dimensionalizing and assuming small values of α, one obtains:

$$(C_{Mw1})_{ref} \approx \left[(C_{Mac})_1 \hat{c}_1 - (C_L)_{w1}(\hat{l}_{w1} + \hat{h}_{w1}\alpha) + (C_D)_{w1}(\hat{h}_{w1} - \hat{l}_{w1}\alpha) \right] S_{w1}/S$$

$$(C_{Mw2})_{ref} \approx \left[(C_{Mac})_2 \hat{c}_2 - (C_L)_{w2}(\hat{l}_{w2} + \hat{h}_{w2}\alpha) + (C_D)_{w2}(\hat{h}_{w2} - \hat{l}_{w2}\alpha) \right] S_{w2}/S$$

$$(C_{Ms})_{ref} \approx \left[(C_{Mac})_s \hat{c}_s - (C_L)_s(\hat{l}_s + \hat{h}_s\alpha) + (C_D)_s(\hat{h}_s - \hat{l}_s\alpha) \right] S_s/S$$

$$(C_{Madd})_{ref} \approx (C_D)_{add} \hat{h}_{excres}$$

The pitching-moment coefficient about the wings' aerodynamic center is the same as that for the 7.5% circular-arc airfoil:

$$(C_{Mac})_{w1} = (C_{Mac})_{w2} = (C_{mac})_{airfoil} = \pi \alpha_{ZLL}/2 = -0.2356, (C_{Mac})_s = 0$$

The reference length and area are $\bar{c} = 9.0''$ and $S = 864.0$ inch^2 and the other terms are:

$$c_1 = c_2 = 9.0'' \rightarrow \hat{c}_1 = c_1/\bar{c} = 1.0; \quad \hat{c}_2 = c_2/\bar{c} = 1.0; \quad \hat{c}_s = 11.0/9.0 = 1.222$$

$$l_{w1} = l_{w2} = 2.25'' \rightarrow \hat{l}_{w1} = l_{w1}/\bar{c} = 0.25; \quad \hat{l}_{w2} = l_{w2}/\bar{c} = 0.25$$

$$l_s = 30.63'' \rightarrow \hat{l}_s = l_s/\bar{c} = 3.403$$

$$h_{w1} = 12.0'' \rightarrow \hat{h}_{w1} = 1.333; \quad h_{w2} = 0 \rightarrow \hat{h}_{w2} = h_{w2}/\bar{c} = 0$$

$$h_s = 12.0'' \rightarrow \hat{h}_s = h_s/\bar{c} = 1.333; \quad h_{excres} \approx 7.0'' \rightarrow \hat{h}_{excres} = h_{excres}/\bar{c} = 0.778$$

$$S_{w1} = S_{w2} = 432.0 \text{ inch}^2, \quad S_s = 159.6 \text{ inch}^2$$

Now, the summation of these contributions gives:

$$(C_M)_{ref} = (C_{Mw1})_{ref} + (C_{Mw2})_{ref} + (C_{Ms})_{ref} + (C_{Madd})_{ref}$$

Canard Airplanes and Biplanes

And, as before, the computer is used to obtain the tabulated values of $(C_M)_{ref}$ vs. α, as well as other values of interest:

α (deg)	C_L	C_D	C_L/C_D	$C_L^{1.5}/C_D$	C_R	$(C_M)_{ref}$	$l_R/\bar{c}$	l_R (inch)
-4	0.2566	0.2095	1.225	0.621	0.3313	0.0492	0.1487	1.338
-2	0.3997	0.1915	2.087	1.319	0.4432	-0.0227	-0.0512	-0.461
0	0.5427	0.1831	2.964	2.184	0.5728	-0.0941	-0.1643	-1.478
2	0.6858	0.1842	3.723	3.083	0.7101	-0.1653	-0.2328	-2.096
4	0.8289	0.1948	4.254	3.873	0.8515	-0.2368	-0.2781	-2.503
6	0.9719	0.2149	4.522	4.458	0.9954	-0.3089	-0.3104	-2.793
8	1.1150	0.2446	4.559	4.814	1.1415	-0.3821	-0.3347	-3.012
10	1.2581	0.2837	4.434	4.974	1.2897	-0.4566	-0.3540	-3.186
12	1.4011	0.3324	4.216	4.990	1.4400	-0.5329	-0.3700	-3.330

and the plotted values of $(C_M)_{ref}$ vs. α are shown below:

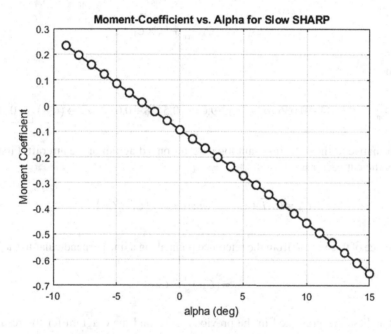

Moment-Coefficient vs. Alpha for Slow SHARP

It is seen that this is very linear and it curve-fitted to the following equation:

$$(C_M)_{ref} = -0.0944 - 0.0365\alpha, \quad (\alpha \text{ in degrees})$$
$$(C_M)_{ref} = -0.0944 - 2.0913\alpha, \quad (\alpha \text{ in radians})$$

For low values of α, and high-enough values of L/D, the pitching moment about any other point, $\hat{l}$, along the reference axis is approximately given by:

$$C_M \approx (C_M)_{ref} + \hat{l} \times C_L$$

And the derivative of this is:

$$C_{M\alpha} \approx (C_{M\alpha})_{ref} + \hat{l} \times C_{L\alpha}$$

If $C_{M\alpha} = 0$, then that value of $\hat{l}$ is the non-dimensional distance from the nose to the neutral point:

$$\hat{l}_{np} \approx -(C_{M\alpha})_{ref}/C_{L\alpha} \to l_{np} \approx -(C_{M\alpha})_{ref}\bar{c}/C_{L\alpha}$$

From the above equations, it is derived that:

$$(C_{M\alpha})_{ref} = -0.0365/\deg = -2.0913/\mathrm{rad}$$

Further, from before, $C_{L\alpha} = 4.0985/\mathrm{rad}$, so:

$$l_{np} \approx 2.0913 \times 9.0/4.0984 \to l_{np} \approx 4.59\,\mathrm{inches}$$

The moment about the neutral point is given by:

$$(C_M)_{np} \approx (C_M)_{ref} + \hat{l}_{np} \times C_L$$

which becomes:

$$(C_M)_{np} \approx -0.0944 - 0.0365\alpha + (4.59/9.0) \times (0.5428 + 0.0715\alpha) \to (C_M)_{np} \approx 0.182$$

It is also informative to show the resultant force vector on a diagram of the aircraft. First, the resultant force-coefficient vector is:

$$\vec{C}_R \equiv C_D \vec{i} + C_L \vec{j} \quad \text{where } C_R = (C_L^2 + C_D^2)^{1/2}$$

Also, the distance of this vector, from the reference point along a line perpendicular to C_R, is given by:

$$l_R = (C_M)_{ref}\,\bar{c}/C_R$$

The values for these are presented in the previous table, and the diagram for the resultant force-vector coefficients are shown below. The length of each vector is proportional to the magnitude of C_R, and its perpendicular distance from the reference point is the moment arm about that point, l_R. Therefore, if the mass-center lies somewhere along a vector, that would be a trim position for the α associated with that vector. For example, consider the vector for $\alpha = 4°$. Any mass-center location along that vector would be a trim position. However, the diagram also shows that the mass-center should be placed low for static stability (observe the pitching moment produced by the $\alpha = 8°$ vector for either low or high cm location on the $\alpha = 4°$ vector). This illustrates the limitation of the neutral-point concept for such a tall configuration. That is, if the mass-center is placed higher, the pitching-stability characteristics will change. In fact, if the mass-center is placed too high, then the aircraft would become unstable.

Such force-vector diagrams were once popular for airplane design before the concept of neutral points was introduced. As seen, though, for this peculiar case where the aircraft has a large vertical dimension compared with its length, the vector-diagram representation is useful and informative.

Canard Airplanes and Biplanes

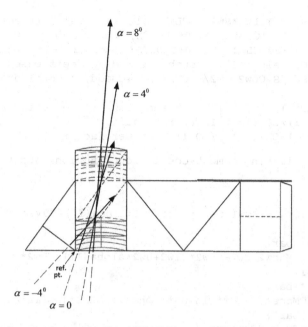

The Matlab program for calculating the Slow SHARP's aerodynamic characteristics is given below:

```
% Polar Curve and Equation for Slow SHARP(3), 17 June 2020
clc
clear all

% All dimensions in inches
S=864.0; % reference area
cbar=9.0; % reference length
Sw1=432.0; % area of wing1
Sw2=432.0; % area of wing2
Ss=159.6; % area of stabilizer
cw1=9.0; % wing 1 mean aerodynamic chord
cw2=9.0; % wing 2 mean aerodynamic chord
cs=11.0; % stabilizer mean aerodynamic chord
lw1=2.25; % long. dist. of wing1 aero center from ref. point
lw2=2.25; % long. dist. of wing2 aero center from ref. point
ls=30.63; % long. dist. of stabilizer aero center from ref. point
hw1=12.0; % vertical dist. of wing1 aero center from ref. point.
hw2=0; % vertical dist. of wing2 aero center from ref. point
hs=12.0; % vertical dist. of stabilizer aero center from ref. point
CDadd=0.0493; % additional drag coefficient (based on S)
hadd=7.0; % vertical distance of additional drag from ref. point
CMacw1=-0.2356; % CMac for wing1
CMacw2=-0.2356; % CMac for wing2
CMacs=0.0; % CMac for stabilizer

alphamax=15.0; % airplane maximum angle of attack (deg)
alphamin=-10.0; % airplane minimum angle of attack (deg)
deltalpha=1.0; % step size for alpha variation (deg)
N=(alphamax-alphamin)/deltalpha;
for I=1:1:N
    alpha(I)=(alphamin+I*deltalpha)*pi/180; % radians
    CLw1=0.5890+3.9266*alpha(I); % wing1 lift coeff. based on Sw1
    CLw2=0.5890+3.9266*alpha(I); % wing2 lift coeff. based on Sw2
    CLs=-0.2504+0.9308*alpha(I); % stabilizer lift coeff. based on Ss
```

```
        CL(I)=CLw1*Sw1/S+CLw2*Sw2/S+CLs*Ss/S;   % total lift coeff. based on S
        CDw1=0.2216-0.300*CLw1+0.2440*CLw1^2; % wing1 drag coeff. based on Sw1
        CDw2=0.2216-0.300*CLw2+0.2400*CLw2^2; % wing2 drag coeff. based on Sw2
        CDs=0.0269+CLs*alpha(I); % stabilizer drag coeff based on Ss
        CD(I)=CDw1*Sw1/S+CDw2*Sw2/S+CDs*Ss/S+CDadd; % total drag coeff. based
on S
        CR(I)=sqrt(CL(I)^2+CD(I)^2); % resultant force coeff. based on S
        Ratio(I)=CL(I)/CD(I); % lift/drag ratio
        PowerFactor(I)=CL(I)^1.5/CD(I); % power factor

    % component pitching moment-coeff. contributions about the ref. pt.,
based on S
    % and cbar:

        CMw1ref(I)=(CMacw1*cw1-CLw1*(lw1+hw1*alpha(I))+CDw1*
(hw1-lw1*alpha(I)))...
            *Sw1/(S*cbar);
        CMw2ref(I)=(CMacw2*cw2-CLw2*(lw2+hw2*alpha(I))+CDw2*
(hw2-lw2*alpha(I)))...
            *Sw2/(S*cbar);
        CMsref(I)=(CMacs*cs-CLs*(ls+hs*alpha(I))+CDs*(hs-ls*alpha(I)))...
            *Ss/(S*cbar);
        CMadd(I)=CDadd*hadd/cbar;

    % total pitching-moment coeff.about the ref. pt.

        CMref(I)=CMw1ref(I)+CMw2ref(I)+CMsref(I)+CMadd(I);
        lrbar(I)=CMref(I)/CR(I);
        lr(I)=lrbar(I)*cbar; % perpendicular dist. from ref. pt. to resultant
force vector
end
figure
I=[1:1:N];
plot(CD(I),CL(I),'-bo','linewidth',1.5,'markersize',8,...
    'markeredgecolor','b','markerfacecolor','w')
grid on
xlabel('Drag Coefficient')
ylabel('Lift Coefficient')
title('Drag Polar for Slow SHARP')

figure
I=[1:1:N];
alphadeg(I)=alpha(I)*180/pi;
plot(alphadeg(I),CL(I),'-bo','linewidth',1.5,'markersize',8,...
    'markeredgecolor','b','markerfacecolor','w')
grid on
xlabel('alpha (deg)')
ylabel('Lift Coefficient')
title('Lift-Coefficient vs. Alpha for Slow SHARP')

figure
I=[1:1:N];
alphadeg(I)=alpha(I)*180/pi;
plot(alphadeg(I),CD(I),'-bo','linewidth',1.5,'markersize',8,...
    'markeredgecolor','b','markerfacecolor','w')
grid on
xlabel('alpha (deg)')
ylabel('Drag Coefficient')
title('Drag-Coefficient vs. Alpha for Slow SHARP')
```

Canard Airplanes and Biplanes

```
figure
I=[1:1:N];
alphadeg(I)=alpha(I)*180/pi;
plot(alphadeg(I),CMref(I),'-bo','linewidth',1.5,'markersize',8,...
    'markeredgecolor','b','markerfacecolor','w')
grid on
xlabel('alpha (deg)')
ylabel('Moment Coefficient')
title('Moment-Coefficient vs. Alpha for Slow SHARP')

% Curve fitting of CL vs. Alpha (deg)
fprintf('\n')
fprintf('Linear Curve Fit for CL vs. alpha\n')
p=polyfit(alphadeg,CL,1);
xp=alphamin:deltalpha:alphamax;
yp=polyval(p,xp);
fprintf('\n')
P0=p(2);
P1=p(1);
CL0=P0;
CLa=P1;
CLaRad=CLa*180/pi;
fprintf('P0=%6.4f, P1=%6.4f\n',p(2),p(1))

figure
plot(alphadeg,CL,'o',xp,yp)
xlabel('alpha(deg)');ylabel('CL')
title('Linear Curve-Fit Check')

% Curve fitting of CD vs. CL
fprintf('\n')
fprintf('Polynomial Curve Fit for Polar\n')
p=polyfit(CL,CD,2);
alphaminrad=alphamin*pi/180;
alphamaxrad=alphamax*pi/180;
CLmin=CL(1);
CLmax=CL(N);
DeltCL=(CLmax-CLmin)/N;
xp=CLmin:DeltCL:CLmax;
yp=polyval(p,xp);
fprintf('\n')
P0=p(3);
P1=p(2);
P2=p(1);
fprintf('P0=%6.4f, P1=%6.4f, P2=%6.4f\n',p(3),p(2),p(1))

% Performance
fprintf('\n')
CLMaxLD=sqrt(P0/P2);
fprintf('CL for Max L/D=%6.4f\n',CLMaxLD)
fprintf('\n')
LDMax=(P0/CLMaxLD+P1+P2*CLMaxLD)^-1;
fprintf('Max L/D=%6.4f\n',LDMax)
fprintf('\n')
AlphaMaxLDrad=(CLMaxLD-CL0)/CLaRad;
AlphaMaxLDdeg=AlphaMaxLDrad*180/pi;
fprintf('Alpha for Max L/D (deg)=%6.4f\n',AlphaMaxLDdeg)
```

```
figure
plot(CL,CD,'o',xp,yp)
xlabel('CL');ylabel('CD')
title('Polar Cubic Curve-Fit Check')

% Polynomial curve fitting of CMref vs. alpha
fprintf('\n')
fprintf('Polynomial Curve Fit for Moment Coefficient\n')
p=polyfit(alphadeg,CMref,2);
xp=alphamin:deltalpha:alphamax;
yp=polyval(p,xp);
fprintf('\n')
P0=p(3);
P1=p(2);
P2=p(1);
fprintf('P0=%6.4f, P1=%6.4f, P2=%6.4f\n',p(3),p(2),p(1))

figure
plot(alphadeg,CMref,'o',xp,yp)
xlabel('alpha(deg)');ylabel('CMref')
title('Cubic Curve-Fit Check')

% Linear curve fitting of CMref vs. alpha
fprintf('\n')
fprintf('Linear Curve Fit for Moment Coefficient\n')
p=polyfit(alphadeg,CMref,1);
xp=alphamin:deltalpha:alphamax;
yp=polyval(p,xp);
fprintf('\n')
P0=p(2);
P1=p(1);
fprintf('P0=%6.4f, P1=%6.4f\n',p(2),p(1))

figure
plot(alphadeg,CMref,'o',xp,yp)
xlabel('alpha(deg)');ylabel('CMref')
title('Linear Curve-Fit Check')

% Print out aero terms
fprintf('\n')
disp('Tabulated Aerodynamic Characteristics')
fprintf('\n')
fprintf(' Alpha(deg)   Lift Coeff.   Drag Coeff.    Lift/Drag    CL^1.5/CD\n')
for ID=1:1:N
fprintf('%8.4f        %8.4f        %8.4f        %8.4f
%8.4f\n',alphadeg(ID),...
CL(ID),CD(ID),Ratio(ID),PowerFactor(ID))
end

fprintf('\n')
fprintf(' Alpha(deg)   Result. Coeff.  Moment Coeff.    lr/cbar
lr(inch)\n')
for ID=1:1:N
fprintf('%8.4f        %8.4f         %8.4f        %8.4f
%8.4f\n',alphadeg(ID),...
CR(ID),CMref(ID),lrbar(ID),lr(ID))
end
```

Canard Airplanes and Biplanes

FURTHER CONSIDERATIONS OF BIPLANE ANALYSIS

The above example was for an equal-span biplane, $b_1 = b_2$. However, the previous induced-drag equation explicitly accounts for different spans and areas.

$$\left(C_{Di}\right)_{biplane} = \frac{S_1}{S}\frac{(C_L)_{w1}^2}{\pi(AR)_{w1}} + \sigma\frac{2}{S}\frac{(C_L)_{w1}(C_L)_{w2}S_1S_2}{\pi b_1 b_2} + \frac{S_2}{S}\frac{(C_L)_{w2}^2}{\pi(AR)_{w2}}$$

For the equal-span equal-area case, the middle term was equally divided between the upper and lower wing. However, that's not so the unequal-span situation. Therefore, it will be assumed that the apportionment will be based upon the relative spans of the two wings, assuming the *mean chords* of the two wings are *identical*:

$$\left(C_{Di}\right)_{biplane}^{w1} \approx \frac{S_1}{S}\frac{(C_L)_{w1}^2}{\pi(AR)_{w1}} + \sigma\frac{2}{S}\frac{(C_L)_{w1}(C_L)_{w2}S_1S_2}{\pi b_1 b_2}\frac{b_1}{(b_1+b_2)}$$

$$\left(C_{Di}\right)_{biplane}^{w2} \approx \sigma\frac{2}{S}\frac{(C_L)_{w1}(C_L)_{w2}S_1S_2}{\pi b_1 b_2}\frac{b_2}{(b_1+b_2)} + \frac{S_2}{S}\frac{(C_L)_{w2}^2}{\pi(AR)_{w2}}$$

The induced drag coefficients of the individual wings, with respect to their reference areas, S_1 and S_2 are given by:

$$\left(C_{Dw1}\right)_{induced} = \frac{S}{S_1}\left(C_{Di}\right)_{biplane}^{w1}, \quad \left(C_{Dw2}\right)_{induced} = \frac{S}{S_2}\left(C_{Di}\right)_{biplane}^{w2}$$

Thus, from the above equations, one obtains:

$$\left(C_{Dw1}\right)_{induced} \approx \frac{(C_L)_{w1}^2}{\pi(AR)_{w1}} + 2\sigma\frac{(C_L)_{w1}(C_L)_{w2}S_2}{\pi b_2}\frac{1}{(b_1+b_2)}$$

$$\left(C_{Dw2}\right)_{induced} \approx \frac{(C_L)_{w2}^2}{\pi(AR)_{w2}} + 2\sigma\frac{(C_L)_{w1}(C_L)_{w2}S_1}{\pi b_1}\frac{1}{(b_1+b_2)}$$

Now, as was done before, these functions are equated to the induced drag coefficients of the wings given by:

$$\left(C_{Dw1}\right)_{induced} = \frac{(C_L)_{w1}^2}{\pi(AR)_{w1}} + (C_L)_{w1}(\varepsilon_{w1})_{w2}, \quad \left(C_{Dw2}\right)_{induced} = \frac{(C_L)_{w2}^2}{\pi(AR)_{w2}} + (C_L)_{w2}(\varepsilon_{w2})_{w1}$$

Therefore, upon comparison, one obtains that:

$$(\varepsilon_{w1})_{iw2} = 2\sigma\frac{(C_L)_{w2}S_2}{\pi b_2}\frac{1}{(b_1+b_2)}, \quad (\varepsilon_{w2})_{w1} = 2\sigma\frac{(C_L)_{w1}S_1}{\pi b_1}\frac{1}{(b_1+b_2)}$$

As for application to the lift-coefficient equations for the wings, there are two considerations:

1. Note that the downwash from the longer span will completely cover the shorter wing, that from the shorter wing will only cover the inner portion of the longer wing (as was the case for the Canard airplane). However, since the inner part of a wing produces most of the lift, this approximate method should be valid for $b_2/b_1 \geq 0.75$.
2. For the equal-span biplane with elliptic loadings on both wings, the earlier-defined mutual-downwash values, ε_1 and ε_2 were considered to be constant along the spans. For non-elliptical span loadings, this approximate method assumes that the average values are close to the ideal values, so that $\bar{\varepsilon}_1 \approx \varepsilon_1$ and $\bar{\varepsilon}_2 \approx \varepsilon_2$. The lift-coefficient equations are now written as:

$$(C_L)_{w1} = (C_{L\alpha})_{w1}^*\left[-(\alpha_{ZLL})_{w1} + \alpha_{w1} + i_{w1} - \frac{(\bar{\varepsilon}_{w1})_{w2}}{(C_L)_{w2}}(C_L)_{w2}\right]$$

$$(C_L)_{w2} = (C_{L\alpha})_{w2}^*\left[-(\alpha_{ZLL})_{w2} + \alpha_{w2} + i_{w2} - \frac{(\bar{\varepsilon}_{w2})_{w1}}{(C_L)_{w1}}(C_L)_{w1}\right]$$

where,

$$\frac{(\varepsilon_{w1})_{w2}}{(C_L)_{w2}} = 2\sigma \frac{S_2}{\pi b_2 (b_1 + b_2)}, \quad \frac{(\varepsilon_{w2})_{w1}}{(C_L)_{w1}} = 2\sigma \frac{S_1}{\pi b_1 (b_1 + b_2)}$$

Finally, the σ values for different span ratios and gap ratios are obtained from Walter S. Diehl (*Engineering Aerodynamics*, Roland Press, 1936, now owned by Wiley) and are shown below. Observe that when the span ratio, $\mu = b_2/b_1 \rightarrow 0$, all values of $\sigma \rightarrow 0$, which is what one would expect for a monoplane.

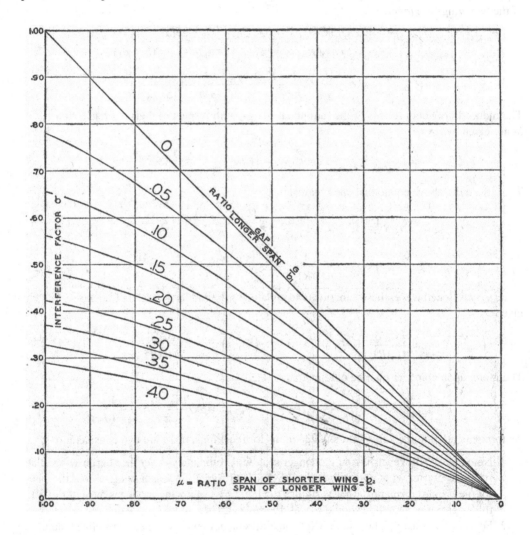

ASIDE

Biplanes, where the lower wing has less than half the span of the upper wing are called "Sesquiplanes". These are rare since they do not generally seem to offer any particular aerodynamic or structural advantages.

END ASIDE

Canard Airplanes and Biplanes

Another geometric variable to consider is stagger (defined earlier in this chapter). From Max Munk (NACA TR-151, 1923, p. 32), moderate values of stagger have only a secondary effect on the biplane's total induced drag coefficient. From before, for a biplane with identical wings ($b_1 = b_2 = b$, $S_1 = S_2 = S \rightarrow (AR)_{w1} = (AR)_{w2} = AR'$), this is given by:

$$\left(C_{Di}\right)_{biplane} = \frac{(C_L)_{w1}^2}{2\pi AR'} + \sigma \frac{(C_L)_{w1}(C_L)_{w2}}{\pi AR'} + \frac{(C_L)_{w2}^2}{2\pi AR'}$$

However, Munk shows that stagger does have an effect on the individual lift coefficients of the two wings, giving an increment of lift coefficient:

$$\Delta C_L = 2\frac{C_L}{AR}\left(\frac{1}{k^2} - 0.5\right)\frac{s/b}{\bar{R}}$$

where C_L is the lift coefficient for the un-staggered wing and

$$s/b = stagger/span$$

Also, k and $\bar{R}$ are functions of $gap/span = g/b$. The tabulated values given by Munk are:

g/b	k	$\bar{R}$
0	1.00	0.20
0.05	1.02	0.27
0.10	1.05	0.32
0.15	1.09	0.37
0.20	1.15	0.42
0.30	1.21	0.50
0.40	1.24	0.57
0.50	1.27	0.65

Upon assuming a positive stagger (top wing ahead of the bottom wing), the ΔC_L value for the upper wing is positive and the ΔC_L value for the lower wing is negative:

$$\frac{\Delta C_{Lw1}}{C_{Lw1}} = \frac{2}{AR'}\left(\frac{1}{k^2} - 0.5\right)\frac{s/b}{\bar{R}}, \quad \frac{\Delta C_{Lw2}}{C_{Lw2}} = \frac{-2}{AR'}\left(\frac{1}{k^2} - 0.5\right)\frac{s/b}{\bar{R}}$$

These may be added to the wings' lift coefficients as:

$$\left(C_{Lw1}\right)_{staggered} = \left(C_{Lw1}\right)_{un\text{-}staggered}\left(1 + \frac{\Delta C_{Lw1}}{C_{Lw1}}\right) \equiv \left(C'_L\right)_{w1}$$

$$\left(C_{Lw2}\right)_{staggered} = \left(C_{Lw2}\right)_{un\text{-}staggered}\left(1 + \frac{\Delta C_{Lw2}}{C_{Lw2}}\right) \equiv \left(C'_L\right)_{w2}$$

where $(C'_L)_{w1}$ and $(C'_L)_{w2}$ are the individual lift coefficients for each staggered wing.

Although, as stated before, the total induced drag coefficient of both wings is relatively invariant with moderate stagger, it is still useful to know the individual induced-drag coefficients for both wings. This may be estimated from a variation of the un-staggered equations:

$$\left(C'_{Dw1}\right)_{induced} \approx \frac{(C'_L)_{w1}^2}{\pi AR'} + \sigma \frac{(C'_L)_{w1}(C'_L)_{w2}}{\pi AR'}, \quad \left(C'_{Dw2}\right)_{induced} \approx \frac{(C'_L)_{w2}^2}{\pi AR'} + \sigma \frac{(C'_L)_{w1}(C'_L)_{w2}}{\pi AR'}$$

If $\Delta C_L/C_L$ is small, then the sum of the individual induced-drag coefficients will approximate the total induced-drag coefficient of the un-staggered case. For the Stagger-SHARP airplane, where $s/\bar{c} = 1.0 \rightarrow s/b = 9/48 = 0.1875$ and $g/b = 12/48 = 0.25$,

One obtains:

$$\frac{\Delta C_L}{C_L} = \frac{2}{5.333}\left(\frac{1}{1.18^2} - 0.5\right)\frac{0.188}{0.46} = 0.0334$$

EXAMPLE, BIPLANE GLIDER

The figure below shows a drawing of a 12-inch span balsa-wood biplane glider built by the author as a classroom demonstration model. This is a "cleaner" design than the Slow SHARP and has a half-chord positive stagger.

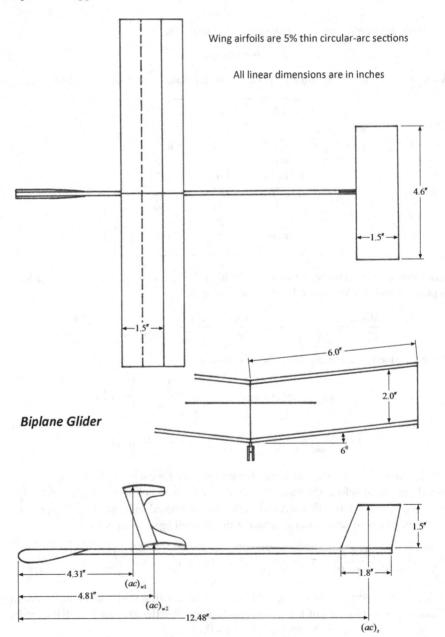

Biplane Glider

Canard Airplanes and Biplanes

As before, the lift-curve slope for a single wing is obtained from the Lowry and Polhamus equation, as described in the "Wings" section of Chapter 2:

$$C_{L\alpha} = \frac{2\pi\, AR}{2+\left[\left(\frac{AR}{\kappa}\right)^2 \left(1+\tan^2 \Lambda_{c/2}\right)+4\right]^{1/2}}$$

where $\Lambda_{c/2} = 0$, $AR = 12.0/1.5 = 8.0$, $\kappa = 1$ (assumed), so that each wing's lift-curve slope is:

$$\left(C_{L\alpha}^*\right)_{w1} = \left(C_{L\alpha}^*\right)_{w2} = \frac{50.27}{2+(64.0+4)^{1/2}} = 4.906, \quad (1/\text{rad})$$

The stabilizer has an aspect ratio $(4.6)^2/(1.5 \times 4.6) = 21.16/6.9 = (AR)_s = 3.07$. Also, $\Lambda_{c/2} = 0$ and $\kappa = 1$ (assumed), so that the Lowry and Polhamus equation gives:

$$\left(C_{L\alpha}^*\right)_s = \frac{19.29}{2+\left[9.42+4\right]^{1/2}} = \frac{19.29}{2+3.664} = \left(C_{L\alpha}^*\right)_s = 3.406, \quad (1/\text{rad})$$

Note that these values for $(C_{L\alpha}^*)$ assume no mutual interference, and each is referenced to its own area.

The wing airfoils are 5% thin circular arcs, as was the case for the Canard Glider example. Therefore, $\alpha_{ZLL} = -0.10\,\text{rad}\,(-5.73°)$, so $(\alpha_{ZLL})_{w1} = (\alpha_{ZLL})_{w2} = -0.10$. Also, the incidence angle of both wings with respect to the airplane's reference line is zero, so that $i_{w1} = i_{w2} = 0$. Further, $\alpha_{w1} = \alpha_{w2} = \alpha$, the angle-of-attack of the airplane.

For this example, the two wings are of equal spans and areas, $b_1 = b_2 = 12''$, $S_1 = S_2 = 18.0\,in^2 \rightarrow (AR)_{w1} = (AR)_{w2} = AR' = 8.0$. Further, the gap ratio is $g/b = 2/12 = 0.17 \rightarrow \sigma = 0.53$. Therefore the mutual-downwash terms from the previously-derived equations are:

$$\frac{(\varepsilon_{w1})_{w2}}{(C_L)_{w2}} = \frac{(\varepsilon_{w2})_{w1}}{(C_L)_{w1}} \rightarrow \frac{\sigma}{\pi\, AR'} = \frac{0.53}{\pi \times 8.0} = 0.021$$

Also, the Munk stagger-lift factor is:

$$\frac{\Delta C_{Lw1}}{C_{Lw1}} = \frac{2}{(AR)_{w1}}\left(\frac{1}{k^2} - 0.5\right)\frac{s/b_1}{\bar{R}} = -\frac{\Delta C_{Lw2}}{C_{Lw2}}$$

where $s/b = 0.75/12 = 0.0625$ and $g/b = 0.17 \rightarrow k = 1.114$, $\bar{R} = 0.390$, so that:

$$\frac{\Delta C_{Lw1}}{C_{Lw1}} = -\frac{\Delta C_{Lw2}}{C_{Lw2}} = \frac{2}{8}\left(\frac{1}{1.114^2} - 0.5\right)\frac{0.0625}{0.390} = 0.0123$$

Now, the un-staggered lift coefficients are given by:

$$\left(C_{Lw1}\right)_{\text{un-staggered}} = 4.906\left[0.10 + \alpha - 0.021(C_L)_{w2}\right]$$

$$\left(C_{Lw2}\right)_{\text{un-staggered}} = 4.906\left[0.10 + \alpha - 0.021(C_L)_{w1}\right]$$

From these, one may solve for $(C_{Lw1})_{\text{un-staggered}}$ and $(C_{Lw2})_{\text{un-staggered}}$ as functions of α:

$$(C_{Lw1})_{\text{un-staggered}} = 0.4448 + 4.4477\alpha, \quad (C_{Lw2})_{\text{un-staggered}} = 0.4448 + 4.4477$$

Upon applying the stagger correction:

$$(C_L)_{w1} = (0.4448 + 4.4477\alpha)(1 + 0.0123), \quad (C_L)_{w2} = (0.4448 + 4.4477\alpha)(1 - 0.0123)$$

One obtains the lift-coefficient equations for both wings:

$$\underline{(C_L)_{w1} = 0.450 + 4.502\alpha, \quad (C_L)_{w2} = 0.439 + 4.393\alpha}$$

Unlike the earlier un-staggered example, both wings produce slightly different lifts. This difference is much less than what would be implied by Hiscock's equations, given earlier in this chapter. However, the author's approximate methodology is consistent with Munk's analysis.

Next, the stabilizer's aerodynamic contribution will be estimated. The lift coefficient of the stabilizer is given by:

$$(C_L)_s = (C_{L\alpha})_s^* \left[\alpha + i_s - (\varepsilon_s)_{w1} - (\varepsilon_s)_{w2}\right]$$

where $i_s = 0$, $(\varepsilon_s)_{w1} = (\varepsilon_s / C_L)_{w1} (C_L)_{w1}$, $(\varepsilon_s)_{w2} = (\varepsilon_s / C_L)_{w2} (C_L)_{w2}$

The downwash parameters will be calculated from the USAF DATCOM methodology described in the "Complete-Aircraft Aerodynamics" section of Chapter 2, where:

$$\left(\frac{\varepsilon}{C_L}\right)_w = \frac{(\partial \varepsilon / \partial \alpha)_w}{(C_{L\alpha})_w}, \quad \left(\frac{\partial \varepsilon}{\partial \alpha}\right)_w = 4.44\left[K_A K_\lambda K_H (\cos \Lambda_{c/4})_w^{1/2}\right]^{1.19}$$

For the downwash on the stabilizer due to *wing 1*, one has that:

$$(K_A)_{w1} = \frac{1}{(AR)_{w1}} - \frac{1}{1+(AR)_{w1}^{1.7}} = \frac{1}{8.0} - \frac{1}{1+(8.0)^{1.7}} = 0.125 - 0.028 = 0.097$$

Next,

$$(K_\lambda)_{w1} = \frac{10 - 3\lambda_{w1}}{7} = \frac{10-3}{7} = 1.0$$

The next term in the downwash equation is:

$$(K_H)_{w1} = \frac{1 - |(h_H)_{w1}/b_{w1}|}{\left[2(l_H)_{w1}/b_{w1}\right]^{1/3}} = \frac{1 - 0.5/12}{(2 \times 8.17/12)^{1/3}} = \frac{0.958}{1.108} = 0.865$$

Therefore, with $(\Lambda_{c/4})_{w1} = 0$, one obtains that:

$$(\partial \varepsilon / \partial \alpha)_{w1} = 4.44(0.097 \times 1.0 \times 0.865)^{1.19} = 0.233$$

Canard Airplanes and Biplanes

Furthermore, from previous work, $(C_{L\alpha})_{w1} = 4.502$, so that:

$$(\varepsilon_s/C_L)_{w1} = 0.233/4.502 = 0.052$$

For *wing* 2, all the terms stay the same except for K_H which, in this case, becomes:

$$(K_H)_{w2} = \frac{1-|(h_H)_{w1}/b_{w1}|}{[2(l_H)_{w1}/b_{w1}]^{1/3}} = \frac{1-1.5/12}{(2\times 7.67/12)^{1/3}} = \frac{0.875}{1.085} = 0.806$$

Therefore,

$$(\partial\varepsilon/\partial\alpha)_{w1} = 4.44(0.097\times 1.0\times 0.806)^{1.19} = 0.214$$

Because $(C_{L\alpha})_{w2} = 4.393$, one obtains that:

$$(\varepsilon_s/C_L)_{w2} = 0.214/4.393 = 0.049$$

Finally, the downwash values are given by:

$$(\varepsilon_s)_{w1} = (\varepsilon_s/C_L)_{w1}(C_L)_{w1} = 0.052\times(0.450+4.502\alpha) = 0.023+0.234\alpha$$

$$(\varepsilon_s)_{w2} = (\varepsilon_s/C_L)_{w2}(C_L)_{w2} = 0.049\times(0.439+4.393\alpha) = 0.022+0.215\alpha$$

Therefore,

$$(C_L)_s = 3.406\left[\alpha-(0.023+0.022)-(0.234+0.215)\alpha\right] = 3.406\alpha - 0.153 - 1.529\alpha \rightarrow$$

$$\underline{\underline{(C_L)_s = -0.153+1.877\alpha}}$$

Now, $S_{w1} = S_{w2} = 18.0 \text{ in}^2$ and $S_s = 6.9 \text{ in}^2$. Choose a reference area of $S = S_{w1} + S_{w2} = 36.0 \text{ in}^2$, so that the total lift coefficient is given by:

$$C_L = (C_L)_{w1}\frac{S_{w1}}{S} + (C_L)_{w2}\frac{S_{w2}}{S} + (C_L)_s\frac{S_s}{S} =$$

$$C_L = (0.450+4.502\alpha)\times 0.5 + (0.439+4.393\alpha)\times 0.5 + (-0.153+1.877\alpha)\times 0.192$$

$$= (0.225+0.220-0.029) + (2.251+2.197+0.360)\alpha$$

$$\underline{\underline{C_L = 0.416+4.808\alpha, \quad \alpha \text{ in radians}}}$$

$$\underline{\underline{C_L = 0.416+0.0839\alpha, \quad \alpha \text{ in degrees}}}$$

The body, being just a stick, is assumed to produce negligible lift compared with the wings and stabilizer. A graph of lift coefficient vs. angle of attack is shown below:

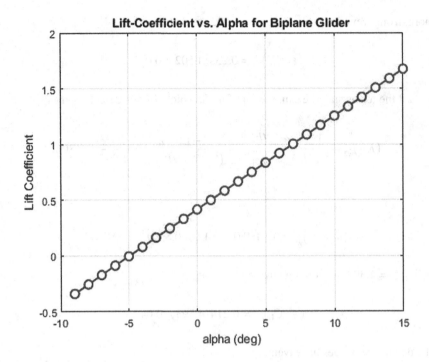

Now, for the drag coefficient, the first step is to obtain the profile drag of the circular-arc airfoil. From the "Airfoils" section in Chapter 2, one has:

$$C_d \approx (C_d)_{\text{fric}} + (1-\eta_{\text{le}})\left[\, 2\pi\alpha_{\text{ZLL}}^2 + 2\alpha_{\text{ZLL}}C_l + C_l^2/(2\pi)\,\right]$$

The sharp leading edge of this airfoil gives that $\eta_{\text{le}} \approx 0$. So, the equation becomes:

$$C_d \approx (C_d)_{\text{fric}} + 2\pi\alpha_{\text{ZLL}}^2 + 2\alpha_{\text{ZLL}}C_l + C_l^2/(2\pi)$$

From before, for this 5% circular-arc section, $\alpha_{\text{ZLL}} = -0.10\,\text{radian}$. So, the equation further becomes:

$$C_d \approx (C_d)_{\text{fric}} + 0.0628 - 0.20\,C_l + 0.1592\,C_l^2$$

From glide tests, the glider flight speed is approximately 8 ft/sec (2.44 m/sec). So, the Reynolds number for the wings is found from the equation in Imperial units given in the "Airfoils" section of Chapter 2:

$$RN = 6.36\,V\,\bar{c}\times 10^3 = 6.36\times 8.0\times 0.125\times 10^3 = 6360.0$$

and the skin-friction drag coefficient is given by:

$$C_{\text{fric}} = \frac{0.455}{\left[\log(RN)\right]^{2.58}} = 0.01450 \rightarrow (C_d)_{\text{fric}} = 2C_{\text{fric}} = 0.0290$$

So, the airfoil's profile drag coefficient is:

$$C_d = 0.0290 + 0.0628 - 0.20\,C_l + 0.1592\,C_l^2 = 0.0918 - 0.20\,C_l + 0.1592\,C_l^2$$

As assumed for the canard example, this will be the profile drag for each wing:

$$\left(C_{\text{Dw}}\right)_{\text{profile}} = 0.0918 - 0.20(C_L)_w + 0.1592(C_L)_w^2$$

Canard Airplanes and Biplanes

Next, the induced drag coefficient for each wing will draw upon the earlier equations for a biplane with wings of equal spans and areas:

$$\left(C_{Dw1}\right)_{induced} = \frac{(C_L)_{w1}^2}{\pi\,AR'} + \sigma\frac{(C_L)_{w1}(C_L)_{w2}}{\pi\,AR'}$$

$$\left(C_{Dw2}\right)_{induced} = \frac{(C_L)_{w2}^2}{\pi\,AR'} + \sigma\frac{(C_L)_{w1}(C_L)_{w2}}{\pi\,AR'}$$

From before, $\sigma \approx 0.53$ and $AR' = 8.0$. Therefore, one obtains that:

$$\left(C_{Dw1}\right)_{induced} = 0.0398(C_L)_{w1}^2 + 0.0211(C_L)_{w1}(C_L)_{w2}$$

$$\left(C_{Dw2}\right)_{induced} = 0.0398(C_L)_{w2}^2 + 0.0211(C_L)_{w1}(C_L)_{w2}$$

Therefore, the total drag of each wing is $(C_D)_w = \left(C_{Dw}\right)_{profile} + \left(C_{Dw}\right)_{induced}$:

$$(C_D)_{w1} = 0.0918 - 0.20(C_L)_{w1} + 0.1592(C_L)_{w1}^2 + 0.0398(C_L)_{w1}^2 + 0.0211(C_L)_{w1}(C_L)_{w2}$$

$$(C_D)_{w2} = 0.0918 - 0.20(C_L)_{w2} + 0.1592(C_L)_{w2}^2 + 0.0398(C_L)_{w2}^2 + 0.0211(C_L)_{w1}(C_L)_{w2}$$

$$\underline{(C_D)_{w1} = 0.0918 - 0.20(C_L)_{w1} + 0.0211(C_L)_{w1}(C_L)_{w2} + 0.1990(C_L)_{w1}^2}$$

$$\underline{(C_D)_{w2} = 0.0918 - 0.20(C_L)_{w2} + 0.0211(C_L)_{w1}(C_L)_{w2} + 0.1990(C_L)_{w2}^2}$$

The stabilizer has a smooth single-surface flat-plate airfoil with the same chord as the wings'. Therefore, its RN and C_f is the same as those for the wings' and its zero-angle drag coefficient is given by:

$$\left(C_{Ds}\right)_{\alpha=0} \approx \left(C_{Ds}\right)_{fric} = 2C_{fric} = 0.0290$$

As was the case for the wing, the stabilizer will be assumed to have zero leading-edge suction. Therefore, the stabilizer's drag can simply be obtained by observing that, with $\eta_{le} = 0$, the lift force is perpendicular to the stabilizer's surface. Thus, the drag force, relative to the airplane's reference line, is given by:

$$D_s \approx D_{fric} + L\alpha \rightarrow (C_D)_s \approx \left(C_{Ds}\right)_{\alpha=0} + (C_L)_s \alpha$$

$$\underline{(C_D)_s = 0.0290 + (C_L)_s \alpha}$$

As for the body and excrescence drag, the wetted area of the body stick is:

$$\left(S_{wet}\right)_{body} \approx 8.5\,in^2\,(0.059\,ft^2)$$

and the Reynolds number, upon noting the body length of 13.4 inches (1.12 ft), is:

$$(RN)_{body} = 6.36 \times 8.0 \times 1.12 \times 10^3 = 56986 \rightarrow (C_{fric})_b = 0.0081$$

So,

$$(C_D)_b\,S_b = (C_{fric})_b(S_{wet})_b = 0.0082 \times 8.5 = 0.0692\,in^2 \rightarrow (C_D)_b\,S_b/S = 0.0019$$

The additional drag includes the fin, wing struts, nose fairing, and riblets, etc., which totals a wetted area of:

$$(S_{\text{wet}})_{\text{add}} = (S_{\text{wet}})_{\text{fin}} + (S_{\text{wet}})_{\text{struts}} + (S_{\text{wet}})_{\text{fairing}} + (S_{\text{wet}})_{\text{riblets}} \rightarrow$$

$$(S_{\text{wet}})_{\text{add}} \approx 4.95 + 3 \times 3.6 + 1.5 + 8 \times 0.4 = 20.45 \text{ in}^2$$

Also, the average Reynolds number is $RN_{\text{ex}} \approx 10{,}000$, which gives a skin-friction coefficient of $(C_{\text{fric}})_{\text{add}} = 0.0127 \rightarrow$

$$(C'_D)_{\text{add}} S_{\text{add}} = (C_{\text{fric}})_{\text{add}} (S_{\text{wet}})_{\text{add}} = 0.260 \text{ in}^2 \rightarrow (C'_D)_{\text{add}} S_{\text{add}}/S = 0.0072$$

There is also a contribution from the interference drag between the struts and wings, etc. For the "Interference Drag" chapter in S. Hoerner's **Fluid Dynamic Drag**, this is estimated to increase the added drag by 10%:

$$(C_D)_{\text{add}} S_{\text{add}}/S \approx 1.10 \, (C'_D)_{\text{add}} S_{\text{add}}/S = 0.0079$$

So, the total drag coefficient is:

$$C_D = (C_D)_{w1} \frac{S_1}{S} + (C_D)_{w2} \frac{S_2}{S} + (C_D)_s \frac{S_s}{S} + (C_D)_b \frac{S_b}{S} + (C_D)_{\text{add}} \frac{S_{\text{add}}}{S}$$

$$\underline{\underline{C_D = 0.5(C_D)_{w1} + 0.5(C_D)_{w2} + 0.192(C_D)_s + 0.0098}}$$

where $(C_D)_{w1}$, $(C_D)_{w2}$ and $(C_D)_s$ are obtained from the previously-derived equations.

As for the Slow-SHARP example, a short program is used to calculate and plot C_D as a function of α, as well as obtaining the C_D vs. C_L plot for which a polynomial curve-fit equation is:

$$\underline{\underline{C_D = 0.1012 - 0.1795 C_L + 0.2040 C_L^2}}$$

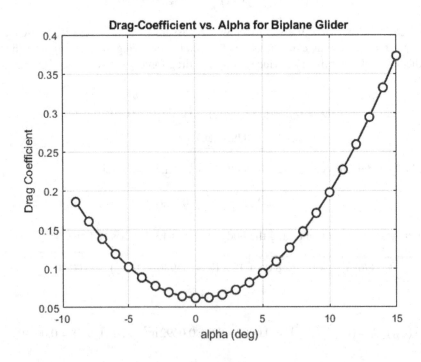

Canard Airplanes and Biplanes

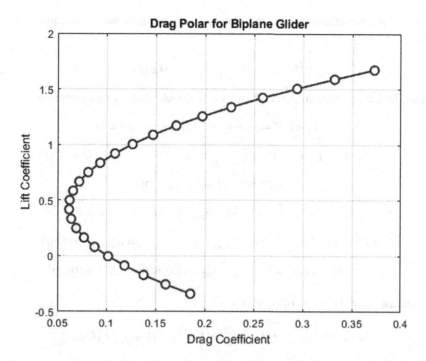

From the *Complete Airplane Lift and Drag Coefficients* subsection of the "Numerical Example" section in Chapter 2, one has that for a polynomial polar equation expressed as:

$$C_D = P_0 + P_1 C_L + P_2 C_L^2$$

Then

$$(C_L)_{\text{Max L/D}} = (P_0/P_2)^{1/2} = (0.1012/0.2040)^{1/2} = 0.704$$

$$(L/D)_{\text{Max}} = \left[\frac{P_0}{(C_L)_{\text{Max L/D}}} + P_1 + P_2 \times (C_L)_{\text{Max L/D}} \right]^{-1} = 9.271$$

Further, the α value at which this occurs is given by:

$$(C_L)_{\text{Max L/D}} = 0.416 + 4.808(\alpha)_{\text{Max L/D}} \rightarrow (\alpha)_{\text{Max L/D}} = (0.704 - 0.416)/4.808 \rightarrow$$

$$(\alpha)_{\text{Max L/D}} = 0.060 \text{ rad} = 3.43°$$

The moment coefficient equation is adapted from that previously derived for the Slow SHARP example. The reference point is the nose of the glider:

$$(C_{Mw1})_{\text{ref}} \approx \left[(C_{Mac})_1 \hat{c}_1 - (C_L)_{w1} (\hat{l}_{w1} + \hat{h}_{w1}\alpha) + (C_D)_{w1} (\hat{h}_{w1} - \hat{l}_{w1}\alpha) \right] S_{w1}/S$$

$$(C_{Mw2})_{\text{ref}} \approx \left[(C_{Mac})_2 \hat{c}_2 - (C_L)_{w2} (\hat{l}_{w2} + \hat{h}_{w2}\alpha) + (C_D)_{w2} (\hat{h}_{w2} - \hat{l}_{w2}\alpha) \right] S_{w2}/S$$

$$(C_{Ms})_{\text{ref}} \approx \left[(C_{Mac})_s \hat{c}_s - (C_L)_s (\hat{l}_s + \hat{h}_s\alpha) + (C_D)_s (\hat{h}_s - \hat{l}_s\alpha) \right] S_s/S$$

$$(C_{Madd})_{\text{ref}} \approx (C_D)_{\text{add}} S_{\text{add}}/S \times \hat{h}_{\text{add}}, \quad (C_{Mb})_{\text{ref}} \approx (C_D)_b S_b/S \times \hat{h}_b$$

The pitching-moment coefficient about the wings' aerodynamic center is the same as that for the 5% circular-arc airfoil:

$$\left(C_{Mac}\right)_{w1} = \left(C_{Mac}\right)_{w2} = \left(C_{mac}\right)_{airfoil} = \pi\alpha_{ZLL}/2 = -0.1571$$

The reference length and area are $\bar{c} = 1.5''$ and $S = 36.0$ inch^2 and the other terms are:

$$c_1 = c_2 = 1.5'' \rightarrow \hat{c}_1 = c_1/\bar{c} = 1.0; \quad \hat{c}_2 = c_2/\bar{c} = 1.0$$

$$l_{w1} = 4.31'', \quad l_{w2} = 4.81'' \rightarrow \hat{l}_{w1} = l_{w1}/\bar{c} = 2.873; \quad \hat{l}_{w2} = l_{w2}/\bar{c} = 3.207$$

$$l_s = 12.48'' \rightarrow \hat{l}_s = l_s/\bar{c} = 8.320$$

$$h_{w1} = 2.0'' \rightarrow \hat{h}_{w1} = 1.333; \quad h_{w2} = 0 \rightarrow \hat{h}_{w2} = h_{w2}/\bar{c} = 0$$

$$h_s = 1.5'' \rightarrow \hat{h}_s = h_s/\bar{c} = 1.0; \quad h_{excres.} \approx 1.0 \rightarrow \hat{h}_{excres} = h_{excres}/\bar{c} = 0.667,$$

$$h_b = 0 \rightarrow \hat{h}_b = h_b/\bar{c} = 0; \quad S_{w1} = S_{w2} = 18.0\,\text{inch}^2, \quad S_s = 6.9\,\text{inch}^2$$

Now, the summation of these contributions gives:

$$\left(C_M\right)_{ref} = \left(C_{Mw1}\right)_{ref} + \left(C_{Mw2}\right)_{ref} + \left(C_{Ms}\right)_{ref} + \left(C_{Mb}\right)_{ref} + \left(C_{Madd}\right)_{ref}$$

And, as before, a computer is used to calculate values of $(C_M)_{ref}$ vs. α, which plot out as shown:

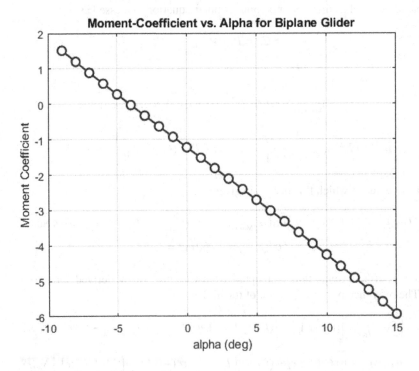

A linear curve-fit equation to this is:

$$\left(C_M\right)_{ref} = -1.2317 - 0.3065\alpha, \quad \alpha \text{ in degrees}$$

$$\left(C_M\right)_{ref} = -1.2317 - 17.561\alpha, \quad \alpha \text{ in radians}$$

Canard Airplanes and Biplanes

For low values of α, and reasonably high values of L/D, the pitching moment about any other point, $\hat{l}$, along the reference axis is approximately given by:

$$C_M \approx (C_M)_{ref} + \hat{l} \times C_L$$

And the derivative of this is:

$$C_{M\alpha} \approx (C_{M\alpha})_{ref} + \hat{l} \times C_{L\alpha}$$

If $C_{M\alpha} = 0$, then that value of $\hat{l}$ is the non-dimensional distance from the nose to the neutral point:

$$\hat{l}_{np} \approx -(C_{M\alpha})_{ref} / C_{L\alpha} \rightarrow l_{np} \approx -(C_{M\alpha})_{ref} \bar{c}/C_{L\alpha}$$

As seen from the above plot, the variation of $(C_M)_{ref}$ with α is nearly straight, and the averaged slope represents the typical condition for the glider's flight:

$$(C_{M\alpha})_{ref} = -0.3065/\text{deg} = -17.561/\text{rad}$$

Further, from before, $C_{L\alpha} = 4.808/\text{rad}$, so:

$$l_{np} \approx 17.561 \times 1.5/4.808 \rightarrow \underline{l_{np} \approx 5.48 \text{ inches}}$$

Further, the moment about the neutral point at is given by:

$$(C_M)_{np} \approx (C_M)_{ref} + \hat{l}_{np} \times C_L$$
$$(C_M)_{np} \approx -1.2317 - 17.561\alpha + (5.48/1.5) \times (0.416 + 4.808\alpha) \rightarrow \underline{(C_M)_{np} \approx 0.288}$$

Finally, the author built the Biplane Glider in accordance with the drawing shown earlier.

The mass-center location for the stable trimmed flight was at $l_{cm} = 4.70''$, so the Static Margin is:

$$\text{Static Margin} \equiv \hat{x}_{np} = \frac{l_{np} - l_{cm}}{\bar{c}} = \frac{5.48 - 4.70}{1.5} = 0.52 \rightarrow 52\%$$

This is a very strong value. The pitching moment about the mass-center is approximately given by:

$$M_{cm} \approx M_{np} - L(l_{np} - l_{cm}) \rightarrow (C_M)_{cm} \approx (C_M)_{np} - C_L \times \hat{x}_{np}$$

The trim state is when $(C_M)_{cm} = 0$, so that:

$$(C_M)_{cm} = 0 = (C_M)_{np} - (C_L)_{trim} \times \hat{x}_{np} \rightarrow (C_L)_{trim} = (C_M)_{np}/\hat{x}_{np} \approx 0.288/0.52$$

$$(C_L)_{trim} \approx 0.554$$

As before, with the drag polar expressed as:

$$C_D = P_0 + P_1 C_L + P_2 C_L^2$$

the L/D ratio for that lift coefficient is given by:

$$(L/D)_{trim} = \left[\frac{P_0}{(C_L)_{trim}} + P_1 + P_2 \times (C_L)_{trim}\right]^{-1} = \left[\frac{0.1012}{0.554} - 0.1795 + 0.2040 \times 0.554\right]^{-1}$$

$$(L/D)_{trim} \approx 8.607$$

Also, for the lift coefficient expressed as:

$$C_L = C_{L0} + C_{L\alpha}\alpha$$

the α value this occurs at is given by:

$$(\alpha)_{trim} = \frac{(C_L)_{trim} - C_{L0}}{C_{L\alpha}} = \frac{(C_L)_{trim} - 0.416}{4.808} = \frac{0.554 - 0.416}{4.808} = 0.0287 \,\text{rad} = 1.64°$$

It's clear that the model can be trimmed for a glide closer to that for $(L/D)_{max}$ by moving its mass-center further aft, but in its present state, the L/D ratio is only 7.2% less than the maximum value of 9.271, and is on the safe side of stalling.

The lift coefficient of the stabilizer for this trim state (referenced to S_s) is found from:

$$(C_L)_s = -0.153 + 1.873\alpha = -0.153 + 1.873 \times 0.0287 = -0.099$$

Usually, the desired trim state for a tail-aft airplane is to have the stabilizer operating at a zero-lift coefficient. This case is reasonably close to that, which explains the excellent L/D for a model of this size.

Big Caveat
It is important to emphasize that despite the numerical examples giving answers to the third decimal place, this is an approximate methodology. Although it draws upon more rigorous, and referenced, methodologies, the author has made simplifying assumptions and adaptations. However, what can

Canard Airplanes and Biplanes

be said is that in comparison with treating the two wings as acting independently, this methodology brings their interactions closer to the correct values.

A more exacting treatment of biplane theory may be found in the references cited in this chapter. Additionally, two excellent sources are:

W.S. Diehl, "Relative Loading on Biplane Wings", NACA TR 458, 1933.
W.S. Diehl, "Relative Loading on Biplane Wings of Unequal Chord", NACA TR 501, 1934.

Of course, one could also use a modern vortex-lattice methodology, such as offered by Mark Drela's AVL program: http://web.mit.edu/drela/Public/web/avl/. This would give results justifying a third decimal place. However, one would then miss the opportunity to appreciate the elegant derivation of Ludwig Prandtl's lifting-line analysis extended to a biplane configuration.

FINAL BIPLANE COMMENT

The geometrical parameter, "decalage", was presented on page 203 as the difference between the lower wing's incidence angle and the upper wing's incidence angle.

$$Decalage \equiv i_{w1} - i_{w2}$$

More precisely, in accordance to Hiscocks, it is the angular difference between the zero-lift lines of the wings. For identical wings, decalage defined either way will be identical.

Some references, however, give an opposite sign to decalage. Namely, they state that positive decalage means that the upper wing has a larger incidence angle than that for the lower wing. The author has chosen Hiscock's definition, which is also in accordance to that given by Karl Wood in "Technical Aerodynamics" (McGraw-Hill, 1935, Page 270).

Also, note that "decalage" has sometimes been used to express the difference in incidence angles between the wing and stabilizer for monoplane configurations. However, an alternative terminology, in this case, is "longitudinal dihedral".

6 Flight Dynamics

INTRODUCTION

This chapter will describe an aircraft's flight-dynamic equations, which involve the vehicle's motion responses to imposed forces and moments. Previously, the concept of *static* longitudinal stability had been discussed. It is an interesting fact that if a normally proportioned airplane has positive static stability, then it is likely to have longitudinal dynamic stability. However, note that this is not always so. Also, no static criteria have been established for lateral stability. Therefore, in order to be assured that the aircraft has the desired stability and response to control inputs, one must perform a dynamic analysis. Also, it will be seen that the static-stability criterion is a special case from the dynamic-stability solutions.

An introduction to the concept of flight-dynamic stability is given by a simple, balanced, wind-vane model; and the idea is to find the system's dynamic response to a small discrete perturbation. That is, imagine that the vane is held at a small θ angle in a smooth, steady flow and then released.

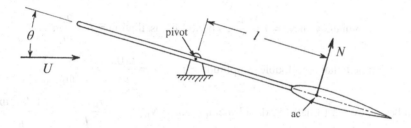

In this case, the angle of attack at the vane's aerodynamic center is given by:

$$\alpha \approx \theta + \frac{l\dot\theta}{U}$$

This gives a moment about the pivot point (positive in the θ direction):

$$M_{\text{pivot}} = -Nl = -q_{\text{dyn}} C_{N_\alpha} \alpha S l$$

where S is the area of the vane, q_{dyn} is the dynamic pressure and C_{N_α} is the derivative of the normal-force coefficient with respect to α.

The inertial reaction to this is given by:

$$M_{\text{pivot}} = I_{yy} \ddot\theta$$

where I_{yy} is the pitching moment of inertia about the pivot. So, the complete dynamic equation is:

$$q_{\text{dyn}} C_{N_\alpha} S l \theta + \frac{q_{\text{dyn}} C_{N_\alpha} S l^2}{U} \dot\theta + I_{yy} \ddot\theta = 0$$

Note that small angles have been assumed ($\theta \ll 1$, $\dot{\theta} \ll U$). Upon defining:

$$\frac{q_{\text{dyn}} C_{N_\alpha} S l^2}{U} \equiv A, \quad q_{\text{dyn}} C_{N_\alpha} S l \equiv B$$

the dynamic equation becomes:

$$I_{yy}\ddot{\theta} + A\dot{\theta} + B\theta = 0$$

A solution for this may be obtained by assuming that $\theta = \Theta e^{\omega t}$. Upon substitution into the dynamic equations and cancellation of like terms, one obtains the "characteristic equation":

$$I_{yy}\omega^2 + A\omega + B = 0$$

The solution of this gives the "dimensional stability root", which is generally of the form: $\omega = \omega_r + \omega_j$. It is seen that there is a real part, ω_r and an imaginary part, ω_j. These terms characterize the dynamic behavior of the system in that ω_j is the oscillatory frequency of the wind vane's response to a discrete perturbation (in rad/sec) and ω_r quantifies the damping or divergence of the wind vane's motion. In particular, the useful parameters for defining the wind vane's dynamic behavior are:

$$\text{frequency in } Hz = f = \frac{\omega_j}{2\pi}, \quad \text{period of oscillation} = T = \frac{2\pi}{\omega_j},$$

$$\text{time to half (or double) amplitude} = T_{1/2(or2)} = \frac{\ln 0.5}{\omega_r} = \frac{-0.69315}{\omega_r}$$

$$\text{number of cycles to half (or double) amplitude} = N_{1/2(or2)} = \frac{T_{1/2(or2)}}{T} = \frac{0.11032\,\omega_j}{\omega_r}$$

Numerical Example 1:

Consider a wind vane with the following characteristics:

$S = 0.03$ m^2, $C_{N_\alpha} = 4.5$, $l = 0.5$ m, $U = 10$ m/sec, $I_{yy} = 0.065$ kg – m^2, $\rho = 1.225$ kg/m^3 $\to q = 61.25$ N/m^2

So,

$$A = \frac{61.25 \times 4.5 \times 0.03 \times 0.5^2}{10} = 0.2067, \quad B = 61.25 \times 4.5 \times 0.03 \times 0.5 = 4.1344$$

The characteristic equation is, therefore:

$$0.065\omega^2 + 0.2067\omega + 4.1344 = 0 \to \omega = -1.590 \pm i7.815$$

which gives $\omega_r = -1.590$ sec, $\omega_j = 7.815$ sec. Therefore, the wind vane is oscillating at a frequency of 1.244 Hz and the time to half amplitude from a discrete perturbation is 0.436 sec. Clearly, this is a very stable system.

End Example

Flight Dynamics

Next, consider a wind vane that is free to translate vertically:

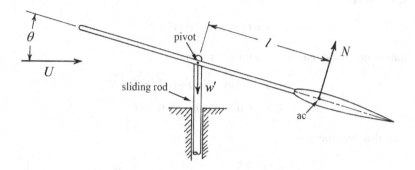

Observe that gravity is not accounted for. Physically, this can be envisioned as the vane being sideways to the flow.

The angle of attack of the airfoil's aerodynamic center is now given as:

$$\alpha = \theta + \frac{l}{U}\dot\theta + \frac{w'}{U}$$

where w' is the vertical plunging velocity (positive down). Therefore, the moment about the pivot point is:

$$M_{pivot} = -Nl = -q_{dyn}\,C_{N_\alpha}\,\alpha\,S\,l = I_{yy}\ddot\theta \rightarrow$$

$$q_{dyn}\,C_{N_\alpha}\,Sl\,\theta + \frac{q_{dyn}\,C_{N_\alpha}\,Sl^2}{U}\dot\theta + q_{dyn}\,C_{N_\alpha}\frac{Sl}{U}w' + I_{yy}\ddot\theta = 0.$$

Upon defining:

$$C \equiv q_{dyn}\,C_{N_\alpha}\frac{Sl}{U}$$

this becomes:

$$I_{yy}\ddot\theta + A\dot\theta + B\theta + C\,w' = 0 \tag{6.1}$$

An additional equation is obtained from the vertical acceleration. Assume that the vane's mass-center is at the pivot point and that the sliding rod has negligible mass by comparison. One thus obtains:

$$N = q_{dyn}\,S\,C_{N_\alpha}\,\alpha = -m\dot w' \rightarrow q_{dyn}\,S\,C_{N_\alpha}\,\theta + \frac{q_{dyn}\,S\,C_{N_\alpha}\,l}{U}\dot\theta + \frac{q_{dyn}\,S\,C_{N_\alpha}}{U}w' + m\dot w' = 0$$

Define:

$$E \equiv q_{dyn}\,S\,C_{N_\alpha}, \quad D \equiv \frac{q_{dyn}\,S\,C_{N_\alpha}\,l}{U}, \quad F \equiv \frac{q_{dyn}\,S\,C_{N_\alpha}}{U}$$

so, the second equation becomes:

$$m\dot w' + D\dot\theta + E\theta + F\,w' = 0 \tag{6.2}$$

One now has two equations for the θ and w' motions. Again, a solution may be found by assuming the following motions:

$$\theta = \Theta e^{\omega t}, \quad w' = W' e^{\omega t} \qquad (6.3)$$

Upon substitution into the dynamic equations, one obtains:

$$(I_{yy}\omega^2 + A\omega + B)\Theta + CW' = 0 \qquad (6.4)$$

$$(D\omega + E)\Theta + (m\omega + F)W' = 0 \qquad (6.5)$$

In matrix form, this becomes:

$$\begin{bmatrix} (I_{yy}\omega^2 + A\omega + B) & C \\ (D\omega + E) & (m\omega + F) \end{bmatrix} \begin{bmatrix} \Theta \\ W' \end{bmatrix} = 0$$

The characteristic equation is found by multiplying out the determinant of the square matrix:

$$(I_{yy}\omega^2 + A\omega + B)(m\omega + F) - C(D\omega + E) = 0 \rightarrow$$
$$I_{yy}m\omega^3 + (I_{yy}F + Am)\omega^2 + (AF + Bm - CD)\omega + (BF - CE) = 0$$

Numerical Example 2:

In addition to the parameters previously defined, the vane's mass is $m = 0.26$ kg. Therefore, one obtains:

$$A = 0.2067, \ B = 4.1344, \ C = 0.4134, \ D = C = 0.4134, \ E = 8.2688, \ F = 0.8269$$

which gives the terms in the characteristic equation:

$I_{yy}m = 0.065 \times 0.26 = 0.01690$, $(I_{yy}F + Am) = 0.10749$, $(AF + Bm - CD) = 1.07497$, $(BF - CE) = 0$

Thus, the characteristic equation is:

$$(0.01690\,\omega^2 + 0.10749\,\omega + 1.07497)\omega = 0$$

from which the first root of this cubic equation is $\omega_1 = 0$ and the remaining two roots are found from the quadratic equation inside the parentheses:

$$\omega_{2,3} = -3.1802 \pm i7.3140 \rightarrow \omega_r = -3.1802, \ \omega_j = 7.3140$$

So, the frequency of oscillation is 1.164 Hz and the time to half amplitude is 0.2180 sec. Again, this is a very stable system.

End Example

In comparison with the first example, there are two motions: pitching and plunging (the vertical motion at the pivot, defined as positive down). It is of interest to compare the relative magnitudes of these motions, which may be determined from either equation 6.4 or 6.5. Upon choosing equation 6.5, this may be rewritten as:

$$\frac{W'}{\Theta} = \frac{-(D\omega + E)}{(m\omega + F)}$$

Flight Dynamics

with ω_2, this equation gives that:

$$\frac{W'}{\Theta} = \frac{-[0.4134(-3.1802+i7.3140)+8.2688]}{0.26(-3.1802+i7.3140)+0.8269} = \frac{-6.9541-i3.0236}{0.0+i1.9016}$$

$$= \frac{-i1.9016}{-i1.9016} \times \frac{-6.9541-i3.0236}{0.0+i1.9016} = \frac{i13.2239+5.7497}{3.6162} = -1.5900+i3.6569$$

This may also be expressed in the form:

$$W'/\Theta = 3.9876 e^{i(113.5 \text{deg})}$$

The result gives the relative values of the plunging motion magnitude, W', to the pitching motion magnitude, Θ. However, comparing a velocity to an angle is not particularly informative; but if W' is divided by the free-stream velocity, U, then the comparison becomes more meaningful:

$$\frac{W'/U}{\Theta} = 0.3988 e^{i(113.5 \text{deg})}$$

This states that the magnitude of the relative angle of attack caused by the plunging oscillation is about 40% of that from the pitching oscillation. A convenient way of visualizing this is provided by a modal-vector diagram, where relative magnitudes and phase angles are readily seen and compared.

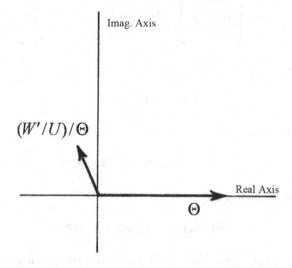

Now the dynamic equations may be expressed in a different way, for which an eigenvalue solution may be found. If one defines the *pitch-rate* $\dot{\theta}$ as q, then one obtains three first-order differential equations:

$$q - \dot{\theta} = 0 \tag{6.6}$$

$$I_{yy}\dot{q} + Aq + B\theta + Cw' = 0 \tag{6.7}$$

$$m\dot{w}' + Dq + E\theta + Fw' = 0 \tag{6.8}$$

The assumed solutions, equation (6.3), must be extended to include:

$$q = Qe^{\omega t} \tag{6.9}$$

Therefore, upon substitution of equations (6.3) and (6.9) into equations (6.6), (6.7) and (6.8), one obtains:

$$Q - \omega\Theta = 0$$

$$I_{yy}Q\omega + AQ + B\Theta + CW' = (I_{yy}\omega + A)Q + CW' + B\Theta = 0$$

$$mW'\omega + DQ + E\Theta + FW' = DQ + (m\omega + F)W' + E\Theta = 0$$

In matrix form these become:

$$\begin{bmatrix} 1 & 0 & -\omega \\ (I_{yy}\omega + A) & C & B \\ D & (m\omega + F) & E \end{bmatrix} \begin{bmatrix} Q \\ W' \\ \Theta \end{bmatrix} = \begin{bmatrix} 1 & 0 & 0 \\ A & C & B \\ D & F & E \end{bmatrix} \begin{bmatrix} Q \\ W' \\ \Theta \end{bmatrix}$$

$$-\omega \begin{bmatrix} 0 & 0 & 1 \\ -I_{yy} & 0 & 0 \\ 0 & -m & 0 \end{bmatrix} \begin{bmatrix} Q \\ W' \\ \Theta \end{bmatrix} = 0$$

upon defining

$$M_1 \equiv \begin{bmatrix} 1 & 0 & 0 \\ A & C & B \\ D & F & E \end{bmatrix}, \quad M_2 \equiv \begin{bmatrix} 0 & 0 & 1 \\ -I_{yy} & 0 & 0 \\ 0 & -m & 0 \end{bmatrix}$$

The equations become:

$$[M_1 - \omega M_2] = 0$$

The MATLAB solution for the eigenvalues of this is obtained from:

$$[\text{modes, stabroots}] = \text{eig}(M_1, M_2)$$

For the numerical example, the stability roots are:

$$\omega_1 = 0, \quad \omega_{2,3} = -3.1800 \pm j7.3139$$

This is the same as before (accounting for round-off error from the previous hand calculations) and the values for the oscillatory mode are thus:

$$\omega_r = -3.1800, \quad \omega_j = 7.3139$$

Also, the mode for the second root is $W'/\Theta = -1.5902 + i3.6569$, which is essentially the same as before. Besides the convenience of an eigenvalue solution, what is also acquired is a new modal vector:

$$Q/\Theta = -3.1800 + i7.3139$$

Modal-vector results will be particularly useful for interpreting the motions of an aircraft from the non-dimensional form of its equations, as described later in this chapter.

Flight Dynamics

The final introductory example will be a free-flying wind vane (or arrow) as shown below.

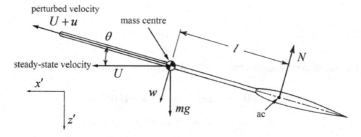

Observe that this now has the freedom to move horizontally as well as vertically and in pitch. Also, this vane is assumed to be flying along a horizontal path and the motions are small perturbations from this.

> **ASIDE**
>
> Note that this is a rather special "cooked-up" case for the purpose of introducing the notion of a free-flying vehicle with three degrees of freedom. Imagine that the vane is describing straight and level flight, with the weight, mg, balanced by an equilibrium aerodynamic lift (say, N_0) and with a net equilibrium moment of zero about the mass-center. Therefore, the motions, forces and moment described below are *perturbations* from that equilibrium state.
>
> **END ASIDE**

The perturbation velocities are now chosen to be defined in a body-fixed reference frame as shown, so the dynamic equations will be defined accordingly.

First of all, the moment equation about the mass-center is the same as before (note that $w \approx w'$ from the previous example):

$$M_{cm} = -Nl = -q_{dyn} C_{N_\alpha} \alpha S l = I_{yy} \ddot{\theta} \rightarrow$$

$$q_{dyn} C_{N_\alpha} S l \theta + \frac{q_{dyn} C_{N_\alpha} S l^2}{U} \dot{\theta} + q_{dyn} C_{N_\alpha} \frac{S l}{U} w + I_{yy} \ddot{\theta} = 0 \rightarrow$$

$$I_{yy} \ddot{\theta} + A \dot{\theta} + B \theta + C w = 0 \tag{6.10}$$

For the linear-acceleration terms, first observe that the mass-center's velocity components in the inertial reference frame, defined by the coordinates *fixed in that frame*, x', z', are given by:

$$v_{x'} = \frac{dx'}{dt} = \dot{x}', \quad v_{z'} = \frac{dz'}{dt} = \dot{z}'$$

If these velocity components are expressed in the *body-fixed axes system*, then one has that:

$$v_{x'} = (U+u)\cos\theta + w\sin\theta, \quad v_{z'} = w\cos\theta - (U+u)\sin\theta$$

The time derivatives of these give the acceleration components in the inertial reference frame:

$$a_{x'} = \frac{d^2 x'}{dt^2} = \ddot{x}' = \dot{v}_{x'} = \dot{u}\cos\theta - (U+u)\dot{\theta}\sin\theta + w\dot{\theta}\cos\theta + \dot{w}\sin\theta$$

$$a_{z'} = \frac{d^2 z'}{dt^2} = \ddot{z}' = \dot{v}_{z'} = \dot{w}\cos\theta - w\dot{\theta}\sin\theta - (U+u)\dot{\theta}\cos\theta - \dot{u}\sin\theta$$

With the small-perturbation assumptions of $\theta \ll 1$, u/U, $w/U \ll 1$, $\dot\theta \ll U/l$, these acceleration terms become:

$$a_{x'} \approx \dot u, \quad a_{z'} \approx \dot w - U\dot\theta$$

Further, the acceleration components in u and w directions are:

$$a_u = a_{x'}\cos\theta - a_{z'}\sin\theta \approx a_{x'} - (\dot w - U\dot\theta)\theta \approx a_{x'} \rightarrow$$

$$a_u \approx \dot u$$

$$a_w = a_{z'}\cos\theta + a_{x'}\sin\theta \approx a_{z'} + \dot u\theta \approx a_{z'} \rightarrow$$

$$a_w \approx \dot w - U\dot\theta$$

Note again that these are accelerations due to perturbations of the vane from a straight and level flight path, and the *perturbation* forces that cause these (in u and w body-fixed directions) are:

$$(\Delta F)_u = -mg\sin\theta \approx -mg\theta, \quad (\Delta F)_w = -N + mg(1-\cos\theta) \approx -N$$

so, the force-acceleration equations are:

$$mg\theta + m\dot u = 0$$

$$N + m(\dot w - U\dot\theta) = 0 \rightarrow$$

$$q_{\text{dyn}}SC_{N_\alpha}\theta + \frac{q_{\text{dyn}}SC_{N_\alpha}l}{U}\dot\theta + \frac{q_{\text{dyn}}SC_{N_\alpha}}{U}w + m(\dot w - U\dot\theta) = 0 \rightarrow$$

$$q_{\text{dyn}}SC_{N_\alpha}\theta + \left(\frac{q_{\text{dyn}}SC_{N_\alpha}l}{U} - mU\right)\dot\theta + \frac{q_{\text{dyn}}SC_{N_\alpha}}{U}w + m\dot w = 0 \tag{6.11}$$

upon defining:

$$G \equiv \left(\frac{q_{\text{dyn}}SC_{N_\alpha}l}{U} - mU\right) = D - mU$$

the equation becomes:

$$m\dot w + G\dot\theta + E\theta + Fw = 0 \tag{6.12}$$

As with the previous example, the introduction of $q = \dot\theta$ gives a set of first-order linear differential equations:

$$q - \dot\theta = 0$$

$$I_{yy}\dot q + Aq + B\theta + Cw = 0$$

$$mg\theta + m\dot u = 0$$

$$m\dot w + Gq + E\theta + Fw = 0$$

the assumed solutions are:

$$\theta = \Theta e^{\omega t}, \quad w = We^{\omega t}, \quad q = Qe^{\omega t}, \quad u = \tilde U e^{\omega t}$$

Flight Dynamics

so, the equations become:

$$Q - \omega\Theta = 0$$
$$I_{yy}Q\omega + AQ + B\Theta + CW = (I_{yy}\omega + A)Q + CW + B\Theta = 0$$
$$mg\Theta + m\tilde{U}\omega = 0$$
$$mW\omega + GQ + E\Theta + FW = GQ + (m\omega + F)W + E\Theta = 0$$

In matrix form these become:

$$\begin{bmatrix} 0 & 0 & 1 & -\omega \\ 0 & C & (I_{yy}\omega + A) & B \\ m\omega & 0 & 0 & mg \\ 0 & (m\omega + F) & G & E \end{bmatrix} \begin{bmatrix} \tilde{U} \\ W \\ Q \\ \Theta \end{bmatrix} = 0$$

This expands to give:

$$\begin{bmatrix} 0 & 0 & 1 & 0 \\ 0 & C & A & B \\ 0 & 0 & 0 & mg \\ 0 & F & G & E \end{bmatrix} \begin{bmatrix} \tilde{U} \\ W \\ Q \\ \Theta \end{bmatrix} - \omega \begin{bmatrix} 0 & 0 & 0 & 1 \\ 0 & 0 & -I_{yy} & 0 \\ -m & 0 & 0 & 0 \\ 0 & -m & 0 & 0 \end{bmatrix} \begin{bmatrix} \tilde{U} \\ W \\ Q \\ \Theta \end{bmatrix} = 0.$$

Define the two matrices:

$$M_3 \equiv \begin{bmatrix} 0 & 0 & 1 & 0 \\ 0 & C & A & B \\ 0 & 0 & 0 & mg \\ 0 & F & G & E \end{bmatrix}, \quad M_4 \equiv \begin{bmatrix} 0 & 0 & 0 & 1 \\ 0 & 0 & -I_{yy} & 0 \\ -m & 0 & 0 & 0 \\ 0 & -m & 0 & 0 \end{bmatrix}$$

The equation then becomes:

$$[M_3 - \omega M_4] = 0$$

One may, again, obtain a solution for the eigenvalues and eigenvectors:

$$[\text{modes, stabroots}] = \text{eig}(M_3, M_4)$$

Numerical Example 3:

For this case, all the terms remain as before. However, the new term is:

$$G = \left(\frac{q_{dyn} S C_{N\alpha} l}{U} - mU\right) = D - mU = 0.4134 - 0.26 \times 10.0 = -2.1866$$

Also, $mg = 0.26 \times 9.81 = 2.5506$
The eigenvalues are:

$$\omega_1 = 0.0, \quad \omega_{2,3} = -3.1801 \pm j10.8210, \quad \omega_4 = 0.0$$

So, the motion consists of two neutrally stable non-oscillatory roots and a stable oscillatory complex-conjugate root. The modal vectors for the oscillatory root are:

$$\tilde{U}/\Theta = 0.2452 + i\,0.8345 = 0.8698\,e^{i(73.63\,\text{deg})} \rightarrow (\tilde{U}/U)/\Theta = 0.0870\,e^{i(73.63\,\text{deg})}$$

$$W/\Theta = 8.4099 + i\,5.4108 = 10.0000\,e^{i(32.76\,\text{deg})} \rightarrow (W/U)/\Theta = 1.0000\,e^{i(32.76\,\text{deg})}$$

$$Q/\Theta = -3.1801 + i\,10.8210 = 11.2786\,e^{i(106.38\,\text{deg})}$$

The modal vector diagram for this example shows that the magnitudes of pitch and non-dimensional plunge are identical, even though their phase angles differ by 32.76°. Also, the non-dimensional surging motion, u/U, is virtually negligible as compared with the pitching.

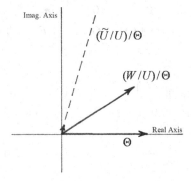

End Example

ASIDE

It is important to take a closer look at how the force and acceleration vectors are represented in the body-fixed reference frame.

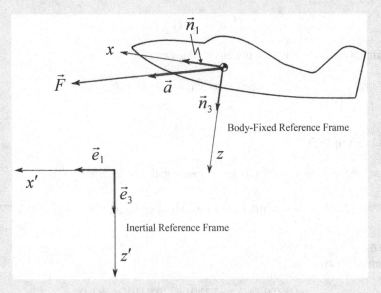

First of all, note that:

$$\vec{F} = m\vec{a}, \quad \text{where} \quad \vec{a} = \frac{d\vec{V}}{dt}$$

The velocity vector, $\vec{V}$, may be represented in both reference frames as:

$$\vec{V} = V_1 \vec{e}_1 + V_3 \vec{e}_3 = U \vec{n}_1 + W \vec{n}_3$$

so that the acceleration is given by:

$$\vec{a} = \frac{d\vec{V}}{dt} = \frac{dV_1}{dt}\vec{e}_1 + \frac{dV_3}{dt}\vec{e}_3 = \frac{dU}{dt}\vec{n}_1 + U\frac{d\vec{n}_1}{dt} + \frac{dW}{dt}\vec{n}_3 + W\frac{d\vec{n}_3}{dt}$$

Because the unit vectors $\vec{e}_1$ and $\vec{e}_3$ are fixed in the inertial reference frames, their time derivatives are zero. This is not the case for the $\vec{n}_1$ and $\vec{n}_3$ unit vectors, and their time derivatives must be determined.

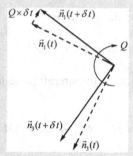

A vector is characterized by its magnitude and orientation. The magnitude of a unit vector is fixed, so its orientation is what changes with time. In the case of the body-fixed unit vectors, $\vec{n}_1$ and $\vec{n}_3$, their change in orientation is due to the pitch rate of the body, Q.

Consider $\vec{n}_1$ at time t. After a very small time increment, δt, it has rotated to a new position given by:

$$\vec{n}_1(t+\delta t) = \vec{n}_1(t) + \delta \vec{n}_1 \approx \vec{n}_1(t) - (Q \times \delta t)\vec{n}_3 \to \frac{\delta \vec{n}_1}{\delta t} \approx -Q\vec{n}_3 \to \frac{d\vec{n}_1}{dt} = -Q\vec{n}_3$$

A similar exercise gives:

$$\frac{d\vec{n}_3}{dt} = Q\vec{n}_1$$

Therefore, one has:

$$\vec{a} = (\dot{U} + WQ)\vec{n}_1 + (\dot{W} - UQ)\vec{n}_3$$

For the small motion perturbations, such as those assumed in the previous examples, one has:

$$U = U_0 + u, \quad W = W_0 + w, \quad Q = q$$

The parameters U_0 and W_0 are the body's steady-state equilibrium velocity components. Equilibrium also means that there are no rotation rates (there is no Q_0). Also, in aircraft flight dynamics, the orientation of the steady-state $\vec{n}_1$ vector is in the direction of the body's steady-state equilibrium velocity, so $W_0 = 0$.

Therefore, upon observing that wq and uq are vanishingly small, one obtains the small-perturbation form of the acceleration equation:

$$\vec{a} = \dot{u}\,\vec{n}_1 + (\dot{w} - U_0\,q)\,\vec{n}_3$$

Finally, the imposed force on the body may be written in both coordinate systems:

$$\vec{F} = F_1\,\vec{e}_1 + F_3\,\vec{e}_3 = X\,\vec{n}_1 + Z\,\vec{n}_3$$

Further, the imposed force, $\vec{F}$, on the aircraft consists of an equilibrium part and a perturbation part:

$$\vec{F} = \vec{F}_0 + \Delta\vec{F}$$

Therefore, the perturbation force on the aircraft is:

$$\Delta\vec{F} = \Delta X\,\vec{n}_1 + \Delta Z\,\vec{n}_3$$

In the same way, the pitching moment on the aircraft (nose-up positive) consists of an equilibrium part and a perturbation part:

$$\vec{M} = \vec{M}_0 + \Delta\vec{M} \rightarrow \Delta\vec{M} = \Delta M\,\vec{n}_2$$

Therefore, the force-acceleration equations for the body-fixed coordinate system are:

$$\Delta X = m\dot{u},\ \Delta Z = m(\dot{w} - U_0 q)$$

When one compares these equations with the force-acceleration equations of the previous example, it is seen that these are the same, where:

$$\Delta X \equiv (\Delta F)_u, \quad \Delta Z \equiv (\Delta F)_w \quad \text{and} \quad \Delta M \equiv M_{cm}$$

END ASIDE

AIRCRAFT LONGITUDINAL SMALL-PERTURBATION DYNAMIC EQUATIONS

The small-perturbation dynamic equations for an aircraft are similar to those for the previous "arrow" example, except that the imposed *perturbation* forces and moment now consist of the following contributions:

thrust: ΔT_x, ΔT_z
control: ΔX_c, ΔZ_c
gross buoyancy: B, which is based on the molded volume of the flight vehicle
gravity minus buoyancy: $(mg - B)$
aerodynamics: ΔX_{aero}, ΔZ_{aero}, ΔM_{aero}

It is also important to define what is meant by the mass, m. This includes not only the structural mass and other such components but also any enclosed air and gas because these likewise move with the flight vehicle. However, with the exception of airships and human-powered aircraft, the buoyancy and enclosed air and gas effects are generally negligible.

Upon noting that mg acts in the z' direction in the inertial reference frame, and B acts opposite to this, the resulting small-perturbation longitudinal dynamic equations are:

$$\Delta X_{aero} + \Delta X_c - (mg - B)\cos\Theta_0\,\theta + \Delta T_x = m\dot{u}$$

$$\Delta Z_{aero} + \Delta Z_c - (mg - B)\sin\Theta_0\,\theta + \Delta T_z = m(\dot{w} - U_0 q)$$

$$\Delta M_{aero} + \Delta M_c - x_B B \sin\Theta_0\,\theta + z_B B \cos\Theta_0\,\theta - x_T \Delta T_z + z_T \Delta T_x = I_{yy}\dot{q}$$

where Θ_0 is the aircraft's climb angle (negative for descent), x_T, z_T are the coordinates for the thruster's location (e.g., propeller hub) in the body-fixed axis system, and x_B, z_B are the coordinates for the vehicle's gross-buoyancy location.

ASIDE

The gross buoyancy of the aircraft is defined to be:

$$B = \rho g (Vol)_{mold}$$

where $(Vol)_{mold}$ is the aircraft's "molded volume". This would be the volume of a solid object of the same size and shape as the aircraft. This term could also be called the "Archimedean buoyancy". The "net buoyancy" is then given by:

$$B_{net} = B - mg$$

where m, again, is the total mass as defined above.

END ASIDE

The x and z body-fixed coordinates have been defined in a previous illustration. However, their equilibrium orientation depends upon the aircraft's equilibrium climb angle. If the equilibrium flight condition is level ($\Theta_0 = 0$), then the non-perturbed orientation of x is level, with z pointing directly down. However, if the aircraft has an equilibrium climb angle Θ_0, then the non-perturbed x-axis is in that direction.

A further point to note is that, in general, the aircraft's equilibrium angle of attack will vary with its equilibrium flight speed. For example, a higher C_L is needed at slow speeds, requiring a higher angle of attack. So, the orientation of the aircraft's reference line (say, the center line) relative to the x-axis will differ for various equilibrium flight conditions. This influences certain terms in the equations, as will be seen subsequently.

Finally, in comparison with the similar equations of the previous "arrow" example, a notational change has been made where the equilibrium flight speed in the inertial reference frame, U, is now called U_0. This serves to emphasize its meaning as an equilibrium (non-changing) value.

Now, the aerodynamic forces are assumed to be functions of the motion variables, u, w, q and their time derivatives $\dot{u}, \dot{w}, \dot{q}$ respectively. Therefore, for example, ΔX_{aero} may be expressed by the following Taylor series expansion:

$$\Delta X_{aero} = \left(\frac{\partial(\Delta X)_{aero}}{\partial u}\right)_0 u + \left(\frac{\partial(\Delta X)_{aero}}{\partial w}\right)_0 w + \left(\frac{\partial(\Delta X)_{aero}}{\partial q}\right)_0 q +$$

$$\left(\frac{\partial(\Delta X)_{aero}}{\partial \dot{u}}\right)_0 \dot{u} + \left(\frac{\partial(\Delta X)_{aero}}{\partial \dot{w}}\right)_0 \dot{w} + \left(\frac{\partial(\Delta X)_{aero}}{\partial \dot{q}}\right)_0 \dot{q} + \text{Higher-Order Terms}$$

where $(\ )_0$ denotes the derivative evaluated at the reference equilibrium state. Also, because this is a small-perturbation analysis, where the motion variables are assumed to be small, one has that:

$$u, v, w \ll U_0, \dot{u}, \dot{v}, \dot{w} \ll 2U_0^2/c, \ q \ll 2U_0/c, \ \dot{q} \ll 4U_0^2/c^2$$

c is a parameter called the "characteristic length" of the aircraft, which will be discussed in greater detail; and the higher-order terms are negligible.

For notational convenience, the following terms will be defined:

$$\left(\frac{\partial(\Delta X)_{\text{aero}}}{\partial u}\right)_0 \equiv X_u, \quad \left(\frac{\partial(\Delta X)_{\text{aero}}}{\partial w}\right)_0 \equiv X_w, \quad \left(\frac{\partial(\Delta X)_{\text{aero}}}{\partial \dot{u}}\right)_0 \equiv X_{du}, \text{ etc.}$$

and these will be called *dimensional stability derivatives*. Therefore,

$$\Delta X_{\text{aero}} = X_u u + X_w w + X_q q + X_{du} \dot{u} + X_{dw} \dot{w} + X_{dq} \dot{q}$$

The other *aero* terms may be expressed in the same way, so that:

$$\begin{bmatrix} \Delta X_{\text{aero}} \\ \Delta Z_{\text{aero}} \\ \Delta M_{\text{aero}} \end{bmatrix} = \begin{bmatrix} X_u & X_w & X_q & X_{du} & X_{dw} & X_{dq} \\ Z_u & Z_w & Z_q & Z_{du} & Z_{dw} & Z_{dq} \\ M_u & M_w & M_q & M_{du} & M_{dw} & M_{dq} \end{bmatrix} \begin{bmatrix} u \\ w \\ q \\ \dot{u} \\ \dot{w} \\ \dot{q} \end{bmatrix}$$

Similarly, one may assume that the propulsion terms may be expressed as:

$$\Delta T_X = T_{Xu} u + T_{Xw} w + T_{Xq} q, \quad \Delta T_Z = T_{Zu} u + T_{Zw} w + T_{Zq} q,$$
$$\Delta T_M = (z_T T_{Xu} - x_T T_{Zu})u + (z_T T_{Xw} - x_T T_{Zw})w + T_{Mq} q$$

where x_T and z_T are the distances along the x and z axes from the aircraft's mass-center to the point of application of the propulsive thrust vector. For a propeller-driven aircraft, this point would be the propeller's hub.

Thus, upon taking note, again, that $q = \dot{\theta}$, one may obtain the complete dimensional longitudinal stability equations:

$$\left[(m - X_{du})\frac{d}{dt} - X_u - T_{Xu}\right]u - \left[X_{dw}\frac{d}{dt} + X_w + T_{Xw}\right]w - \left[X_{dq}\frac{d}{dt} + X_q + T_{Xq}\right]q$$
$$-\left[(B - mg)\cos\Theta_0\right]\theta = \Delta X_c$$

$$-\left[Z_{du}\frac{d}{dt} + Z_u + T_{Zu}\right]u + \left[(m - Z_{dw})\frac{d}{dt} - Z_w - T_{Zw}\right]w - \left[Z_{dq}\frac{d}{dt} + (mU_0 + Z_q + T_{Zq})\right]q$$
$$-\left[(B - mg)\sin\Theta_0\right]\theta = \Delta Z_c$$

$$-\left[M_{du}\frac{d}{dt} + M_u - x_T T_{Zu} + z_T T_{Xu}\right]u - \left[M_{dw}\frac{d}{dt} + M_w - x_T T_{Zw} + z_T T_{Xw}\right]w$$
$$+\left[(I_{yy} - M_{dq})\frac{d}{dt} - M_q - T_{Mq}\right]q + (x_B B \sin\Theta_0 - z_B B \cos\Theta_0)\theta = \Delta M_c$$

$$q - \dot{\theta} = 0$$

Flight Dynamics

These linear first-order equations may also be represented in matrix form:

$$[A]\begin{bmatrix} u \\ w \\ q \\ \theta \end{bmatrix} + [B]\begin{bmatrix} \dot{u} \\ \dot{w} \\ \dot{q} \\ \dot{\theta} \end{bmatrix} = \begin{bmatrix} \Delta X_c \\ \Delta Z_c \\ \Delta M_c \\ 0 \end{bmatrix}$$

where the components of matrices $[A]$ and $[B]$ are:

$a_{1,1} = -X_u - T_{Xu}, \quad a_{1,2} = -X_w - T_{Xw}, \quad a_{1,3} = -X_q - T_{Xq}, \quad a_{1,4} = (mg - B)\cos\Theta_0$

$a_{2,1} = -Z_u - T_{Zu}, \quad a_{2,2} = -Z_w - T_{Zw}, \quad a_{2,3} = -(mU_0 + Z_q + T_{Zq}), \quad a_{2,4} = (mg - B)\sin\Theta_0$

$a_{3,1} = -M_u + x_T T_{Zu} - z_T T_{Xu}, \quad a_{3,2} = -M_w - x_T T_{Zw} + z_t T_{Xw}, \quad a_{3,3} = -M_q - T_{Mq}$

$a_{3,4} = x_B B\sin\Theta_0 - z_B B\cos\Theta_0, \quad a_{4,1} = 0, \quad a_{4,2} = 0, \quad a_{4,3} = 1, \quad a_{4,4} = 0$

$b_{1,1} = m - X_{du}, \quad b_{1,2} = -X_{dw}, \quad b_{1,3} = -X_{dq}, \quad b_{1,4} = 0$

$b_{2,1} = -Z_{du}, \quad b_{2,2} = m - Z_{dw}, \quad b_{2,3} = -Z_{dq}, \quad b_{2,4} = 0$

$b_{3,1} = -M_{du}, \quad b_{3,2} = -M_{dw}, \quad b_{3,3} = I_{yy} - M_{dq}, \quad b_{3,4} = 0$

$b_{4,1} = 0, \quad b_{4,2} = 0, \quad b_{4,3} = 0, \quad b_{4,4} = -1$

NON-DIMENSIONAL FORM OF THE EQUATIONS

It has traditionally been convenient to non-dimensionalize the dynamic equations used for aircraft analysis. The reasons are:

1. The non-dimensional stability derivatives would generally be constant, or simple functions of angle-of-attack, throughout the aircraft's performance envelope.
2. The non-dimensional parameters (including stability derivatives) for one aircraft may be directly compared with those for other aircraft, even though these aircraft may differ greatly in size and flight regime (example: comparisons of C_L and C_D for wings).
3. The non-dimensional stability results for a given aircraft may be meaningfully compared with those for other aircraft (although, for a particular aircraft, dimensional results, such as $T_{1/2}$ may have the most meaning).

There are situations, though, where the use of the dimensional dynamic equations make the most sense, for example:

1. Aircraft with large rotors or propellers may have stability derivatives that vary greatly throughout their performance envelope, whether they're non-dimensionalized or not.
2. For special "one-off" aircraft, comparisons with other aircraft may be meaningless.
3. Pilot-in-the-loop analysis necessarily requires working with dimensional physical variables (time, in particular).

It should be noted, though, that whether one chooses to use the dimensional or non-dimensional dynamic equations, the dimensional stability derivatives are usually calculated from the non-dimensional values, methods for which are readily found in publications, such as the previously cited USAF Stability and Control DATCOM.

The non-dimensional terms are defined as follows:

force: $C_X, C_Z, \hat{B}, \hat{mg} \equiv \dfrac{X, Z, B, mg}{\rho V_{cm}^2 S/2}$ moment: $C_M \equiv \dfrac{M}{\rho V_{cm}^2 S\bar{c}/2}$

distance: $\hat{x}_B, \hat{z}_B, \hat{x}_T, \hat{z}_T \equiv \dfrac{x_B, z_B, x_T, z_T}{\bar{c}}$

velocity: $\hat{u}, \alpha \equiv \dfrac{u, w}{U_0}$ angular velocity: $\hat{q} \equiv \dfrac{q}{2U_0/\bar{c}}$

time derivatives: $D(\ ) \equiv \dfrac{d(\ )}{d\hat{t}} = \dfrac{\bar{c}}{2U_0}\dfrac{d(\ )}{dt}, \quad D^2(\ ) \equiv \dfrac{d^2(\ )}{d\hat{t}^2} = \dfrac{\bar{c}^2}{4U_0^2}\dfrac{d^2(\ )}{dt^2}$

(note that time is non-dimensionalized by $\bar{c}/(2U_0)$)

mass: $\mu \equiv \dfrac{2m}{\rho S \bar{c}}$ moment of inertia: $i_{yy} \equiv \dfrac{I_{yy}}{\rho S(\bar{c}/2)^3}$

where V_{cm} is the instantaneous velocity of the aircraft's mass-center (not necessarily the equilibrium velocity, U_0), and S & $\bar{c}$ are the defined reference area and length for the aircraft. For a conventional airplane design, this is usually the wing's planform area and mean aerodynamic chord, respectively.

Now, for the dimensional stability derivative X_u, one may draw upon the above definitions to obtain:

$$X_u = \left(\dfrac{\partial X}{\partial u}\right)_0 = \left[\dfrac{\partial}{\partial u}\left(\dfrac{C_X \rho V_{cm}^2 S}{2}\right)\right]_0 = \left(\dfrac{\partial C_X}{\partial u}\right)_0 \dfrac{\rho V_{cm}^2 S}{2} + \dfrac{(C_X)_0 \rho S}{2}\left(\dfrac{\partial V_{cm}^2}{\partial u}\right)_0$$

Further,

$$V_{cm}^2 = (U_o + u)^2 + v^2 + w^2 \approx U_0^2$$

where v is the lateral velocity component, along a y-axis that is perpendicular to the x- and z-axes. This will be important for the lateral dynamic equations to be discussed later in this chapter.

So,

$$\dfrac{\partial V_{cm}^2}{\partial u} = 2(U_0 + u) \rightarrow \left(\dfrac{\partial V_{cm}^2}{\partial u}\right)_0 = 2U_0$$

ASIDE

Recall that $(\ )_0$ denotes that the derivative is evaluated at the reference equilibrium state.

END ASIDE

Upon the definition of a non-dimensional stability derivative:

$$C_{Xu} \equiv \left(\dfrac{\partial C_X}{\partial \hat{u}}\right)_0 = \left(\dfrac{\partial C_X}{\partial u}\right)_0 U_0 \rightarrow \left(\dfrac{\partial C_X}{\partial u}\right)_0 = \dfrac{1}{U_0}\left(\dfrac{\partial C_X}{\partial \hat{u}}\right)_0$$

one obtains, from the above equations, that:

$$X_u = \left[C_{xu} + 2(C_X)_0\right]\rho U_0 S/2$$

Similarly, the other $(\ )_u$ stability derivatives are:

$$Z_u = \left[C_{Zu} + 2(C_Z)_0\right]\rho U_0 S/2$$

$$M_u = \left[C_{Mu} + 2(C_M)_0\right]\rho U_0 S\bar{c}/2$$

Flight Dynamics

where:

$$C_{Zu} \equiv \left(\frac{\partial C_Z}{\partial \hat{u}}\right)_0, \quad C_{Mu} \equiv \left(\frac{\partial C_M}{\partial \hat{u}}\right)_0$$

also, from the above equation for V_{cm}^2, one has that:

$$\left(\frac{\partial V_{cm}^2}{\partial v}\right)_0 = 0, \quad \left(\frac{\partial V_{cm}^2}{\partial w}\right)_0 = 0$$

so, the "w" derivatives may be directly obtained as:

$$X_w = C_{X\alpha}\rho U_0 S/2, \quad Z_w = C_{Z\alpha}\rho U_0 S/2, \quad M_w = C_{M\alpha}\rho U_0 S\bar{c}/2$$

where:

$$C_{X\alpha}, C_{Z\alpha}, C_{M\alpha} \equiv \left(\frac{\partial C_X}{\partial \hat{w}}\right)_0, \left(\frac{\partial C_Z}{\partial \hat{w}}\right)_0, \left(\frac{\partial C_M}{\partial \hat{w}}\right)_0$$

for the "q" derivatives, one has:

$$X_q = \left(\frac{\partial X}{\partial q}\right)_0 = \left[\frac{\partial}{\partial q}\left(\frac{C_X \rho V_{cm}^2 S}{2}\right)\right]_0 = \left(\frac{\partial C_X}{\partial q}\right)_0 \frac{\rho U_0^2 S}{2}$$

upon defining:

$$C_{Xq} \equiv \left(\frac{\partial C_X}{\partial \hat{q}}\right)_0$$

and when recalling the non-dimensional definition of q, one obtains:

$$X_q = C_{Xq}\rho U_0 S\bar{c}/4$$

In the same way the remaining "q" derivatives become:

$$Z_q = C_{Zq}\rho U_0 S\bar{c}/4, \quad M_q = C_{Mq}\rho U_0 S\bar{c}^2/4$$

where:

$$C_{Zq}, C_{Mq} \equiv \left(\frac{\partial C_Z}{\partial \hat{q}}\right)_0, \left(\frac{\partial C_M}{\partial \hat{q}}\right)_0$$

Next, the non-dimensional "acceleration derivatives" will be defined. These are derivatives of $\dot{u}$ and $\dot{w}$. The X_{du} expression is:

$$X_{du} = \left(\frac{\partial X}{\partial \dot{u}}\right)_0 = \left[\frac{\partial}{\partial \dot{u}}\left(\frac{C_X \rho V_{cm}^2 S}{2}\right)\right]_0 = \left(\frac{\partial C_X}{\partial \dot{u}}\right)_0 \frac{\rho U_0^2 S}{2}$$

Define:

$$C_{XDu} \equiv \left(\frac{\partial C_X}{\partial (D\hat{u})}\right)_0$$

so that, with the previous non-dimensional definitions of u and the time derivative, one obtains:

$$X_{du} = C_{XDu}\,\rho\,\bar{c}\,S/4$$

In a similar fashion, the other derivatives become:

$$Z_{du} = C_{ZDu}\,\rho\,\bar{c}\,S/4, \quad M_{du} = C_{MDu}\,\rho\,\bar{c}^2\,S/4$$

$$X_{dw} = C_{XD\alpha}\,\rho\,\bar{c}\,S/4, \quad Z_{dw} = C_{ZD\alpha}\,\rho\,\bar{c}\,S/4, \quad M_{dw} = C_{MD\alpha}\,\rho\,\bar{c}^2\,S/4$$

where:

$$C_{ZDu},\ C_{MDu} \equiv \left(\frac{\partial C_Z}{\partial(D\hat{u})}\right)_0,\ \left(\frac{\partial C_M}{\partial(D\hat{u})}\right)_0$$

$$C_{XD\alpha},\ C_{ZD\alpha},\ C_{MD\alpha} \equiv \left(\frac{\partial C_X}{\partial(D\hat{w})}\right)_0,\ \left(\frac{\partial C_Z}{\partial(D\hat{w})}\right)_0,\ \left(\frac{\partial C_M}{\partial(D\hat{w})}\right)_0$$

Next, one may obtain the $D\hat{q}$ derivative from the following expression:

$$X_{dq} \equiv \left(\frac{\partial X}{\partial \dot{q}}\right)_0 = \left[\frac{\partial}{\partial \dot{q}}\left(\frac{C_X \rho V_{cm}^2 S}{2}\right)\right]_0 = \left(\frac{\partial C_X}{\partial \dot{q}}\right)_0 \frac{\rho U_0^2 S}{2}$$

upon defining:

$$C_{XDq} \equiv \left(\frac{\partial C_X}{\partial(D\hat{q})}\right)_0 = \frac{4U_0^2}{\bar{c}^2}\left(\frac{\partial C_X}{\partial \dot{q}}\right)_0$$

one obtains:

$$X_{dq} = C_{XDq}\,\rho\,S\,\bar{c}^2/8$$

similarly,

$$Z_{dq} = C_{ZDq}\,\rho\,S\,\bar{c}^2/8, \quad M_{dq} = C_{MDq}\,\rho\,S\,\bar{c}^3/8$$

where:

$$C_{ZDq} \equiv \left(\frac{\partial C_Z}{\partial(D\hat{q})}\right)_0, \quad C_{MDq} \equiv \left(\frac{\partial C_M}{\partial(D\hat{q})}\right)_0$$

Finally, the thrust derivatives are expressed in a somewhat different form than the previous derivatives. For example, for T_{X_u}, one has:

$$T_X = \hat{T}_X \frac{\rho U_0^2 S}{2} \rightarrow T_{Xu} \equiv \left(\frac{\partial T_X}{\partial u}\right)_0 = \hat{T}_{Xu}\frac{\rho \dot{U}_0 S}{2}$$

Similarly, the other thrust-component terms are expressed as:

$$T_{Xu}, T_{Zu}, T_{Xw}, T_{Zw} = (\hat{T}_{Xu}, \hat{T}_{Zu}, \hat{T}_{X\alpha}, \hat{T}_{Z\alpha})\rho U_0 S/2$$

$$T_{Xq}, T_{Zq} = (\hat{T}_{Xq}, \hat{T}_{Zq})\rho U_0 S\bar{c}/4, \quad T_{Mq} = \hat{T}_{Mq}\,\rho U_0 S\bar{c}^2/4$$

Now, when these terms are substituted into the stability-axes dynamic equations, one may obtain the non-dimensional equations of motion. For example, the X-force equation becomes:

$$\left[\frac{\rho S \bar{c}}{4}(2\mu - C_{XDu})\left(\frac{2U_0}{\bar{c}}\right)D - \frac{\rho S U_0}{2}\left(C_{Xu} + 2(C_X)_0 + \hat{T}_{Xu}\right)\right]U_0 \hat{u}$$

$$-\left[\frac{\rho S \bar{c}}{4}C_{XD\alpha}\left(\frac{2U_0}{\bar{c}}\right)D + \frac{\rho S U_0}{2}\left(C_{X\alpha} + \hat{T}_{X\alpha}\right)\right]U_0 \alpha$$

$$-\left[\frac{\rho S \bar{c}^2}{8}C_{XDq}\left(\frac{2U_0}{\bar{c}}\right)D + \frac{\rho S U_0 \bar{c}}{4}\left(C_{Xq} + \hat{T}_{Xq}\right)\right]\frac{2U_0}{\bar{c}}\hat{q}$$

$$-\left[\frac{\rho U_0^2 S}{2}\left(\hat{B} - \hat{m}\hat{g}\right)\cos\Theta_0\right]\theta = \frac{\rho U_0^2 S}{2}\Delta C_{X_C} \rightarrow$$

$$\frac{\rho U_0^2 S}{2}\left[\left(2\mu - C_{XDu}\right)D - C_{Xu} - 2(C_X)_0 - \hat{T}_{Xu}\right]\hat{u} - \frac{\rho U_0^2 S}{2}\left(C_{XD\alpha}D + C_{X\alpha} + \hat{T}_{X\alpha}\right)\alpha$$

$$-\frac{\rho U_0^2 S}{2}\left(C_{XDq}D + C_{Xq} + \hat{T}_{Xq}\right)\hat{q} - \frac{\rho U_0^2 S}{2}\left[\left(\hat{B} - \hat{m}\hat{g}\right)\cos\Theta_0\right]\theta = \frac{\rho U_0^2 S}{2}\Delta C_{X_C}$$

As may be seen, the $\rho U_0^2 S/2$ term cancels out and one is left with the non-dimensional X-force equation. The same procedure may be used for the other equations, and one thus obtains the following set of first-order non-dimensional equations of motion:

$$\left[\left(2\mu - C_{XDu}\right)D - C_{Xu} - 2(C_X)_0 - \hat{T}_{Xu}\right]\hat{u} - \left(C_{XD\alpha}D + C_{X\alpha} + \hat{T}_{X\alpha}\right)\alpha$$

$$-\left(C_{XDq}D + C_{Xq} + \hat{T}_{Xq}\right)\hat{q} - \left(\hat{B} - \hat{m}\hat{g}\right)\cos\Theta_0 \theta$$

$$= \Delta C_{X_C} - \left[C_{ZDu}D + C_{Zu} + 2(C_Z)_0 + \hat{T}_{Zu}\right]\hat{u} + \left[\left(2\mu - C_{ZD\alpha}\right)D - C_{Z\alpha} - \hat{T}_{Z\alpha}\right]\alpha$$

$$-\left[C_{ZDq}D + \left(2\mu + C_{Zq} + \hat{T}_{Zq}\right)\right]\hat{q} - \left(\hat{B} - \hat{m}\hat{g}\right)\sin\Theta_0 \theta$$

$$= \Delta C_{Z_C} - \left[C_{MDu}D + C_{Mu} + 2(C_M)_0 - \hat{x}_T \hat{T}_{Zu} + \hat{z}_T \hat{T}_{Xu}\right]\hat{u} - \left(C_{MD\alpha}D + C_{M\alpha} - \hat{x}_T \hat{T}_{Z\alpha} + \hat{z}_T \hat{T}_{X\alpha}\right)\alpha$$

$$+\left[\left(i_{yy} - C_{MDq}\right)D - C_{Mq} - \hat{T}_{Mq}\right]\hat{q} + \left(\hat{x}_B \hat{B}\sin\Theta_0 - \hat{z}_B \hat{B}\cos\Theta_0\right)\theta = \Delta C_{M_C}$$

$$\hat{q} - D\theta = 0$$

As with the dimensional equations, these may be expressed in matrix form as:

$$\left[\hat{A}\right]\begin{bmatrix}\hat{u} \\ \alpha \\ \hat{q} \\ \theta\end{bmatrix} + \left[\hat{B}\right]\begin{bmatrix}D\hat{u} \\ D\alpha \\ D\hat{q} \\ D\theta\end{bmatrix} = \begin{bmatrix}\Delta C_{X_C} \\ \Delta C_{Z_C} \\ \Delta C_{M_C} \\ 0\end{bmatrix}$$

where the components of the $\left[\hat{A}\right]$ and $\left[\hat{B}\right]$ matrices are:

$\hat{a}_{1,1} = -C_{Xu} - 2(C_X)_0 - \hat{T}_{Xu}, \quad \hat{a}_{1,2} = -C_{X\alpha} - \hat{T}_{X\alpha}, \quad \hat{a}_{1,3} = -C_{Xq} - \hat{T}_{Xq}, \quad \hat{a}_{1,4} = -\left(\hat{B} - \hat{m}\hat{g}\right)\cos\Theta_0$

$\hat{a}_{2,1} - C_{Zu} - 2(C_Z)_0 - \hat{T}_{Zu}, \quad \hat{a}_{2,2} = -C_{Z\alpha} - \hat{T}_{Z\alpha}, \quad \hat{a}_{2,3} = -2\mu - C_{Zq} - \hat{T}_{Zq},$

$$\hat{a}_{2,4} = -\left(\hat{B} - \hat{m}\hat{g}\right)\sin\Theta_0$$
$$\hat{a}_{3,1} = -C_{Mu} - 2(C_M)_0 + \hat{x}_T\hat{T}_{Zu} - \hat{z}_T\hat{T}_{Xu}, \quad \hat{a}_{3,2} = -C_{M\alpha} + \hat{x}_T\hat{T}_{Z\alpha} - \hat{z}_T\hat{T}_{X\alpha}, \quad \hat{a}_{3,3} = -C_{Mq} - \hat{T}_{Mq},$$
$$\hat{a}_{3,4} = \hat{x}_B\hat{B}\sin\Theta_0 - \hat{z}_B\hat{B}\cos\Theta_0$$
$$\hat{a}_{4,1} = 0, \quad \hat{a}_{4,2} = 0, \quad \hat{a}_{4,3} = 1, \quad \hat{a}_{4,4} = 0$$
$$\hat{b}_{1,1} = 2\mu - C_{XDu}, \quad \hat{b}_{1,2} = -C_{XD\alpha}, \quad \hat{b}_{1,3} = -C_{XDq}, \quad \hat{b}_{1,4} = 0$$
$$\hat{b}_{2,1} - C_{ZDu}, \quad \hat{b}_{2,2} = 2\mu - C_{ZD\alpha}, \quad \hat{b}_{2,3} = -C_{ZDq}, \quad \hat{b}_{2,4} = 0$$
$$\hat{b}_{3,1} = -C_{MDu}, \quad \hat{b}_{3,2} = -C_{MD\alpha}, \quad \hat{b}_{3,3} = i_{yy} - C_{MDq}, \quad \hat{b}_{3,4} = 0$$
$$\hat{b}_{4,1} = 0, \quad \hat{b}_{4,2} = 0, \quad \hat{b}_{4,3} = 0, \quad \hat{b}_{4,4} = -1$$

ESTIMATION OF THE LONGITUDINAL STABILITY DERIVATIVES

The crux of an aircraft's flight-dynamic analysis is the determination of the stability derivatives. The described methodologies are generally based upon the individual contributions of the components, which are then added together, accounting for mutual interference factors. This is similar to what was described in Chapter 2 for the derivation of the aircraft's static lift, drag and pitching-moment equations.

Now, the way to envision the determination of the stability derivatives is to consider the aerodynamic forces and moments caused by the perturbation of only one of the motion variables, with all others held fixed. The first case below is for surging motion *only*, u, with α, q, $\dot{u}$, $\dot{\alpha}$, and $\dot{q}$ equal to zero.

The ( )$_0$ Terms

The value of these terms is dependent only on the aerodynamic configuration of the aircraft. Propeller contributions are accounted for by separate terms in the dynamic equations. The drag-coefficient equation remains the same, which gives:

$$(C_X)_0 = -(C_D)_{equil} \rightarrow \underline{\underline{(C_X)_0 = -\left(C_{D0} + C_1\alpha_{equil} + C_2\alpha_{equil}\right)}}$$

However, the lift-coefficient equation is that for the aircraft minus the propeller (the "glider" mode):

$$C'_L = C'_{L0} + C'_{L\alpha}\alpha$$

Thus,

$$(C_Z)_0 = -(C'_L)_{equil} \rightarrow \underline{\underline{(C_Z)_0 = -\left(C'_{L0} + C'_{L\alpha}\alpha_{equil}\right)}}$$

Also,

$$(C_M)_0 = \left(C'_{Mcm}\right)_{equil} = \left(C'_{M0}\right)_{np} + C'_{M\alpha}\alpha_{total}, \quad \text{where} \quad C'_{M\alpha} = C'_{L\alpha}\frac{\left(l_{cm} - l'_{np}\right)}{\bar{c}}, \quad \text{and}$$

$$\alpha_{total} = \alpha'_0 + \alpha_{equil}, \quad \text{where} \quad \alpha'_0 = C'_{L0}/C'_{L\alpha}$$

So,

$$\underline{\underline{(C_M)_0 = \left(C'_{M0}\right)_{np} + C'_{L\alpha}\frac{\left(l_{cm} - l'_{np}\right)}{\bar{c}}\left(\frac{C'_{L0}}{C'_{L\alpha}} + \alpha_{equil}\right)}}$$

Flight Dynamics

The $\hat{u}$ Derivatives

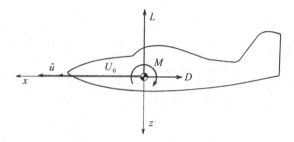

If the aircraft is disturbed by a u perturbation only, then:

$$X = -D, \quad Z = -L \to C_X = -C_D, \quad C_Z = -C_L$$

therefore,

$$C_{X_u} = -\left(\frac{\partial C_D}{\partial \hat{u}}\right)_0$$

The variation of C_D with speed is greatest at extremely low Reynolds Numbers, at the transition RN (from laminar to turbulent boundary layers) and transonic speeds. Thus,

$$\left(\frac{\partial C_D}{\partial \hat{u}}\right)_0 = \left(\frac{\partial C_D}{\partial (RN)}\right)_0 \left(\frac{\partial (RN)}{\partial \hat{u}}\right)_0 + \left(\frac{\partial C_D}{\partial \hat{M}}\right)_0 \left(\frac{\partial \hat{M}}{\partial \hat{u}}\right)_0 \to$$

where $\hat{M}$ is the Mach Number. Also, note that:

$$\left(\frac{\partial (RN)}{\partial \hat{u}}\right) = \frac{\partial (Const \times U)}{\partial \hat{u}} = \frac{\partial (Const \times (U_0 + U_0 \hat{u}))}{\partial \hat{u}} = Const \times U_0 = (RN)_0$$

$$\left(\frac{\partial \hat{M}}{\partial \hat{u}}\right)_0 = \frac{\partial (U/a)}{\partial \hat{u}} = \frac{\partial ((U_0 + U_0 \hat{u})/a)}{\partial \hat{u}} = \frac{\partial (U_0/a)}{\partial \hat{u}} = \hat{M}_0$$

where a is the speed of sound. Therefore, the expression for C_{X_u} is:

$$\underline{\underline{C_{X_u} = -\left(\frac{\partial C_D}{\partial (RN)}\right)_0 (RN)_0 - \left(\frac{\partial C_D}{\partial \hat{M}}\right)_0 \hat{M}_0}}$$

In a similar fashion, one obtains:

$$\underline{\underline{C_{Z_u} = -\left(\frac{\partial C_L}{\partial (RN)}\right)_0 (RN)_0 - \left(\frac{\partial C_L}{\partial \hat{M}}\right)_0 \hat{M}_0}}, \quad \underline{\underline{C_{M_u} = -\left(\frac{\partial C_M}{\partial (RN)}\right)_0 (RN)_0 - \left(\frac{\partial C_M}{\partial \hat{M}}\right)_0 \hat{M}_0}}$$

Observe that in most cases, $C_{X_u}, C_{Z_u}, C_{M_u} \approx 0$.

The α Derivatives

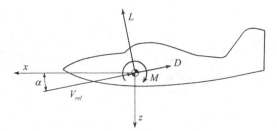

Again, these derivatives pertain to the aerodynamics of the aircraft without the propeller (the aircraft is a "glider" with the propeller removed).

If the aircraft is disturbed by an α perturbation due to heaving, w, one has:

$$X = -D\cos\alpha + L'\sin\alpha \rightarrow C_X \approx C_L'\alpha - C_D \rightarrow \frac{\partial C_X}{\partial \alpha} = C_L' + \alpha\frac{\partial C_L'}{\partial \alpha} - \frac{\partial C_D}{\partial \alpha}$$

For the reference flight condition, $(\partial C_X/\partial \alpha)_0$, this becomes:

$$\underline{\underline{C_{X\alpha} = C_{L0}' - (C_{D\alpha})_0}}$$

Observe that for those aircraft whose drag-polar equation may be represented by:

$$C_D = (C_D)_{C_L=0} + A'C_L' + B'(C_L')^2$$

then,

$$(C_{D\alpha})_0 = A'(C_{L\alpha}')_0 + 2B'C_{L0}'(C_{L\alpha}')_0$$

for $C_{Z\alpha}$, one similarly obtains:

$$Z = -L'\cos\alpha - D\sin\alpha \rightarrow C_Z \approx -C_L' - C_D\alpha \rightarrow \frac{\partial C_Z}{\partial \alpha} = -\frac{\partial C_L'}{\partial \alpha} - C_D - \alpha\frac{\partial C_D}{\partial \alpha}$$

For the reference flight condition, $(\partial C_Z/\partial \alpha)_0$, this becomes:

$$\underline{\underline{C_{Z\alpha} = -(C_{L\alpha}')_0 - C_{D0}}}$$

For $C_{M\alpha}$, one must first consider the C_M equation for the aircraft (with respect to the mass-center, cm). As discussed in Chapter 2, an aircraft generally has a longitudinal "neutral point" of location l_{np} and h_{np} with respect to the mass-center, cm, which gives

$$x_{np} = (l_{cm} - l_{np})\cos\alpha + (h_{cm} - h_{np})\sin\alpha$$

$$z_{np} = (h_{cm} - h_{np})\cos\alpha - (l_{cm} - l_{np})\sin\alpha$$

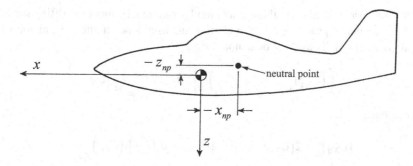

As likewise discussed in Chapter 2, the neutral point is analogous to the aerodynamic center on an airfoil. Namely, it's that point about which the pitching-moment coefficient stays constant with α (accurate for totally attached flow and relatively small angles). That constant-moment coefficient is denoted by C_{Mnp}, and thus one has that the pitching-moment coefficient about the mass-center is given by:

$$C_{M\,aero} = C_{X\,aero}\frac{z_{np}}{\bar{c}} - C_{Z\,aero}\frac{x_{np}}{\bar{c}} + C_{Mnp}$$

where $C_{X\,aero}$ and $C_{Z\,aero}$ are the aerodynamic forces acting through the neutral point.

Therefore,

$$C_{M\alpha} = C_{X\alpha}\frac{z_{np}}{\bar{c}} - C_{Z\alpha}\frac{x_{np}}{\bar{c}} - C_{D0}\frac{d(z_{np}/\bar{c})}{d\alpha} + C_{L0}\frac{d(x_{np}/\bar{c})}{d\alpha}$$

For most configurations, the neutral point stays fixed for small perturbations with respect to a reference position. Therefore, if $l_{np} = l'_{np}$ and $h_{np} = h'_{np}$, the equation becomes:

$$\underline{C_{M\alpha} = C_{X\alpha}\, z'_{np}/\bar{c} - C_{Z\alpha}\, x'_{np}/\bar{c}}$$

Also, for most aircraft configurations, $C_{X\alpha}\, z_{np}/\bar{c} \ll C_{Z\alpha}\, x_{np}/\bar{c}$, so that the equation simplifies to:

$$\underline{C_{M\alpha} \approx C_{L\alpha}\, x'_{np}/\bar{c}}$$

The $\hat{q}$ derivatives

Consider an *isolated lifting surface* (wing, stabilizer, etc.) as shown:

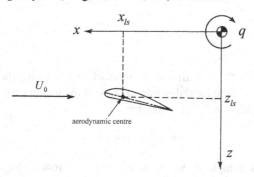

If it is subjected to a q perturbation about the cm, then the perturbation velocities at the surface's aerodynamic center are:

$$u_{ls} = q\, z_{ls} \rightarrow \hat{u}_{ls} = 2\hat{q}\, z_{ls}/\bar{c}, \quad w_{ls} = -q\, x_{ls} \rightarrow \alpha_{ls} = -2\hat{q}\, x_{ls}/\bar{c}$$

Further, note that besides $\hat{u}_{ls}$ and α_{ls}, there is an angular rotation, q, about the lifting surface's aerodynamic center, ac, which produces forces and moments itself since it alters the airfoil's boundary conditions along its chord. These will be denoted as:

$$\left[\left(C^*_{Xq}\right)_{ac}\right]_{ls} \hat{q}, \quad \left[\left(C^*_{Zq}\right)_{ac}\right]_{ls} \hat{q}, \quad \left[\left(C^*_{Mq}\right)_{ac}\right]_{ls} \hat{q}$$

Therefore, one has:

$$\left(C^*_{Xq}\right)_{ls} = 2\left(C^*_{Xu}\right)_{ls} z_{ls}/\bar{c} - 2\left(C^*_{X\alpha}\right)_{ls} x_{ls}/\bar{c} + \left[\left(C^*_{Xq}\right)_{ac}\right]_{ls}$$

$$\left(C^*_{Zq}\right)_{ls} = 2\left(C^*_{Zu}\right)_{ls} z_{ls}/\bar{c} - 2\left(C^*_{Z\alpha}\right)_{ls} x_{ls}/\bar{c} + \left[\left(C^*_{Zq}\right)_{ac}\right]_{ls}$$

The terms $\left(C^*_{Xu}\right)_{ls}$, $\left(C^*_{Z\alpha}\right)_{ls}$, etc. are determined in a similar fashion as the corresponding derivatives for the entire aircraft:

$$\left(C^*_{Xu}\right)_{ls} = -\left[\left(\frac{\partial\left(C^*_D\right)_{ls}}{\partial(RN)}\right)_0 (RN)_0 + \left(\frac{\partial\left(C^*_D\right)_{ls}}{\partial \hat{M}}\right)_0 \hat{M}_0\right]\frac{S_{ls}}{S}$$

$$\left(C^*_{Zu}\right)_{ls} = -\left[\left(\frac{\partial\left(C^*_L\right)_{ls}}{\partial(RN)}\right)_0 (RN)_0 + \left(\frac{\partial\left(C^*_L\right)_{ls}}{\partial \hat{M}}\right)_0 \hat{M}_0\right]\frac{S_{ls}}{S}$$

$$\left(C^*_{X\alpha}\right)_{ls} = \left[\left(C^*_L\right)_0 - \left(C^*_{D\alpha}\right)_0\right]_{ls}\frac{S_{ls}}{S}, \quad \left(C^*_{Z\alpha}\right)_{ls} = -\left[\left(C^*_{L\alpha}\right)_0 + \left(C^*_D\right)_0\right]_{ls}\frac{S_{ls}}{S}$$

where the $\left(C^*_L\right)_0$, $\left(C^*_{D\alpha}\right)_0$, $\left(C^*_{L\alpha}\right)_0$ and $\left(C^*_D\right)_0$ terms are those equilibrium-flight values for the isolated lifting surface (no effects from the rest of the aircraft).

Now, consider the *local flow* at the lifting surface as illustrated below:

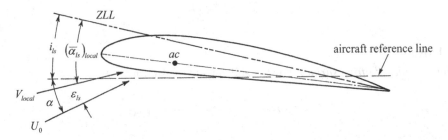

As defined in Chapter 2, α is the aircraft's angle of attack and i_{ls} is total the lifting surface's built-in incidence angle. Further, ε_{ls} is the local downwash angle at the lifting surface. Therefore, the lifting surface's total angle of attack, $\left(\bar{\alpha}_{ls}\right)_{local}$ is given by:

$$\left(\bar{\alpha}_{ls}\right)_{local} = \alpha + i_{ls} - \varepsilon_{ls}$$

Further, the local downwash angle, ε_{ls} is due not only to the aircraft's angle of attack, as described in Chapter 2, but can also be due to the pitch rate q for this dynamic case:

$$\varepsilon_{ls} = \left(\varepsilon_{ls}\right)_0 + \frac{\partial \varepsilon}{\partial \alpha}\alpha + \frac{\partial \varepsilon}{\partial \hat{q}}\hat{q}$$

Therefore, the lifting surface's perturbation angle of attack, α_{ls}, due to $\hat{q}$ motion alone, is given by:

$$\alpha_{ls} = \hat{q}\left[-2\frac{x_{ls}}{\bar{c}} - \left(\frac{\partial \varepsilon}{\partial \hat{q}}\right)_{ls}\right]$$

The lifting surface's aerodynamic behavior is also modified by being in conjunction with the rest of the aircraft (as described in the "Wing-Body Combination" section of Chapter 2). This is accounted for by an efficiency factor, η_{ls}. Further, the dynamic pressure at the lifting surface may differ somewhat from that for the free stream. This effect is accounted for by another efficiency factor $(\eta_q)_{ls}$, where

$$(\eta_q)_{ls} = q_{ls}/q$$

In summary then, upon ignoring the $\hat{u}$ contributions, the expressions for $\left(C_{Xq}\right)_{ls}$ and $\left(C_{Zq}\right)_{ls}$ becomes:

$$\left(C_{Xq}\right)_{ls} = -\left(C_{X\alpha}^*\right)_{ls}\left[2\frac{x_{ls}}{\bar{c}} + \left(\frac{\partial \varepsilon}{\partial \hat{q}}\right)_{ls}\right](\eta_q)_{ls}\,\eta_{ls} + \left[\left(C_{Xq}^*\right)_{ac}\right]_{ls}(\eta_q)_{ls}\,\eta_{ls}$$

$$\left(C_{Zq}\right)_{ls} = -\left(C_{Z\alpha}^*\right)_{ls}\left[2\frac{x_{ls}}{\bar{c}} + \left(\frac{\partial \varepsilon}{\partial \hat{q}}\right)_{ls}\right](\eta_q)_{ls}\,\eta_{ls} + \left[\left(C_{Zq}^*\right)_{ac}\right]_{ls}(\eta_q)_{ls}\,\eta_{ls}$$

ASIDE

It is clear that the downwash and efficiency effects are most pertinent to those lifting surfaces that are aft of larger lifting surfaces (such as the stabilizer on a tail-aft configuration). Also, in the author's experience, no published values of $\left(\partial \varepsilon / \partial \hat{q}\right)_{ls}$ are readily available. This effect may be negligible, but is included for completeness.

END ASIDE

Now, as discussed previously, the Reynolds Number and Mach Number can produce a significant u-derivative effect under certain circumstances:

$$\left(C_{Mu}^*\right)_{ls} \approx -\left[\left(\frac{\partial \left(C_M^*\right)_{ls}}{\partial (RN)}\right)_0 (RN)_0 + \left(\frac{\partial \left(C_M^*\right)_{ls}}{\partial \hat{M}}\right)_0 \hat{M}_0\right]\frac{S_{ls}}{S}$$

However, for most subsonic aircraft, this is considered negligible. Thus, the moment derivative is given by,

$$\left(C_{Mq}\right)_{ls} = \left(C_{Xq}\right)_{ls}\frac{z_{ls}}{\bar{c}} - \left(C_{Zq}\right)_{ls}\frac{x_{ls}}{\bar{c}} + \left[\left(C_{Mq}^*\right)_{ac}\right]_{ls}(\eta_q)_{ls}\,\eta_{ls}$$

where, as stated before, for most aircraft the $\hat{u}$ derivatives are assumed to be negligible.

For the evaluation of the $(\ )_{ac}$ terms, note first that, from linear airfoil theory, the effects of the different boundary conditions are additive. That is, the general airfoil problem may be subdivided into several distinct sub-problems:

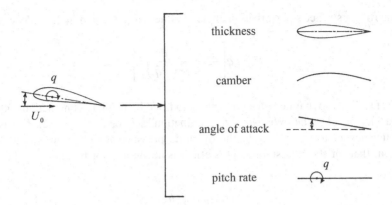

because only the q perturbation is being considered; attention is focused on the last sub-problem:

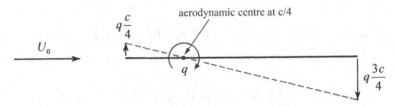

The pitch rate, q, gives a linearly varying normal (perpendicular) perturbation velocity along the length of the chord. To first order, this is mathematically equivalent to a static circular-arc airfoil at an angle of attack, α, as shown:

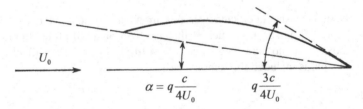

Also, from thin-airfoil theory, the Zero-Lift Line (*ZLL*) of a circular-arc airfoil passes through the camber's midpoint ($\alpha_0 = -\alpha_{ZLL}$):

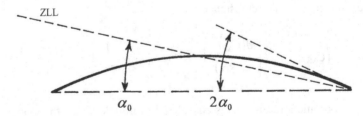

so, for the pitching airfoil,

$$\alpha_0 = \frac{1}{2}\left(q\frac{3c}{4U_0} - \alpha\right) = \frac{1}{2}\left(q\frac{3c}{4U_0} - q\frac{c}{4U_0}\right) = q\frac{c}{4U_0}$$

so, the total effective angle of attack is:

$$\bar{\alpha} = \alpha + \alpha_0 = q\frac{c}{2U_0} = \hat{q}\frac{c}{\bar{c}}$$

Therefore, the increment of lift due to the airfoil's pitch rate about its aerodynamic center is

$$\Delta C_L = C_{L\alpha}\,\bar{\alpha} = C_{L\alpha}\,\hat{q}\,c/\bar{c}$$

This result may be directly applied to rectangular-planform wings with no sweep angle to give the contributions to the entire aircraft:

$$\left[\left(C^*_{Xq}\right)_{ac}\right]_{ls} \approx 0, \quad \left[\left(C^*_{Zq}\right)_{ac}\right]_{ls} \approx \left(C^*_{Z\alpha}\right)_{ls}\,\bar{c}_{ls}/\bar{c}$$

If c_{ls} is replaced with the mean aerodynamic chord of the lifting surface, $\bar{c}_{ls}$, the above equations may be approximately applied to un-swept wings of arbitrary planform that are pitching about that planform's aerodynamic center.

The pitching moment due to q, $[(C_{Mq})_{ac}]_{ls}$, may be treated in the same fashion. Because this derivative is referenced to the aerodynamic center, it is independent of α and dependent only on the effective camber. From thin-airfoil theory, the $(C_M)_{ac}$ for a circular-arc airfoil is:

$$\left(C_M\right)_{ac} = -\frac{\pi}{2}\alpha_0, \quad \text{which gives}\left(C_M\right)_{ac} = -\frac{\pi}{4}\frac{c}{\bar{c}}\hat{q}$$

This may be directly applied to rectangular-planform wings with no taper to give the lifting surface's contribution to the entire aircraft:

$$\left[\left(C^*_{Mq}\right)_{ac}\right]_{ls} = -\frac{\pi}{4}\left(\frac{c_{ls}}{\bar{c}}\right)^2 \frac{S_{ls}}{S}$$

For swept wings with uniform taper, the USAF Stability and Control DATCOM gives:

$$\left[\left(C^*_{Mq}\right)_{ac}\right]_{ls} = -\frac{\pi}{4}\cos\Lambda\left[\frac{1}{3}\left(\frac{AR^3 \tan^2 \Lambda_{1/4}}{AR + 6\cos\Lambda_{1/4}}\right)+1\right]\left(\frac{\bar{c}_{ls}}{\bar{c}}\right)^2 \frac{S_{ls}}{S}$$

where $\Lambda_{1/4}$ is the sweep angle of the aerodynamic-center line.

ASIDE

Note that for the aircraft polar equation used for deriving the $\hat{u}$ and α derivatives, as well as the $\hat{q}$ derivatives, the propeller contribution has been omitted. These contributions are calculated separately, later in this Section.

END ASIDE

The $D\hat{u}$ Derivatives

The primary contribution to the $D\hat{u}$ derivatives is the "apparent-mass" effect. This concept is the mathematical consequence of accelerating a body in a fluid field. As the body accelerates, all the fluid particles in the field likewise accelerate. This increase in fluid-particle velocity means that the field is gaining kinetic energy, and thus a reaction force is acting on the body. It is important to note two things:

1. This applies even in potential fluid, and no viscosity is involved.
2. The "apparent mass" is not a mass but a mathematical consequence of this derivation. It acquired this name because the force necessary to accelerate the body through the fluid

field can be expressed as a constant times the acceleration. That constant has been referred to as the "apparent mass" or "virtual mass".
3. The "apparent mass" generally varies with the direction of the acceleration.

ASIDE

A further, and more complete, exposition on this topic may be found in **Appendix D: Apparent-Mass Effects**.

END ASIDE

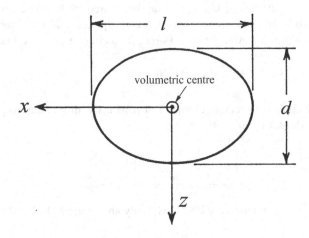

Lifting surfaces that are thin and have chords that are closely aligned with the x-direction have negligible $D\hat{u}$ contributions. However, this is not necessarily so for bodies. For example, an ellipsoid of revolution whose x-axis is aligned with the acceleration vector $\vec{a}_{cm}$ has an apparent mass in the x-direction of

$$\left(m_{app}\right)_{xx} = \rho k_1 \, Vol$$

where:
$\rho \equiv$ density of the air (or other surrounding fluid medium),
$k_1 \equiv$ apparent-mass coefficient, which is a function of the body's fineness ratio, $f = l/d$,
$Vol \equiv$ volume of the ellipsoid.

When $\left(m_{app}\right)_{xx}$ is non-dimensionalized in accordance with the previous work, one has:

$$\left(C_{XDu}\right)_{body} = \frac{-4 k_1 \rho \, Vol}{\rho S \bar{c}} = -\frac{4 k_1 \, Vol}{S \bar{c}}$$

also, for this situation, $\left(C_{ZDu}\right)_{body}$ and $\left(C_{MDu}\right)_{body}^{cv}$ are zero, where cv refers to the volumetric center.
When the body either has principal axes that are not aligned with $\vec{a}_{cm}$ or has a general shape, then there are non-zero values of $\left(C_{ZDu}\right)_{body}$ and $\left(C_{MDu}\right)_{body}^{cv}$. However, if the body is a streamlined fuselage

or hull (for the case of an airship) that is nearly aligned with $\vec{a}_{cm}$, then good approximate equations are given by:

$$\left(C_{XDu}\right)_{body} = -\frac{4k_1 Vol}{S\bar{c}}, \quad \left(C_{ZDu}\right)_{body} \approx 0, \quad \left(C_{MDu}\right)_{body} = \frac{z_{cv}}{\bar{c}}\left(C_{XDu}\right)_{body}$$

where z_{cv} is the z-coordinate distance from the mass-center, cm, to the body's volumetric center, cv:

$$z_{cv} = \left(h_{cm} - h_{cv}\right)\cos\alpha_{equil} - \left(l_{cm} - l_{cv}\right)\sin\alpha_{equil}$$

An equation to find k_1 is given by:

$$k_1 = \frac{\gamma}{(2-\gamma)}, \quad \text{where} \quad \gamma = 2\left(\frac{1-e^2}{e^3}\right)\left[\frac{1}{2}\ln\left(\frac{1+e}{1-e}\right)-e\right], \quad e = \left(\frac{f^2-1}{f^2}\right)^{1/2}$$

Recall that $f = l/d$ is the fineness ratio.

The $D\alpha$ Derivatives

In a similar manner as for the $D\hat{u}$ derivatives, the apparent mass contributes to the $D\hat{w}$ (denoted as $\dot{\alpha}$) derivatives. In this case,

$$\left(m_{app}\right)_{zz} = \rho k_3 Vol$$

where k_3 is the apparent-mass coefficient of the ellipsoid described before, with its acceleration in the z-direction. This non-dimensionalizes to give:

$$\left(C_{ZD\alpha}\right)_{body} = \frac{-4k_3\rho Vol}{\rho S\bar{c}} = -\frac{4k_3 Vol}{S\bar{c}}$$

Again, for arbitrarily shaped bodies, $\left(C_{XD\alpha}\right)_{body}^{cv}$ and $\left(C_{MD\alpha}\right)_{body}^{cv}$ are non-zero. However, if the body is a streamlined fuselage or hull whose major axis is nearly perpendicular with $\vec{a}_{cm}$, then $\left(C_{XD\alpha}\right)_{body}$ and $\left(C_{MD\alpha}\right)_{body}^{cv}$ are approximately equal to zero. Therefore, the body's contribution to the $D\alpha$ derivatives are:

$$\left(C_{ZD\alpha}\right)_{body} = -\frac{4k_3 Vol}{S\bar{c}}, \quad \left(C_{XD\alpha}\right)_{body} \approx 0, \quad \left(C_{MD\alpha}\right)_{body} = -\frac{x_{cv}}{\bar{c}}\left(C_{ZD\alpha}\right)_{body}$$

where x_{cv} is the x-coordinate distance from the mass-center, cm, to the volumetric center, cv:

$$x_{cv} = \left(l_{cm} - l_{cv}\right)\cos\alpha_{equil} + \left(h_{cm} - h_{cv}\right)\sin\alpha_{equil}$$

An equation to find k_3 is given by:

$$k_3 = \frac{\delta}{(2-\delta)}, \quad \text{where} \quad \delta = \frac{1}{e^2} - \frac{1-e^2}{2e^3}\ln\left(\frac{1+e}{1-e}\right), \quad e = \left(\frac{f^2-1}{f^2}\right)^{1/2}$$

As described in Appendix E, lifting surfaces (wings, stabilizers, canards, etc.), are a more complicated situation because the normal force caused by the plunging acceleration, $\dot{w}$, is due to:

1. The apparent-mass of the planform and
2. The influence of the shed vortex.

These two effects generally oppose with each other, as illustrated in the figure below:

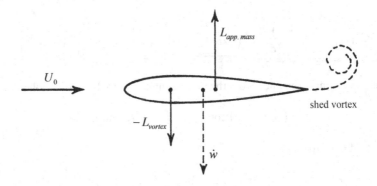

Thus,

$$Z_{ls} \approx -\left(L_{app} + L_{circ}\right) \rightarrow$$

$$\left(C_{ZD\alpha}\right)_{ls} D\alpha \rho S U_0^2/2 = -\left[\left(C_{LD\alpha}\right)_{app} + \left(C_{LD\alpha}\right)_{cir}\right]_{ls} (D\alpha)_{ls} \rho S_{ls} U_0^2/2 \rightarrow$$

$$\left(C_{ZD\alpha}\right)_{ls} D\alpha S = -\left[\left(C_{LD\alpha}\right)_{app} + \left(C_{LD\alpha}\right)_{cir}\right]_{ls} (D\alpha)_{ls} S_{ls} \rightarrow$$

$$\left(C_{ZD\alpha}\right)_{ls} \frac{\bar{c}}{2U_0} \dot{\alpha} S = -\left[\left(C_{LD\alpha}\right)_{app} + \left(C_{LD\alpha}\right)_{cir}\right]_{ls} \frac{(c_0)_{ls}}{2U_0} \dot{\alpha}_{ls} S_{ls}$$

Since $\dot{\alpha} = \dot{\alpha}_{ls}$, one obtains:

$$\left(C_{ZD\alpha}\right)_{ls} = -\left[\left(C_{LD\alpha}\right)_{app} + \left(C_{LD\alpha}\right)_{cir}\right]_{ls} \frac{(c_0)_{ls}}{\bar{c}} \frac{S_{ls}}{S} = -\left(C_{LD\alpha}\right)_{ls} \frac{(c_0)_{ls}}{\bar{c}} \frac{S_{ls}}{S}$$

where $(c_0)_{ls}$ is the root chord of the lifting surface.

ASIDE

Note that the derivations for above terms in the Appendix were based upon the assumption of an elliptical-planform wing, and that the characteristic wing chord for the derivation was the root chord of that wing, c_0. For application to other planforms, an approximation will be made that this criterion also applies. In other words, for rectangular, tapered, or multiple-tapered wings, that planform's root chord will be the characteristic chord for its $C_{LD\alpha}$ value.

END ASIDE

Observe that, in general,

$$\left(C_{XD\alpha}\right)_{ls} \approx 0$$

Flight Dynamics

Now, it is important to note that whereas $(C_{LD\alpha})_{cir}$ acts at the lifting surface's aerodynamic center, as previously defined, $(C_{L\dot\alpha})_{app\,mass}$ acts at the area center of the lifting surface, as given by $(x_{ca})_{ls}$ and $(z_{ca})_{ls}$:

$$(x_{ca})_{ls} = \left[l_{cm} - (l_{ca})_{ls}\right]\cos\alpha_{equil} + \left[h_{cm} - (h_{ca})_{ls}\right]\sin\alpha_{equil}$$

$$(z_{ca})_{ls} = \left[h_{cm} - (h_{ca})_{ls}\right]\cos\alpha_{equil} - \left[l_{cm} - (l_{ca})_{ca}\right]\sin\alpha_{equil}$$

Further, define:

$$\left[(C_{ZD\alpha})_{app}\right]_{ls} \equiv -\left[(C_{LD\alpha})_{app}\right]_{ls}\frac{(c_0)_{ls}}{\bar c}\frac{S_{ls}}{S},\quad \left[(C_{ZD\alpha})_{cir}\right]_{ls} \equiv -\left[(C_{LD\alpha})_{cir}\right]_{ls}\frac{(c_0)_{ls}}{\bar c}\frac{S_{ls}}{S}$$

Therefore, the pitching-moment equation is:

$$(C_{MD\alpha})_{ls} = -\frac{(x_{ca})_{ls}}{\bar c}\left[(C_{ZD\alpha})_{app}\right]_{ls} - \frac{x_{ls}}{\bar c}\left[(C_{ZD\alpha})_{cir}\right]_{ls}$$

Values for $\left[(C_{LD\alpha})_{app}\right]_{ls}, \left[(C_{LD\alpha})_{cir}\right]_{ls}$ and $(C_{LD\alpha})_{ls}$ may be obtained from the Appendix. Below are curve-fit equations to the results:

$$\left[(C_{LD\alpha})_{app}\right]_{ls} = 0.8905 + 0.8297\,AR - 0.1638\,AR^2 + 0.0158\,AR^3 - 0.00073\,AR^4 + 0.000013\,AR^5$$

$$\left[(C_{LD\alpha})_{cir}\right]_{ls} = 0.7749 - 0.9488\,AR - 0.0810\,AR^2 + 0.0155\,AR^3 - 0.00085\,AR^4 + 0.000016\,AR^5$$

$$(C_{LD\alpha})_{ls} = 1.6654 - 0.1192\,AR - 0.2447\,AR^2 + 0.0313\,AR^3 - 0.0016\,AR^4 + 0.00003\,AR^5$$

Finally there is an effect, for tail-aft designs, due to "downwash lag".

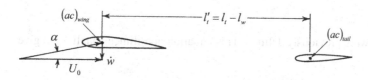

When α changes, the wing's downwash angle, ε, changes. However, the tail isn't instantaneously influenced by this because it takes a finite time, Δt, for the effect to travel downstream from the wing's aerodynamic center to the tail's aerodynamic center. Note that $\Delta t = l'_t/U_0$, so that the change of ε at the tail's ac, at time t, corresponds to that for the wing's α at time $(t - \Delta t)$. The downwash at the tail is thus:

$$\varepsilon_{tail} = \frac{\partial\varepsilon}{\partial\alpha}\frac{d\alpha}{dt}(t - \Delta t) = \varepsilon(\alpha) + \varepsilon(\dot\alpha),\quad\text{where}\quad \varepsilon(\dot\alpha) = -\frac{\partial\varepsilon}{\partial\alpha}\dot\alpha\,\Delta t \to -\frac{\partial\varepsilon}{\partial\alpha}\frac{l'_t}{U_0}\dot\alpha$$

Now, a negative downwash angle corresponds to a positive angle-of-attack at the tail, so one obtains:

$$(\Delta C_Z)_t \approx -(\Delta C_L)_t = -(C_{L\alpha})_t\,\varepsilon(\dot\alpha)\frac{S_t}{S} = -(C_{L\alpha})_t\frac{\partial\varepsilon}{\partial\alpha}\frac{l'_t}{U_0}\frac{S_t}{S}\dot\alpha =$$

$$(\Delta C_Z)_t = -(C_{L\alpha})_t\frac{\partial\varepsilon}{\partial\alpha}\frac{l'_t}{U_0}\frac{S_t}{S}\left(\frac{2U_0}{\bar c}\right)D\alpha$$

Note that the tail's efficiency factor is already incorporated in $(C_{L\alpha})_t$. Therefore, upon accounting for the efficiency factors, one obtains that:

$$\left(C_{ZD\alpha}\right)_t^{lag} = -2\left(C_{LD\alpha}^*\right)_t \frac{\partial \varepsilon}{\partial \alpha} \frac{l_t'}{\bar{c}} \frac{S_t}{S}(\eta_q)_t \eta_t$$

In a similar fashion, one finds:

$$(\Delta C_M)_t = -\frac{x_t}{\bar{c}}(\Delta C_Z)_t \rightarrow \left(C_{MD\alpha}\right)_t^{lag} = -\frac{x_t}{\bar{c}}\left(C_{ZD\alpha}\right)_t^{lag}$$

The $D\hat{q}$ Derivatives

With reference to the ellipsoid again, there is an apparent moment of inertia, $\left(I_{yy}\right)_{app}$, about the body's volumetric center:

$$\left(I_{yy}\right)_{app} = \rho k_2' \frac{\left(l^2 + d^2\right)}{20} Vol$$

where k_2' is a coefficient given by

$$k_2' = \frac{e^4(\delta - \gamma)}{(2-e^2)\left[2e^2 - (2-e^2)(\delta - \gamma)\right]}$$

and e, γ and δ were previously defined. This equation non-dimensionalizes to give

$$\left(C_{MDq}\right)_{body}^{cv} = -\frac{2k_2'\left(l^2 + d^2\right) Vol}{5 S \bar{c}^3}$$

Further, for generally configured bodies, $\left(C_{XDq}\right)_{body}^{cv}$ and $\left(C_{ZDq}\right)_{body}^{cv}$ are non-zero. However, for most axisymmetric bodies these terms are negligible. Therefore, one has:

$$\left(C_{XDq}\right)_{body} = 2\frac{z_{cv}}{\bar{c}}\left(C_{XDu}\right)_{body}, \quad \left(C_{ZDq}\right)_{body} = -2\frac{x_{cv}}{\bar{c}}\left(C_{ZD\alpha}\right)_{body}$$

$$\left(C_{MDq}\right)_{body} = \frac{z_{cv}}{\bar{c}}\left(C_{XDq}\right)_{body} - \frac{x_{cv}}{\bar{c}}\left(C_{ZDq}\right)_{body} + \left(C_{MDq}\right)_{body}^{cv}$$

> **ASIDE**
>
> Very few aircraft bodies are ellipsoids of revolution, even for airships. Therefore, in order to apply these equations to a fuselage (or airship hull), an "equivalent ellipsoid of revolution" is defined, the criteria being that both the ellipsoid and body have equal volumes, $(Vol)_{body} = (Vol)_{ellipsoid}$ and equal cross-sectional areas, S_{max}. Therefore, the length of the ellipsoid will generally differ from the body's total length and is given by:
>
> $$l = L_{ellipsoid} = 1.5(Vol)_{ellipsoid}/S_{max}$$
>
> Further, the diameter of the equivalent ellipsoid is given by:
>
> $$d \equiv (dia)_{max} = (4S_{max}/\pi)^{1/2}$$
>
> The numerical example will show how this is applied, in addition to an extension to account for bodies of unequal maximum depth and height.
>
> **END ASIDE**

The author is not aware of any analyses for a lifting surface that accounts for such terms as $\left[(C_{XDq})_{ac}\right]_{ls}$, $\left[(C_{ZDq})_{ac}\right]_{ls}$ and $\left[(C_{MDq})_{ac}\right]_{ls}$. Therefore, for this methodology, these will be assumed to be generally negligible. One thus obtains the following equations for the lifting-surface's contribution:

$$(C_{XDq})_{ls} \approx 0, \quad (C_{ZDq})_{ls} = -2\frac{(x_{ca})_{ls}}{\bar{c}}\left[(C_{ZD\alpha})_{app}\right]_{ls} - 2\frac{x_{ls}}{\bar{c}}\left[(C_{ZD\alpha})_{circ}\right]_{ls}$$

$$(C_{MDq})_{ls} = -\frac{(x_{ca})_{ls}}{\bar{c}}\left[(C_{ZDq})_{app}\right]_{ls} - \frac{x_{ls}}{\bar{c}}\left[(C_{ZDq})_{circ}\right]_{ls}$$

The Propulsion Derivatives

When the propulsive thrust vector is approximately aligned with U_0, then $\hat{T}_{Zu} \approx 0$ and T_{Xu} is obtained as follows:

a. Gliding flight: $T = 0$, so $\hat{T}_{Xu} = 0$.
b. Jet and rocket propulsion: the thrust variation with speed is usually negligible, so $\hat{T}_{Xu} \approx 0$.
c. Propeller-driven aircraft with fixed throttle: constant thrust-power output is assumed during the perturbation motion, so that:

$$T(U_0 + u) = T_0 U_0 = \text{const.} \rightarrow T = \frac{T_0 U_0}{(U_0 + u)} \rightarrow \left(\frac{\partial T}{\partial u}\right) = \frac{-T_0 U_o}{(U_0 + u)^2} \rightarrow \left(\frac{\partial T}{\partial u}\right)_0 = -\frac{T_0}{U_0}$$

This non-dimensionalizes to give:

$$\hat{T}_{Xu} = \frac{-2T_0}{\rho S U_0^2} \equiv -\hat{T}_0$$

also,

$$\hat{T}_{Zu} \approx 0$$

For the propulsive α derivatives, recall the propeller side-force estimation in the "Complete Aircraft Aerodynamics" section of Chapter 2.

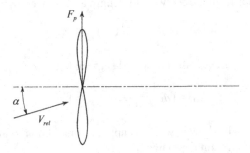

From this, one obtains that

$$F_{prop} = F_{P\alpha}\alpha \approx \frac{\rho V_{rel}^2}{2} S_{prop}(C_{L\alpha})_{prop} \alpha \approx \frac{\rho U_0^2}{2} S_p(C_{L\alpha})_{prop} \alpha \quad \text{when} \quad \alpha \ll 1$$

Therefore,

$$T_Z \approx -F_{P\alpha}\alpha \rightarrow \hat{T}_{Z\alpha} = \frac{-2F_{P\alpha}}{\rho U_0^2 S} = -(C_{L\alpha})_{prop}\frac{S_{prop}}{S}$$

Next, one may upon the previous derivation of $C_{T\alpha}$ in the "Pitching Moment (with non-extended undercarriage)" subsection of the "Complete Aircraft Aerodynamics (for a tail-aft monoplane)" section of Chapter 2 to obtain:

$$C_{T\alpha} = \frac{dC_T}{d\alpha} = \frac{3}{2} K_p \eta_p C_L^{1/2} C_{L\alpha}, \quad \text{where} \quad K_p \eta_p \equiv \frac{2\rho^{1/2} P \eta_p}{(2W/S)^{3/2} \tilde{S}}$$

Thus

$$\hat{T}_{X\alpha} = C_{T\alpha} \tilde{S}/S = C_{T\alpha}(prop.dia.)^2/S$$

Next, the pitching terms may be written as

$$\hat{T}_{Xq} = 2\hat{T}_{Xu}\hat{z}_T - 2\hat{T}_{X\alpha}\hat{x}_T \quad \text{and} \quad \hat{T}_{Zq} = 2\hat{T}_{Zu}\hat{z}_T - 2\hat{T}_{Z\alpha}\hat{x}_T$$

However, as seen above, $\hat{T}_{X\alpha}$ is a function of the plunging of the entire aircraft and not the localized motion due to pitching. Therefore, these equations may be written as:

$$\hat{T}_{Xq} = 2\hat{T}_{Xu}\hat{z}_T, \quad \hat{T}_{Zq} = -2\hat{T}_{Z\alpha}\hat{x}_T$$

Further, the moment contribution due to pitching is given by

$$\hat{T}_{Mq} = \hat{T}_{Xq}\hat{z}_T - \hat{T}_{Zq}\hat{x}_T \rightarrow \hat{T}_{Mq} = 2\hat{T}_{Xu}\hat{z}_T^2 + 2\hat{T}_{Z\alpha}\hat{x}_T^2$$

Flight Dynamics

where:

$$\hat{x}_T = \left[(l_{cm} - l_T)\cos\alpha_{equil} + (h_{cm} - h_T)\sin\alpha_{equil} \right]/\bar{c}$$

$$\hat{z}_T = \left[(h_{cm} - h_T)\cos\alpha_{equil} - (l_{cm} - l_T)\sin\alpha_{equil} \right]/\bar{c}$$

The application of these equations is demonstrated with the following numerical example:

LONGITUDINAL NUMERICAL EXAMPLE ("SCHOLAR" TAIL-AFT MONOPLANE)

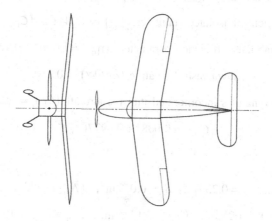

"Scholar", Tail-Aft Monoplane

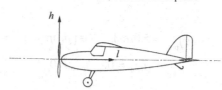

The geometric, physical and aerodynamic properties of this example airplane are presented in Chapters 2 and 7, and are summarized below:

Complete aircraft:

$$b = 2.0\,\text{m}, \quad \bar{c} = 0.30\,\text{m}, \quad S = 0.556\,\text{m}^2, \quad l_{cm} = 0.497\,\text{m}, \quad h_{cm} = 0.03$$

total length of airplane (excluding propeller): $l = 1.462\,\text{m}$

complete aerodynamics: $C_L = 0.418 + 4.842\alpha$, $C_D = 0.0509 + 0.2001\alpha + 1.3439\alpha^2$

complete aerodynamics without propeller: $C'_L = 0.418 + 4.788\alpha$

Wing:

$$b_w = 2.0\,\text{m}, \quad \bar{c}_w = 0.30\,\text{m}, \quad S_w = 0.556\,\text{m}^2, \quad (\Lambda_{c/4})_w = 8^0, \quad l_w = 0.494\,\text{m}, \quad h_w = 0.156\,\text{m}$$

$$(l_{ca})_w = 0.559\,\text{m}, \quad (h_{ca})_w = 0.16\,\text{m}, \quad (c_0)_w \equiv c_{root} = 0.30\,\text{m}, \quad \Gamma_w = 5^0$$

lift coefficient of isolated wing: $(C_L^*)_w = 0.518 + 4.636\alpha$, $(C_{L\alpha}^*) = 4.636$

wing efficiency parameters: $(\eta_q)_w \approx 1.0$, $\eta_w = 0.877$

wing aerodynamics incorporated into the aircraft: $(C_L)_w = 0.454 + 4.066\alpha$

$(C_D)_w = 0.0243 + 0.2196\alpha + 1.1490\alpha^2$, $(C_D)_w = 0.0141 + 0.0091(C_L)_w + 0.0695(C_L)_w^2$

Stabilizer (horizontal tail):

$b_s = 0.75\,\text{m}$, $\bar{c}_s = 0.20\,\text{m}$, $S_s = 0.140\,\text{m}^2$, $(\Lambda_{c/4})_s = 0$, $l_s = 1.30\,\text{m}$, $h_s = 0$

$(l_{ca})_s = 1.350\,\text{m}$, $(h_{ca})_s = 0$, $(c_0)_s \equiv c_{root} = 0.20\,\text{m}$, $l'_s = 0.806\,\text{m}$

lift coefficient of isolated stabilizer: $(C_L^*)_s = 4.244\alpha$, $(C_{L\alpha}^*)_s = 4.244$,

stabilizer efficiency parameters: $(\eta_q)_s \approx 1.0$, $\eta_s = 0.96$,

downwash derivative: $(\partial\varepsilon/\partial\alpha)_s = 0.310$

stabilizer aerodynamics incorporated into the aircraft: $(C_L)_s = -0.141 + 2.815\alpha$

$(C_D)_s = 0.0081 + 0.0977(C_L)_s^2$

Vertical fin:

$b_f = 0.261\,\text{m}$, $\bar{c}_f = 0.233\,\text{m}$, $S_f = 0.057\,\text{m}^2$, $(\Lambda_{c/2})_f = 12°$, $l_f = 1.225\,\text{m}$

$h_f = 0.110\,\text{m}$, $(l_{ca})_f = 1.283\,\text{m}$, $(h_{ca})_f = 0.110\,\text{m}$, $(c_0)_f \equiv c_{root} = 0.275\,\text{m}$, $\eta_f = 0.882$

Propeller:

$S_{prop} \approx 0.006\,\text{m}^2$, $(C_{L\alpha})_{prop} \approx 5.0/\text{rad}$, $\tilde{S} \equiv (prop.\ dia.)^2 = (0.37)^2 = 0.1369\,\text{m}^2$,

$l_{prop} = 0$, $h_{prop} = 0$

Body:

$c_b = 1.40\,\text{m}$, $S_b = 0.04175\,\text{m}^2$ (max. cross-sectional area), $(C_{M\alpha})_{body} = 0.470$

$(Vol)_{body} = 0.0352\,\text{m}^3$, $l_{cv} = 0.60\,\text{m}$, $h_{cv} = 0.0$

Equilibrium-flight conditions:

$V = 14.458\,\text{m/sec}$, $\alpha = 0.01897\,\text{rad}$ (1.0871°), $\Theta_0 = 0.001248\,\text{rad}$ (0.0715°)

$T_0 = 3.974\,\text{N}$, $m = 3.70\,\text{kg}$

Example Values of $(C_X)_0$, $(C_Z)_0$ and $(C_M)_0$

For an equilibrium angle of attack, $\alpha_{equil} = 0.01897\,\text{rad}$, one obtains:

$(C_X)_0 = -(C_D)_{equil} = -0.0509 - 0.2001 \times 0.01897 - 1.3439 \times 0.01897^2 \to \underline{\underline{(C_X)_0 = -0.0552}}$

From the "Example, Tail-Aft Monoplane" section of Chapter 2, it was found that the no-propeller lift coefficient is:

$$C'_L = 0.418 + 4.788\alpha$$

Flight Dynamics

which gives that:

$$(C_Z)_0 = -(C'_L)_{equil} = -0.418 - 4.788 \times 0.01897 \rightarrow (C_Z)_0 = -0.5088$$

Also, from the Section, $l'_{np} = 0.603\,\text{m}$ and $(C'_{M0})_{np} = 0.1770$, therefore one obtains:

$$(C'_{M0})_{np} = 0.1770 + 4.788 \times \frac{(0.497 - 0.603)}{0.30} \times \left(\frac{0.418}{4.788} + 0.01897\right) \rightarrow (C_M)_0 = -0.0028$$

> **ASIDE**
>
> For most conventionally configured airplanes, the trim condition is such that the aerodynamic moment about the mass-center is essentially zero. An exception is when the thrust vector provides a significant moment contribution, in which case the values of l_{np}, $C_{M\alpha}$ and C_{M0} will vary with different equilibrium flight conditions. However, as seen below, their values remain relatively close to those for the "glider" configuration calculated in the "Example, Tail-Aft Monoplane" section of Chapter 2.
>
> Another case that will give a non-zero value for $(C_M)_0$ is when the magnitude of the buoyancy vector is significant relative to the weight vector and doesn't pass through the mass-center. This would be a possible situation for airships.
>
> **END ASIDE**

Example Values of $C_{X\alpha}$, $C_{Z\alpha}$ and $C_{M\alpha}$

The equilibrium values of C_L, $C_{D\alpha}$, $C_{L\alpha}$ and C_D are given by:

$$(C'_L)_0 = 0.418 + 4.788 \times 0.01897 = 0.5088$$

$$(C_{D\alpha})_0 = 0.2001 + 2 \times 1.3439 \times 0.01897 = 0.2511$$

$$(C'_{L\alpha})_0 = 4.788, \quad (C_D)_0 = 0.0509 + 0.2001 \times 0.01897 + 1.3439 \times (0.01897)^2 = 0.05518$$

Therefore,

$$C_{X\alpha} = (C'_L)_0 - (C_{D\alpha})_0 = 0.5088 - 0.2511 \rightarrow C_{X\alpha} = 0.2577$$

$$C_{Z\alpha} = -(C'_{L\alpha})_0 - (C_D)_0 = -4.788 - 0.05518 \rightarrow C_{Z\alpha} = -4.8432$$

Now, the x and z locations of the neutral point relative to the mass-center are given by:

$$x'_{np} = (0.497 - 0.603) \times \cos(1.0871°) + (0.03 - 0.08) \times \sin(1.0871°) \rightarrow x'_{np} = -0.1069\,\text{m}$$

$$z'_{np} = (0.03 - 0.08) \times \cos(1.0871°) - (0.497 - 0.603) \times \sin(1.0871°) \rightarrow z'_{np} = -0.0480\,\text{m}$$

Note that information on the vertical location of h_{np} is sparse; and it seems sufficient, for small angles of attack, to choose a vertical distance midway between the aerodynamic center of the wing and that for the stabilizer. One thus obtains:

$$C_{M\alpha} = C_{X\alpha} \frac{z'_{np}}{\bar{c}} - C_{Z\alpha} \frac{x'_{np}}{\bar{c}} = 0.2577 \times \frac{-0.0480}{0.30} + 4.8432 \times \frac{-0.1069}{0.30} \rightarrow C_{M\alpha} = -1.7670$$

Note that an approximate value, stated earlier, is given by:

$$C_{M\alpha} \approx -C'_{L\alpha}(l'_{np} - l_{cm})/\bar{c} = -4.788 \times (0.603 - 0.497)/0.30 \to C_{M\alpha} \approx -1.6918$$

It's seen that this is a reasonable approximation, within 5%.

Example values of C_{Xq}, C_{Zq} and C_{Mq}

Wing:

$$\left[(C_L^*)_0\right]_w = 0.518 + 4.636 \times 0.01897 = 0.6059$$

$$\left[(C_{L\alpha}^*)_0\right]_w = 4.636$$

The wing's drag coefficient for this value of lift coefficient is found from:

$$(C_D^*)_w = 0.0141 - 0.0091(C_L^*)_w + 0.0695(C_L^*)_w^2$$

Thus, one obtains:

$$(C_D^*)_w = 0.0141 - 0.0091 \times (0.518 + 4.636\alpha) + 0.0695 \times (0.518 + 4.636\alpha)^2 \to$$

$$(C_D^*)_w = 0.0280 + 0.2916\alpha + 1.4937\alpha^2$$

so that,

$$\left[(C_D^*)_0\right]_w = 0.0280 + 0.2916 \times 0.01897 + 1.4937 \times (0.01897)^2 = 0.03407$$

$$\left[(C_{D\alpha}^*)_0\right]_w = 0.2916 + 2 \times 1.4937 \times 0.01897 = 0.3483$$

Therefore,

$$(C_{X\alpha}^*)_w = \left\{\left[(C_L^*)_0\right]_w - \left[(C_{D\alpha}^*)_0\right]_w\right\}\frac{S_w}{S} = (0.6059 - 0.3483)\frac{0.556}{0.556} \to (C_{X\alpha}^*)_w = 0.2576$$

$$(C_{Z\alpha}^*)_w = -\left\{\left[(C_{L\alpha}^*)_0\right]_w + \left[(C_D^*)_0\right]_w\right\}\frac{S_w}{S} = -(4.636 + 0.03407)\frac{0.556}{0.556} \to (C_{Z\alpha}^*)_w = -4.6701$$

$$\left[(C_{Zq}^*)_{ac}\right]_w \approx (C_{Z\alpha}^*)_w \bar{c}_w/\bar{c} = -4.6701 \times 0.30/0.30 \to \left[(C_{Zq}^*)_{ac}\right]_w = -4.6701$$

$$x_w = (l_{cm} - l_w)\cos\alpha_{equil} + (h_{cm} - h_w)\sin\alpha_{equil} \to$$

$$x_w = (0.497 - 0.494) \times \cos(1.0871°) + (0.03 - 0.156) \times \sin(1.0871°) \to x_w = 0.0006\,\text{m}$$

Thus, upon assuming that $(\partial \varepsilon / \partial \hat{q})_w \approx 0$, one obtains:

$$\left(C_{Xq}\right)_w = -2\left(C_{X\alpha}^*\right)_w \frac{x_w}{\bar{c}} (\eta_q)_w \eta_w = -2 \times (0.2576) \times \frac{0.0006}{0.30} \times 1.0 \times 0.877 \rightarrow$$

$$\left(C_{Xq}\right)_w = -0.0009$$

$$\left(C_{Zq}\right)_w = \left\{-2\left(C_{Z\alpha}^*\right)_w \frac{x_w}{\bar{c}} + \left[\left(C_{Zq}^*\right)_{ac}\right]_w\right\}(\eta_q)_w \eta_w$$

$$\left(C_{Zq}\right)_w = \left\{-2 \times (-4.6701) \times \frac{0.0006}{0.30} - 4.6701\right\} \times 1.0 \times 0.877 \rightarrow$$

$$\left(C_{Zq}\right)_w = -4.0793$$

Stabilizer:

$$\left[\left(C_L^*\right)_0\right]_s = 4.244 \times 0.01897 = 0.08051$$

$$\left[\left(C_{L\alpha}\right)_0\right]_s = 4.244$$

The stabilizer's drag coefficient for this value of lift coefficient is found from:

$$\left(C_D^*\right)_s = 0.0081 + 0.0977\left(C_L^*\right)_s^2$$

Thus, one obtains:

$$\left(C_D^*\right)_s = 0.0081 + 0.0977 \times (4.244\alpha)^2 \rightarrow \left(C_D^*\right)_s = 0.0081 + 1.7597\alpha^2$$

so that,

$$\left[\left(C_D^*\right)_0\right]_s = 0.0081 + 1.7597 \times (0.01897)^2 = 0.0087$$

$$\left[\left(C_{D\alpha}^*\right)_0\right]_s = 2 \times 1.7597 \times 0.01897 = 0.0668$$

Therefore,

$$\left(C_{X\alpha}^*\right)_s = \left\{\left[\left(C_L^*\right)_0\right]_s - \left[\left(C_{D\alpha}^*\right)_0\right]_s\right\} \frac{S_s}{S} = (0.08051 - 0.0668)\frac{0.140}{0.556} \rightarrow \left(C_{X\alpha}^*\right)_s = 0.00345$$

$$\left(C_{Z\alpha}^*\right)_s = -\left\{\left[\left(C_{L\alpha}^*\right)_0\right]_s + \left[\left(C_D^*\right)_0\right]_s\right\} \frac{S_s}{S} = -(4.244 + 0.0087)\frac{0.140}{0.556} \rightarrow \left(C_{Z\alpha}^*\right)_s = -1.0708$$

$$\left[\left(C_{Zq}^*\right)_{ac}\right]_s \approx \left(C_{Z\alpha}^*\right)_s \bar{c}_s / \bar{c} = -1.0708 \times 0.2/0.3 \rightarrow \left[\left(C_{Zq}^*\right)_{ac}\right]_s = -0.7139$$

$$x_s = (l_{cm} - l_s)\cos\alpha_{equil} + (h_{cm} - h_s)\sin\alpha_{equil} \rightarrow$$

$$x_s = (0.497 - 1.30) \times \cos(1.0871°) + (0.03 - 0) \times \sin(1.0871°) \rightarrow x_s = -0.8023 \text{ m}$$

Thus, upon assuming that $(\partial\varepsilon/\partial\hat{q})_s \approx 0$, one obtains:

$$\left(C_{Xq}\right)_s = -2\left(C_{X\alpha}^*\right)_s \frac{x_s}{\bar{c}}(\eta_q)_s \eta_s = -2\times(0.00345)\times\frac{-0.8023}{0.30}\times 1.0\times 0.96 \rightarrow$$

$$\underline{\left(C_{Xq}\right)_s = 0.0177}$$

$$\left(C_{Zq}\right)_s = \left\{-2\left(C_{Z\alpha}^*\right)_s \frac{x_s}{\bar{c}} + \left[\left(C_{Z\alpha}^*\right)_{ac}\right]_s\right\}(\eta_q)_s \eta_s$$

$$= \left\{-2\times(-1.0708)\times\frac{-0.8023}{0.30} - 0.7139\right\}\times 1.0\times 0.96 \rightarrow$$

$$\underline{\left(C_{Zq}\right)_s = -6.1836}$$

Therefore, one has that:

$$C_{Xq} = \left(C_{Xq}\right)_w + \left(C_{Xq}\right)_s = -0.0009 + 0.0177 \rightarrow \underline{C_{Xq} = 0.0168}$$

$$C_{Zq} = \left(C_{Zq}\right)_w + \left(C_{Zq}\right)_s = -4.0820 - 6.1836 \rightarrow \underline{C_{Zq} = -10.2629}$$

Now,

$$\left[\left(C_{Mq}^*\right)_{ac}\right]_w = -\frac{\pi}{4}\cos(\Lambda_{c/4})_w \left[\frac{1}{3}\left(\frac{AR_w^3 \tan^2(\Lambda_{c/4})_w}{AR_w + 6\cos(\Lambda_{c/4})_w}\right) + 1\right]\left(\frac{\bar{c}_w}{\bar{c}}\right)^2 \frac{S_w}{S}$$

$$= -\frac{\pi}{4}\times\cos 8^0 \left[\frac{1}{3}\times\left(\frac{7.194^3 \tan^2 8^0}{7.194 + 6\cos 8^0}\right) + 1\right]\times\left(\frac{0.3}{0.3}\right)^2 \times \frac{0.556}{0.556} \rightarrow$$

$$\underline{\left[\left(C_{Mq}^*\right)_{ac}\right]_w = -0.9229}$$

$$\left[\left(C_{Mq}^*\right)_{ac}\right]_s = -\frac{\pi}{4}\left(\frac{\bar{c}_s}{\bar{c}}\right)^2 \frac{S_s}{S} = -\frac{\pi}{4}\times\left(\frac{0.2}{0.3}\right)^2 \times \frac{0.140}{0.556} \rightarrow$$

$$\underline{\left[\left(C_{Mq}^*\right)_{ac}\right]_s = -0.0879}$$

Also,

$$z_w = \left(h_{cm} - h_w\right)\cos\alpha_{equil} - \left(l_{cm} - l_w\right)\sin\alpha_{equil} \rightarrow$$

$$z_w = (0.03 - 0.156)\times\cos(1.0871°) - (0.497 - 0.494)\times\sin(1.0871°) \rightarrow \underline{z_w = -0.1260\,\text{m}}$$

$$z_s = \left(h_{cm} - h_s\right)\cos\alpha_{equil} - \left(l_{cm} - l_s\right)\sin\alpha_{equil} \rightarrow$$

$$z_s = (0.03 - 0)\times\cos(1.0871°) - (0.497 - 1.30)\times\sin(1.0871°) \rightarrow \underline{z_s = 0.0452\,\text{m}}$$

So,

$$(C_{Mq})_w = (C_{Xq})_w \frac{z_w}{\bar{c}} - (C_{Zq})_w \frac{x_w}{\bar{c}} + \left[(C^*_{Mq})_{ac}\right]_w (\eta_q)_w \eta_w \rightarrow$$

$$-0.0009 \times \frac{-0.1260}{0.30} - (-4.0793) \times \frac{0.0006}{0.30} - 0.9229 \times 1.0 \times 0.877 \rightarrow$$

$$\underline{\underline{(C_{Mq})_w = -0.8008}}$$

$$(C_{Mq})_s = (C_{Xq})_s \frac{z_s}{\bar{c}} - (C_{Zq})_s \frac{x_s}{\bar{c}} + \left[(C^*_{Mq})_{ac}\right]_s (\eta_q)_s \eta_s \rightarrow$$

$$0.0177 \times \frac{0.0452}{0.30} - (-6.1836) \times \frac{-0.8023}{0.30} - 0.0879 \times 1.0 \times 0.96 \rightarrow$$

$$\underline{\underline{(C_{Mq})_s = -16.6187}}$$

Therefore, in summary, one has:

$$C_{Mq} = (C_{Mq})_w + (C_{Mq})_s = -0.8008 - 16.6187 \rightarrow \underline{\underline{C_{Mq} = -17.4195}}$$

Example Values of C_{XDU} and C_{MDU}

As presented earlier, the primary contribution is due to the apparent-mass effect of the body, as given by:

$$(C_{XDu})_{body} = -\frac{4k_1 (Vol)_{body}}{S\bar{c}}$$

In order to use this equation, the equivalent ellipsoid of revolution is defined as having the same cross-sectional area and volume as the body:

$$(Vol)_{ellipsoid} = (Vol)_{body} = (2/3) L_{ellipsoid} S_{max} \rightarrow L_{ellipsoid} = 1.5 (Vol)_{body} / S_{max} =$$

$$L_{ellipsoid} = 1.5 \times 0.0352 / 0.04175 \rightarrow L_{ellipsoid} = 1.265\,m$$

Further, the diameter of the equivalent ellipsoid of revolution is obtained from:

$$(dia)_{max} = (4 S_{max} / \pi)^{1/2} = (4 \times 0.04175 / \pi)^{1/2} \rightarrow (dia)_{max} = 0.2306\,m$$

Therefore, the fineness ratio is $f = L_{ellipsoid} / (dia)_{max} = 1.265/0.2306 \rightarrow f = 5.486$. Thus the k_1 value, as well as k_2, k_3, k'_2 and k'_3, may be found from a table of solutions of the Munk equations by S. P. Jones (personal communication to the author):

f	k_1	k_2, k_3	k_2', k_3'
1.00	0.5000	0.5000	0.0
1.50	0.3037	0.6221	0.0951
2.00	0.2100	0.7042	0.2394
2.50	0.1563	0.7619	0.3652
3.00	0.1220	0.8039	0.4657
3.50	0.0985	0.8354	0.5450
4.00	0.0816	0.8598	0.6079
5.00	0.0591	0.8943	0.6999
6.00	0.0452	0.9171	0.7623
8.00	0.0293	0.9447	0.8394
10.00	0.0207	0.9602	0.8835

An interpolation of the values between $f = 5.0$ and $f = 6.0$ gives:

$$k_1 = 0.0523, \ k_2, k_3 = 0.9054, \ k_2', k_3' = 0.7302$$

therefore,

$$\left(C_{XDu}\right)_{body} = -\frac{4 \times 0.0523 \times 0.0352}{0.556 \times 0.30} \rightarrow \left(C_{XDu}\right)_{body} = -0.0442$$

next,

$$\left(C_{MDu}\right)_{body} = \frac{z_{cv}}{\bar{c}}\left(C_{XDu}\right)_{body}, \ \text{where} \ z_{cv} = \left(h_{cm} - h_{cv}\right)\cos\alpha_{equil} - \left(l_{cm} - l_{cv}\right)\sin\alpha_{equil} \rightarrow$$

$$z_{cv} = (0.03 - 0) \times \cos(1.0871°) - (0.497 - 0.60) \times \sin(1.0871°) \rightarrow z_{cv} = 0.0319 \ \text{m} \rightarrow$$

$$\left(C_{MDu}\right)_{body} = \frac{0.0319}{0.30} \times (-0.0442) \rightarrow \left(C_{MDu}\right)_{body} = -0.0047$$

As stated before, the body is the dominant contributor to these stability derivatives. Thus, one has:

$$C_{XDu} = -0.0442, \quad C_{MDu} = -0.0047$$

Note that in the flight-dynamic equations, C_{XDu} appears in the following term:

$$\hat{b}_{1,1} = 2\mu - C_{XDu}$$

where, for sea-level conditions,

$$\mu = 2m/(\rho S \bar{c}) = 2 \times 3.70/(1.225 \times 0.556 \times 0.30) \rightarrow \mu = 36.215$$

therefore,

$$\hat{b}_{1,1} = 72.432 + 0.0442 \rightarrow \hat{b}_{1,1} = 72.476$$

The acceleration stability derivative contributes 0.06% to this term and its effect is thus negligible.

Flight Dynamics

Example Values of $C_{ZD\alpha}$ and $C_{MD\alpha}$

The body's contribution is given by:

$$\left(C_{ZD\alpha}\right)_{body} = -\frac{4 k_3 (Vol)_{body}}{S\bar{c}}$$

where, from the previous table for an ellipsoid of revolution, $k_3 = k_2 = 0.9054$. Therefore,

$$\left(C_{ZD\alpha}\right)_{body} = -\frac{4 \times 0.9054 \times 0.0352}{0.556 \times 0.30} \rightarrow \left(C_{ZD\alpha}\right)_{body} = -0.7642$$

The equation for the wing's contribution is:

$$\left(C_{ZD\alpha}\right)_w = -\left(C_{LD\alpha}\right)_w \frac{S_w}{S} \frac{(c_0)_w}{\bar{c}}$$

where $\left(C_{LD\alpha}\right)_w$ may be found either in Appendix E, or the curve-fit equations presented earlier. For the wing's aspect ratio of 7.194, $\left(C_{LD\alpha}\right)_w = -3.910$. Therefore,

$$\left(C_{ZD\alpha}\right)_w = -(-3.910) \times \frac{0.556}{0.556} \times \frac{0.30}{0.30} \rightarrow \left(C_{ZD\alpha}\right)_w = 3.910$$

The stabilizer's contribution may be found in the same way. With its aspect ratio of 4.02, one obtains that $\left(C_{LD\alpha}\right)_s = -1.119$. Therefore,

$$\left(C_{ZD\alpha}\right)_s = -\left(C_{LD\alpha}\right)_s \frac{S_s}{S} \frac{(c_0)_s}{\bar{c}} = -(-1.119) \times \frac{0.140}{0.556} \times \frac{0.20}{0.30} \rightarrow \left(C_{ZD\alpha}\right)_s = 0.1878$$

The downwash-lag effect on the stabilizer is given by:

$$\left(C_{ZD\alpha}\right)_s^{lag} = -2\left(C_{L\alpha}^*\right)_s \frac{\partial \varepsilon}{\partial \alpha} \frac{l'_s}{\bar{c}} \frac{S_s}{S} (\eta_q)_s \eta_s = -2 \times 4.244 \times 0.310 \times \frac{0.806}{0.30} \times \frac{0.140}{0.556} \times 1.0 \times 0.96 \rightarrow$$

$$\left(C_{ZD\alpha}\right)_s^{lag} = -1.7089$$

ASIDE

In flight-dynamic texts that the author is acquainted with (several of which are referenced in this book) the $D\alpha$ stability derivatives are assumed to be negligible, with the exception of those due to downwash lag. However, it is seen that the other $D\alpha$ derivatives can be significant relative to the downwash-lag contributions, especially from the wing.

END ASIDE

The summary value for $C_{ZD\alpha}$ is given by:

$$C_{ZD\alpha} = \left(C_{ZD\alpha}\right)_{body} + \left(C_{ZD\alpha}\right)_w + \left(C_{ZD\alpha}\right)_s + \left(C_{ZD\alpha}\right)_s^{lag} = -0.7642 + 3.910 + 0.1878 - 1.7089 \rightarrow$$

$$C_{ZD\alpha} = 1.6247$$

In the flight-dynamic equations, $C_{ZD\alpha}$ appears in the following term:

$$\hat{b}_{2,2} = 2\mu + C_{ZD\alpha} = 70.8053$$

In this case, $C_{ZD\alpha}$ constitutes 2.3% of this term. This is still fairly negligible.

The moment contribution from the body may be estimated from:

$$(C_{MD\alpha})_{body} = -\frac{x_{cv}}{\bar{c}}(C_{ZD\alpha})_{body}, \text{ where } x_{cv} = (l_{cm} - l_{cv})\cos\alpha_{equil} + (h_{cm} - h_{cv})\sin\alpha_{equil} \rightarrow$$

$$x_{cv} = (0.497 - 0.60) \times \cos(1.0871°) + (0.03 - 0.0) \times \sin(1.0871°) \rightarrow x_{cv} = -0.1024 \text{ m} \rightarrow$$

$$(C_{MD\alpha})_{body} = -\frac{(-0.1024)}{0.30} \times (-0.7642) \rightarrow \underline{(C_{MD\alpha})_{body} = -0.2608}$$

The moment contribution from the wing is given by:

$$(C_{MD\alpha})_w = -\frac{(x_{ca})_w}{\bar{c}}\left[(C_{ZD\alpha})_{app}\right]_w - \frac{x_w}{\bar{c}}\left[(C_{ZD\alpha})_{cir}\right]_w$$

where the longitudinal area center of the wing, $(x_{ca})_w$, is found from:

$$(x_{ca})_w = \left[l_{cm} - (l_{ca})_w\right]\cos\alpha_{equil} + \left[h_{cm} - (h_{ca})_w\right]\sin\alpha_{equil} \rightarrow$$

$$(x_{ca})_w = (0.497 - 0.559) \times \cos(1.0871°) + (0.03 - 0.156) \times \sin(1.0871°) \rightarrow \underline{(x_{ca})_w = -0.0644 \text{ m}}$$

Also, from the curve-fit equations presented earlier, one obtains:

$$\left[(C_{LD\alpha})_{app}\right]_w = 2.560, \quad \left[(C_{LD\alpha})_{cir}\right]_w = -6.440 \rightarrow$$

$$\left[(C_{ZD\alpha})_{app}\right]_w = -\left[(C_{LD\alpha})_{app}\right]_w \frac{(c_0)_w}{\bar{c}}\frac{S_w}{S} = -2.560 \times \frac{0.30}{0.30} \times \frac{0.556}{0.556} \rightarrow \left[(C_{ZD\alpha})_{app}\right]_w = -2.560$$

$$\left[(C_{ZD\alpha})_{cir}\right]_w = -\left[(C_{LD\alpha})_{cir}\right]_w \frac{(c_0)_w}{\bar{c}}\frac{S_w}{S} = -(-6.440) \times \frac{0.30}{0.30} \times \frac{0.556}{0.556} \rightarrow \left[(C_{ZD\alpha})_{cir}\right]_w = 6.440$$

So that,

$$(C_{MD\alpha})_w = -\frac{-0.0644}{0.30} \times (-2.560) - \frac{0.0006}{0.30} \times 6.440 \rightarrow \underline{(C_{MD\alpha})_w = -0.5624}$$

Likewise, the moment contribution from the stabilizer is given by:

$$(C_{MD\alpha})_s = -\frac{(x_{ca})_s}{\bar{c}}\left[(C_{ZD\alpha})_{app}\right]_s - \frac{x_s}{\bar{c}}\left[(C_{ZD\alpha})_{cir}\right]_s$$

where the longitudinal area center of the stabilizer, $(x_{ca})_s$, is found from:

$$(x_{ca})_s = \left[l_{cm} - (l_{ca})_s\right]\cos\alpha_{equil} + \left[h_{cm} - (h_{ca})_s\right]\sin\alpha_{equil} \rightarrow$$

$$(x_{ca})_s = (0.497 - 1.350) \times \cos(1.0871°) + (0.03 - 0.0) \times \sin(1.0871°) \rightarrow \underline{(x_{ca})_s = -0.8523 \text{ m}}$$

Flight Dynamics

The stabilizer's aspect ratio is 4.018, so the curve-fit equations give that:

$$\left[(C_{L D\alpha})_{app}\right]_s = 2.428, \quad \left[(C_{L D\alpha})_{cir}\right]_s = -3.544 \rightarrow$$

$$\left[(C_{Z D\alpha})_{app}\right]_s = -\left[(C_{L D\alpha})_{app}\right]_s \frac{(c_0)_s}{\bar{c}} \frac{S_{ls}}{S} = -2.428 \times \frac{0.20}{0.30} \times \frac{0.140}{0.556} \rightarrow \left[(C_{Z D\alpha})_{app}\right]_s = -0.4076$$

$$\left[(C_{Z D\alpha})_{cir}\right]_s = -\left[(C_{L D\alpha})_{cir}\right]_s \frac{(c_0)_s}{\bar{c}} \frac{S_s}{S} = -(-3.544) \times \frac{0.20}{0.30} \times \frac{0.140}{0.556} \rightarrow \left[(C_{Z D\alpha})_{cir}\right]_s = 0.5949$$

So that,

$$(C_{MD\alpha})_s = -\frac{(-0.8523)}{0.30} \times (-0.4076) - \frac{(-0.8023)}{0.30} \times 0.5949 \rightarrow (C_{MD\alpha})_s = 0.4330$$

The downwash lag contribution is:

$$(C_{MD\alpha})_s^{lag} = -\frac{x_s}{\bar{c}}(C_{ZD\alpha})_s^{lag} = -\frac{(-0.8023)}{0.30} \times (-1.7089) \rightarrow (C_{MD\alpha})_s^{lag} = -4.5702$$

Therefore, the total value of this stability derivative is given by:

$$C_{MD\alpha} = (C_{MD\alpha})_{body} + (C_{MD\alpha})_w + (C_{MD\alpha})_s + (C_{MD\alpha})_s^{lag} = -0.2608 - 0.5624 + 0.4330 - 4.5702 \rightarrow$$

$$C_{MD\alpha} = -4.9604$$

Example Values of C_{XDq}, C_{ZDq} and C_{MDq}

The body's contribution to C_{XDq} is:

$$(C_{XDq})_{body} = 2\frac{z_{cv}}{\bar{c}}(C_{XDu})_{body} = 2 \times \frac{0.0319}{0.30} \times (-0.0442) \rightarrow (C_{XDq})_{body} = -0.0094$$

Since the values of this derivative for the wing and stabilizer are assumed to be zero, then:

$$C_{XDq} = -0.0094$$

Next, one has that:

$$(C_{ZDq})_{body} = -2\frac{x_{cv}}{\bar{c}}(C_{ZD\alpha})_{body} = -2 \times \frac{(-0.1024)}{0.30} \times (-0.7624) = (C_{ZDq})_{body} = -0.5205$$

The wing's contribution is given by:

$$(C_{ZDq})_w = -2\frac{(x_{ca})_w}{\bar{c}}\left[(C_{ZD\alpha})_{app}\right]_w - 2\frac{x_w}{\bar{c}}\left[(C_{ZD\alpha})_{cir}\right]_w \rightarrow$$

$$(C_{ZDq})_w = -2 \times \frac{(-0.0644)}{0.30} \times (-2.560) - 2 \times \frac{0.0006}{0.30} \times 6.440 \rightarrow (C_{ZDq})_w = -1.1249$$

And the stabilizer's contribution is given by:

$$(C_{ZDq})_s = -2\frac{(x_{ca})_s}{\bar{c}}\left[(C_{ZD\alpha})_{app}\right]_s - 2\frac{x_s}{\bar{c}}\left[(C_{ZD\alpha})_{cir}\right]_s \rightarrow$$

$$(C_{ZDq})_s = -2 \times \frac{(-0.8523)}{0.30} \times (-0.4076) - 2 \times \frac{(-0.8023)}{0.30} \times 0.5949 \rightarrow (C_{ZDq})_s = 0.8659$$

In summary one has:

$$C_{ZDq} = (C_{ZDq})_{body} + (C_{ZDq})_w + (C_{ZDq})_s = -0.5205 - 1.1249 + 0.8659 \rightarrow C_{ZDq} = -0.7795$$

The body's contribution to C_{MDq} is given by:

$$(C_{MDq})_{body} = \frac{z_{cv}}{\bar{c}}(C_{XDq})_{body} - \frac{x_{cv}}{\bar{c}}(C_{ZDq})_{body} + (C_{MDq})_{body}^{cv}$$

where:

$$(C_{MDq})_{body}^{cv} = -\frac{2k_2'\left[L_{ellipsoid}^2 + (dia)_{max}^2\right](Vol)_{body}}{5S\bar{c}^3}$$

$$= -\frac{2 \times 0.7302 \times (1.265^2 + 0.2306^2)}{5 \times 0.556 \times 0.30^3} \times 0.0352 \rightarrow (C_{MDq})_{body}^{cv} = -1.1324 \rightarrow$$

$$(C_{MDq})_{body} = \frac{0.0319}{0.30} \times (-0.0094) - \frac{(-0.1024)}{0.30} \times (-0.5205) - 1.1324 \rightarrow (C_{MDq})_{body} = -1.3111$$

The wing's contribution is given by:

$$(C_{MDq})_w = -\frac{(x_{ca})_w}{\bar{c}}\left[(C_{ZDq})_{app}\right]_w - \frac{x_w}{\bar{c}}\left[(C_{ZDq})_{cir}\right]_w$$

where:

$$\left[(C_{ZDq})_{app}\right]_w = -2\frac{(x_{ca})_w}{\bar{c}}\left[(C_{ZD\alpha})_{app}\right]_w = -2 \times \frac{(-0.0644)}{0.30} \times (-2.560) \rightarrow$$

$$\left[(C_{ZDq})_{app}\right]_w = -1.0991$$

$$\left[(C_{ZDq})_{cir}\right]_w = -2\frac{x_w}{\bar{c}}\left[(C_{ZD\alpha})_{cir}\right]_w = -2 \times \frac{0.0006}{0.30} \times 6.440 \rightarrow$$

$$\left[(C_{ZDq})_{cir}\right]_w = -0.0258$$

therefore,

$$(C_{MDq})_w = -\frac{(-0.0644)}{0.30} \times (-1.0991) - \frac{0.0006}{0.30} \times (-0.0258) \rightarrow (C_{MDq})_w = -0.2359$$

and the stabilizer's contribution is given by:

$$\left(C_{ZDq}\right)_s = -\frac{(x_{ca})_s}{\bar{c}}\left[\left(C_{ZDq}\right)_{app}\right]_s - \frac{x_s}{\bar{c}}\left[\left(C_{ZDq}\right)_{cir}\right]_s$$

where:

$$\left[\left(C_{ZDq}\right)_{app}\right]_s = -2\frac{(x_{ca})_s}{\bar{c}}\left[\left(C_{ZD\alpha}\right)_{app}\right]_s = -2\times\frac{(-0.8523)}{0.30}\times(-0.4076) \rightarrow$$

$$\left[\left(C_{ZDq}\right)_{app}\right]_s = -2.3160$$

$$\left[\left(C_{ZDq}\right)_{cir}\right]_s = -2\frac{x_s}{\bar{c}}\left[\left(C_{ZD\alpha}\right)_{cir}\right]_s = -2\times\frac{(-0.8023)}{0.30}\times 0.5949 \rightarrow$$

$$\left[\left(C_{ZDq}\right)_{cir}\right]_s = 3.1819$$

therefore,

$$\left(C_{MDq}\right)_s = -\frac{(-0.8523)}{0.30}\times(-2.3160) - \frac{(-0.8023)}{0.30}\times(3.1819) \rightarrow \left(C_{MDq}\right)_s = 1.9297$$

The summary value is:

$$C_{MDq} = \left(C_{MDq}\right)_{body} + \left(C_{MDq}\right)_w + \left(C_{MDq}\right)_s = -1.3111 - 0.2359 + 1.9297 \rightarrow C_{MDq} = 0.3827$$

This stability derivative appears in the dynamic equations in combination with the non-dimensional pitching moment of inertia:

$$\hat{b}_{3,3} = i_{yy} - C_{MDq}, \quad \text{where} \quad i_{yy} = I_{yy}/\left[\rho S(\bar{c}/2)^3\right]$$

The pitching moment of inertia may be expressed in terms of its radius of gyration, r_y:

$$I_{yy} = m\left(r_y\, l/2\right)^2$$

where a representative value of r_y may be found at www.eng-tips.com for various airplane configurations. For a high-wing monoplane, the reference gives that $r_y \approx 0.397$. Therefore,

$$I_{yy} \approx 3.70\times(0.397\times 1.462/2)^2 \rightarrow I_{yy} \approx 0.3116\,\text{kg-m}^2$$

One then obtains, for sea-level conditions, that:

$$i_{yy} = 0.3116/\left[1.225\times 0.556\times(0.30/2)^3\right] \rightarrow i_{yy} = 135.56$$

It is seen that $C_{MD\hat{q}}$ makes a negligible 0.28% of the contribution to the $\hat{b}_{3,3}$ term.

Example Values of $\hat{T}_{Xu}$, $\hat{T}_{X\alpha}$ and $\hat{T}_{Z\alpha}$

For sea-level conditions ($\rho = 1.225 \text{ kg/m}^3$), one has that:

$$\hat{T}_{Xu} = \frac{-2T_0}{\rho S U_0^2} = \frac{-2 \times 3.974}{1.225 \times 0.556 \times 14.458^2} \rightarrow \hat{T}_{Xu} - 0.0558 \equiv -\hat{T}_0$$

Next, the thrust term, $C_{T\alpha}$, has to be evaluated. From the "Equilibrium Flight" section of Chapter 7, the power output required is given by:

$$P\,\eta_p = T_0 V = 3.974 \times 14.458 = 57.456 \text{ Watts}$$

Also, the weight of the aircraft is $W = mg = 3.70 \times 9.81 = 36.297 \text{ N}$, and the propeller diameter is $dia = 0.37 \text{ m}$. Therefore, from the "Complete-Aircraft Aerodynamics" section of Chapter 2, one has that:

$$C_{T\alpha} = \frac{3}{2} K_p \eta_p (C_L)_0^{1/2} (C_{L\alpha})_0$$

where, for sea-level conditions,

$$K_p \eta_p = \frac{2\rho^{1/2} P \eta_p}{(2W/S)^{3/2} \tilde{S}} = \frac{2 \times (1.225)^{1/2} \times 57.456}{(2 \times 36.297/0.556)^{3/2} \times 0.1369} = \frac{127.184}{204.241} = 0.623$$

so, the result is:

$$C_{T\alpha} = 1.5 \times 0.623 \times (0.5099)^{1/2} \times 4.842 \rightarrow C_{T\alpha} = 3.231$$

Therefore, one obtains:

$$\hat{T}_{X\alpha} = C_{T\alpha} \frac{\tilde{S}}{S} = 3.231 \times \frac{0.1369}{0.556} \rightarrow \hat{T}_{X\alpha} = 0.7955$$

Also,

$$\hat{T}_{Z\alpha} = -(C_{L\alpha})_{\text{prop}} \frac{S_p}{S} = -5.0 \times \frac{0.006}{0.556} \rightarrow \hat{T}_{Z\alpha} = -0.0540$$

Next, for $l_T = l_{\text{prop}}$ and $h_T = h_{\text{prop}}$, one has that:

$$\hat{x}_T = \left[(l_{\text{cm}} - l_{\text{prop}}) \cos \alpha_{\text{equil}} + (h_{\text{cm}} - h_{\text{prop}}) \sin \alpha_{\text{equil}} \right] / \bar{c}$$

$$\hat{x}_T = \left[(0.497 - 0.0) \times \cos(1.0871°) + (0.03 - 0.0) \times \sin(1.0871°) \right] / 0.30 \rightarrow \hat{x}_T = 1.658$$

$$\hat{z}_T = \left[(h_{\text{cm}} - h_{\text{prop}}) \cos \alpha_{\text{equil}} - (l_{\text{cm}} - l_{\text{prop}}) \sin \alpha_{\text{equil}} \right] / \bar{c}$$

$$\hat{z}_T = \left[(0.03 - 0.0) \times \cos(1.0871°) - (0.497 - 0.0) \times \sin(1.0871°) \right] / 0.30 \rightarrow \hat{z}_T = 0.069$$

$$\hat{T}_{Xq} = 2 \hat{T}_{Xu} \hat{z}_T = 2 \times (-0.0558) \times 0.069 \rightarrow \hat{T}_{Xq} = -0.0077$$

$$\hat{T}_{Zq} = -2 \hat{T}_{Z\alpha} \hat{x}_T = -2 \times (-0.0540) \times 1.658 \rightarrow \hat{T}_{Zq} = 0.1791$$

Flight Dynamics

$$\hat{T}_{Mq} = \hat{T}_{Xq}\,\hat{z}_T - \hat{T}_{Zq}\,\hat{x}_T \rightarrow \hat{T}_{Mq} = 2\left(\hat{T}_{Xu}\,\hat{z}_T^2 + \hat{T}_{Z\alpha}\,x_T^2\right) \rightarrow$$

$$\hat{T}_{Mq} = 2\times\left(-0.0558\times 0.069^2 - 0.0540\times 1.658^2\right) \rightarrow \underline{\underline{\hat{T}_{Mq} = -0.2974}}$$

It is seen that, with the exceptions of $\hat{T}_{Xu}$ and $\hat{T}_{X\alpha}$, the propulsion derivatives are relatively small. This is especially so in the cases for $\hat{T}_{Zq}$ compared to C_{Zq}, and $\hat{T}_{Mq}$ compared to C_{Mq}.

To conclude this section, example longitudinal stability derivatives are given for three different aircraft:

1. Piper Comanche: a subsonic general-aviation airplane of tail-aft configuration.
2. Pazmany PL-4: a subsonic home-built airplane of tail-aft configuration.
3. Present numerical example
4. Modern semi-rigid airship described in Chapter 8.

Stability derivative	PA 24-250 (piper comanche)	Pazmany PL-4 home-built airplane	"Scholar" model airplane	Goodyear "Wingfoot 2" airship*
$(C_X)_0$	−0.0308	−0.0510	−0.0552	−0.02051
$(C_Z)_0$	−0.2889	−0.374	−0.5088	−0.04131
$(C_M)_0$	0.0	0.0	−0.0028	0.01024
C_{Xu}	0.0	0.0	0.0	0.0
$C_{X\alpha}$	0.1052	0.156	0.2577	0.00263
C_{Xq}	−0.0484	0.242	0.0168	−0.0067
C_{Zu}	0.0	0.0	0.0	0.0
$C_{Z\alpha}$	−4.829	−5.17	−4.8432	−0.8155
C_{Zq}	−5.777	−9.00	−10.2629	−0.5046
C_{Mu}	0.0	0.0	0.0	0.0
$C_{M\alpha}$	−2.600	−1.13	−1.7670	0.17626
C_{Mq}	−28.63	−11.2	−17.4195	−0.2047
C_{XDu}	≈0.0	≈0.0	−0.0442	−0.0636
$C_{XD\alpha}$	0.0	0.0	0.0	−0.0454
C_{XDq}	≈0.0	≈0.0	−0.0094	0.0091
C_{ZDu}	0.0	0.0	0.0	−0.0454
$C_{ZD\alpha}$	−4.829	−3.5	1.6247	−0.9360
C_{ZDq}	≈0.0	≈0.0	−0.7795	0.0718
C_{MDu}	≈0.0	≈0.0	−0.0047	0.0046
$C_{MD\alpha}$	−5.259	−3.9	−4.9604	0.0359
C_{MDq}	≈0.0	≈0.0	0.3820	−0.0883
$\hat{T}_{Xu}$	−0.0652	−0.0509	−0.0558	≈0.0
$\hat{T}_{X\alpha}$	neglected	neglected	0.7955	≈0.0
$\hat{T}_{Zu}$	"	"	0.0	≈0.0
$\hat{T}_{Z\alpha}$	"	"	−0.0540	≈0.0
$\hat{T}_{Xq}$	"	"	−0.0077	≈0.0
$\hat{T}_{Zq}$	"	"	0.1791	≈0.0
$\hat{T}_{Mq}$	"	"	−0.2974	≈0.0

*Note that, for this particular case, the characteristic length, $\bar{c}$, is the body length: and the characteristic area, S, is the body's molded volume to the 2/3 power.

The derivatives for the Piper Comanche were calculated as a special project by Alex Markov, while a student at the University of Toronto.

The derivatives for the Pazmany PL-4 were calculated by Jeremy M. Harris, as a special-interest project, while a Research Engineer at Battelle Memorial Institute.

The derivatives for the airship were calculated by the author. These are with respect to the aircraft's mass-center, which includes the internal air and gas as well as the structure, etc.

AIRCRAFT LATERAL SMALL-PERTURBATION DYNAMIC EQUATIONS

The lateral dynamic-stability case is more complex, as illustrated by the figure below.

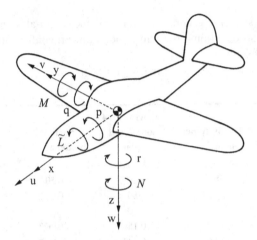

This shows that the small-perturbation motions include a sideways velocity, v, along the y-axis, a yawing angular velocity, r, about the z-axis, and a rolling angular velocity, p, about the x-axis. Further, the aircraft responds to a force, Y, along the y-axis, a yawing moment, N, about the z-axis, and a rolling moment, $\tilde{L}$, about the x-axis (this term has a tilde, so as not be confused with the lift force, L).

The small-perturbation dynamic equations of motion are given in **Appendix C: Rigid-Body Equations of Motion**. Therefore, with the inclusion of the lateral stability derivatives, along with the weight and buoyancy effects, the dimensional lateral-stability equations are:

$$[C]\begin{bmatrix} v \\ p \\ r \\ \varphi \end{bmatrix} + [D]\begin{bmatrix} \dot{v} \\ \dot{p} \\ \dot{r} \\ \varphi \end{bmatrix} = \begin{bmatrix} \Delta Y_C \\ \Delta \tilde{L}_C \\ \Delta N_C \\ 0 \end{bmatrix}$$

where:

$$c_{1,1} = -Y_v, \quad c_{1,2} = -Y_p, \quad c_{1,3} = -Y_r + m, \quad c_{1,4} = (B - mg)\cos\Theta_0,$$

$$c_{2,1} = -\tilde{L}_v, \quad c_{2,2} = -\tilde{L}_p, \quad c_{2,3} = -\tilde{L}_r, \quad c_{2,4} = -B z_B \cos\Theta_0,$$

$$c_{3,1} = -N_v, \quad c_{3,2} = -N_p, \quad c_{3,3} = -N_r, \quad c_{3,4} = x_B B \cos\Theta_0,$$

$$c_{4,1} = 0, \quad c_{4,2} = 1, \quad c_{4,3} = \tan\Theta_0, \quad c_{4,4} = 0$$

$$d_{1,1} = m - Y_{dv}, \quad d_{1,2} = -Y_{dp}, \quad d_{1,3} = -Y_{dr}, \quad d_{1,4} = 0,$$

Flight Dynamics

$$d_{2,1} = -\tilde{L}_{dv}, \quad d_{2,2} = I_{xx} - \tilde{L}_{dp}, \quad d_{2,3} = -\tilde{L}_{dr} - I_{xz}, \quad d_{2,4} = 0,$$

$$d_{3,1} = -N_{dv}, \quad d_{3,2} = -I_{xz} - N_{dp}, \quad d_{3,3} = I_{zz} - N_{dr}, \quad d_{3,4} = 0$$

$$d_{4,1} = 0, \quad d_{4,2} = 0, \quad d_{4,3} = 0, \quad d_{4,4} = -1$$

Notice that, in comparison with the longitudinal dynamic equations, the thrust terms are not explicitly called out. Instead, the lateral forces from the thruster (in particular, the propeller) are most conveniently incorporated in the stability derivatives, as will be subsequently demonstrated.

Non-Dimensional Form of the Equations

As for the longitudinal dynamic equations, a set of non-dimensional lateral equations may be obtained by using the following definitions:

$$\text{force: } C_Y, \hat{B}, \hat{mg} \equiv \frac{Y, B, mg}{\rho V_{cm}^2 S/2} \quad \text{moments: } C_N, C_{\tilde{L}} \equiv \frac{N, \tilde{L}}{\rho V_{cm}^2 Sb/2}$$

$$\text{distance: } \hat{x}_B, \hat{z}_B \equiv \frac{x_B, z_B}{b}$$

$$\text{velocity: } \beta \equiv \hat{v} = \frac{v}{U_0} \quad \text{angular velocities: } \hat{r}, \hat{p} \equiv \frac{r, p}{2U_0/b}$$

$$\text{time derivatives: } D(\ ) \equiv \frac{b}{2U_0}\frac{d(\ )}{dt}, \quad D^2(\ ) \equiv \frac{b^2}{4U_0^2}\frac{d^2(\ )}{dt^2}$$

Note that time is non-dimensionalized by $b/(2U_0)$

$$\text{mass: } \mu \equiv \frac{2m}{\rho S b} \quad \text{moments of inertia: } i_{zz}, i_{xx}, i_{xz} \equiv \frac{I_{zz}, I_{xx}, I_{xz}}{\rho S(b/2)^3}$$

Again, V_{cm} is the instantaneous velocity of the aircraft's mass-center (not necessarily the equilibrium velocity, U_0) and S & b are the defined reference area and lateral reference length, respectively, for the aircraft. For a conventional airplane design, this is usually the wing's planform area and total (flattened) wing span.

With these definitions, one may obtain expressions for the non-dimensional lateral stability derivatives, in the same fashion as for the longitudinal stability derivatives:

$$Y_v = C_{Y\beta} \rho U_0 S/2, \quad \tilde{L}_v = C_{\tilde{L}\beta} \rho U_0 b S/2, \quad N_v = C_{N\beta} \rho U_0 b S/2,$$

where:

$$C_{Y\beta}, C_{\tilde{L}\beta}, C_{N\beta} \equiv \left(\frac{\partial C_Y}{\partial \hat{v}}\right)_0, \left(\frac{\partial C_{\tilde{L}}}{\partial \hat{v}}\right)_0, \left(\frac{\partial C_N}{\partial \hat{v}}\right)_0$$

for the angular velocities, observe that:

$$Y_p = \left(\frac{\partial Y}{\partial p}\right)_0 = \left[\frac{\partial}{\partial p}\left(\frac{C_Y \rho\, V_{cm}^2 S}{2}\right)\right]_0 = \left(\frac{\partial C_Y}{\partial p}\right)_0 \frac{\rho U_0^2 S}{2}$$

Further, define:

$$C_{Yp} \equiv \left(\frac{\partial C_Y}{\partial \hat{p}}\right)_0$$

so, one obtains:

$$Y_p = C_{Yp}\, \rho\, U_0\, S\, b/4$$

In a similar fashion, the other angular-velocity derivatives become:

$$\tilde{L}_p, N_p = (C_{\tilde{L}p}, C_{Np})\rho U_0 S b^2/4, \quad Y_r = C_{Yr}\, \rho U_0 S b/4,$$

$$\tilde{L}_r, N_r = (C_{\tilde{L}r}, C_{Nr})\rho U_0 S b^2/4$$

where:

$$C_{\tilde{L}p}, C_{Np} \equiv \left(\frac{\partial C_{\tilde{L}}}{\partial \hat{p}}\right)_0, \left(\frac{\partial C_N}{\partial \hat{p}}\right)_0$$

$$C_{Yr}, C_{\tilde{L}r}, C_{Nr} \equiv \left(\frac{\partial C_Y}{\partial \hat{r}}\right)_0, \left(\frac{\partial C_{\tilde{L}}}{\partial \hat{r}}\right)_0, \left(\frac{\partial C_N}{\partial \hat{r}}\right)_0$$

For the $\dot{v}$ derivatives, observe that:

$$Y_{dv} = \left(\frac{\partial Y}{\partial \dot{v}}\right)_0 = \left[\frac{\partial}{\partial \dot{v}}\left(\frac{C_Y \rho V_{cm}^2 S}{2}\right)\right]_0 = \left(\frac{\partial C_Y}{\partial \dot{v}}\right)_0 \frac{\rho U_0^2 S}{2}$$

also, define:

$$C_{YD\beta} \equiv \left[\frac{\partial C_Y}{\partial (D\hat{v})}\right]_0$$

So, one obtains:

$$Y_{dv} = C_{YD\beta}\, \rho\, b\, S/4$$

In a similar fashion the other derivatives are:

$$\tilde{L}_{dv}, N_{dv} = \left(C_{\tilde{L}D\beta}, C_{ND\beta}\right)\rho b^2 S/4$$

where:

$$C_{\tilde{L}D\beta}, C_{ND\beta} \equiv \left(\frac{\partial C_{\tilde{L}}}{\partial (D\hat{v})}\right)_0, \left(\frac{\partial C_N}{\partial (D\hat{v})}\right)_0$$

Further, for the $\dot{p}$ and $\dot{r}$ derivatives, observe that:

$$Y_{dp} = \left(\frac{\partial Y}{\partial \dot{p}}\right)_0 = \left[\frac{\partial}{\partial \dot{p}}\left(\frac{C_Y \rho V_{cm}^2 S}{2}\right)\right]_0 = \left(\frac{\partial C_Y}{\partial \dot{p}}\right)_0 \frac{\rho U_0^2 S}{2}$$

Define:

$$C_{YDp} \equiv \left[\frac{\partial C_Y}{\partial (D\hat{p})}\right]_0 = \frac{4U_0^2}{b^2}\left(\frac{\partial C_Y}{\partial \dot{p}}\right)_0$$

This, into the previous equation, gives:

$$Y_{dp} = C_{YDp}\, \rho S b^2/8$$

Finally, the other derivatives are:

$$\tilde{L}_{dp},\ \tilde{L}_{dr},\ N_{dp},\ N_{dr} = \left(C_{\tilde{L}Dp},\ C_{\tilde{L}Dr},\ C_{NDp},\ C_{NDr}\right)\rho b^3 S/8, \quad Y_{dr} = C_{YDr}\,\rho S b^2/8$$

where:

$$C_{\tilde{L}Dp},\ C_{NDp} \equiv \left(\frac{\partial C_{\tilde{L}}}{\partial(D\hat{p})}\right)_0, \left(\frac{\partial C_N}{\partial(D\hat{p})}\right)_0$$

$$C_{YDr},\ C_{\tilde{L}Dr},\ C_{NDr} \equiv \left(\frac{\partial C_Y}{\partial(D\hat{r})}\right)_0, \left(\frac{\partial C_{\tilde{L}}}{\partial(D\hat{r})}\right)_0, \left(\frac{\partial C_N}{\partial(D\hat{r})}\right)_0$$

When these definitions are substituted into the previously given linearized lateral dynamic equations, one obtains the non-dimensional form of these equations:

$$[\hat{C}]\begin{bmatrix}\beta\\ \hat{p}\\ \hat{r}\\ \varphi\end{bmatrix} + [\hat{D}]\begin{bmatrix}D\beta\\ D\hat{p}\\ D\hat{r}\\ D\varphi\end{bmatrix} = \begin{bmatrix}\Delta C_{Yc}\\ \Delta C_{\tilde{L}c}\\ \Delta C_{Nc}\\ 0\end{bmatrix}$$

where:

$$\hat{c}_{1,1} = -C_{Y\beta},\quad \hat{c}_{1,2} = -C_{Yp},\quad \hat{c}_{1,3} = -C_{Yr} + 2\mu,\quad \hat{c}_{1,4} = \left(\hat{B} - \hat{m}\hat{g}\right)\cos\Theta_0,$$

$$\hat{c}_{2,1} = -C_{\tilde{L}\beta},\quad \hat{c}_{2,2} = -C_{\tilde{L}p},\quad \hat{c}_{2,3} = -C_{\tilde{L}r},\quad \hat{c}_{2,4} = -\hat{B}\,\hat{z}_B \cos\Theta_0,$$

$$\hat{c}_{3,1} = -C_{N\beta},\quad \hat{c}_{3,2} = -C_{Np},\quad \hat{c}_{3,3} = -C_{Nr},\quad \hat{c}_{3,4} = \hat{x}_B\,\hat{B}\cos\Theta_0,$$

$$\hat{c}_{4,1} = 0,\quad \hat{c}_{4,2} = 1,\quad \hat{c}_{4,3} = \tan\Theta_0,\quad \hat{c}_{4,4} = 0$$

$$\hat{d}_{1,1} = 2\mu - C_{YD\beta},\quad \hat{d}_{1,2} = -C_{YDp},\quad \hat{d}_{1,3} = -C_{YDr},\quad \hat{d}_{1,4} = 0,$$

$$\hat{d}_{2,1} = -C_{\tilde{L}D\beta},\quad \hat{d}_{2,2} = i_{xx} - C_{\tilde{L}Dp},\quad \hat{d}_{2,3} = -C_{\tilde{L}Dr} - i_{xz},\quad \hat{d}_{2,4} = 0,$$

$$\hat{d}_{3,1} = -C_{ND\beta},\quad \hat{d}_{3,2} = -i_{xz} - C_{NDp},\quad \hat{d}_{3,3} = i_{zz} - C_{NDr},\quad \hat{d}_{3,4} = 0,$$

$$\hat{d}_{4,1} = 0,\quad \hat{d}_{4,2} = 0,\quad \hat{d}_{4,3} = 0,\quad \hat{d}_{4,4} = -1$$

ESTIMATION OF THE LATERAL STABILITY DERIVATIVES

As with the longitudinal stability derivatives, the way to envision the determination of the lateral stability derivatives is to consider the aerodynamic forces and moments caused by the perturbation of only one of the motion variables, with all others held fixed. The first case below is for sideslip motion *only*, β, with r, p, $\dot{\beta}$, $\dot{r}$, and $\dot{p}$ equal to zero.

The β Derivatives:

A body in yaw has a side force, Y_{body}, that depends greatly on the body's geometry. For a bare streamlined body of revolution with no appendages (wings, tail, etc.), non-viscous potential theory gives that $(C_{Y\beta})_{body} = 0$. However, the reality of viscous and flow-separation effects give a small, but finite, value for $(C_{Y\beta})_{body}$. Brian Thwaites ("Incompressible Aerodynamics", Oxford University Press, 1960, p. 408) provides an equation for this, when $\beta < 5°$:

$$(C_{Y\beta})_{body} = \frac{-1.75}{(Re)^{1/5}} \frac{A}{S_{max}} \frac{S_{body}}{S}$$

where:
- $A \equiv$ the projected lateral area of the body (the "shadow area")
- $S_{max} \equiv$ the maximum cross-sectional area
- $Re \equiv$ the Reynolds number based on the body length

$$S_{body} \equiv (Vol_{body})^{2/3}$$

For a body incorporating a wing and tail, another method of estimation, at subsonic speeds, comes from the USAF Stability and Control DATCOM (Section 5.2.1, 1969 edition):

$$(C_{Y\beta})_{body} = -K_{int} (\hat{C}_{L\alpha})_{body} \frac{S_b}{S}$$

where K_{int} is the wing-body interference factor, given by:
- $K_{int} = 1.85 (2 h_w / d)$ for high-winged airplanes,
- $K_{int} = -1.48 (2 h_w / d)$ for low-winged airplanes,
- $h_w \equiv$ the vertical distance from the body's center line (or loft line) to the quarter-chord point of the exposed wing root (positive for that point being above the body center line),
- $d \equiv$ the maximum body height at the wing-body intersection region.

ASIDE

Note that S_b is the body's characteristic area. As stated in the "Bodies" section of Chapter 2, this may either be defined as the body's maximum cross-sectional area (usually chosen for airplane analysis), or the body's molded volume to the 2/3 power (usually chosen for airship analysis). The important thing is to choose one and stay consistent with it throughout the analysis.

END ASIDE

Further, $(\hat{C}_{L\alpha})_{body}$ is a "lateral lift-coefficient slope" for the body with no wings. If the body has a round cross-section, this value may be obtained from:

$$(\hat{C}_{L\alpha})_{body} = \frac{2(k_2 - k_1) S_{base}}{S_b}$$

where k_1 and k_2 are the longitudinal and lateral apparent-mass coefficients defined in Chapter 2, and S_{base} is the body's cross-sectional area where the vertical tail begins. Also, if the body's overall cross-sectional shape is rectangular, this value should be doubled.

These suggested methods may seem rather imprecise but, in truth, for most aircraft the body contribution is small compared with those from the wing(s) and tail. Airships, of course, are an exception.

Next, consider the lateral forces on the vertical fin, as illustrated below:

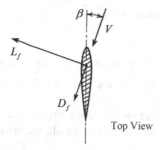

Top View

the Y_f force is given by:

$$Y_f = -L_f \cos\beta - D_f \sin\beta$$

If one considers the fin to be in "isolation" (in the presence of the stabilizer and portion of the body under the fin, but with no effect from the wing, side-wash angle, dynamic-pressure change, or propulsion system), then the isolated derivative, $\left(C_{Y\beta}^*\right)_f$, is given by:

$$\left(C_{Y\beta}^*\right)_f = -\left[\left(C_{L\alpha}\right)_f + \left(C_{D0}\right)_f\right] S_f/S$$

where, of course, $\left(C_{L\alpha}\right)_f$ and $\left(C_{D0}\right)_f$ are coefficients calculated for the "isolated" fin. When the fin is positioned over a horizontal tail (stabilizer), which is significantly larger than the fin, one may conjecture that the stabilizer forms an image plane, so that the effective aspect ratio is that of the fin passing through the fuselage and continuing to its image, as illustrated:

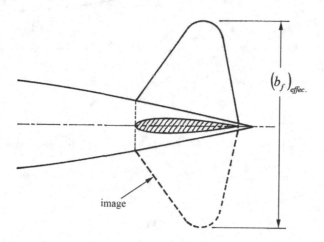

image

If one defines the fin's area as shown in the figure below:

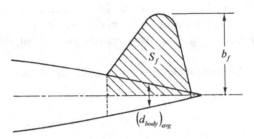

then the aspect ratio is $(b_f)^2_{\text{effec.}}/(2S_f)$, and thus $(C_{L\alpha})_f$ and $(C_{D0})_f$ may be obtained from the "Wings" section of Chapter 2. Further, the effect of the body may be estimated by the method described in the "Wing-Body Combination" section.

However, the effect of the stabilizer on the fin is more complex than that, and DATCOM (Section 5.3.1.1-2) offers a more refined methodology. First of all, the effective aspect ratio of the vertical fin is given by:

$$(AR)_{\text{effec.}} = K_1 \frac{b_f^2}{S_f}\left[1 + K_H(K_2 - 1)\right]$$

where b_f is illustrated above, as well as $(d_{\text{body}})_{\text{avg}}$, which is the body's average diameter within the region under the fin. Also, K_1 is a function of $X_1 \equiv b_f/(d_{\text{body}})_{\text{avg}}$, as shown in the plot below.

K1 Function

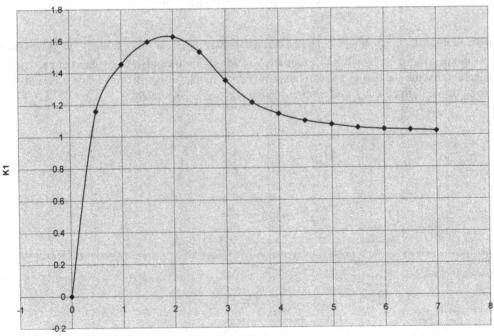

Further, K_H is a function of the ratio of the stabilizer area to the fin area, S_s/S_f, as also shown below.

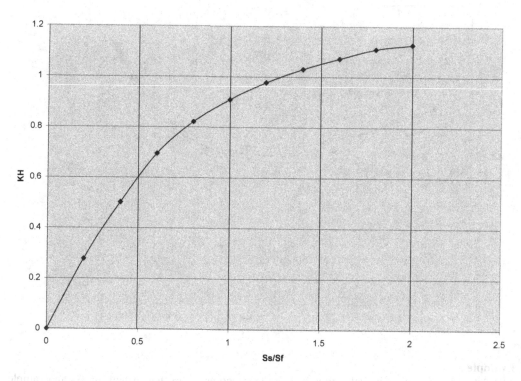

KH Function

Although curve-fit equations may be obtained for these plots, it is sufficiently accurate to obtain the desired values directly from the plots: visually interpolating and extrapolating as required. The methodology is approximate, at best, and does not warrant extreme precision in the values obtained.

Finally, K_2 is a function of the position of the stabilizer relative to the fin. The figure below defines these position parameters, where:

$h_s \equiv$ the vertical distance from the body's reference line (or loft line) to the stabilizer's aerodynamic center.
$\bar{c}_f \equiv$ the mean aerodynamic chord of the fin, positioned upon its aerodynamic center.
$\Delta l_s \equiv$ the longitudinal distance from the front of the fin's mean aerodynamic chord to the aerodynamic center of the stabilizer.

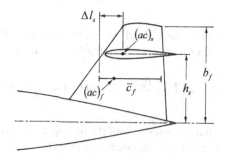

K_2, then, is dependent on and $X_2 \equiv h_s/b_f$ and $X_3 \equiv \Delta l_s/\bar{c}_f$, which is plotted below:

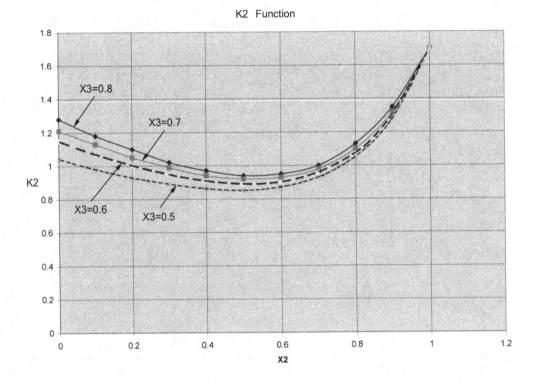

Example

All of the terms are now available to calculate the effectiveness of the vertical fin. As an example, consider a case where the fin is mounted on a very slender body, the stabilizer is on the body's reference line (or loft line), and the fin is positioned directly over the stabilizer. Further, $S_s/S_f = 3.0$. For this case $K_1 \approx 1.0$ and $K_H \approx 1.16$ (visual extrapolation of the plots). Also $X_2 = 0$, and choose $X_3 \approx 0.5$, therefore $K_2 \approx 1.03$. This gives:

$$(AR)_{\text{effec.}} \approx 1.0 \times \frac{b_f^2}{S_f}\left[1 + 1.16(1.03 - 1)\right] = 1.03 \frac{b_f^2}{S_f}$$

So, the effective aspect ratio is 1.03 times greater than the geometric value.

In comparison with the image-plane method for estimating $(AR)_{\text{effec.}}$, where $\left(b_f\right)_{\text{effec.}} = 2b_f$ and $\left(S_f\right)_{\text{effec.}} = 2S_f$, one obtains:

$$(AR)_{\text{effec.}} \approx \frac{(2b_f)^2}{(2S_f)} = 2\frac{b_f^2}{S_f}$$

So, the image-plane method, in this case, clearly over-estimates the fin's effectiveness.

However, if the body's diameter is increased so that $X_1 \approx 2$, then $K_1 \approx 1.62$ and one obtains that:

$$(AR)_{\text{effec.}} \approx 1.62 \times \frac{b_f^2}{S_f}\left[1 + 1.16(1.03 - 1)\right] = 1.68 \frac{b_f^2}{S_f}$$

Therefore, with a body radius that occupies 25% of the fin's span, the effective aspect ratio increases by 63%. Upon looking at the K_1 vs. X_1 graph, one can see that the body itself acts as an image plane.

Of course, as the graph also shows, if the body radius occupies too much of the span, the value of K_1, and hence the effective aspect ratio, precipitously drops to zero.

Now, if one moves the stabilizer further aft on the fin, so that $X_3 = 0.8$, then $K_2 = 1.28$ and the effective fin's aspect ratio is now:

$$(AR)_{\text{effec.}} \approx 1.62 \times \frac{b_f^2}{S_f}\left[1 + 1.16(1.28 - 1)\right] = 2.15 \frac{b_f^2}{S_f}$$

In this case, the result is close to that from the simple image-plane model.

As a final note on this topic, Perkins and Hage ("Airplane Performance Stability and Control", John Wiley & Sons, 1949 Edition, p. 325) state that for most typical fin designs, where the stabilizer is mounted on the body, an approximate effective aspect ratio is given by:

$$(AR)_{\text{effec.}} \approx 1.55 \frac{b_f^2}{S_f}$$

It is interesting that this value is midway between the extreme DATCOM examples. This shows what an inexact methodology this is, which is better served by wind-tunnel tests or vortex-lattice/CFD analyses. However, in the spirit of the piecemeal aerodynamic modeling in this text, the author recommends the DATCOM method (though the Perkins & Hage equation should serve as a reality check).

With the determination of the fin's effective aspect ratio, $(C_{L\alpha})_f$ and $(C_{D0})_f$ may be obtained by using the methods described in Chapter 2. When using the Lowry and Polhamus equation for calculating $(C_{L\alpha})_f$, the sweep angle of the half-chord line, $\Lambda_{c/2}$, may be estimated from the conceptual construction of an equivalent trapezoidal planform:

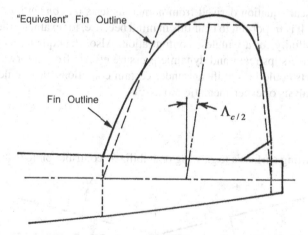

Although one could formulate an elaborate methodology for identifying the equivalent fin, in truth it suffices to "eyeball" the configuration, arranging it so that the area of the equivalent fin is the same as S_f.

ASIDE

The vertical fin is one portion of the aircraft where the designer can exercise a bit of "artistic license". It can be a graceful elliptical planform as shown above, or a trapezoid or a rectangle topped with a semi-circle. All of these shapes may be made to work well and there is no obvious "best" shape.

END ASIDE

The complete equation for $(C_{Y\beta})_f$, for the usual aft-mounted fin, is now given by:

$$(C_{Y\beta})_f = K_3 (C^*_{Y\beta})_f \left(1 - \frac{\partial \sigma_f}{\partial \beta}\right) \frac{q_t}{q}$$

where K_3 is an empirical "sideslip" factor obtained from the DATCOM section. Upon recalling that $X_1 \equiv b_f/(d_{body})_{avg}$, one has that:

$$K_3 = 0.75 \quad \text{for} \quad 0 \leq X_1 \leq 2$$
$$K_3 = 0.75 + 0.1667 \times (X_1 - 2) \quad \text{for} \quad 2 \leq X_1 \leq 3.5$$
$$K_3 = 1.0 \quad \text{for} \quad 3.5 \leq X_1 \leq 6 \text{ (and on)}$$

$(C^*_{Y\beta})_f$ is the sideslip derivative for the "isolated" fin as defined earlier.

$(1 - \partial \sigma_f / \partial \beta)$ is the attenuation of the flow angle, β, acting on the fin due to the side-wash angle, σ. This is the lateral equivalent to the downwash angle for the stabilizer, ε. DATCOM (5.4.1-1) combines this effect with q_t/q, which is the ratio of dynamic pressure at the fin to that of the free stream, to give:

$$\left(1 - \frac{\partial \sigma_f}{\partial \beta}\right) \frac{q_t}{q} = 0.724 + 3.06 \frac{(S_f/S_w)}{(1 + \cos(\Lambda_{c/4})_w)} - 0.4 \frac{h_w}{(d_{body})_{max}} + 0.009(AR)_w$$

As defined in Chapter 2, $(\Lambda_{c/4})_w$ is the sweepback angle of the wing's quarter-chord line. Also, h_w is the vertical distance from the body reference line to the wing's root quarter-chord point (positive up) and $(d_{body})_{max}$ is the maximum vertical *depth* of the body.

ASIDE

This is an empirical equation derived from numerous tests on *conventionally proportioned* tail-aft airplanes. It is important to bear this in mind because, for example, the equation would give a value of infinity for a wingless configuration. Also, the equation does not appear to account for possible slipstream and dynamic-pressure effects from a forward-mounted propeller. Such effects could be significant under certain conditions and would require a much more detailed analysis or experimentation to assess.

END ASIDE

Next, consider the lifting surface's (e.g., wing or stabilizer) contribution to $C_{Y\beta}$:

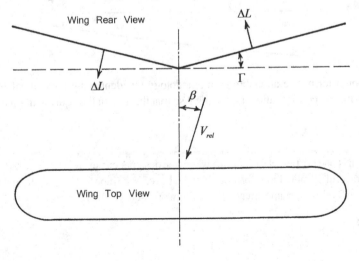

Flight Dynamics

The dihedral angle gives rise to equal and opposite lift increments, ΔL, when the lifting surface has a side-slip angle, β. Namely, the sideslip adds an angle-of-attack increment $\Delta\alpha$ to the right panel and subtracts an equal $\Delta\alpha$ from the left panel, where:

$$|\Delta\alpha| = \beta \sin \Gamma$$

Therefore, the differential lift on each panel is:

$$|\Delta L| = (C_{L\alpha})_{ls} |\Delta\alpha| \frac{\rho V_{rel}^2 S_{ls}}{4} = (C_{L\alpha})_{ls} \beta \sin \Gamma \frac{\rho V_{rel}^2 S_{ls}}{4}$$

also, since the lateral force $Y = 2|\Delta L| \sin \Gamma$, then one has that:

$$(C_{Y\beta})_{ls} = -(C_{L\alpha})_{ls} \sin^2 \Gamma (S_{ls}/S)$$

However, there is a mutual interference between the lift distributions on both halves of the lifting surface, so that this value is reduced. This is accounted for by a correction factor, K_4, which is a function of the taper ratio and the aspect ratio, and may be found in NACA Report 823 by Purser and Campbell:

$$\underline{(C_{Y\beta})_{ls}^{dihedral} = -K_4 (C_{L\alpha})_{ls} \frac{S_{ls}}{S} \sin^2 \Gamma}$$

typically, K_4 ranges from 0.7 to 0.8.

Further, if the lifting surface is swept, there is an *additional* effect given by DATCOM (5.1.1.1-1):

$$\underline{(C_{Y\beta})_{ls}^{sweep} = \left[\frac{6 \tan \Lambda_{c/4} \sin \Lambda_{c/4}}{\pi (AR)(AR + 4\cos \Lambda_{c/4})} \right] (C_L)_{ls}^2}$$

Observe how this is a function of the lifting surface's lift coefficient. So, this sweep contribution to the derivative is not constant throughout the aircraft's flight envelope.

In total, one has:

$$\underline{(C_{Y\beta})_{ls} = \left[(C_{Y\beta})_{ls}^{dihedral} + (C_{Y\beta})_{ls}^{sweep} \right] \tilde{\eta}_{ls}}$$

where $\tilde{\eta}_{ls}$ is a *lateral* wing-body interference factor for the lifting surface. This value is usually close to unity.

> **ASIDE**
>
> Note that if the airplane has a V-Tail instead of the usual cruciform configuration of fin and stabilizer, then these equations for the dihedralled lifting surface may be applied to obtain $(C_{Y\beta})_t$.
>
> **END ASIDE**

If the aircraft has a propeller, this also has a lateral-force contribution, simply given by:

$$\underline{(C_{Y\beta})_{prop} \approx -(C_{L\alpha})_{prop} S_{prop}/S}$$

where the estimation of $(C_{L\alpha})_{prop}$ is described in the "Complete-Aircraft Aerodynamics" section of Chapter 2.

Next, consider the yawing moment due to sideslip, $C_{N\beta}$:

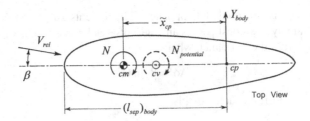

An isolated body-of-revolution (no wing, tail, etc.) has a very strong upsetting moment, as discussed in Chapter 2. This value may be estimated from:

$$N_{body} = N_{potential} + \tilde{x}_{sep} Y_{body}$$

where $N_{potential} = -\rho U_0^2 Vol(k_2 - k_1)\beta$ and $\tilde{x}_{sep}$ is the center-of-action of the flow-separation force on the body (note that this is relative to the aircraft's mass-center, positive forward from that point). Therefore,

$$\left(C_{N\beta}\right)_{body} = \frac{-2(k_2 - k_1)(Vol)_{body}}{Sb} + \frac{\tilde{x}_{sep}}{b}\left(C_{Y\beta}\right)_{body}$$

Upon referring to the above figure, $\left(l_{sep}\right)_{body}$ is approximately 70% of the total body length.

A more accurate methodology accounting for general fuselage shapes with wings was presented by Perkins and Hage which, in turn, was based on unpublished data from the North American Aviation Company (now part of Boeing). The empirical equation is:

$$\left(C_{N\beta}\right)_{body} = -0.96 K_\beta \frac{S_{side}}{S} \frac{l_{body}}{b} \left(\frac{h_1}{h_2}\right)^{1/2} \left(\frac{w_2}{w_1}\right)^{1/3}$$

where S_{side} is the sideways "shadow" area of the fuselage, b is the wing span and the other parameters are defined by the figure below, which is adapted from Figure 8-4 by Perkins & Hage in "Airplane Performance Stability and Control", John Wiley & Sons, 1960 Edition:

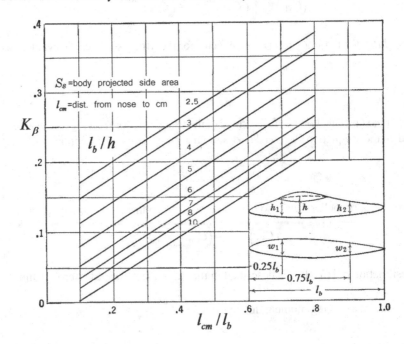

Flight Dynamics

What this methodology attempts to do is to account for how the wing affects the lateral force produced by the body. This is still, in the spirit of the piece-meal method, the body's contribution alone; and the wing's contribution is described later.

Finally, the most refined version of this is presented in DATCOM (5.2.3.1):

$$\left(C_{N\beta}\right)_{body} = -K_N K_{RN} \frac{S_{side}}{S} \frac{l_b}{b} \times 57.3$$

K_N is an empirical parameter obtained from DATCOM's Figure 5.2.3.1-8, which is presented below. Note that this type of plot is called a nomograph.

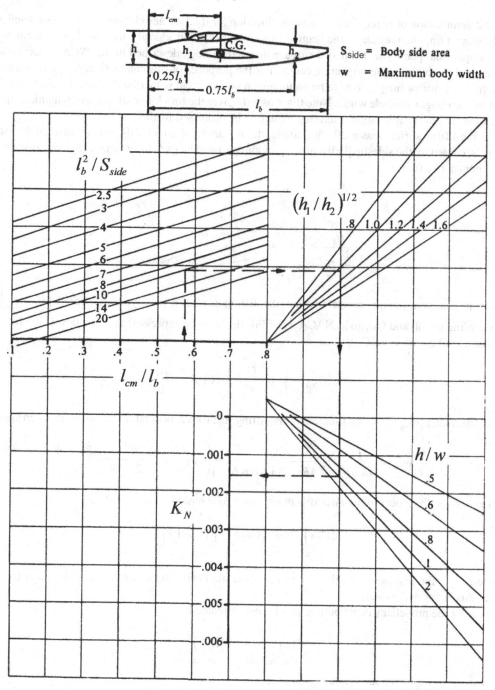

Further, a curve fit to DATCOM Figure 5.2.3.1-9 gives an equation for K_{RN}:

$$K_{RN} = 0.2041 \times \ln(RN)_{body} - 1.82$$

where "ln" is the natural logarithm and $(RN)_{body}$ is the Reynolds Number based on the body's length.

Next, the vertical fin's contribution is simply obtained from its lateral force acting upon its moment arm from the mass-center:

$$\left(C_{N\beta}\right)_f = \frac{x_f}{b}\left(C_{Y\beta}\right)_f$$

The determination of x_f requires an accurate location for the fin's aerodynamic center. For a generally shaped fin, one may apply the "equivalent-fin" concept introduced earlier to define an equivalent trapezoidal planform. Then, one can use the methodology described in the "Wings" section of Chapter 2 to obtain the aerodynamic center. For the purposes of this calculation, one assumes that the fin has a mirror image, so as to be an imagined symmetrical wing (the AR for *this* calculation is that of the imagined whole wing). Note that this only gives the fin's longitudinal aerodynamic-center location. The vertical location is another matter, to be discussed further on.

If the lifting surface has a dihedral angle, the $\Delta\alpha$ angle-of-attack differential between the two panels, caused by the sideslip β (discussed previously), introduces a differential drag and horizontal component of lift.

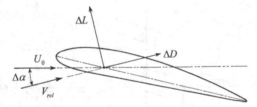

This produces a yawing moment about the lifting surface's aerodynamic center, $\left[\left(C_{N\beta}\right)_{ac}\right]_{ls}$. According to Toll and Queijo in NACA TN-1581, this may be expressed as a function of the lifting surface's lift coefficient, $(C_L)_{ls}$:

$$\left[\left(C_{N\beta}\right)_{ac}\right]_{ls} = \left(\frac{C_{N\beta}}{C_L^2}\right)_{ls}(C_L)_{ls}^2 \frac{S_{ls}}{S}\frac{b_{ls}}{b}$$

The parameter $\left(C_{N\beta}/C_L^2\right)_{ls}$ is a function of the lifting surface's aspect ratio and ¼-chord sweep angle:

$$\left(\frac{C_{N\beta}}{C_L^2}\right)_{ls} = \frac{1}{4\pi AR} - \frac{\tan \Lambda_{c/4}}{\pi AR(AR + 4\cos \Lambda_{c/4})}\left(\cos \Lambda_{c/4} - \frac{AR}{2} - \frac{AR^2}{8\cos \Lambda_{c/4}}\right)$$

Therefore, with respect to the aircraft's mass-center, one has:

$$\left(C_{N\beta}\right)_{ls} = \frac{x_{ls}}{b}\left(C_{Y\beta}\right)_{ls} + \left[\left(C_{N\beta}\right)_{ac}\right]_{ls}$$

where x_{ls} is the distance from the wing's aerodynamic center to the aircraft's mass-center (recall that this is positive forward).

As for the propeller's contribution, this is approximated by:

$$\left(C_{N\beta}\right)_{prop} = \frac{x_{prop}}{b}\left(C_{Y\beta}\right)_{prop}$$

Next considered is the aircraft's rolling moment with respect to sideslip, $C_{\tilde{L}\beta}$. The body's contribution is usually negligible. However, an estimate may be obtained from:

$$\left(C_{\tilde{L}\beta}\right)_{body} \approx -\frac{z_{ref}}{b}\left(C_{Y\beta}\right)_{body}$$

where z_{ref} is measured from the mass-center to the body reference line (positive down).

In a similar fashion, the vertical location of the fin's aerodynamic center contributes a rolling moment:

$$\left(C_{\tilde{L}\beta}\right)_{fin} = -\frac{z_f}{b}\left(C_{Y\beta}\right)_{fin}$$

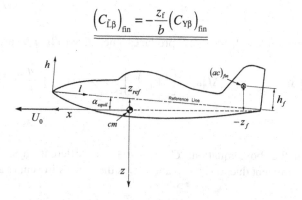

ASIDE

With respect to the figure above, recall that the body-fixed l and h coordinates are oriented with respect to the aircraft's reference line (or "loft line"), with an origin affixed to a point along this line (usually the "nose"). At different flight speeds, the reference line will generally have different equilibrium angles, α_{equil}, relative to the equilibrium velocity vector, $\vec{U}_0$. Therefore, the locations of the fin's aerodynamic center, x_f and z_f, will also generally vary with flight speed. These values may be found from the following equations:

$$x_f = \left(l_{cm} - l_f\right)\cos\alpha_{equil} + \left(h_{cm} - h_f\right)\sin\alpha_{equil}$$

$$z_f = \left(h_{cm} - h_f\right)\cos\alpha_{equil} - \left(l_{cm} - l_f\right)\sin\alpha_{equil}$$

Similar equations apply to the other components of the aircraft, such as the aerodynamic centers of the lifting surfaces.

END ASIDE

For the location of the aerodynamic center, it is assumed that the fin constitutes a "half wing" so that $\left(h_{ac}\right)_{fin}$ may be estimated from a method described by A. H. Yates ("Notes on the Mean Aerodynamic Chord and the Mean Aerodynamic Centre of a Wing", *Journal of the Royal Aeronautical Society*, Vol. 56, June 1952, page 461). Namely, if the leading edge of the illustrated fin is extended to the reference line, so that the planform is a swept trapezoid with constant taper, as illustrated earlier, then:

$$\left(h_{ac}\right)_f \approx 0.42\, b_f$$

where b_f is the span of the fin, measured from the reference line to its tip again, as shown earlier. Of course, this is only an approximation because this value is influenced by the size of the body below the fin (as was the case for the estimation of $\left(C_{Y\beta}\right)_{fin}$), as well as the position of the stabilizer. However, barring more precise analytical or experimental values, this suffices for an initial analysis. Also, for most airplanes, the main lifting surface (e.g., "wing") is a significantly greater contributor to $C_{\tilde{l}\beta}$.

In particular, the lifting surface's contribution is due to its dihedral angle, Γ, its sweepback angle, $\Lambda_{c/4}$ and its vertical location, z_{ls}.

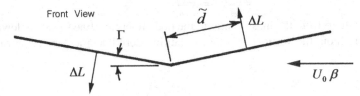

For example, the figure readily shows that Γ produces a negative rolling moment for a positive β:

$$(\Delta \tilde{L})_{dihedral} = -2|\Delta L|\tilde{d} \rightarrow \left[(C_{\tilde{l}\beta})_{ls}\right]_{dihedral} \approx -(C_{L\alpha})_{ls} \frac{\tilde{d}}{b} \frac{S_{ls}}{S} \Gamma$$

ASIDE

Recall again that in the above equation, "C_L" is used in two different ways. The term $C_{\tilde{l}\beta}$ is the rolling-moment coefficient due to sideslip and $C_{L\alpha}$ is the panel's lift-curve slope.

END ASIDE

Now, accurate methods exist for determining $\left[(C_{\tilde{l}\beta})_{ls}\right]_{dihedral}$, which may be represented by an extension of the previous equation to give:

$$\left[(C_{\tilde{l}\beta})_{ls}\right]_{dihedral} = \left[(C_{\tilde{l}\beta})_{\Gamma}\right]_{ls} \frac{S_{ls}}{S} \Gamma_{ls}$$

Campbell and McKinney in NACA TR-1098 give graphs for estimating $\left[(C_{\tilde{l}\beta})_{\Gamma}\right]_{ls}$ for straight-tapered un-swept wings of different taper ratios and a range of aspect ratios, as shown below:

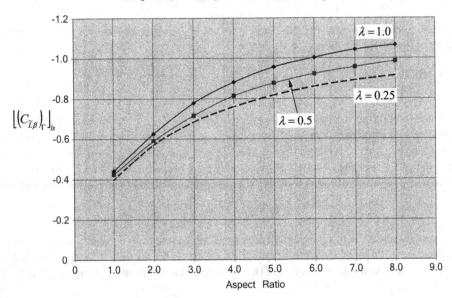

Flight Dynamics

Curve-fit equations from these graphs are:

$$\lambda = 1.0: \quad \left[(C_{\tilde{l}\beta})_\Gamma\right]_{ls} \approx -0.1941 - 0.2743\,AR + 0.0306(AR)^2 - 0.0012(AR)^3$$

$$\lambda = 0.5: \quad \left[(C_{\tilde{l}\beta})_\Gamma\right]_{ls} \approx -0.2080 - 0.2415\,AR + 0.0274(AR)^2 - 0.0012(AR)^3$$

$$\lambda = 0.25: \quad \left[(C_{\tilde{l}\beta})_\Gamma\right]_{ls} \approx -0.1844 - 0.2505\,AR + 0.0326(AR)^2 - 0.0016(AR)^3$$

Note that this is for dihedral angle Γ measured in *radians*. Also, the equations are only valid within the aspect-ratio range shown on the graphs. For lifting surfaces with larger aspect ratios, it is best to "eyeball" an extrapolation of the curves. Clearly, they are monotonically approaching some limiting value at infinite AR.

There is also a moderate attenuation of $(C_{\tilde{l}\beta})_\Gamma$ due to sweep, for which NACA TR-1098 gives the following equation:

$$\left[(C_{\tilde{l}\beta})_\Gamma\right]_{ls} = \frac{(AR+4)\cos\Lambda_{c/4}}{AR + 4\cos\Lambda_{c/4}}\left[(C_{\tilde{l}\beta})_\Gamma\right]_{ls}^{\text{no sweep}}$$

ASIDE

The methodology used for obtaining these graphs in NACA TR-1098 was the best for the time (1950), using lifting-line theory. Modern methodologies, using vortex-lattice or CFD analyses, would no doubt give more accurate values over any range of planform geometries desired. However, it has been the author's experience that this simpler method suffices for most initial-design purposes.

END ASIDE

Sweep can also have an effect in conjunction with the lifting-surface's lift coefficient. This is because the $C_{l\alpha}$ of a given section varies as the $\cos(\Lambda - \beta)$:

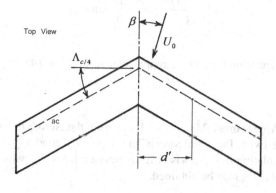

$$(C_{l\alpha})_{\text{section}} = (C_{l\alpha})_0 \cos(\Lambda - \beta)$$

This result comes from infinite-wing theory, but it may be approximately applied to large aspect-ratio wings. Note that:

$$(Rolling\ Moment)_{\text{sweep}} = \Delta\tilde{L}_{\text{sweep}} \approx -(\Delta L_{\text{right}} - \Delta L_{\text{left}})\,d'$$

where d' is the spanwise center-of-lift distance on the lifting-surface's half panel. In non-dimensional form, this equation becomes:

$$\left[(C_{\tilde{L}})_{ls}\right]_{sweep} \approx -(C_{L\alpha})_0 \left[\cos(\Lambda_{c/4}-\beta) - \cos(\Lambda_{c/4}+\beta)\right]\alpha_{ZLL} \frac{1}{2}\frac{\hat{d}}{b_{ls}}\frac{S_{ls}}{S}\frac{b_{ls}}{b} \rightarrow$$

$$\left[(C_{\tilde{L}})_{ls}\right]_{sweep} \approx -(C_{L\alpha})_0 \frac{d'}{b_{ls}}\frac{S_{ls}}{S}\frac{b_{ls}}{b}(\sin\Lambda_{c/4})\alpha_{ZLL}\beta$$

where $(C_{L\alpha})_0$ is the lift-curve slope of the *un-swept* lifting surface's half panel.

Now, the lift coefficient of the lifting surface when $\beta = 0$ is given by:

$$(C_{L0})_{ls} = \left[(C_{L\alpha})_0 \cos\Lambda_{c/4}\right]\alpha_{ZLL}, \text{ from which one obtains } (C_{L\alpha})_0 = \frac{(C_{L0})_{ls}}{\cos\Lambda_{c/4}\,\alpha_{ZLL}}$$

Therefore, the rolling-moment equation becomes:

$$\left[(C_{\tilde{L}\beta})_{ls}\right]_{sweep} \approx -(C_{L0})_{ls}\frac{d'}{b_{ls}}\frac{S_{ls}}{S}\frac{b_{ls}}{b}(\tan\Lambda_{c/4})$$

upon using terminology from NACA TR-1098, this may be written as:

$$\left[(C_{\tilde{L}\beta})_{ls}\right]_{sweep} = \left[\frac{(C_{\tilde{L}\beta})_w}{C_L}\right]_{ls}(C_{L0})_{ls}\frac{S_{ls}}{S}\frac{b_{ls}}{b}$$

where values for $\left[(C_{\tilde{L}\beta})_w / C_L\right]_{ls}$ may be obtained from Figure 8 in NACA TR-1098, as functions of taper ratio and sweep angle (including negative sweep). However, if the span-wise lift distribution of the wing is known, then d' may be calculated and the sweep-effect parameter can be estimated from:

$$\left[\frac{(C_{\tilde{L}\beta})_w}{C_L}\right]_{ls} \approx -\frac{d'}{b_{ls}}\tan\Lambda_{c/4}$$

Pending an accurate calculation for d'/b_{ls}, a representative value of 0.43 may be chosen.

ASIDE

In the section "An Approximate Method for Estimating the Aerodynamic Characteristics of Wings with Variable Twist, Taper and Sweep" in Chapter 4, a good approximate methodology is presented for calculating the spanwise lift distribution for wings with arbitrary geometry. From this, a value of d'/b_{ls} may be obtained.

END ASIDE

It is important to observe that the sweep effect on the rolling moment is linearly dependent on the lifting surface's lift coefficient. The greater the value of $(C_{L0})_{ls}$, the stronger the effect is. Some low-AR highly-swept airplanes can achieve lateral stability with no dihedral angle at all.

Now, in some cases, a wing may have multiple dihedral angles:

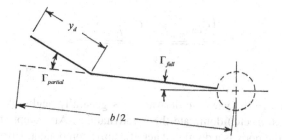

In this instance, from TR-1098, $(C_{L\beta})_\Gamma$ is modified by,

$$\left[\left(C_{\tilde{L}\beta}\right)_\Gamma\right]_{ls}^{4-panel} = \left[\left(C_{\tilde{L}\beta}\right)_\Gamma\right]_{ls}^{2-panel} \left[\Gamma_{full} + K_d \Gamma_{partial}\right]$$

where K_d is given by the following graph:

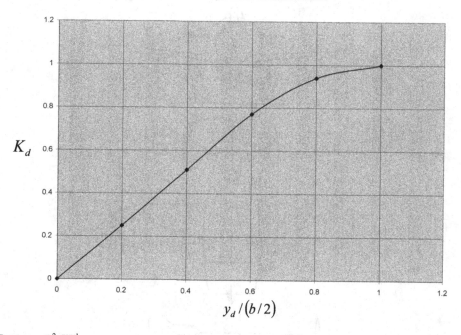

And $\left[\left(C_{\tilde{L}\beta}\right)_\Gamma\right]_{ls}^{2-panel}$ is the previously derived $\left[\left(C_{\tilde{L}\beta}\right)_\Gamma\right]_{ls}$ (modified by the sweep angle if appropriate).

Also, the vertical position of the lifting surface relative to the mass-center likewise has a contribution to $(C_{\tilde{L}\beta})_{ls}$, as given by:

$$(Rolling\ Moment)_{position} = -\Delta Y_{ls}\, z_{ls} \rightarrow \left[\left(C_{\tilde{L}\beta}\right)_{ls}\right]_{position} = -\left(C_{Y\beta}\right)_{ls} z_{ls}/b$$

Therefore, the total rolling-moment derivative from the lifting surface, due to sideslip β, is given by:

$$\underline{\underline{\left(C_{\tilde{L}\beta}\right)_{ls} = \left[\left(C_{\tilde{L}\beta}\right)_{ls}\right]_{position} + \left[\left(C_{\tilde{L}\beta}\right)_{ls}\right]_{dihedral} + \left[\left(C_{\tilde{L}\beta}\right)_{ls}\right]_{sweep}}}$$

Finally, the propeller's contribution may be approximated by:

$$\left(C_{\tilde{L}\beta}\right)_{prop} = -\left(C_{Y\beta}\right)_{prop} z_{prop}/b$$

The $\hat{r}$ Derivatives

The body's contribution to the yaw-rate derivatives is generally negligible for aircraft where the lifting surfaces (wing, tail, etc.) dominate the aerodynamics. An exception would be blimp-like configurations, where the body is a dominant aerodynamic component. However, for this chapter it will be assumed that the body's contribution is nil.

ASIDE

The figure below shows a drawing of the 1932 Gee Bee R-1 racer:

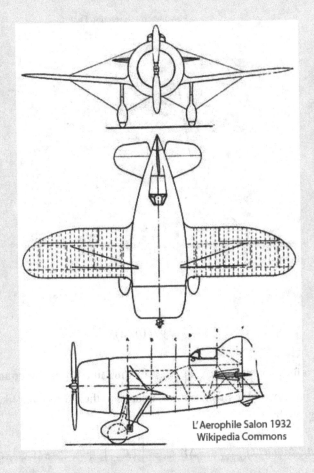

L'Aerophile Salon 1932
Wikipedia Commons

This is a rare example of an airplane where the body's contribution definitely cannot be ignored!

END ASIDE

Flight Dynamics

The fin's contribution is illustrated below:

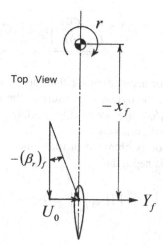

It is seen that, for a fin-aft configuration, a positive yaw rate produces a negative β angle at the fin's aerodynamic center:

$$(\beta_r)_f = \frac{x_f r}{U_0}$$

Therefore, the isolated fin's side-force coefficient is:

$$\left(C_Y^*\right)_f = \left(C_{Y\beta}^*\right)_f (\beta_r)_f = \left(C_{Y\beta}^*\right)_f \frac{x_f}{U_0} r = 2\left(C_{Y\beta}^*\right)_f \frac{x_f}{b} \hat{r}$$

However, $(\beta_r)_f$ is usually modified by the rest of the aircraft. This may be accounted for by a sidewash effect, $\partial \sigma / \partial \hat{r}$, so that:

$$(\beta_r)_f = \left(2\frac{x_f}{b} + \frac{\partial \sigma_f}{\partial \hat{r}}\right) \hat{r}$$

ASIDE

Note the sign of $\partial \sigma_f / \partial \hat{r}$. In this aft-fin case, when it is positive, it acts to attenuate the net lateral angle-of-attack at the fin. Compare this with the previous side-slip case, where $\partial \sigma_f / \partial \beta$ was assigned a negative sign in order to provide attenuation.

END ASIDE

Further, as with $\left(C_{Zq}\right)_{ls}$, there is a contribution from the fin's rotation about its aerodynamic center, which may be estimated by:

$$\left[\left(C_{Yr}\right)_{ac}\right]_f \approx -\left(C_{Y\beta}^*\right)_f \frac{\bar{c}_f}{b} \eta_f$$

Therefore, one obtains the total fin contribution as:

$$(C_{Yr})_f = (C_{Y\beta}^*)_f \left(2\frac{x_f}{b} + \frac{\partial \sigma_f}{\partial \hat{r}}\right)\eta_f + \left[(C_{Yr})_{ac}\right]_f$$

where η_f is the fin-efficiency factor accounting for the portion of the body covering the fin, and the attenuation or augmentation of the dynamic pressure at the fin. Also, note that values for the rotational side-slip parameter, $\partial \sigma_f/\partial \hat{r}$, are not readily available and choosing this to be zero is a reasonable assumption for most configurations.

The lifting-surface's contribution is obtained in the same way. For low to moderate dihedral angles, the planform contribution is negligible:

$$\left[(C_{Yr})_{ac}\right]_{ls} \approx 0$$

so,

$$(C_{Yr})_{ls} = (C_{Y\beta})_{ls}\left(2\frac{x_{ls}}{b} + \frac{\partial \sigma_{ls}}{\partial \hat{r}}\right)$$

Note that the lateral lifting-surface efficiency, $\tilde{\eta}_{ls}$, is already incorporated into $(C_{Y\beta})_{ls}$. Further, the rotational side-slip parameter, $\partial \sigma_{ls}/\partial \hat{r}$, is also assumed to be negligible.

The propeller's contribution is approximated by:

$$(C_{Yr})_{prop} = 2(C_{Y\beta})_{prop} x_{prop}/b$$

Next, the fin's yaw rate gives rise to a yawing moment:

$$N_f = Y_f x_f$$

Further, as with $(C_{Mq})_{ls}$, there is a contribution from the fin's rotation about its aerodynamic center. This may be approximated by:

$$\left[(C_{Nr})_{ac}\right]_f \approx -\frac{\pi}{4}\left(\frac{\bar{c}_f}{b}\right)^2 \frac{S_f}{S}\eta_f$$

Therefore, one obtains:

$$(C_{Nr})_f = \frac{x_f}{b}(C_{Yr})_f + \left[(C_{Nr})_{ac}\right]_f$$

The lifting surface acts in the same way, as well as having a value about its aerodynamic center due to its planform. This planform effect is due primarily to the difference in drag between the two halves, caused by yaw rate, and NACA TR-1098 expresses it as:

$$\left[(C_{Nr})_{ac}\right]_{ls} = \left[\frac{(\Delta C_{Nr})_1}{C_L^2}\right]_{ls} \frac{S_{ls}}{S}\left(\frac{b_{ls}}{b}\right)^2 (C_L)_{ls}^2 + \left[\frac{(\Delta C_{Nr})_2}{C_{D0}}\right]_{ls} \frac{S_{ls}}{S}\left(\frac{b_{ls}}{b}\right)^2 (C_{Dprofile})_{ls}$$

where $\left(C_{\text{D profile}}\right)_{\text{ls}}$ is the profile drag of the lifting surface, given by:
$$C_{\text{D profile}} = C_D - C_L^2/\pi AR$$
and the coefficients are functions of the taper ratio λ, sweepback angle $\Lambda_{c/4}$, and aspect ratio AR. Observe that the induced-drag effect is accounted for by the first term, being a function of the lift coefficient squared, and the profile-drag effect is accounted for by the second term. The curves for these are obtained from NASA TR-1098, which were redrawn in "Dynamics of Flight" by Bernard Etkin (John Wiley & Sons, 1959) and are adapted and reproduced below.

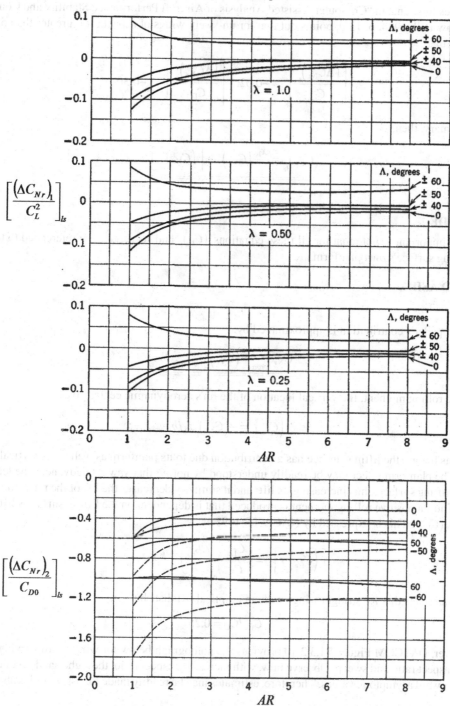

> **ASIDE**
>
> Recall from the "Wings" section of Chapter 2 that the profile-drag coefficient is also, generally, a function of C_L and C_L^2.
>
> **END ASIDE**

Frederick Smetana, in "Computer Assisted Analysis of Aircraft Performance Stability and Control" (McGraw-Hill, 1984, p. 135), points out that for un-swept wings of aspect ratio greater than 5, one obtains that:

$$\left[\frac{(\Delta C_{Nr})_1}{C_L^2}\right]_{ls} \approx -0.02, \quad \left[\frac{(\Delta C_{Nr})_2}{C_{D0}}\right]_{ls} \approx -0.30$$

In summary, then,

$$(C_{Nr})_{ls} = \frac{x_{ls}}{b}(C_{Yr})_{ls} + \left[(C_{Nr})_{ac}\right]_{ls}$$

> **ASIDE**
>
> It is important to note that for all these equations, $(C_L)_{ls}$ and $(C_{Dprofile})_{ls}$ are referenced to the lifting surface's *own* planform area.
>
> **END ASIDE**

The propeller's contribution is approximated by:

$$(C_{Nr})_{prop} \approx (C_{Yr})_{prop} x_{prop}/b$$

For the rolling moment, the vertical location of the fin's aerodynamic center gives:

$$(C_{\tilde{L}r})_f = -(C_{Yr})_f z_f/b$$

Also, as before, the lifting surface has a contribution due to its planform as well as its vertical location. The planform effect may be readily understood by noting that yaw rate advances the left half of the lifting surface, thus increasing its lift; and it similarly decreases the lift of the retreating right half. Thus, a positive rolling moment is produced that is dependent on the lifting surface's lift coefficient, $(C_L)_{ls}$. This is expressed by NACA TR-1098 as:

$$\left[(C_{\tilde{L}r})_{ac}\right]_{ls} = \left(\frac{C_{\tilde{L}r}}{C_L}\right)_{ls}\left(\frac{b_{ls}}{b}\right)^2 \frac{S_{ls}}{S}(C_L)_{ls}$$

Smetana states that for straight wings with an elliptical lift distribution,

$$C_{\tilde{L}r}/C_L = 0.25$$

However, DATCOM Figure 7.1.3.2-10 provides the nomograph below for wings with varying taper ratio, aspect ratio and sweep. Observe how, with increasing aspect ratio, the right-hand set of curves trend toward asymptotic values. Therefore, estimates may be readily made for aspect ratios above 10.

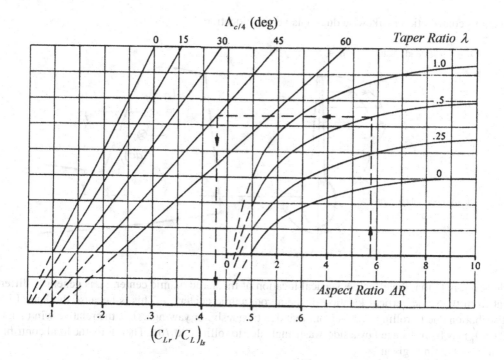

In summary, then, the contribution of the yawing lifting surface to rolling moment is:

$$\left(C_{\tilde{L}r}\right)_{ls} = -\frac{z_{ls}}{b}\left(C_{Yr}\right)_{ls} + \left[\left(C_{\tilde{L}r}\right)_{ac}\right]_{ls}$$

Also, the propeller's contribution is approximated by:

$$\left(C_{\tilde{L}r}\right)_{prop} = -\left(C_{Yr}\right)_{prop} z_{prop}/b$$

The $\hat{p}$ Derivatives:

The rolling-moment derivatives about a body's reference line may be taken as zero. Therefore, it may be assumed that the only effect is due to the vertical location of the reference line from the mass-center:

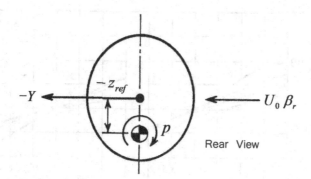

Rear View

$$\left(C_Y\right)_{body} = -\left(C_{Y\beta}\right)_{body}\frac{z_{ref}}{U_0}p = -\left(C_{Yp}\right)_{body}\frac{z_{ref}}{U_0}\frac{2U_0}{b}\hat{p} \rightarrow \left(C_{Yp}\right)_{body} = -2\frac{z_{ref}}{b}\left(C_{Y\beta}\right)_{body}$$

The fin's contribution is likewise due to its vertical location:

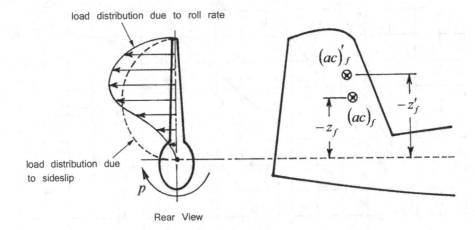

However, note that the effective vertical location of the aerodynamic center, z'_f, is generally different from the static vertical location of the aerodynamic center, z_f. This is because the lateral lift distribution due to rolling differs from that due to sideslip or yawing. Also, the relative wind angle at $(ac)'_f$ may be influenced by a side-wash angle due to rolling, $\partial \sigma_f / \partial \hat{p}$. Therefore, the total contribution from the fin is given by:

$$(C_{Yp})_f = (C^*_{Y\beta})_f \left(-2\frac{z'_f}{b} + \frac{\partial \sigma_f}{\partial \hat{p}} \right) \eta_f$$

The references offer no clear methodology for obtaining z'_f, so it will suffice to choose this value to be 15% greater than z_f

$$z'_f \approx 1.15 z_f$$

When the fin is mounted aft of the wing, as for conventional configurations, swirl in the wake from the wing may give a significant side-wash effect. This was studied by William H. Michael Jr. in NACA Report 1086, "Analysis of the Effects of Wing Interference on The Tail Contributions to the Rolling Derivatives", and the graphs below give summary results from this research for wings of aspect-ratio 3.5 and 6.0, as functions of $b_f/(b_w/2)$ and wing angle of attack.

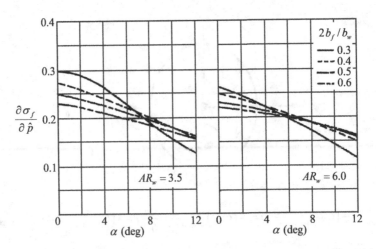

These theoretical results ignored fuselage efforts and, when compared with experiments on models with fuselages, the author concluded that for such cases the value of $\partial \sigma_f/\partial \hat{p}$ at $\alpha = 0$ may be used throughout the angle-of-attack range. Therefore, for $AR_w = 6.0$, approximate values are given by the following curve-fit equation:

$$\partial \sigma_f/\partial \hat{p} \approx \left[0.178 + 0.827(2b_f/b_w) - 2.35(2b_f/b_w)^2 + 1.833(2b_f/b_w)^3 \right] \times \text{sign}|z_f'|$$

This effect is by no means insignificant, as shown in the numerical example presented further in this chapter. It will be assumed that the values for $AR_w = 6.0$ are also representative for wings of moderate aspect ratios above this. However, it would be a useful exercise to program Michael's equations for a wider range of aspect ratios.

The lifting surface's contribution is due to its vertical location and its planform. The planform effect was found from experiment (NACA TR-968 by Goodman and Fisher) to be a function of AR and $\Lambda_{c/4}$, and NACA TR-1098 expresses this as:

$$\left[(C_{Yp})_{ac}\right]_{ls} = \left[\left(\frac{AR + \cos \Lambda_{c/4}}{AR + 4\cos \Lambda_{c/4}}\right)\tan \Lambda_{c/4} + \frac{1}{AR}\right]\frac{b_{ls}}{b}\frac{S_{ls}}{S}(C_L)_{ls}$$

> **ASIDE**
>
> The $1/AR$ term accounts for differential wing-tip suction. For example, at a positive lift coefficient, $(C_L)_{ls}$, the additional angle-of-attack at the right wing tip, due to rolling, increases the wing-tip suction in that area, whereas the opposite effect occurs at the left wing tip. This results in an additional force in the y-direction. Clearly, this depends on smoothly rounded wing tips. If the tips are sharp, the 1/AR term should be ignored.
>
> **END ASIDE**

Therefore, the total contribution from the lifting surface is:

$$(C_{Yp})_{ls} = (C_{Y\beta})_{ls}\left(-2\frac{z_{ls}}{b} + \frac{\partial \sigma_{ls}}{\partial \hat{p}}\right) + \left[(C_{Yp})_{ac}\right]_{ls}$$

If the lifting surface is a V-Tail aft of the wing, then one may determine the side-wash parameter, $\partial \sigma_{ls}/\partial \hat{p}$, from the same equation as for the aft-located fin. In this case, $b_{tail} \sin \Gamma_{tail}/2$ is substituted for b_f, where b_{tail} is the tip-to-tip flattened span of the V-Tail. However, for the wing alone, this is negligible.

For the propeller, its contribution is approximated by:

$$(C_{Yp})_{prop} = -2(C_{Y\beta})_{prop} z_{prop}/b$$

Next, the body's rolling moment produces a yawing moment:

$$(C_{Np})_{body} = -2\frac{z_{ref}}{b}(C_{N\beta})_{body}$$

Similarly, the fin's contribution is:

$$(C_{Np})_f = \frac{x_f}{b}(C_{Yp})_f$$

As before, the lifting surface's contribution is due to its location and its planform. The planform's effect may be readily seen by noting that p increases the relative angle of attack of the right half of the lifting surface, while decreasing that of the left half. This angle change has two effects:

1. The distributed profile-drag coefficients, C_d, along the lifting surface changes with the local lift coefficients, C_l. This integrates along the span to give a yawing moment.
2. This differential change in angle of attack will change each section's leading-edge suction force. When the lifting surface is at a positive lift coefficient, $(C_L)_{ls}$, this generally gives a negative yawing moment when integrated along the span.

NACA TR-1098 expresses these two contributions in the following equation:

$$\left[(C_{Np})_{ac}\right]_{ls} = \left[\left(\frac{(\Delta C_{Np})_1}{C_L}\right)_{ls}(C_L)_{ls} + \left(\frac{(\Delta C_{Np})_2}{(C_{D0})_\alpha}\right)_{ls}\left[(C_{D\text{profile}})_\alpha\right]_{ls}\right]\left(\frac{b_{ls}}{b}\right)^2\frac{S_{ls}}{S}$$

The graph from TR-1098 for the parameters is reproduced below (as redrawn by Etkin):

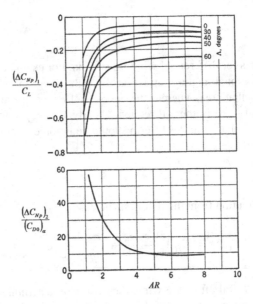

The derivative of the profile drag with the angle of attack, $(C_{D\text{profile}})_\alpha$, may be obtained from the methods described in Chapter 2 where, according to TR-1098,

$$(C_{D\text{profile}})_\alpha = \frac{\partial}{\partial \alpha}\left(C_D - \frac{C_L^2}{\pi AR}\right)$$

However, the authors of TR-1098 point out that, for wings with smooth leading edges and attached flow, this may generally be assumed to be zero.

Alternatively, an equation is offered in NACA TN-1581 (in which it is clear that the $(C_{D\text{profile}})_\alpha$ effect is ignored):

$$\left[(C_{Np})_{ac}\right]_{ls} = (C_L)_{ls}\left(\frac{AR+4}{AR+4\cos\Lambda_{c/4}}\right)\left[1+6\left(1+\frac{\cos\Lambda_{c/4}}{AR}\right)\frac{\tan^2\Lambda_{c/4}}{12}\right]\left(\frac{b_{ls}}{b}\right)^2\frac{S_{ls}}{S}\left(\frac{C_{Np}}{C_L}\right)_{\Lambda=0}$$

where $(C_{Np}/C_L)_{\Lambda=0}$ is approximated by:

$$(C_{Np}/C_L)_{\Lambda=0} \approx -0.0004 - 0.0099\,AR + 0.0003(AR)^2$$

Therefore, in summary, the yawing moment produced by the lifting surface undergoing a roll rate is:

$$(C_{Np})_{ls} = \frac{x_{ls}}{b}(C_{Yp})_{ls} + \left[(C_{Np})_{ac}\right]_{ls}$$

Further, the propeller's contribution is approximated by:

$$(C_{Np})_{prop} = (C_{Yp})_{prop}\, x_p/b$$

The rolling moment due to roll rate will be considered next. The body's contribution may be estimated from:

$$(C_{\tilde{L}p})_{body} \approx -\frac{z_{ref}}{b}(C_{Yp})_{body}$$

Similarly, the fin's contribution may be obtained from:

$$(C_{\tilde{L}p})_{fin} = -\frac{\hat{z}'_f}{b}(C_{Yp})_f$$

And, again, the lifting surface's contribution is due to its location and planform. The planform's contribution is generally very significant, which may be understood by noting that the $\Delta\alpha$ angles generated along the span by roll rate give rise to $\Delta(lift)$ values that, in turn, give an opposing rolling moment. The coefficient of this moment, $(C_{\tilde{L}p})_{planform}$, is generally expressed as a function of AR, $\Lambda_{c/4}$ and taper ratio λ, and are given by TR-1098 as

$$(C_{\tilde{L}p})_{planform} = (C_{\tilde{L}p})_0 \frac{(C_{L\alpha})_{CL}}{(C_{L\alpha})_{CL=0}} - \frac{1}{8}\frac{(C_L^2)_{ls}}{\pi\, AR\cos^2\Lambda_{c/4}}\left(1 + 2\sin^2\Lambda_{c/4}\frac{AR + 2\cos\Lambda_{c/4}}{AR + 4\cos\Lambda_{c/4}}\right) - \frac{1}{8}C_{Dprofile}$$

where $(C_{L\alpha})_{CL}/(C_{L\alpha})_{CL=0}$ is the ratio of the lift-curve slope at lift coefficient, C_L, to that at $C_L = 0$. Also, $C_{Dprofile}$ is the profile-drag coefficient defined earlier; and there is a correction accounting for the airfoil's two-dimensional lift-curve slope, $C_{l\alpha}$:

$$(C_{\tilde{L}p})_0 = (C_{\tilde{L}p})'_0 \frac{AR + 4\cos\Lambda_{c/4}}{(2\pi/C_{l\alpha})AR + 4\cos\Lambda_{c/4}}$$

The values of $(C_{Lp})'_0$ are given by the following plots from Figure 11 in NACA TR-1098:

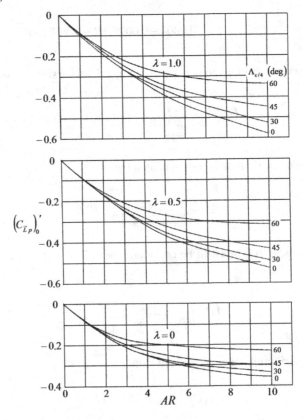

Observe that for a reasonable initial approximation, one may assume that:

$$\left(C_{\bar{L}p}\right)_{planform} \approx \left(C_{\bar{L}p}\right)'_0$$

Now, upon converting this to the aircraft's reference length and area, one obtains:

$$\left[\left(C_{\bar{L}p}\right)_{ac}\right]_{ls} = \left(C_{\bar{L}p}\right)_{planform} \left(\frac{b_{ls}}{b}\right)^2 \frac{S_{ls}}{S}$$

and the total contribution of the lifting surface is thus:

$$\underline{\underline{\left(C_{\bar{L}p}\right)_{ls} = -\frac{z_{ls}}{b}\left(C_{Yp}\right)_{ls} + \left[\left(C_{\bar{L}p}\right)_{ac}\right]_{ls}}}$$

Further, the propeller's contribution is approximated by:

$$\underline{\underline{\left(C_{\bar{L}p}\right)_{prop} = -\left(C_{Yp}\right)_{prop} z_{prop}/b}}$$

The $D\beta$ Derivatives:

Generally, the largest contribution is due to the body, and one may draw upon the previous equation for the body's $D\alpha$ derivatives to obtain:

$$\underline{\underline{\left(C_{YD\beta}\right)_{body} = -4k_2 \frac{(Vol)_{body}}{Sb}}}$$

ID: flight-dynamics-325

where k_2 is the lateral apparent-mass coefficient, as defined in the "Bodies" section of Chapter 2. Further, note that lateral force, Y_f on the vertical fin is equal to the negative value of its lateral "lift":

$$Y_f = -(L_{\text{lateral}})_f \rightarrow (C_{Y D\beta})_f D\beta \rho S U_0^2/2 = -(C_{L D\alpha})_f (D\alpha)_{\text{lateral}} \rho S_f U_0^2/2 \rightarrow$$

$$(C_{Y D\beta})_f D\beta\, S = -(C_{L D\alpha})_f (D\alpha)_{\text{lateral}} S_f \rightarrow (C_{Y D\beta})_f \frac{b}{2U_0} \dot{\beta}\, S = -(C_{L D\alpha})_f \frac{\bar{c}_f}{2U_0} \dot{\alpha}_{\text{lateral}} S_f$$

Since, in this case, $\dot{\beta} = \dot{\alpha}_{\text{lateral}}$, one obtains:

$$(C_{Y D\beta})_f = -(C_{L D\alpha})_f \frac{S_f}{S} \frac{\bar{c}_f}{b}$$

Upon introducing the possibility of a "swirl" effect modifying $\dot{\alpha}_{\text{lateral}}$, as well as an efficiency factor, the equation becomes:

$$(C_{Y D\beta})_f = -(C_{L D\alpha})_f \frac{S_f}{S} \frac{\bar{c}_f}{b}\left(1 - \frac{\partial \sigma_f}{\partial D\beta}\right)\eta_f$$

A similar equation is obtained for the contribution of a lifting surface:

$$(C_{Y D\beta})_{ls} = -K_4 (C_{L D\alpha})_{ls} \frac{S_{ls}}{S} \frac{\bar{c}_{ls}}{b} \tilde{\eta}_{ls} \sin^2 \Gamma$$

where K_4 is the lateral interference parameter defined previously. Likewise, the $C_{LD\alpha}$ terms may be obtained from the equation presented in "The $D\alpha$ Derivatives" subsection. However, the side-wash acceleration parameter, $\partial \sigma_f / \partial D\beta$, may generally be chosen as zero.

Now, the body's apparent-mass force is assumed to act at the body's volumetric center, given by:

$$x_{cv} = (l_{cm} - l_{cv})\cos\alpha_{\text{equil.}} + (h_{cm} - h_{cv})\sin\alpha_{\text{equil.}}$$

$$z_{cv} = (h_{cm} - h_{cv})\cos\alpha_{\text{equil.}} - (l_{cm} - l_{cv})\sin\alpha_{\text{equil.}}$$

For the moment contributions from the fin and lifting surfaces, recall that the circulation term in $C_{LD\alpha}$ acts at the aerodynamic center, as previously defined. However, the apparent-mass term acts at the area center. Therefore, this difference must be accounted for in the moment equations:

$$(C_{N D\beta})_{\text{body}} = \frac{x_{cv}}{b}(C_{Y D\beta})_{\text{body}}, \quad (C_{N D\beta})_f = \frac{(x_{ca})_f}{b}\left[(C_{Y D\beta})_{\text{app}}\right]_f + \frac{x_f}{b}\left[(C_{Y D\beta})_{\text{cir}}\right]_f,$$

where:

$$\left[(C_{Y D\beta})_{\text{app}}\right]_f = -\left[(C_{L D\alpha})_{\text{app}}\right]_f \frac{S_f}{S} \frac{\bar{c}_f}{b}\left(1 - \frac{\partial \sigma_f}{\partial D\beta}\right)\eta_f$$

$$\left[(C_{Y D\beta})_{\text{cir}}\right]_f = -\left[(C_{L D\alpha})_{\text{cir}}\right]_f \frac{S_f}{S} \frac{\bar{c}_f}{b}\left(1 - \frac{\partial \sigma_f}{\partial D\beta}\right)\eta_f$$

$$(C_{N D\beta})_{ls} = \frac{(x_{ca})_{ls}}{b}\left[(C_{Y D\beta})_{\text{app}}\right]_{ls} + \frac{x_{ls}}{b}\left[(C_{Y D\beta})_{\text{cir}}\right]_{ls}$$

where:

$$\left[\left(C_{Y D \beta}\right)_{app}\right]_{ls} = -K_4\left[\left(C_{LD\alpha}\right)_{app}\right]_{ls}\frac{S_{ls}}{S}\frac{\bar{c}_{ls}}{b}\tilde{\eta}_{ls}\sin^2\Gamma_{ls}$$

$$\left[\left(C_{Y D \beta}\right)_{cir}\right]_{ls} = -K_4\left[\left(C_{LD\alpha}\right)_{cir}\right]_{ls}\frac{S_{ls}}{S}\frac{\bar{c}_{ls}}{b}\tilde{\eta}_{ls}\sin^2\Gamma_{ls}$$

ASIDE

These equations for the lifting surface assume that it acts with minimal effect from side-wash, $\partial\sigma_{ls}/\partial D\beta$. However, if the "lifting surface" is an aft-mounted V-Tail, such an effect might be significant.

END ASIDE

$$\left(C_{\tilde{L}D\beta}\right)_{body} = -\frac{z_{cv}}{b}\left(C_{YD\beta}\right)_{body}, \quad \left(C_{\tilde{L}D\beta}\right)_f = -\frac{(z_{ca})_f}{b}\left[\left(C_{YD\beta}\right)_{app}\right]_f - \frac{z_f}{b}\left[\left(C_{YD\beta}\right)_{cir}\right]_f$$

$$\left(C_{\tilde{L}D\beta}\right)_{ls} = -\frac{(z_{ca})_{ls}}{b}\left[\left(C_{YD\beta}\right)_{app}\right]_{ls} - \frac{z_{ls}}{b}\left[\left(C_{YD\beta}\right)_{cir}\right]_{ls}$$

Observe that the lifting-surface's planform-geometry effects have been neglected, which is generally a reasonable assumption.

The $D\hat{r}$ Derivatives

Again, one may directly obtain from the previous derivation of the r derivatives that,

$$\left(C_{YDr}\right)_{body} = 2\frac{x_{cv}}{b}\left(C_{YD\beta}\right)_{body}$$

$$\left(C_{YDr}\right)_f = \left[\left(C_{YDr}\right)_{app}\right]_f + \left[\left(C_{YDr}\right)_{cir}\right]_f$$

where:

$$\left[\left(C_{YDr}\right)_{app}\right]_f = -\left[\left(C_{LD\alpha}\right)_{app}\right]_f\left(\frac{2(x_{ca})_f}{b} + \frac{\partial\sigma_f}{\partial(D\hat{r})}\right)\frac{S_f}{S}\frac{\bar{c}_f}{b}\eta_f$$

$$\left[\left(C_{YDr}\right)_{cir}\right]_f = -\left[\left(C_{LD\alpha}\right)_{cir}\right]_f\left(\frac{2x_f}{b} + \frac{\partial\sigma_f}{\partial(D\hat{r})}\right)\frac{S_f}{S}\frac{\bar{c}_f}{b}\eta_f$$

and

$$\left(C_{YDr}\right)_{ls} = \left[\left(C_{YDr}\right)_{app}\right]_{ls} + \left[\left(C_{YDr}\right)_{cir}\right]_{ls}$$

where:

$$\left[\left(C_{YDr}\right)_{app}\right]_{ls} = 2\frac{(x_{ca})_{ls}}{b}\left[\left(C_{YD\beta}\right)_{app}\right]_{ls}, \quad \left[\left(C_{YDr}\right)_{cir}\right]_{ls} = 2\frac{x_{ls}}{b}\left[\left(C_{YD\beta}\right)_{cir}\right]_{ls}$$

Again, pending accurate information on the rotary side-wash acceleration parameters, $\partial\sigma_f/\partial(D\hat{r})$ and $\partial\sigma_{ls}/\partial(D\hat{r})$, these may be chosen to be negligible.

The angular acceleration, $\dot{r}$, about the body's volumetric center gives rise to a moment coefficient, $\left(C_{NDr}\right)^{cv}_{body}$, which may be obtained in the same way as for $\left(C_{MDq}\right)^{cv}_{body}$ obtained previously:

$$\left(C_{NDr}\right)^{cv}_{body} = \frac{-2k'\left(length^2 + diameter^2\right)}{5Sb^3}(Vol)_{body}$$

Therefore, the C_{NDr} derivatives are:

$$\left(C_{NDr}\right)_{body} = \frac{x_{cv}}{b}\left(C_{YDr}\right)_{body} + \left(C_{NDr}\right)^{cv}_{body},$$

$$\left(C_{NDr}\right)_f = \frac{(x_{ca})_f}{b}\left[\left(C_{YDr}\right)_{app}\right]_f + \frac{x_f}{b}\left[\left(C_{YDr}\right)_{cir}\right]_f$$

$$\left(C_{NDr}\right)_{ls} = \frac{(x_{ca})_{ls}}{b}\left[\left(C_{YDr}\right)_{app}\right]_{ls} + \frac{x_{ls}}{b}\left[\left(C_{YDr}\right)_{cir}\right]_{ls}$$

Generally, $\left(C_{\tilde{L}Dr}\right)^{cv}_{body} \approx 0$, so the $C_{\tilde{L}Dr}$ derivatives are:

$$\left(C_{\tilde{L}Dr}\right)_{body} = -\frac{z_{cv}}{b}\left(C_{YDr}\right)_{body}, \quad \left(C_{\tilde{L}Dr}\right)_f = -\frac{(z_{ca})_f}{b}\left[\left(C_{YDr}\right)_{app}\right]_f - \frac{z_f}{b}\left[\left(C_{YDr}\right)_{cir}\right]_f$$

$$\left(C_{\tilde{L}Dr}\right)_{ls} = -\frac{(z_{ca})_{ls}}{b}\left[\left(C_{YDr}\right)_{app}\right]_{ls} - \frac{z_{ls}}{b}\left[\left(C_{YDr}\right)_{cir}\right]_{ls}$$

Again, note that the lifting surface's planform-geometry effects are neglected.

The $D\hat{p}$ Derivatives

The development of the equations for the Dp derivatives closely follows that for the p derivatives, so that:

$$\left(C_{YDp}\right)_{body} = -2\frac{z_{cv}}{b}\left(C_{YD\beta}\right)_{body},$$

$$\left(C_{YDp}\right)_f = \left[\left(C_{YDp}\right)_{app}\right]_f + \left[\left(C_{YDp}\right)_{cir}\right]_f$$

where:

$$\left[\left(C_{YDp}\right)_{app}\right]_f = \left[\left(C_{LD\alpha}\right)_{app}\right]_f \left(\frac{2(z_{ca})_f}{b} + \frac{\partial \sigma_f}{\partial(D\hat{p})}\right)\frac{S_f}{S}\frac{\bar{c}_f}{b}\eta_f$$

$$\left[\left(C_{YDp}\right)_{cir}\right]_f = \left[\left(C_{LD\alpha}\right)_{cir}\right]_f \left(\frac{2z'_f}{b} + \frac{\partial \sigma_f}{\partial(D\hat{p})}\right)\frac{S_f}{S}\frac{\bar{c}_f}{b}\eta_f$$

$$\left(C_{YDp}\right)_{ls} = \left[\left(C_{YDp}\right)_{app}\right]_{ls} + \left[\left(C_{YDp}\right)_{cir}\right]_{ls}$$

where:

$$\left[\left(C_{YDp}\right)_{app}\right]_{ls} = -2\frac{(z_{ca})_f}{b}\left[\left(C_{YD\beta}\right)_{app}\right]_{ls}, \quad \left[\left(C_{YDp}\right)_{cir}\right]_{ls} = -2\frac{x_f}{b}\left[\left(C_{YD\beta}\right)_{cir}\right]_{ls}$$

$$\left(C_{NDp}\right)_{body} = \frac{x_{cv}}{b}\left(C_{YDp}\right)_{body}$$

$$\left(C_{NDp}\right)_f = \frac{(x_{ca})_f}{b}\left[\left(C_{YDp}\right)_{app}\right]_f + \frac{x_f}{b}\left[\left(C_{YDp}\right)_{cir}\right]_f$$

$$\left(C_{NDp}\right)_{ls} = \frac{(x_{ca})_{ls}}{b}\left[\left(C_{YDp}\right)_{app}\right]_{ls} + \frac{x_{ls}}{b}\left[\left(C_{YDp}\right)_{cir}\right]_{ls}$$

$$\left(C_{\tilde{L}Dp}\right)_{body} = -\frac{z_{cv}}{b}\left(C_{YDp}\right)_{body}$$

$$\left(C_{\tilde{L}Dp}\right)_f = -\frac{(z_{ca})_f}{b}\left[\left(C_{YDp}\right)_{app}\right]_f - \frac{z'_f}{b}\left[\left(C_{YDp}\right)_{cir}\right]_f$$

The swirl terms, $\partial \sigma / \partial(D\hat{p})$, are again not documented and may be assumed to be negligible.

Also, the lifting surface undergoing roll acceleration may have a contribution from its planform, $\left[\left(C_{\tilde{L}Dp}\right)_{ac}\right]_{ls}$, which must be calculated from the $C_{LD\alpha}$ variation along the span. One approach is to choose the $C_{LD\alpha}$ for the lifting-surface's aspect ratio, from the equation presented earlier. Then the rolling moment per unit roll rate may be obtained by integrating along the span:

$$\Delta \tilde{L} = -y\Delta(Lift), \text{ where}$$

$$\Delta(Lift) = \rho\frac{U_0^2}{2}(C_{LD\alpha})(D\alpha)c\,dy \rightarrow \rho\frac{U_0^2}{2}(C_{LD\alpha})\frac{c\dot{\alpha}}{2U_0}c\,dy$$

Also, $\dot{\alpha} = \dot{w}/U_0 = y\dot{p}/U_0$, so $\Delta(Lift) = \rho\frac{U_0^2}{2}(C_{LD\alpha})\frac{c^2 y}{2U_0^2}\dot{p}\,dy$, and therefore, one has:

$$\Delta \tilde{L} = -\rho\frac{U_0^2}{2}(C_{LD\alpha})\frac{c^2 y^2}{2U_0^2}\dot{p}\,dy$$

Flight Dynamics

Now, recall the non-dimensional definitions of $\tilde{L}$ and $\dot{p}$:

$$\tilde{L} \equiv \rho \frac{U_0^2}{2} S b C_{\tilde{L}} \rightarrow \Delta \tilde{L} = \rho \frac{U_0^2}{2} S b \Delta C_{\tilde{L}}, \quad \dot{p} \equiv \frac{4 U_0^2}{b^2}(D\hat{p})$$

The rolling-moment equation thus becomes:

$$\Delta C_{\tilde{L}} = -\frac{(C_{LD\alpha})}{Sb} \frac{c^2 y^2}{2U_0^2} \frac{4U_0^2}{b^2}(D\hat{p}) dy = -2\frac{(C_{LD\alpha})}{Sb^3}(D\hat{p}) c^2 y^2 dy$$

Upon integration along the span, one has:

$$C_{\tilde{L}} = \int_{span} \Delta C_{\tilde{L}} = -2\frac{(D\hat{p})}{Sb^3} \int_{span}(C_{LD\alpha}) c^2 y^2 dy \rightarrow \left[\left(C_{\tilde{L}Dp}\right)_{ac}\right]_{ls} = -\frac{2}{Sb^3}\int_{span}(C_{LD\alpha}) c^2 y^2 dy$$

If, as stated above, $C_{LD\alpha}$ is chosen to be a constant value appropriate to the lifting surface's aspect ratio, then the equation becomes:

$$\left[\left(C_{\tilde{L}Dp}\right)_{ac}\right]_{ls} = -\frac{2 C_{LD\alpha}}{Sb^3}\int_{span} c^2 y^2 dy$$

where the integration is over the complete span of the wing.

The contribution of the lifting surface to the aircraft's rolling acceleration may now be written as:

$$\left(C_{\tilde{L}Dp}\right)_{ls} = -\frac{(z_{ca})_{ls}}{b}\left[\left(C_{YDp}\right)_{app}\right]_{ls} - \frac{z_{ls}}{b}\left[\left(C_{YDp}\right)_{cir}\right]_{ls} + \left[\left(C_{\tilde{L}Dp}\right)_{ac}\right]_{ls}$$

It is important to note that, for most aircraft, the acceleration derivatives are negligible (including those due to downwash lag, for tail-aft airplanes). The exceptions are very ultra-light aircraft, like human-powered airplanes. Also, aircraft whose average structural density is comparable to the air density, like airships, have very significant contributions from the acceleration derivatives. These derivatives are included in this analysis for completeness and, because in the author's experience, you never know what strange aircraft you may be called upon to analyze.

EXAMPLE LATERAL STABILITY DERIVATIVES FOR THE "SCHOLAR" TAIL-AFT MONOPLANE

The pertinent geometric and aerodynamic properties of the example are given previously in the "Example Longitudinal Stability Derivatives for the 'Scholar' Tail-Aft Airplane" section.

Example Value of $C_{Y\beta}$

The DATCOM method previously described will be used, where:

$$\left(C_{Y\beta}\right)_{body} = -K_{int}\left(\hat{C}_{L\alpha}\right)_{body}\frac{S_b}{S}, \quad \text{with } K_{int} = 1.85\left(2 h_w/d\right) \text{ for high-wing airplanes}$$

For this particular example, $h_w = -0.156 \text{ m}$ and $d = 0.250 \text{ m}$. Also,

$$\left(\hat{C}_{L\alpha}\right)_{body} = \frac{2(k_2 - k_1) S_{base}}{S_b}$$

where k_2 and k_1 generally refer to an ellipsoid-of-revolution. In this case, the equivalent ellipsoid is assumed to have the same maximum cross-sectional area as that for the fuselage, which is given by $S_{max} = d \times w_{max}$ where the maximum width, $w_{max} = 0.167\,\text{m}$. So, $S_{max} = 0.250 \times 0.167 = 0.04175\,\text{m}^2 = S_b$. Further, from the "Example, Tail-Aft Monoplane" section in Chapter 2, the volume of the fuselage is $(Vol)_{body} = 0.0352\,\text{m}^3$. Now, the volume of an ellipsoid-of-revolution is given by:

$$(Vol)_{ellipsoid} = (2/3) L\, S_{max}$$

Therefore, the length of the equivalent ellipsoid-of-revolution is as follows:

$$L = 1.5 (Vol)_{ellipsoid} / S_{max} = 1.5 \times 0.0352 / 0.04175 \to 1.265\,\text{m}$$

Further, the diameter of the equivalent ellipsoid-of-revolution is obtained from:

$$S_{max} = \pi (dia)_{max}^2 / 4 \to (dia)_{max} = \left[4 S_{max} / \pi \right]^{1/2} \to 0.2306\,\text{m}$$

And the fineness ratio is $f = L/(dia)_{max} = 1.265/0.2306 = 5.486$. Thus the k_1 and k_2 values may be obtained from interpolations of the solutions for the Munk equations by S.P. Jones (personal communication to the author):

f	k_1	k_2, k_3	k'_2, k'_3
1.00	0.5000	0.5000	0.0
1.50	0.3037	0.6221	0.0951
2.00	0.2100	0.7042	0.2394
2.50	0.1563	0.7619	0.3652
3.00	0.1220	0.8039	0.4657
3.50	0.0985	0.8354	0.5450
4.00	0.0816	0.8598	0.6079
5.00	0.0591	0.8943	0.6999
6.00	0.0452	0.9171	0.7623
8.00	0.0293	0.9447	0.8394
10.00	0.0207	0.9602	0.8835

$$k_1 = 0.0523, \quad k_2 = 0.9054, \quad k' = 0.7302$$

For this case, S_{base} is the truncated cross-sectional area where the fin leading edge intersects the fuselage. This value is $S_{base} = 0.1135 \times 0.1009 = 0.01145\,\text{m}^2$

Altogether, then:

$$\left(\hat{C}_{L\alpha} \right)_{body} = \frac{2 \times (0.9054 - 0.0523) \times 0.01145}{0.04175} = \frac{0.01954}{0.04175} = 0.4679,$$

$$K_{int} = -\frac{1.85 \times 2 \times (0.156)}{0.250} = 2.309$$

So,

$$\left(\hat{C}_{Y\beta} \right)_{body} = -2.309 \times 0.4679 \times 0.04175 / 0.556 \to \left(\hat{C}_{Y\beta} \right)'_{body} = -0.0811$$

Flight Dynamics

This would be the value if the fuselage had a circular cross section. However, because the overall cross section is rectangular, this must be corrected by a factor of 2.0:

$$\left(\hat{C}_{Y\beta}\right)_{body} = 0.1622$$

An Extension to the Analysis

The example fuselage is deeper than it is wider, so a more representative ellipsoid would have an oval cross section. At its deepest and widest part, the ratio of depth to width is $0.250/0.167 = 1.497$. Now for an ellipse of this ratio, its area and width are given by:

$$S_{max} = \frac{1.497}{4}\pi w^2 \rightarrow w_{equiv.ellipse} = \sqrt{0.8505\, S_{max}} \rightarrow w_{equiv.ellipse} = 0.1884\, m$$

Also, of course, the height of the equivalent ellipse is:

$$h_{equiv.\,elipse} = 1.497\, w_{equiv.ellipse} = 0.2821\, m$$

The volume of this ellipsoid is given by:

$$(Vol)_{equiv.ellipsoid} = \frac{4}{3}\pi \frac{L}{2}\frac{w}{2}\frac{h}{2} = \frac{\pi}{6}Lwh \rightarrow L_{equiv.ellipsoid} = \frac{6\,(Vol)_{equiv.ellipsoid}}{\pi\, w_{equiv.ellipse}\, h_{equiv.ellipse}} \rightarrow$$

$$L_{equiv.ellipsoid} = \frac{6 \times 0.0352}{\pi \times 0.1884 \times 0.2821} = 1.2649\, m$$

Now, $L/h = 4.484$, and $h/w = 1.497$, and fairly lengthy extensions of the Munk apparent-mass calculations give coefficients for ellipsoids with three unequal axes. Charts for these values may be found in Appendix D. For this case,

$$k_1 = 0.087, \quad k_2 = 1.30, \quad k_3 = 0.56$$

Upon assuming the same base area, $S_{base} = 0.01145\, m^2$, one obtains that:

$$\left(\hat{C}_{L\alpha}\right)_{body} = \frac{2 \times (1.30 - 0.087) \times 0.01145}{0.04175} = \frac{0.02778}{0.04175} = 0.6653$$

This is 42% larger than the round-fuselage value, which is what one would expect from such a fish-shaped body. K_{int} is the same value, so one obtains that:

$$\left(\hat{C}_{Y\beta}\right)_{body} = -2.309 \times 0.6653 \times 0.04175/0.556 \rightarrow \left(\hat{C}_{Y\beta}\right)'_{body} = -0.11535$$

Upon applying the rectangular cross section correction factor of 2, the final result is:

$$\left(\hat{C}_{Y\beta}\right)_{body} = -0.2307$$

In the author's opinion this value is closer to being correct, and it will be what is chosen for subsequent calculations.

Now, the fin's contribution will be considered.

In accordance with the fin-area definition given earlier, the value of $S_f = 0.057\,\text{m}^2$. Further, the geometry of the equivalent trapezoidal shape is shown below:

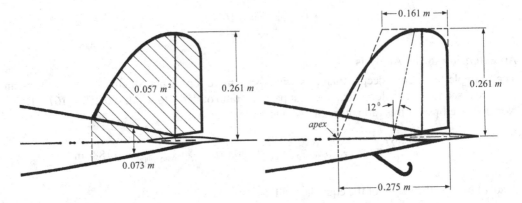

It should be noted that the definition of an "equivalent trapezoidal planform" can be somewhat arbitrary. In this case, the author chose to match the root chord, height and area. This gave the tip chord. Also, this particular planform has a straight trailing edge, which was matched to give the half-chord sweep angle. Upon using the equations presented in the "Wings" section of Chapter 2, the mean aerodynamic chord for the fin's equivalent trapezoidal planform is given by:

$$\bar{c}_f = \frac{2}{3}(c_{\text{root}})_f \left(\frac{1 + \lambda_f + \lambda_f^2}{1 + \lambda_f} \right), \quad \text{where} \quad \lambda_f = \frac{(c_{\text{tip}})_f}{(c_{\text{root}})_f} \rightarrow$$

$$\lambda_f = \frac{0.161}{0.275} = 0.585 \rightarrow \bar{c}_f = \frac{2}{3} \times 0.275 \times \left(\frac{1 + 0.585 + 0.585^2}{1 + 0.585} \right) \rightarrow \bar{c}_f = 0.223\,\text{m}$$

Next, the longitudinal location of the aerodynamic center is given by:

$$(l_{ac})_f = l_{\text{apex}} + \frac{\bar{c}_f}{4} + \chi, \quad \text{where} \quad \chi = \frac{1 + 2\lambda_f}{12}(c_{\text{root}})_f (AR)_{\text{image}} \tan \Lambda_{le}$$

where the fin's image-plane aspect ratio, $(AR)_{\text{image}}$ is given by:

$$(AR)_{\text{image}} = \frac{(2b_f)^2}{2S_f} = \frac{2b_f^2}{S_f} = 2 \times \frac{0.261^2}{0.057} = 2.39$$

also,

$$\tan \Lambda_{le} = \frac{0.5\left[(c_{\text{root}})_f - (c_{\text{tip}})_f\right] + b_f \tan \Lambda_{c/2}}{b_f} = \frac{0.5 \times (0.275 - 0.161) + 0.261 \times \tan 12°}{0.261} \rightarrow$$

$$\tan \Lambda_{le} = 0.431 \quad (\Lambda_{le} = 23.3°)$$

therefore,

$$\chi = \frac{1 + 2 \times 0.585}{12} \times 0.275 \times 2.39 \times 0.431 \rightarrow \chi = 0.051\,\text{m}$$

Observe that this is the longitudinal distance from the apex to the leading edge of $\bar{c}_f$.
Further,

$$l_{apex} = 1.1187\,\text{m} \to (l_{ac})_f = 1.1187 + \frac{0.223}{4} + 0.051 \to l_f \equiv (l_{ac})_f = 1.225\,\text{m}$$

One may now find the longitudinal distance from the nose to the leading edge of $\bar{c}_f$:

$$(l_{cbar})_f = l_{apex} + \chi = 1.1187 + 0.051 = 1.170\,\text{m}$$

Also, recall that $l_s = 1.30\,\text{m}$. Therefore, in accordance with the previously presented methodology, one has that,

$$\Delta l_s = l_s - (l_{cbar})_f = 1.30 - 1.170 \to \Delta l_3 = 0.13\,\text{m} \to X_3 = \Delta l_s/\bar{c}_f = 0.13/0.223 \to$$

$$X_3 = 0.58; \quad \text{also,} \quad X_2 = h_s/b_f = 0$$

ASIDE

The question arises about configurations where the fin is aft of the stabilizer such that $X_3 < 0.5$. The graphs from DATCOM do not account for this. A certain extrapolation is possible, but this eventually gets into uncomfortably unknown territory. It may be that the given range of X_3 is considered good design practice, since K_2 diminishes as X_3 diminishes. Also, one consideration that the author had heard of is that, in a situation where the stabilizer is stalled, its separated flow could blanket the fin when it's needed most. However, successful fin-aft airplanes have been made, so a different methodology (or wind-tunnel testing) will have to be used for such configurations.

END ASIDE

Now, $X_1 = b_f/(d_{body})_{avg}$. The DATCOM method assumed a circular body *under the fin*. For this case, where the body is rectangular, $(d_{body})_{avg}$ will be assumed to be the diameter of the circular area equal to the area of the rectangle formed by average height and width under the fin:

$$(d_{body})_{avg} = 2\left(\frac{h_{avg}w_{avg}}{\pi}\right)^{1/2} = 2\times\left(\frac{0.073\times 0.069}{\pi}\right)^{1/2} \to (d_{body})_{avg} \approx 0.080\,\text{m}$$

therefore,

$$X_1 = 0.261/0.080 \to X_1 = 3.26$$

So, the information is now obtained to calculate the fin's effective aspect ratio, $(AR)_{effec.}$, where:

$$(AR)_{effec.} = K_1 \frac{b_f^2}{S_f}\left[1 + K_H(K_2 - 1)\right]$$

From the graphs presented earlier, $K_1 \approx 1.27$. Also, $S_s/S_f = 0.140/0.057 = 2.46$. So, $K_H \approx 1.14$. Finally, $K_2 \approx 1.14$. Therefore, one obtains:

$$(AR)_{effec.} = 1.27\times\frac{0.261^2}{0.057}\times\left[1 + 1.14\times(1.14 - 1)\right] \to (AR)_{effec.} = 1.760$$

This shows that the effective aspect ratio of the fin has been augmented by 1.473 times its geometrical aspect ratio of $b_f^2/S_f = 1.195$. This compares well with the previously mentioned Perkins & Hage augmentation factor of 1.55.

From this, one may now obtain the lateral lift-coefficient, $(C_{L\alpha})_f$, of the fin, isolated from the effects of side-wash and dynamic pressure augmentation or reduction:

$$(C_{L\alpha})_f = \frac{2\pi(AR)_{\text{effec.}}}{2+\left\{\frac{(AR)^2_{\text{effec.}}}{\kappa^2}\left[1+\tan^2(\Lambda_{c/2})\right]+4\right\}^{1/2}}$$

Upon assuming that $\kappa \approx 1$, this becomes:

$$(C_{L\alpha})_f = \frac{2\pi \times 1.760}{2+\left\{1.760^2 \times \left[1+\tan^2(12°)\right]+4\right\}^{1/2}} = \frac{11.058}{4.690} \rightarrow (C_{L\alpha})_f = 2.358/\text{rad}$$

Now, the sea-level Reynolds Number for the fin will be based upon its mean aerodynamic chord:

$$RN = 6.85\, V\, \bar{c}_f \times 10^4 = 6.85 \times 14.458 \times 0.223 \times 10^4 \rightarrow 2.2085 \times 10^5$$

So, the turbulent boundary-layer skin-friction coefficient is given by:

$$C_{\text{fric}} = \frac{0.455}{(\log_{10} RN)^{2.58}} \rightarrow C_{\text{fric}} = 0.006027$$

Further, as described in the "Airfoils" section of Chapter 2, an airfoil's friction-drag coefficient is given by:

$$(C_d)_{\text{fric}} = 2 C_{\text{fric}} \left[1 + 2(t/c) + 60(t/c)^4\right]$$

For this example, the thickness/chord ratio, t/c, will be assumed to be 0.09. So, one obtains that:

$$(C_d)_{\text{fric}} = 2 \times 0.006027 \times \left(1 + 2 \times 0.09 + 60 \times 0.09^4\right) \rightarrow (C_d)_{\text{fric}} = 0.01427$$

Further, it will be assumed that the airfoil is same throughout the fin. Therefore, the zero-yaw drag of the fin is:

$$(C_{D0})_f = 0.0143$$

Thus, as presented earlier, the "isolated" value of $(C_{Y\beta})_f$ is given by:

$$(C^*_{Y\beta})_f = -\left[(C_{L\alpha})_f + (C_{D0})_f\right] S_f/S = -(2.358 + 0.0143) \times 0.057/0.556 \rightarrow$$

$$(C^*_{Y\beta})_f = -0.2432$$

ASIDE

Observe that the contribution of the profile-drag term, $(C_{D0})_f$ is negligible as compared to the lateral lift coefficient, $(C_{L\alpha})_f$. In fact, it seems downright fussy to include this, in light of all of the other assumptions made to this point. However, its inclusion illustrates the important fact that, for certain configurations where the fin is particularly draggy, it can have a significant effect on $C_{Y\beta}$, as well as the important "directional derivative" $C_{N\beta}$. The author became aware of this while analyzing the stability of tethered aerostats with rather bulbous inflatable fins.

END ASIDE

Now, as presented previously, DATCOM accounts for wake and sideslip effects on $\left(C_{Y\beta}^*\right)_f$ from the following equation:

$$\left(C_{Y\beta}\right)_f = K_3 \left(C_{Y\beta}^*\right)_f \left(1 - \frac{\partial \sigma_f}{\partial \beta}\right) \frac{q_f}{q}$$

Since $X_1 = 3.26$, DATCOM gives that:

$$K_3 = 0.75 + 0.1667 \times (X_1 - 2) \rightarrow K_3 = 0.9600, \quad \text{and}$$

$$\left(1 - \frac{\partial \sigma_f}{\partial \beta}\right) \frac{q_t}{q} = 0.724 + 3.06 \frac{(S_f/S_w)}{\left(1 + \cos(\Lambda_{c/4})_w\right)} - 0.4 \frac{h_w}{(d_{body})_{max}} + 0.009(AR)_w$$

For this application, $(d_{body})_{max}$ will be chosen to be the maximum height of the fuselage, which is 0.25 m. Therefore, one obtains:

$$\left(1 - \frac{\partial \sigma_f}{\partial \beta}\right) \frac{q_t}{q} = 0.724 + 3.06 \frac{0.057/0.556}{(1 + \cos 8.0°)} - 0.4 \frac{0.156}{0.25} + 0.009 \times 7.194$$

$$= 0.724 + 0.1576 - 0.2496 + 0.06475 \rightarrow \left(1 - \frac{\partial \sigma_f}{\partial \beta}\right) \frac{q_t}{q} = 0.6968$$

So, finally, one obtains:

$$\left(C_{Y\beta}\right)_f = 0.9600 \times (-0.2432) \times 0.6968 \rightarrow \underline{\left(C_{Y\beta}\right)_f = -0.1627}$$

It's interesting that this value is less than that for the body itself, for this particular case. This is due to the fact that the body has a slab-sided rectangular cross section with a high-mounted wing.

The wing's contribution is now considered, which has contributions from dihedral angle and sweep:

$$\left(C_{Y\beta}\right)_w^{dihedral} = -K_4 \left(C_{L\alpha}\right)_w \frac{S_w}{S} \sin^2 \Gamma_w$$

K_4 is assumed to be 0.8. Also,

$$\left(C_{L\alpha}\right)_w = \left(C_{L\alpha}^*\right)_w \eta_w = 4.636 \times 0.877 = 4.066$$

So, one obtains:

$$\left(C_{Y\beta}\right)_w^{dihedral} = -0.8 \times 4.066 \times \frac{0.556}{0.556} \times \sin^2(5.0°) \rightarrow \left(C_{Y\beta}\right)_w^{dihedral} = -0.0247$$

Further,

$$\left(C_{Y\beta}\right)_w^{\text{sweep}} = \left[\frac{6\tan\Lambda_{c/4}\sin\Lambda_{c/4}}{\pi\,AR(AR+4\cos\Lambda_{c/4})}\right](C_L)_w^2$$

From the "Example, Tail-Aft Monoplane" section in Chapter 2, one has that:

$$(C_L)_w = 0.454 + 4.066\alpha \rightarrow 0.454 + 4.066 \times 0.018973 = (C_L)_w = 0.5311$$

Therefore, one has:

$$\left(C_{Y\beta}\right)_w^{\text{sweep}} = \left[\frac{6\times\tan(8°)\times\sin(8°)}{\pi\times 7.194\times(7.194+4\times\cos(8°))}\right]\times 0.5311^2 = \frac{0.1174}{252.11}\times 0.5311^2 \rightarrow$$

$$\left(C_{Y\beta}\right)_w^{\text{sweep}} = 0.00013 \approx 0$$

The total value for the wing contribution is thus:

$$\left(C_{Y\beta}\right)_w = -0.0247$$

The stabilizer is also a "lifting surface". But, because it has no dihedral or sweep, its contribution is zero.

ASIDE

DATCOM and other sources, give an approximate equation for $\left(C_{Y\beta}\right)_{1s}^{\text{dihedral}}$:

$$\left(C_{Y\beta}\right)_{1s}^{\text{dihedral}} = -0.00573\,\Gamma_{1s}, \quad (\Gamma_{1s} \text{ in degrees})$$

In this case, the equation gives that $\left(C_{Y\beta}\right)_w^{\text{dihedral}} = -0.0287$, which is close to the previously calculated value.

END ASIDE

The final contribution considered for this example is that from the propeller. As given previously, the equation for this is:

$$\left(C_{Y\beta}\right)_{\text{prop}} \approx -\left(C_{L\alpha}\right)_{\text{prop}} S_{\text{prop}}/S \rightarrow -5.0\times 0.006/0.556 \rightarrow \left(C_{Y\beta}\right)_{\text{prop}} = -0.0540$$

The sum total of these contributions gives the value for $C_{Y\beta}$:

$$C_{Y\beta} = \left(C_{Y\beta}\right)_{\text{body}} + \left(C_{Y\beta}\right)_f + \left(C_{Y\beta}\right)_w + \left(C_{Y\beta}\right)_{\text{prop}} = -0.2307 - 0.1627 - 0.0247 - 0.0540$$

$$C_{Y\beta} = -0.4721$$

Observe that the contributions of the undercarriage are assumed to be negligible. For a large-wheeled "bush plane", or an airplane with floats, that would not be the case at all.

Flight Dynamics

Example Value of $C_{N\beta}$

For the body contribution, the previously presented Perkins & Hage equation is:

$$\left(C_{N\beta}\right)_{body} = -0.98 \, K_\beta \, \frac{S_{side}}{S} \, \frac{l_{body}}{b} \left(\frac{h_1}{h_2}\right)^{1/2} \left(\frac{w_2}{w_1}\right)^{1/3}$$

where,

$$S_{side} = 0.265 \, m^2, \quad l_{body} = 1.40 \, m, \quad h_1 = 0.246 \, m, \quad h_2 = 0.139 \, m, \quad w_1 = 0.167 \, m,$$
$$w_2 = 0.114 \, m, \quad h = 0.25 \, m, \quad w = 0.167 \, m$$

Also, K_β is found from the graph presented earlier, where $l_{body}/h = 1.40/0.25 = 5.60$ and $l_{cm}/l_{body} = 0.497/1.40 = 0.355 \rightarrow K_\beta \approx 0.14$. So, the equation gives:

$$\left(C_{N\beta}\right)_{body}^{P\&H} = -0.98 \times 0.14 \times \frac{0.265}{0.556} \times \frac{1.40}{2.0} \left(\frac{0.246}{0.139}\right)^{1/2} \left(\frac{0.114}{0.167}\right)^{1/3} \rightarrow$$

$$\underline{\left(C_{N\beta}\right)_{body}^{P\&H} = -0.0536}$$

For the DATCOM method, the previously presented equation is:

$$\left(C_{N\beta}\right)_{body}^{DATCOM} = -57.3 \, K_N \, K_{RN} \, \frac{S_{side}}{S} \, \frac{l_b}{b}$$

In this case, K_N is found from the previously-presented graph, where $h/w = 0.25/0.167 = 1.497$, $\left(h_1/h_2\right)^{1/2} = (0.246/0.139)^{1/2} = 1.330$, $l_{body}^2/S_{side} = 1.40^2/0.265 = 7.396 \rightarrow$

$$K_N = 0.0019$$

Also, the sea-level Reynolds Number of the body is given by:

$$(RN)_{body} = 6.85 \, V \, l_{body} \times 10^4 = 6.85 \times 14.458 \times 1.40 \times 10^4 = 1.386 \times 10^6$$

so that,

$$K_{RN} = 0.2041 \times \ln(RN)_{body} - 1.82 \rightarrow K_{RN} = 1.0664$$

therefore,

$$\left(C_{N\beta}\right)_{body}^{DATCOM} = -57.3 \times 0.0019 \times 1.0664 \times \frac{0.265}{0.556} \times \frac{1.40}{2.0} \rightarrow \underline{\left(C_{N\beta}\right)_{body}^{DATCOM} = -0.0387}$$

Both values are comparable, but the author chooses the DATCOM value because of its Reynolds Number dependence. Note that if $RN = 10.3 \times 10^6$, which is within the range of fuselage RN's for full-scale airplanes from the era that the Perkins & Hage text was published (1949), then the DATCOM result would match the P & H result.

DATCOM makes no comment about the effect of a prismatic body (rectangular cross section). Therefore, barring further information, the value will be taken as given from DATCOM.

Further, observe that this value of $\left(C_{N\beta}\right)_{body}$ is unstable. That is, the yawing moment is negative with a positive sideslip. This has to be corrected with the fin.

The fin's contribution is given by:

$$(C_{N\beta})_f = \frac{x_f}{b}(C_{Y\beta})_f, \quad \text{where} \quad x_f = (l_{cm} - l_f)\cos\alpha_{equil} + (h_{cm} - h_f)\sin\alpha_{equil}$$

Also, as presented earlier,

$$(h_{ac})_{fin} \approx 0.42\, b_{fin} = 0.42 \times 0.261 \rightarrow h_f \equiv (h_{ac})_f = 0.110\,\text{m}$$

So,

$$x_f = (0.497 - 1.225) \times \cos(1.0871°) + (0.03 - 0.110) \times \sin(1.0871°) \rightarrow x_f = -0.7294\,\text{m}$$

So,

$$(C_{N\beta})_f = \frac{-0.7294}{2.0} \times (-0.1627) \rightarrow (C_{N\beta})_f = 0.0593$$

As presented earlier, the wing's contribution to its aerodynamic center is given by:

$$\left[(C_{N\beta})_{ac}\right]_w = \left(\frac{C_{N\beta}}{C_L^2}\right)_w (C_L)_w^2 \frac{S_w}{S} \frac{b_w}{b}, \quad \text{where}$$

$$\left(\frac{C_{N\beta}}{C_L^2}\right)_w = \frac{1}{4\pi AR_w} - \frac{\tan\Lambda_{c/4}}{\pi AR_w(AR_w + 4\cos\Lambda_{c/4})}\left(\cos\Lambda_{c/4} - \frac{AR_w}{2} - \frac{AR_w^2}{8\cos\Lambda_{c/4}}\right)$$

$$= \frac{1}{4\pi \times 7.194} - \frac{\tan 8°}{\pi \times 7.194 \times (7.194 + 4 \times \cos 8°)} \times \left(\cos 8° - \frac{7.194}{2} - \frac{7.194^2}{8 \times \cos 8°}\right)$$

$$= 0.011062 - 0.000557 \times (0.9903 - 3.597 - 6.5328) \rightarrow \left(\frac{C_{N\beta}}{C_L^2}\right)_w = 0.01615$$

Therefore,

$$\left[(C_{N\beta})_{ac}\right]_w = 0.01615 \times (C_L)_w^2 \times \frac{0.556}{0.556} \times \frac{2}{2} = 0.01615 \times (C_L)_w^2$$

It is seen that this contribution is stable, and dependent on the lift coefficient. For this particular example, from before,

$$(C_L)_w = 0.5311 \rightarrow \left[(C_{N\beta})_{ac}\right]_w = 0.0046$$

The wing's stability derivative with respect to the aircraft's mass-center, as previously stated, is:

$$(C_{N\beta})_w = (C_{Y\beta})_w \frac{x_w}{b} + \left[(C_{N\beta})_{ac}\right]_w$$

where,

$$x_w = (l_{cm} - l_w)\cos\alpha_{equil} + (h_{cm} - h_w)\sin\alpha_{equil} \rightarrow$$

$$x_w = (0.497 - 0.494) \times \cos(1.0871°) + (0.03 - 0.156) \times \sin(1.0871°) \rightarrow x_w = 0.0006\,\text{m}$$

thus,

$$\left(C_{N\beta}\right)_w = \left(C_{Y\beta}\right)_w \frac{x_w}{b} + \left[\left(C_{N\beta}\right)_{ac}\right]_w = -0.0247 \times \frac{0.0006}{2.0} + 0.0046 \rightarrow \left(C_{N\beta}\right)_w = 0.0046$$

The stabilizer's contribution to this stability derivative is negligible, so the final contribution is from the propeller:

$$\left(C_{N\beta}\right)_{prop} = \left(C_{Y\beta}\right)_{prop} \frac{x_{prop}}{b}, \quad \text{where}$$

$$x_{prop} = \left(l_{cm} - l_{prop}\right)\cos\alpha_{equil} + \left(h_{cm} - h_{prop}\right)\sin\alpha_{equil} \rightarrow$$

$$x_{prop} = \left(0.497 - 0\right) \times \cos\left(1.0871°\right) + \left(0.03 - 0\right) \times \sin\left(1.0871°\right) \rightarrow x_{prop} = 0.4975\,m$$

so,

$$\left(C_{N\beta}\right)_{prop} = -0.0540 \times \frac{0.4975}{2.0} \rightarrow \left(C_{N\beta}\right)_{prop} = -0.0134$$

The sum total of these contributions gives the value for $C_{N\beta}$:

$$C_{N\beta} = \left(C_{N\beta}\right)_{body} + \left(C_{N\beta}\right)_f + \left(C_{N\beta}\right)_w + \left(C_{N\beta}\right)_{prop} = -0.0387 + 0.0593 + 0.0046 - 0.0134$$

$$C_{N\beta} = 0.0118$$

Example Value of $C_{\bar{L}\beta}$

The body's contribution, from before, is estimated by:

$$\left(C_{\bar{L}\beta}\right)_{body} \approx -\frac{z_{ref}}{b}\left(C_{Y\beta}\right)_{body}, \quad \text{where} \quad z_{ref} = \left(h_{cm} - h_{cp}\right)\cos\alpha_{equli} - \left(l_{cm} - l_{cp}\right)\sin\alpha_{equil}$$

It's assumed for this aircraft that the body's center of pressure is on its reference line. Therefore, $h_{cp} = 0$. As for l_{cp}, this may be approximated by:

$$l_{cp} \approx b\left(C_{N\beta}\right)_{body} / \left(C_{Y\beta}\right)_{body} = 2.0 \times (-0.0387)/(-0.2307) \rightarrow l_{cp} = 0.336\,m$$

Therefore,

$$z_{ref} = (0.03 - 0) \times \cos(1.0871°) - (0.497 - 0.336) \times \sin(1.0871°) \rightarrow z_{ref} = 0.027\,m \rightarrow$$

$$\left(C_{\bar{L}\beta}\right)_{body} \approx -\frac{0.027}{2.0} \times (-0.2307) \rightarrow \left(C_{\bar{L}\beta}\right)_{body} \approx 0.0031$$

The fin's contribution is given by:

$$\left(C_{\bar{L}\beta}\right)_f = -\frac{z_f}{b}\left(C_{Y\beta}\right)_f, \quad \text{where} \quad z_f = \left(h_{cm} - h_f\right)\cos\alpha_{equli} - \left(l_{cm} - l_f\right)\sin\alpha_{equil} \rightarrow$$

$$z_f = \left(0.03 - 0.110\right) \times \cos\left(1.0871°\right) - \left(0.497 - 1.225\right) \times \sin\left(1.0871°\right) \rightarrow z_f = -0.0662\,m \rightarrow$$

$$\left(C_{\bar{L}\beta}\right)_f = -\frac{(-0.0662)}{2.0} \times (-0.1627) \rightarrow \left(C_{\bar{L}\beta}\right)_f = -0.0054$$

As presented before, the wing's contribution comes from dihedral and sweep:

$$\left[\left(C_{\bar{L}\beta}\right)_w\right]_{dihedral} = \left[\left(C_{\bar{L}\beta}\right)_\Gamma\right]_w \frac{S_w}{S} \Gamma_w$$

From a previously-presented graph, for $AR_w = 7.194$ and $\lambda = 1.0$, one obtains that $\left[\left(C_{\bar{L}\beta}\right)_\Gamma\right]_{ls} = -1.05$. Also, $\Gamma_w = 5.0° = 0.0873$ rad. Therefore,

$$\left[\left(C_{\bar{L}\beta}\right)_w\right]_{dihedral} = -1.05 \times \frac{0.556}{0.556} \times 0.0873 \rightarrow \left[\left(C_{\bar{L}\beta}\right)_w\right]_{dihedral} = -0.0917$$

$$\left[\left(C_{\bar{L}\beta}\right)_w\right]_{sweep} = \left[\frac{\left(C_{\bar{L}\beta}\right)_w}{C_L}\right]_w \left(C_L\right)_w \frac{S_w}{S} \frac{b_w}{b}, \quad \text{where} \quad \left[\frac{\left(C_{\bar{L}\beta}\right)_w}{C_L}\right]_w \approx -0.43 \tan \Lambda_{c/4} \rightarrow$$

$$\left[\frac{\left(C_{\bar{L}\beta}\right)_w}{C_L}\right]_w \approx -0.43 \times \tan(8°) = -0.0604. \quad \text{Also,} \quad \left(C_L\right)_w = 0.5311, \text{ so}$$

$$\left[\left(C_{\bar{L}\beta}\right)_w\right]_{sweep} = -0.0604 \times 0.5311 \times \frac{0.556}{0.556} \times \frac{2.0}{2.0} = -0.0321$$

There is also a contribution due to the wing's vertical location on the body:

$$\left[\left(C_{\bar{L}\beta}\right)_w\right]_{position} = -\left(C_{Y\beta}\right)_w z_w/b, \quad \text{where} \quad z_w = \left(h_{cm} - h_w\right)\cos\alpha_{equli} - \left(l_{cm} - l_w\right)\sin\alpha_{equil} \rightarrow$$

$$z_w = (0.03 - 0.156) \times \cos(1.0871°) - (0.497 - 0.494) \times \sin(1.0871°) \rightarrow z_w = -0.1260 \, \text{m} \rightarrow$$

$$\left[\left(C_{\bar{L}\beta}\right)_w\right]_{position} = -(-0.0247) \times (-0.1260)/2.0 = -0.0016$$

In total, the wing's contribution is:

$$\left(C_{\bar{L}\beta}\right)_w = \left[\left(C_{\bar{L}\beta}\right)_w\right]_{dihedral} + \left[\left(C_{\bar{L}\beta}\right)_w\right]_{sweep} + \left[\left(C_{\bar{L}\beta}\right)_w\right]_{position}$$

$$= -0.0917 - 0.0321 - 0.0016 \rightarrow \left(C_{\bar{L}\beta}\right)_w = -0.1254$$

The propeller and stabilisers contribution are negligible for this example, so, in summary, one has:

$$C_{\bar{L}\beta} = \left(C_{\bar{L}\beta}\right)_{body} + \left(C_{\bar{L}\beta}\right)_f + \left(C_{\bar{L}\beta}\right)_w = 0.0031 - 0.0054 - 0.1254 \rightarrow$$

$$\underline{\underline{C_{\bar{L}\beta} = -0.1277}}$$

Example Value of C_{Yr}

The body's contribution to this stability derivative is assumed to be negligible for most airplanes, including this example. However, this would certainly not be the case for an airship.

The fin contribution, as previously presented, is given by:

$$\left(C_{Yr}\right)_f = \left(C_{Y\beta}^*\right)_f \left(2\frac{x_f}{b} + \frac{\partial \sigma_f}{\partial \hat{r}}\right)\eta_f + \left[\left(C_{Yr}\right)_{ac}\right]_f$$

Flight Dynamics

As stated earlier, the rotational side-slip effect, $\partial \sigma_f / \partial \hat{r}$, will be assumed to be negligible and $q_f/q \approx 1$. Therefore, η_f can be estimated from:

$$\eta_f = 1 - 1.4(S_{f-b}/S_f) + 0.4(S_{f-b}/S_f)^2$$

where S_f is the portion of the defined fin area, illustrated in a previous figure, that is covered by the fuselage. For this example, $S_f = 0.057 \, m^2$ and $S_{f-b} = 0.008 \, m^2$. So,

$$\eta_f \approx 1 - 1.4 \times (0.008/0.057) + 4.0 \times (0.008/0.057)^2 \rightarrow \eta_f = 0.882$$

also,

$$\left[(C_{Yr})_{ac}\right]_f \approx -(C^*_{Y\beta})_f \eta_f \, \bar{c}_f/b = -(-0.2432) \times 0.882 \times 0.223/2.0 \rightarrow \left[(C_{Yr})_{ac}\right]_f = 0.0239$$

therefore,

$$(C_{Yr})_f = -0.2432 \times 2 \times \frac{-0.7294}{2.0} \times 0.882 + 0.0239 \rightarrow (C_{Yr})_f = 0.1804$$

Further, the wing's contribution is estimated from:

$$(C_{Yr})_w \approx 2(C_{Y\beta})_w \frac{x_w}{b} = 2 \times (-0.0247) \times \frac{0.0006}{2.0} \rightarrow (C_{Yr})_w \approx 0$$

This is seen to be negligible compared with the fin's contribution. This is likewise in the case for the stabilizer's contribution.

Finally, the propeller's contribution is given by:

$$(C_{Yr})_{prop} \approx 2(C_{Y\beta})_{prop} \frac{x_{prop}}{b} = 2 \times (-0.0540) \times \frac{0.4975}{2.0} \rightarrow (C_{Yr})_{prop} = -0.0269$$

Therefore, in summary, one has:

$$C_{Yr} = (C_{Yr})_f + (C_{Yr})_w + (C_{Yr})_{prop} = 0.1804 + 0 - 0.0269 \rightarrow$$

$$C_{Yr} = 0.1535$$

Example Value of C_{Nr}

The body contribution is again assumed to be negligible for most airplanes, with the important caveat that this is definitely not negligible for airships.

The fin contribution is given by:

$$(C_{Nr})_f = \frac{x_f}{b}(C_{Yr})_f + \left[(C_{Nr})_{ac}\right]_f$$

where,

$$\left[(C_{Nr})_{ac}\right]_f \approx -\frac{\pi}{4}\left(\frac{\bar{c}_f}{b}\right)^2 \frac{S_f}{S} \eta_f = -\frac{\pi}{4} \times \left(\frac{0.223}{2.0}\right)^2 \times \frac{0.057}{2.0} \times 0.882 \rightarrow \left[(C_{Nr})_{ac}\right]_f = -0.0003$$

Therefore,

$$(C_{Nr})_f = \frac{-0.7294}{2.0} \times 0.1804 - 0.0003 \rightarrow (C_{Nr})_f = -0.0661$$

The wing contribution is given by:

$$(C_{Nr})_w = \frac{x_w}{b}(C_{Yr})_w + \left[(C_{Nr})_{ac}\right]_w$$

where,

$$\left[(C_{Nr})_{ac}\right]_w = \left[\frac{(\Delta C_{Nr})_1}{C_L^2}\right]_w \frac{S_w}{S}\left(\frac{b_w}{b}\right)^2 (C_L)_w^2 + \left[\frac{(\Delta C_{Nr})_2}{C_{D0}}\right]_w \frac{S_w}{S}\left(\frac{b_w}{b}\right)^2 (C_{D\,profile})_w$$

From the curves previously given, one obtains for this example:

$$\left[\frac{(\Delta C_{Nr})_1}{C_L^2}\right]_w \approx -0.02, \quad \left[\frac{(\Delta C_{Nr})_2}{C_{D0}}\right]_w \approx -0.32$$

also,

$$(C_L)_w = 0.5311, (C_D)_w = 0.0243 + 0.2196\alpha_{equil.} + 1.1490\alpha_{equil}^2 \rightarrow$$

$$(C_D)_w = 0.0243 + 0.2196 \times 0.01897 + 1.1490 \times 0.01897^2 \rightarrow (C_D)_w = 0.0289$$

so,

$$(C_{D\,profile})_w \approx (C_D)_w - (C_L)_w^2/(\pi AR_w) \rightarrow 0.0289 - (0.5311)^2/(\pi \times 7.194) \rightarrow$$

$$(C_{D\,profile})_w = 0.0164$$

Alternately, one may also draw upon the calculations for the example airplane in Chapter 2 to directly obtain:

$$(C_{D\,profile})_w = 0.0141 - 0.0091(C_L)_w + 0.0224(C_L)_w^2 \rightarrow (C_{D\,profile})_w = 0.0156$$

Therefore, the chosen value will be $(C_{D\,profile})_w = 0.016$ and one thus obtains:

$$\left[(C_{Nr})_{ac}\right]_w = -0.02 \times \frac{0.556}{0.556}\left(\frac{2.0}{2.0}\right)^2 0.5311^2 - 0.32 \times \frac{0.556}{0.556}\left(\frac{2.0}{2.0}\right)^2 \times 0.016 \rightarrow$$

$$\left[(C_{Nr})_{ac}\right]_w = -0.0108$$

The complete wing contribution is:

$$(C_{Nr})_w = \frac{0.0006}{2.0} \times 0.0 - 0.0108 \rightarrow (C_{Nr})_w = -0.0108$$

Flight Dynamics

The stabilizer's contribution for this example is negligible.
The propeller's contribution is given by:

$$(C_{Nr})_{prop} = (C_{Yr})_{prop} x_{prop}/b = -0.0269 \times 0.4975/2.0 \rightarrow (C_{Nr})_{prop} = -0.0067$$

In summary, one obtains:

$$C_{Nr} = (C_{Nr})_f + (C_{Nr})_w + (C_{Nr})_{prop} = -0.0673 - 0.0108 - 0.0067 \rightarrow$$

$$C_{Nr} = -0.0848$$

Example Value of $C_{\tilde{L}r}$

Again, the contribution of the body to this stability derivative is assumed to be negligible relative to contributions from other components.

The fin's contribution is given by:

$$(C_{\tilde{L}r})_f = -\frac{z_f}{b}(C_{Yr})_f = -\frac{(-0.0662)}{2.0} \times 0.1804 \rightarrow (C_{\tilde{L}r})_f = 0.0060$$

Next, the wing's contribution is given by:

$$(C_{\tilde{L}r})_w = -\frac{z_w}{b}(C_{Yr})_w + \left[(C_{\tilde{L}r})_{ac}\right]_w, \quad \text{where} \quad \left[(C_{\tilde{L}r})_{ac}\right]_w = \left(\frac{C_{\tilde{L}r}}{C_L}\right)_w \left(\frac{b_w}{b}\right)^2 \frac{S_w}{S}(C_L)_w$$

From a previous graph, $(C_{\tilde{L}r}/C_L)_w = 0.31$. Therefore,

$$\left[(C_{\tilde{L}r})_{ac}\right]_w = 0.31 \times \left(\frac{2.0}{2.0}\right)^2 \times \frac{0.556}{0.556} \times 0.5311 \rightarrow \left[(C_{\tilde{L}r})_{ac}\right]_w = 0.1646$$

So, the total wing contribution is:

$$(C_{\tilde{L}r})_w = -\frac{-0.1260}{2.0} \times 0.0 + 0.1646 \rightarrow (C_{\tilde{L}r})_w = 0.1646$$

Because the stabilizer of this example has no dihedral angle, $(C_{Y\beta})_s = 0 \rightarrow (C_{Yr})_s = 0$. Therefore, its contribution is:

$$(C_{\tilde{L}r})_s = 0 + \left[(C_{\tilde{L}r})_{ac}\right]_s, \quad \text{where} \quad \left[(C_{\tilde{L}r})_{ac}\right]_s = \left(\frac{C_{\tilde{L}r}}{C_L}\right)_s \left(\frac{b_s}{b}\right)^2 \frac{S_s}{S}(C_L)_s$$

From the results of this example airplane in Chapter 2, one has:

$$(C_L)_s = -0.141 + 2.815\alpha_{equil} = -0.141 + 2.815 \times 0.01897 \rightarrow (C_L)_s = -0.0876$$

Also, $(C_{\tilde{L}r}/C_L)_s = 0.25$, therefore,

$$\left[(C_{\tilde{L}r})_{ac}\right]_s = 0.25 \times \left(\frac{0.75}{2.0}\right)^2 \times \frac{0.140}{0.556} \times (-0.0876) \rightarrow (C_{\tilde{L}r})_s = -0.0008$$

It is seen that this is negligible, in this case, relative to the wing's contribution.

The propeller's contribution is given by:

$$\left(C_{\tilde{L}r}\right)_{prop} = -\left(C_{Yr}\right)_{prop} z_{prop}/b, \text{ where}$$

$$z_{prop} = (0.03 - 0) \times \cos(1.0871°) - (0.497 - 0) \times \sin(1.0871°) \rightarrow z_{prop} = 0.0206\,m$$

Therefore,

$$\left(C_{\tilde{L}r}\right)_{prop} = -\left(C_{Yr}\right)_{prop} z_{prop}/b = -(-0.0269) \times 0.0206/2.0 \rightarrow \left(C_{\tilde{L}r}\right)_{prop} = 0.0003$$

The summary result is:

$$C_{\tilde{L}r} = \left(C_{\tilde{L}r}\right)_f + \left(C_{\tilde{L}r}\right)_w + \left(C_{\tilde{L}r}\right)_s + \left(C_{\tilde{L}r}\right)_{prop} = 0.0060 + 0.1646 - 0.0008 + 0.0003 \rightarrow$$

$$C_{\tilde{L}r} = 0.1701$$

Example Value of C_{Yp}

As presented earlier, the body's contribution may be estimated from:

$$\left(C_{Yp}\right)_{body} = -2\frac{z_{ref}}{b}\left(C_{Y\beta}\right)_{body} = -2 \times \frac{0.027}{2.0} \times (-0.2307) \rightarrow \left(C_{Yp}\right)_{body} = 0.0062$$

The vertical fin's contribution is given by:

$$\left(C_{Yp}\right)_f = \left(C_{Y\beta}^*\right)_f \left(-2\frac{z'_f}{b} + \frac{\partial \sigma_f}{\partial \hat{p}}\right)\eta_f, \quad \text{where} \quad z'_f \approx 1.15 \times (-0.0662) = -0.076\,m$$

Also, the swirl from the forward wing is estimated from:

$$\partial \sigma_f/\partial \hat{p} \approx \left[0.178 + 0.827(2b_f/b_w) - 2.35(2b_f/b_w)^2 + 1.833(2b_f/b_w)^3\right] \times sign|z'_f|$$

where $2b_f/b_w = 2 \times 0.261/2.0 = 0.261 \rightarrow$

$$\partial \sigma_f/\partial \hat{p} \approx \left[0.178 + 0.827 \times 0.261 - 2.35 \times 0.261^2 + 1.833 \times 0.261^3\right] \times sign|z'_f|$$

$$\partial \sigma_f/\partial \hat{p} \approx 0.266 \times sign|z'_f| = -0.266$$

Therefore,

$$\left(C_{Yp}\right)_f = (-0.2432) \times \left(-2 \times \frac{-0.076}{2.0} - 0.266\right) \times 0.882 \rightarrow \left(C_{Yp}\right)_f = 0.0408$$

Notice how the swirl effect is greater, and of opposite sign, than the induced effect without swirl.

The wing's contribution is given by:

$$\left(C_{Yp}\right)_w = \left(C_{Y\beta}\right)_w \left(-2\frac{z_w}{b} + \frac{\partial \sigma_w}{\partial \hat{p}}\right) + \left[\left(C_{Yp}\right)_{ac}\right]_w, \quad \text{where}$$

$$\left[\left(C_{Yp}\right)_{ac}\right]_w = \left[\left(\frac{AR + \cos\Lambda_{c/4}}{AR + 4\cos\Lambda_{c/4}}\right)\tan\Lambda_{c/4} + \frac{1}{AR}\right]\frac{b_w}{b}\frac{S_w}{S}(C_L)_w$$

Flight Dynamics

The wing tips are sufficiently rounded on this example so that the $1/AR$ term must be included. Thus, one obtains:

$$\left[(C_{Yp})_{ac}\right]_w = \left[\left(\frac{7.194+\cos 8°}{7.194+4\times\cos 8°}\right)\times\tan 8° + \frac{1}{7.194}\right]\times\frac{2.0}{2.0}\times\frac{0.556}{0.556}\times 0.5311 = 0.1286$$

Also, $\partial\sigma_w/\partial\hat{p} = 0$ for this forward-wing design, so that:

$$(C_{Yp})_w = (-0.0247)\times\left(-2\times\frac{-0.1260}{2.0}\right)+0.1286 \rightarrow (C_{Yp})_w = 0.1255$$

The stabilizer contribution is assumed to be negligible and the propeller contribution is estimated from:

$$(C_{Yp})_{prop} = -2(C_{Y\beta})_{prop} z_{prop}/b = -2\times(-0.0540)\times 0.0206/2.0 \rightarrow (C_{Yp})_{prop} = 0.0011$$

The summary result is:

$$C_{Yp} = (C_{Yp})_{body} + (C_{Yp})_f + (C_{Yp})_w + (C_{Yp})_{prop} = 0.0062 + 0.0408 + 0.1255 + 0.0011 \rightarrow$$

$$C_{Yp} = 0.1736$$

Example Value of C_{Np}

The body's contribution may be estimated from the equation presented earlier:

$$(C_{Np})_{body} = -2\frac{z_{ref}}{b}(C_{N\beta})_{body} = -2\times\frac{0.027}{2.0}\times(-0.0387) \rightarrow (C_{Np})_{body} = 0.0011$$

Likewise, the fin's contribution is given by:

$$(C_{Np})_f = \frac{x_f}{b}(C_{Yp})_f = \frac{-0.7294}{2.0}\times(0.0408) \rightarrow (C_{Np})_f = -0.0149$$

Next, the equation for the wing's contribution is given as:

$$(C_{Np})_w = \frac{x_w}{b}(C_{Yp})_w + \left[(C_{Np})_{ac}\right]_w,$$

The following equation from NACA TN-1581 is chosen for $\left[(C_{Np})_{ac}\right]_w$:

$$\left[(C_{Np})_{ac}\right]_w = (C_L)_w\left(\frac{AR+4}{AR+4\cos\Lambda_{c/4}}\right)\left[1+6\left(1+\frac{\cos\Lambda_{c/4}}{AR}\right)\frac{\tan^2\Lambda_{c/4}}{12}\right]\left(\frac{b_w}{b}\right)^2\frac{S_w}{S}\left(\frac{C_{Np}}{C_L}\right)_{\Lambda=0}$$

where,

$$\left(\frac{C_{Np}}{C_L}\right)_{\Lambda=0} \approx -0.0004 - 0.0099\, AR + 0.0003(AR)^2$$

$$= -0.0004 - 0.0099\times 7.194 + 0.0003\times 7.194^2 \rightarrow \left(\frac{C_{Np}}{C_L}\right)_{\Lambda=0} = -0.0561$$

Therefore,

$$\left[\left(C_{Np}\right)_{ac}\right]_w = 0.5311 \times \left(\frac{7.194+4}{7.194+4\times\cos 8°}\right)$$

$$\times \left[1 + 6 \times \left(1 + \frac{\cos 8°}{7.194}\right) \times \frac{\tan^2 8°}{12}\right] \times \left(\frac{2.0}{2.0}\right)^2 \times \frac{0.556}{0.556} \times (-0.0561)$$

$$= 0.5330 \times 1.011 \times 1.0 \times 1.0 \times (-0.0561) \rightarrow \left[\left(C_{Np}\right)_{ac}\right]_w = -0.0302 \rightarrow$$

$$\left(C_{Np}\right)_w = \frac{0.0006}{2.0} \times 0.1254 - 0.0302 \rightarrow \left(C_{Np}\right)_w = -0.0302$$

The stabilizer's contribution is negligible for this design, so finally the propeller contribution is:

$$\left(C_{Np}\right)_{prop} = \left(C_{Yp}\right)_{prop} x_{prop}/b = 0.0011 \times 0.4975/2.0 = \left(C_{Np}\right)_{prop} = 0.0003$$

In summary, one obtains:

$$C_{Np} = \left(C_{Np}\right)_{body} + \left(C_{Np}\right)_f + \left(C_{Np}\right)_w + \left(C_{Np}\right)_{prop} = 0.0011 - 0.0149 - 0.0302 + 0.0003 \rightarrow$$

$$C_{Np} = -0.0437$$

Example Value of C_{lp}

The body's contribution is estimated from:

$$\left(C_{lp}\right)_{body} \approx -\frac{z_{ref}}{b}\left(C_{Yp}\right)_{body} = -\frac{0.027}{2.0} \times 0.0062 \rightarrow \left(C_{lp}\right)_{body} = -0.0001$$

Similarly, the fin's contribution is given by:

$$\left(C_{lp}\right)_f \approx -\frac{z'_f}{b}\left(C_{Yp}\right)_f = -\frac{(-0.076)}{2.0} \times 0.0408 \rightarrow \left(C_{lp}\right)_f = 0.0016$$

Notice how the swirl from the wing is actually driving the fin's rolling-moment contribution to a positive value.

For the wing itself, its contribution is given by:

$$\left(C_{lp}\right)_w = -\frac{z_w}{b}\left(C_{Yp}\right)_w + \left[\left(C_{lp}\right)_{ac}\right]_w, \quad \text{where} \quad \left[\left(C_{lp}\right)_{ac}\right]_w = \left(C_{lp}\right)_{planform}\left(\frac{b_w}{b}\right)^2\frac{S_w}{S} \rightarrow$$

$$\left(C_{lp}\right)_{planform} = \left(C_{lp}\right)_0 \frac{\left(C_{L\alpha}\right)_{CL}}{\left(C_{L\alpha}\right)_{CL=0}} -$$

$$-\frac{1}{8}\frac{(C_L)_w^2}{\pi AR \cos^2\Lambda_{c/4}}\left(1 + 2\sin^2\Lambda_{c/4}\frac{AR + 2\cos\Lambda_{c/4}}{AR + 4\cos\Lambda_{c/4}}\right) - \frac{1}{8}C_{D\,profile}$$

Also,

$$\left(C_{\tilde{l}p}\right)_0 = \left(C_{\tilde{l}p}\right)_0' \frac{AR + 4\cos\Lambda_{c/4}}{(2\pi/C_{l\alpha})AR + 4\cos\Lambda_{c/4}}$$

where, from the graph presented earlier, $\left(C_{\tilde{l}p}\right)_0' = -0.48$. Also, as presented in the "Example, Tail-Aft Monoplane" section of Chapter 2, the sectional lift-curve slope of the wing's airfoil is:

$$C_{l\alpha} = 6.10/\text{rad}$$

Therefore,

$$\left(C_{\tilde{l}p}\right)_0 = -0.48 \times \frac{7.194 + 4\times\cos 8°}{(2\pi/6.10)\times 7.194 + 4\times\cos 8°} = -0.48 \times \frac{11.1551}{11.3711} \rightarrow \left(C_{\tilde{l}p}\right)_0 = -0.471$$

Also, from before,

$$C_{D\text{profile}} = 0.016, \quad (C_L)_w = 0.5311$$

Thus, upon assuming that $(C_{L\alpha})_{CL} = (C_{L\alpha})_{CL=0}$, one obtains:

$$\left(C_{\tilde{l}p}\right)_{\text{planform}} = -0.471 - \frac{1}{8}\times\frac{0.5311^2}{\pi\times 7.194\times\cos^2 8°}\left(1 + 2\sin^2 8° \times \frac{7.194 + 2\times\cos 8°}{7.194 + 4\times\cos 8°}\right) - \frac{0.016}{8}$$

$$= -0.471 - 0.00159\times 1.0319 - 0.0020 = -0.471 - 0.0016 - 0.0020 \rightarrow \left(C_{\tilde{l}p}\right)_{\text{planform}} = -0.475$$

Further,

$$\left[\left(C_{\tilde{l}p}\right)_{ac}\right]_w = -0.475\times\left(\frac{2}{2}\right)^2 \frac{0.556}{0.556} \rightarrow \left[\left(C_{\tilde{l}p}\right)_{ac}\right]_w = -0.475 \rightarrow$$

$$\left(C_{\tilde{l}p}\right)_w = -\frac{-0.1260}{2.0}\times 0.1254 - 0.475 \rightarrow \underline{\left(C_{\tilde{l}p}\right)_w = -0.467}$$

One may also calculate the contribution from the stabilizer, in the same way. However, in that case, swirl from the wing would have to be accounted for, which is a significant attenuating factor. For this particular configuration, where $b_s/b_w = 0.375$, the stabilizer's contribution may be assumed to be negligible compared with that from the wing.

Finally, the contribution from the propeller may be estimated from:

$$\left(C_{\tilde{l}p}\right)_{\text{prop}} = -\left(C_{Yp}\right)_{\text{prop}} z_{\text{prop}}/b = -0.0011\times 0.0206/2.0 \rightarrow \left(C_{\tilde{l}p}\right)_{\text{prop}} \approx 0$$

In summary, one obtains:

$$C_{\tilde{l}p} = \left(C_{\tilde{l}p}\right)_{\text{body}} + \left(C_{\tilde{l}p}\right)_f + \left(C_{\tilde{l}p}\right)_w + \left(C_{\tilde{l}p}\right)_{\text{prop}} = -0.0001 + 0.0016 - 0.467 + 0 \rightarrow$$

$$\underline{C_{\tilde{l}p} = -0.4655}$$

It's clear, for this stability derivative, that the contribution from the wing is dominant. In fact, simply choosing a value from the graph suffices with the error band of this methodology. This is typical for normally-proportioned tail-aft airplanes.

Now, as mentioned before, the acceleration stability derivatives for most aircraft are assumed to have a negligible contribution to the flight dynamics, with the exception of the downwash-lag effect. That doesn't mean that these acceleration stability derivatives are zero. It's just that they may be very small relative to other terms in the flight-dynamic equations. A measure of the significance of the acceleration stability derivatives is provided by the "wing loading", which is the mass of the aircraft divided by its wing area, m/S (in the Imperial system, this is given as the weight divided by the wing area, lbs/ft^2). A numerical example later in this chapter is the Piper PA-250 "Comanche", which is a single-engine all-metal general-aviation design. This has a wing loading of $79.5 \, kg/m^2$, which is by no means large compared with other full-scale airplanes. The "Scholar", however, has a wing loading of $6.65 \, kg/m^2$, so it will be informative to estimate its acceleration stability derivatives:

Example Value of $C_{YD\beta}$

The body's contribution is given by:

$$\left(C_{YD\beta}\right)_{body} = -4k_2 \frac{(Vol)_{body}}{Sb} = -4 \times 1.30 \times \frac{0.0352}{0.556 \times 2.0} \rightarrow \left(C_{YD\beta}\right)_{body} = -0.1646$$

This equation for the fin's contribution is:

$$\left(C_{YD\beta}\right)_f = -\left(C_{LD\alpha}\right)_f \frac{S_f}{S} \frac{\bar{c}_f}{b} \left(1 - \frac{\partial \sigma_f}{\partial D\beta}\right) \eta_f$$

The acceleration side-wash term will be assumed to be zero, $\partial \sigma_f / \partial D\dot{\beta} \approx 0$. Also, for the fin's effective aspect ratio, $(AR)_{effec.} = 1.760$, Appendix E gives that $\left(C_{LD\alpha}\right)_f = 0.853$. Thus, one obtains:

$$\left(C_{YD\beta}\right)_f = -0.853 \times \frac{0.057}{0.556} \times \frac{0.223}{2.0} \times 0.882 \rightarrow \left(C_{YD\beta}\right)_f = -0.0090$$

The wing's contribution is given by:

$$\left(C_{YD\beta}\right)_w = -K_4 \left(C_{LD\alpha}\right)_w \frac{S_w}{S} \frac{\bar{c}_w}{b} \sin^2 \Gamma$$

where,

$$K_4 \approx 0.8, \quad \tilde{\eta} \approx 1.0, \quad \text{and} \quad \left(C_{LD\alpha}\right)_w = -3.910 \rightarrow$$

$$\left(C_{YD\beta}\right)_w = -0.8 \times (-3.910) \times \frac{0.556}{0.556} \times \frac{0.30}{2.0} \sin^2 5° \rightarrow \left(C_{YD\beta}\right)_w = 0.0036$$

In summary, one has:

$$C_{YD\beta} = \left(C_{YD\beta}\right)_{body} + \left(C_{YD\beta}\right)_f + \left(C_{YD\beta}\right)_w = -0.1646 - 0.0090 + 0.0036 - C_{YD\beta} = 0.1700$$

Now, in the flight-dynamic equations, this derivative appears in conjunction with the non-dimensional mass, μ:

$$\hat{d}_{1,1} = 2\mu - C_{YD\beta}$$

Flight Dynamics

where,

$$\mu = 2m/(\rho S b) = 2 \times 3.70/(1.225 \times 0.556 \times 2.0) \to \mu = 5.432, \quad \text{(standard sea level)}$$

Therefore,

$$\hat{d}_{1,1} = 10.865 + 0.1700 \to \hat{d}_{1,1} = 11.035$$

It is seen that the acceleration stability derivative contributes less than 2% toward the $\hat{d}_{1,1}$ term and, considering the approximations involved in this analysis, might be justifiably ignored.

Example Value of $C_{ND\beta}$

The body's contribution is estimated from:

$$\left(C_{ND\beta}\right)_{body} = \frac{x_{cv}}{b}\left(C_{YD\beta}\right)_{body}, \quad \text{where} \quad x_{cv} = \left(l_{cm} - l_{cv}\right)\cos\alpha_{equil} + \left(h_{cm} - h_{cv}\right)\sin\alpha_{equil} \to$$

$$x_{cv} = (0.497 - 0.60) \times \cos(1.0871°) + (0.03 - 0) \times \sin(1.0871°) \to x_{cv} = -0.1024\,\text{m} \to$$

$$\left(C_{ND\beta}\right)_{body} = \frac{-0.1024}{2.0} \times (-0.1646) \to \left(C_{ND\beta}\right)_{body} = 0.0084$$

The fin's contribution is:

$$\left(C_{ND\beta}\right)_f = \frac{(x_{ca})_f}{b}\left[\left(C_{YD\beta}\right)_{app}\right]_f + \frac{x_f}{b}\left[\left(C_{YD\beta}\right)_{cir}\right]_f$$

where,

$$(x_{ca})_f = \left[l_{cm} - (l_{ca})_f\right]\cos\alpha_{equil} + \left[h_{cm} - (h_{ca})_f\right]\sin\alpha_{equil} \to$$

$$(x_{ca})_f = (0.497 - 1.283) \times \cos(1.0871°) + (0.03 - 0.110) \times \sin(1.0871°) \to (x_{ca})_f = -0.7874\,\text{m}$$

$$x_f = (0.497 - 1.225) \times \cos(1.0871°) + (0.03 - 0.110) \times \sin(1.0871°) \to x_f = -0.7294\,\text{m}$$

also,

$$\left[\left(C_{YD\beta}\right)_{app}\right]_f = -\left[\left(C_{LD\alpha}\right)_{app}\right]_f \frac{S_f}{S}\frac{\bar{c}_f}{b}\left(1 - \frac{\partial\sigma_f}{\partial D\beta}\right)\eta_f$$

$$\left[\left(C_{YD\beta}\right)_{cir}\right]_f = -\left[\left(C_{LD\alpha}\right)_{cir}\right]_f \frac{S_f}{S}\frac{\bar{c}_f}{b}\left(1 - \frac{\partial\sigma_t}{\partial D\beta}\right)\eta_f$$

where,

$$\left[\left(C_{LD\alpha}\right)_{app}\right]_f = 1.9227, \quad \left[\left(C_{LD\alpha}\right)_{cir}\right]_f = -1.0693, \quad \text{and} \quad \partial\sigma_f/\partial D\beta \approx 0 \to$$

$$\left[\left(C_{YD\beta}\right)_{app}\right]_f = -1.9227 \times \frac{0.057}{0.556} \times \frac{0.223}{2.0} \times (1-0) \times 0.882 = -0.0194$$

$$\left[(C_{YD\beta})_{cir}\right]_f = 1.0693 \times \frac{0.057}{0.556} \times \frac{0.223}{2.0} \times (1-0) \times 0.882 = 0.0108 \rightarrow$$

$$(C_{ND\beta})_f = \frac{-0.7874}{2.0} \times (-0.0194) + \frac{-0.7294}{2.0} \times 0.0108 \rightarrow (C_{ND\beta})_f = 0.00370$$

And, the wing's contribution is estimated from:

$$\left[(C_{YD\beta})_{app}\right]_w = -K_4 \left[(C_{LD\alpha})_{app}\right]_w \frac{S_{ls}}{S} \frac{\bar{c}_{ls}}{b} \tilde{\eta}_{ls} \sin^2 \Gamma_w$$

$$= -0.8 \times 2.560 \times \frac{0.556}{0.556} \times \frac{0.30}{2.0} \times 1.0 \times \sin^2(5°) \rightarrow \left[(C_{YD\beta})_{app}\right]_w = -0.0023$$

$$\left[(C_{YD\beta})_{cir}\right]_w = -K_4 \left[(C_{LD\alpha})_{cir}\right]_w \frac{S_{ls}}{S} \frac{\bar{c}_{ls}}{b} \tilde{\eta}_{ls} \sin^2 \Gamma_w$$

$$= -0.8 \times (-6.440) \times \frac{0.556}{0.556} \times \frac{0.30}{2.0} \times 1.0 \times \sin^2(5°) \rightarrow \left[(C_{YD\beta})_{cir}\right]_w = 0.0059$$

therefore,

$$(C_{ND\beta})_w = \frac{(x_{ca})_w}{b} \left[(C_{YD\beta})_{app}\right]_w + \frac{x_w}{b} \left[(C_{YD\beta})_{cir}\right]_w$$

$$(C_{ND\beta})_w = \frac{-0.0644}{2.0} \times (-0.0023) + \frac{0.0006}{2.0} \times 0.0059 \rightarrow (C_{ND\beta})_w \approx 0.0$$

The summary value is:

$$C_{ND\beta} = (C_{ND\beta})_{body} + (C_{ND\beta})_f + (C_{ND\beta})_w = 0.0084 + 0.00370 + 0.0 \rightarrow C_{ND\beta} = 0.01210$$

Example Value of $C_{\tilde{L}D\beta}$

The body's contribution is:

$$(C_{\tilde{L}D\beta})_{body} = -\frac{z_{cv}}{b}(C_{YD\beta})_{body}, \quad \text{where} \quad z_{cv} = (h_{cm} - h_{cv})\cos\alpha_{equli} - (l_{cm} - l_{cv})\sin\alpha_{equil} \rightarrow$$

$$z_{cv} = (0.030 - 0) \times \cos(1.0871°) - (0.497 - 0.60) \times \sin(1.0871°) \rightarrow z_{cv} = 0.0319\,\text{m} \rightarrow$$

$$(C_{\tilde{L}D\beta})_{body} = -\frac{0.0319}{2.0}(-0.1646) \rightarrow (C_{\tilde{L}D\beta})_{body} = 0.0026\,\text{m}$$

The fin's contribution is:

$$(C_{\tilde{L}D\beta})_f = -\frac{(z_{ca})_f}{b}\left[(C_{YD\beta})_{app}\right]_f - \frac{z_f}{b}\left[(C_{YD\beta})_{cir}\right]_f$$

where,

$$(z_{ca})_f = \left[h_{cm} - (h_{ca})_f\right]\cos\alpha_{equli} - \left[l_{cm} - (l_{ca})_f\right]\sin\alpha_{equil} \rightarrow$$

$$(z_{ca})_f = (0.030 - 0.110) \times \cos(1.0871°) - (0.497 - 1.283) \times \sin(1.0871°) \rightarrow (z_{ca})_f = -0.0651\,\text{m} \rightarrow$$

$$(C_{\tilde{L}D\beta})_f = -\frac{(-0.0651)}{2.0} \times (-0.0194) - \frac{(-0.0662)}{2.0} \times (0.0108) \rightarrow (C_{\tilde{L}D\beta})_f = -0.00027$$

And the wing's contribution is estimated from:

$$\left(C_{\tilde{L}D\beta}\right)_w = -\frac{(z_{ca})_w}{b}\left[\left(C_{YD\beta}\right)_{app}\right]_w - \frac{z_w}{b}\left[\left(C_{YD\beta}\right)_{cir}\right]_w$$

where,

$$(z_{ca})_w = \left[h_{cm} - (h_{ca})_w\right]\cos\alpha_{equli} - \left[l_{cm} - (l_{ca})_w\right]\sin\alpha_{equil} \rightarrow$$

$$(z_{ca})_w = (0.030 - 0.16)\times\cos(1.0871°) - (0.497 - 0.559)\times\sin(1.0871°) \rightarrow (z_{ca})_w = -0.1288\,\mathrm{m} \rightarrow$$

$$\left(C_{\tilde{L}D\beta}\right)_w = -\frac{(-0.1288)}{2.0}\times(-0.0023) - \frac{(-0.1260)}{2.0}\times 0.0059 \rightarrow \left(C_{\tilde{L}D\beta}\right)_w = 0.00022$$

The summary value is:

$$C_{\tilde{L}D\beta} = \left(C_{\tilde{L}D\beta}\right)_{body} + \left(C_{\tilde{L}D\beta}\right)_f + \left(C_{\tilde{L}D\beta}\right)_w = 0.0026 - 0.00027 + 0.00022 \rightarrow C_{\tilde{L}D\beta} = 0.0026$$

Example Value of C_{YDr}

The body's contribution is:

$$\left(C_{YDr}\right)_{body} = 2\frac{x_{cv}}{b}\left(C_{YD\beta}\right)_{body} \approx 2\times\frac{(-0.1024)}{2.0}\times(-0.1646) \rightarrow \left(C_{YDr}\right)_{body} = 0.0169$$

The fin's contribution is obtained from:

$$\left(C_{YDr}\right)_f = \left[\left(C_{YDr}\right)_{app}\right]_f + \left[\left(C_{YDr}\right)_{cir}\right]_f$$

where,

$$\left[\left(C_{YDr}\right)_{app}\right]_f = -\left[\left(C_{LD\alpha}\right)_{app}\right]_f\left(\frac{2(x_{ca})_f}{b} + \frac{\partial\sigma_f}{\partial(D\hat{r})}\right)\frac{S_f}{S}\frac{\bar{c}_f}{b}\eta_f$$

$$\left[\left(C_{YDr}\right)_{cir}\right]_f = -\left[\left(C_{LD\alpha}\right)_{cir}\right]_f\left(\frac{2x_f}{b} + \frac{\partial\sigma_f}{\partial(D\hat{r})}\right)\frac{S_f}{S}\frac{\bar{c}_f}{b}\eta_f$$

With $\partial\sigma_f/\partial(D\hat{r})$ assumed to be zero, this gives:

$$\left[\left(C_{YDr}\right)_{app}\right]_f = -1.9227\times\frac{2\times(-0.7874)}{2.0}\times\frac{0.057}{0.556}\times\frac{0.223}{2.0}\times 0.882 \rightarrow$$

$$\left[\left(C_{YDr}\right)_{app}\right]_f = 0.0153$$

$$\left[\left(C_{YDr}\right)_{cir}\right]_f = -(-1.0693)\times\frac{2\times(-0.7294)}{2.0}\times\frac{0.057}{0.556}\times\frac{0.223}{2.0}\times 0.882 \rightarrow$$

$$\left[\left(C_{YDr}\right)_{cir}\right]_f = -0.0079$$

so,

$$\left(C_{YDr}\right)_f = 0.0153 - 0.0079 \rightarrow \left(C_{YDr}\right)_f = 0.0074$$

The wing's contribution is given by:

$$\left(C_{YDr}\right)_w = \left[\left(C_{YDr}\right)_{app}\right]_w + \left[\left(C_{YDr}\right)_{cir}\right]_w$$

where,

$$\left[\left(C_{YDr}\right)_{app}\right]_w = 2\frac{(x_{ca})_w}{b}\left[\left(C_{YD\beta}\right)_{app}\right]_w = 2 \times \frac{-0.0644}{2.0} \times (-0.0023) \rightarrow \left[\left(C_{YDr}\right)_{app}\right]_w = 0.00015$$

$$\left[\left(C_{YDr}\right)_{cir}\right]_w = 2\frac{x_w}{b}\left[\left(C_{YD\beta}\right)_{cir}\right]_w = 2 \times \frac{0.0006}{2.0} \times 0.0059 \rightarrow \left[\left(C_{YDr}\right)_{cir}\right]_w \approx 0.0$$

so,

$$\left(C_{YDr}\right)_w = 0.00015 + 0.0 \rightarrow \left(C_{YDr}\right)_w = 0.00015$$

The summary value is:

$$C_{YDr} = \left(C_{YDr}\right)_{body} + \left(C_{YDr}\right)_f + \left(C_{YDr}\right)_w = 0.0169 + 0.0074 + 0.00015 \rightarrow C_{YDr} = 0.0245$$

Example Value of C_{NDr}

The body's contribution is:

$$\left(C_{NDr}\right)_{body} = \frac{x_{cv}}{b}\left(C_{YDr}\right)_{body} + \left(C_{NDr}\right)_{body}^{cv}, \quad \text{where, from previous work,}$$

$$\left(C_{NDr}\right)_{body}^{cv} = \frac{-2k'\left[l_{body}^2 + (dia)_{body}^2\right]}{5Sb^2} \times (Vol)_{body} \rightarrow$$

$$\frac{-2 \times 0.7302 \times \left(1.265^2 + 0.2306^2\right)}{5 \times 0.556 \times 2.0^2} \times 0.0352 \rightarrow \left(C_{NDr}\right)_{body}^{cv} = -0.0076$$

An Extension to the Analysis

As pointed out earlier, the fuselage is deeper than it is wider, so a representative ellipsoid would have an oval cross section. It was determined that the length, width and height of this equivalent ellipsoid is as follows:

$$L_{equiv.\,ellipse} = 1.2649\,m, \quad w_{equiv.\,ellipse} = 0.1884\,m, \quad h_{equiv.\,ellipse} = 0.2821\,m$$

Therefore, $L/w = 6.7139$ and $h/w = 1.497$. So, from the previously mentioned charts in Appendix D, the apparent moment-of-inertia coefficient for rotation about the z-axis is:

$$k'_z = 1.09$$

Flight Dynamics

The body's stability derivative is then given by:

$$\left(C_{NDr}\right)_{body}^{cv} = \frac{-2k_z'\left(L_{equiv.ellipse}^2 + h_{equiv.ellipse}^2\right)}{5Sb^2} \times (Vol)_{body} \rightarrow$$

$$\frac{-2 \times 1.09 \times \left(1.265^2 + 0.2821^2\right)}{5 \times 0.556 \times 2.0^2} \times 0.0352 \rightarrow \left(C_{NDr}\right)_{body}^{cv} = -0.0116$$

This will be the value chosen for the body's contribution to this derivative. Therefore,

$$\left(C_{NDr}\right)_{body} = \frac{(-0.1024)}{2.0} \times 0.0169 - 0.0116 \rightarrow \left(C_{NDr}\right)_{body} = -0.0125$$

The fin's contribution is:

$$\left(C_{NDr}\right)_f = \frac{(x_{ca})_f}{b}\left[\left(C_{YDr}\right)_{app}\right]_f + \frac{x_f}{b}\left[\left(C_{YDr}\right)_{cir}\right]_f \rightarrow$$

$$\left(C_{NDr}\right)_f = \frac{-0.7874}{2.0} \times 0.0153 + \frac{-0.7294}{2.0} \times (-0.0079) \rightarrow \left(C_{NDr}\right)_f = -0.0031$$

And the wing's contribution is:

$$\left(C_{NDr}\right)_w = \frac{(x_{ca})_w}{b}\left[\left(C_{YDr}\right)_{app}\right]_w + \frac{x_w}{b}\left[\left(C_{YDr}\right)_{cir}\right]_w \rightarrow$$

$$\left(C_{NDr}\right)_w = \frac{-0.0644}{2.0} \times 0.00015 + \frac{0.0006}{2.0} \times 0.0 \rightarrow \left(C_{NDr}\right)_w \approx 0.0$$

The summary value is:

$$C_{NDr} = \left(C_{NDr}\right)_{body} + \left(C_{NDr}\right)_f + \left(C_{NDr}\right)_w = -0.0125 - 0.0031 + 0.0 \rightarrow C_{NDr} = -0.0156$$

Note that in the flight-dynamic equations C_{NDr} appears in the term:

$$\hat{d}_{3,3} = i_{zz} - C_{NDr}, \quad \text{where} \quad i_{zz} = I_{zz}/\left[\rho S(b/2)^3\right]$$

Now, I_{zz} may be estimated from:

$$I_{zz} \approx m\left[r_z(b+l)/4\right]^2, \quad \text{where } r_z \text{ is the non-dimensional yawing radius-of-gyration.}$$

From www.eng-tips.com, a typical value for a single-engine high-wing monoplane is:

$$r_z = 0.393 \rightarrow I_{zz} \approx 3.70 \times \left[0.393 \times (2.0 + 1.462)/4\right]^2 \rightarrow I_{zz} = 0.4281 \, kg-m^2$$

Therefore, one obtains, for standard sea-level conditions, that:

$$i_{zz} = \frac{I_{zz}}{\rho S(b/2)^3} = \frac{0.4281}{1.225 \times 0.556 \times (2.0/2)^3} \rightarrow i_{xx} = 0.6285$$

It is seen that C_{NDr} is less than 3% of this and ignoring its effect is justified.

Example Value of $C_{\tilde{L}Dr}$

The body's contribution is estimated from:

$$\left(C_{\tilde{L}Dr}\right)_{body} = -\frac{z_{cv}}{b}\left(C_{YDr}\right)_{body} = \frac{0.0319}{2.0} \times 0.0169 \rightarrow \left(C_{\tilde{L}Dr}\right)_{body} = -0.0003$$

The fin's contribution is given by:

$$\left(C_{\tilde{L}Dr}\right)_f = -\frac{(z_{ca})_f}{b}\left[\left(C_{YDr}\right)_{app}\right]_f - \frac{z_f}{b}\left[\left(C_{YDr}\right)_{cir}\right]_f \rightarrow$$

$$\left(C_{\tilde{L}Dr}\right)_f = -\frac{(-0.0651)}{2.0} \times 0.0153 - \frac{(-0.0662)}{2.0} \times -0.0079 \rightarrow \left(C_{\tilde{L}Dr}\right)_f = 0.0002$$

And the wing's contribution is:

$$\left(C_{\tilde{L}Dr}\right)_w = -\frac{(z_{ca})_w}{b}\left[\left(C_{YDr}\right)_{app}\right]_w - \frac{z_w}{b}\left[\left(C_{YDr}\right)_{cir}\right]_w \rightarrow$$

$$\left(C_{\tilde{L}Dr}\right)_w = -\frac{-0.1288}{2.0} \times 0.00015 - \frac{-0.1260}{2.0} \times 0.0 \rightarrow \left(C_{\tilde{L}Dr}\right)_f \approx 0.0$$

The summary value is:

$$C_{\tilde{L}Dr} = \left(C_{\tilde{L}Dr}\right)_{body} + \left(C_{\tilde{L}Dr}\right)_f + \left(C_{\tilde{L}Dr}\right)_w = -0.0003 + 0.0002 + 0.0 \rightarrow C_{\tilde{L}Dr} = 0.0001$$

Example Value of C_{YDp}

The body's contribution is estimated from:

$$\left(C_{YDp}\right)_{body} = -2\frac{z_{cv}}{b}\left(C_{YD\beta}\right)_{body} = -2 \times \frac{0.0319}{2.0} \times (-0.1646) \rightarrow \left(C_{YDp}\right)_{body} = 0.0053$$

The fin's contribution is given by:

$$\left(C_{YDp}\right)_f = \left[\left(C_{YDp}\right)_{app}\right]_f + \left[\left(C_{YDp}\right)_{cir}\right]_f$$

where,

$$\left[\left(C_{YDp}\right)_{app}\right]_f = \left[\left(C_{LD\alpha}\right)_{app}\right]_f \left(\frac{2(z_{ca})_f}{b} + \frac{\partial \sigma_f}{\partial(D\hat{p})}\right)\frac{S_f}{S}\frac{\bar{c}_f}{b}\eta_f$$

$$\left[\left(C_{YDp}\right)_{cir}\right]_f = \left[\left(C_{LD\alpha}\right)_{cir}\right]_f \left(\frac{2z'_f}{b} + \frac{\partial \sigma_f}{\partial(D\hat{p})}\right)\frac{S_f}{S}\frac{\bar{c}_f}{b}\eta_f$$

With the swirl effect due to roll acceleration $\partial\sigma_f/\partial(D\hat{p})$ assumed to be negligible (pending further research), one obtains:

$$\left[\left(C_{YDp}\right)_{app}\right]_f = 1.9227 \times \frac{2 \times (-0.0651)}{2.0} \times \frac{0.057}{0.556} \times \frac{0.223}{2.0} \times 0.882 \rightarrow \left[\left(C_{YDp}\right)_{app}\right]_f = -0.0013$$

$$\left[\left(C_{YDp}\right)_{cir}\right]_f = -1.0693 \times \frac{2 \times (-0.076)}{2.0} \times \frac{0.057}{0.556} \times \frac{0.223}{2.0} \times 0.882 \rightarrow \left[\left(C_{YDp}\right)_{cir}\right]_f = 0.0008$$

so,
$$(C_{YDp})_f = -0.0013 + 0.0008 \to (C_{YDp})_f = -0.0005$$

Similarly, the wing's contribution is given by:
$$(C_{YDp})_w = \left[(C_{YDp})_{app}\right]_w + \left[(C_{YDp})_{cir}\right]_w$$

where,
$$\left[(C_{YDp})_{app}\right]_w = -2\frac{(z_{ca})_w}{b}\left[(C_{YD\beta})_{app}\right]_w \to \left[(C_{YDp})_{app}\right]_w = -2 \times \frac{(-0.1288)}{2.0} \times (-0.0023) \to$$

$$\left[(C_{YDp})_{app}\right]_w = -0.0003$$

$$\left[(C_{YDp})_{cir}\right]_w = -2\frac{z_w}{b}\left[(C_{YD\beta})_{cir}\right]_w \to \left[(C_{YDp})_{cir}\right]_w = -2 \times \frac{(-0.1260)}{2.0} \times 0.0059 \to$$

$$\left[(C_{YDp})_{cir}\right]_w = 0.00074$$

so,
$$(C_{YDp})_w = -0.0003 + 0.00074 \to (C_{YDp})_w = 0.00044$$

In this case $\partial \sigma_w / \partial(D\hat{p})$ is also assumed to be negligible. This is more certain because of the wing's forward position.

The summary value is:
$$C_{YDp} = (C_{YDp})_{body} + (C_{YDp})_f + (C_{YDp})_w = 0.0053 - 0.0005 + 0.0004 \to C_{YDp} = 0.0052$$

Example Value of C_{NDp}:

The body's contribution is estimated from:
$$(C_{NDp})_{body} = \frac{x_{cv}}{b}(C_{YDp})_{body} = \frac{-0.1024}{2.0} \times 0.0053 \to (C_{NDp})_{body} = -0.0003$$

The fin's contribution is:
$$(C_{NDp})_f = \frac{(x_{ca})_f}{b}\left[(C_{YDp})_{app}\right]_f + \frac{x_f}{b}\left[(C_{YDp})_{cir}\right]_f \to$$

$$(C_{NDp})_f = \frac{-0.7874}{2.0} \times (-0.0013) + \frac{-0.7294}{2.0} \times (0.0008) \to (C_{NDp})_f = 0.0002$$

And the wing's contribution is:
$$(C_{NDp})_w = \frac{(x_{ca})_w}{b}\left[(C_{YDp})_{app}\right]_w + \frac{x_w}{b}\left[(C_{YDp})_{cir}\right]_w \to$$

$$(C_{NDp})_w = \frac{-0.0644}{2.0} \times (-0.0003) + \frac{0.0006}{2.0} \times 0.00077 \to (C_{NDp})_w \approx 0.0$$

The summary value is:

$$C_{NDp} = (C_{NDp})_{body} + (C_{NDp})_f + (C_{NDp})_w = -0.0003 + 0.0002 + 0.0 \rightarrow \underline{\underline{C_{NDp} = 0.0001}}$$

Example Value of $C_{\tilde{L}Dp}$

The body's contribution is estimated from:

$$(C_{\tilde{L}Dp})_{body} = -\frac{z_{cv}}{b}(C_{YDp})_{body} = -\frac{0.0319}{2.0} \times 0.0053 \rightarrow (C_{\tilde{L}Dp})_{body} \approx 0.0$$

The fin's contribution is:

$$(C_{\tilde{L}Dp})_f = -\frac{(z_{ca})_f}{b}\left[(C_{YDp})_{app}\right]_f - \frac{z'_f}{b}\left[(C_{YDp})_{cir}\right]_f \rightarrow$$

$$(C_{\tilde{L}Dp})_f = -\frac{(-0.0651)}{2.0} \times (-0.0013) - \frac{(-0.076)}{2.0} \times 0.0008 \rightarrow (C_{\tilde{L}Dp})_f \approx 0.0$$

and the wing's contribution is given by:

$$(C_{\tilde{L}Dp})_w = -\frac{(z_{ca})_w}{b}\left[(C_{YDp})_{app}\right]_w - \frac{z_w}{b}\left[(C_{YDp})_{cir}\right]_w + \left[(C_{\tilde{L}Dp})_{ac}\right]_w, \quad \text{where}$$

$$\left[(C_{\tilde{L}Dp})_{ac}\right]_w = -\frac{2(C_{L D\alpha})_w}{Sb^3}\int_{-b/2}^{b/2} c^2 y^2 dy = -\frac{4(C_{L D\alpha})_w}{Sb^3}\int_{0}^{b/2} c^2 y^2 dy$$

Although the wing has curved tips, it will be assumed that the chord is constant along the span. Therefore, one has that:

$$\left[(C_{\tilde{L}Dp})_{ac}\right]_w = -\frac{4(C_{L D\alpha})_w}{Sb^3}c^2\int_0^{b/2} y^2 dy = -\frac{4(C_{L D\alpha})_w}{Sb^3}c^2\left|\frac{y^3}{3}\right|_0^{b/2}$$

$$= -\frac{4(C_{L D\alpha})_w}{Sb^3}c^2\frac{b^3}{24} = -\frac{(C_{L D\alpha})_w}{S}\frac{c^2}{6} = -\frac{-3.910}{0.556}\times\frac{0.30^2}{6} \rightarrow \left[(C_{\tilde{L}Dp})_{ac}\right]_w = 0.1055$$

so,

$$(C_{\tilde{L}Dp})_w = -\frac{-0.1288}{2.0}\times(-0.0003) - \frac{-0.1260}{2.0}\times 0.00074 + 0.1055 \rightarrow \underline{\underline{(C_{\tilde{L}Dp})_w = 0.1055}}$$

An Extension to the Analysis

The wing's actual planform geometry is given by:

y, span from center (m)	c, chord (m)
0	0.30
0.788	0.30
0.833	0.291
0.878	0.274
0.917	0.246
0.957	0.192
1.0	0

From this, one obtains:

$$\int_0^{b/2} c^2 y^2 dy \approx \int_0^{0.788} c^2 y^2 dy + \sum_{n=1}^{5} c^2 y^2 \Delta y$$

$$= \int_0^{0.788} c^2 y^2 dy + \sum_{n=1}^{5} \left[\left(\frac{c_{n+1}+c_n}{2}\right)^2 \times \left(\frac{y_{n+1}+y_n}{2}\right)^2 \times (y_{n+1}-y_n) \right]$$

$$= (0.30)^2 \times \frac{0.788^3}{3} + (0.296)^2 \times (0.811)^2 \times 0.045 + (0.283)^2 \times (0.856)^2$$

$$\times 0.045 + (0.260)^2 \times (0.898)^2 \times 0.039 + (0.219)^2 \times (0.937)^2 \times 0.040$$

$$+ (0.096)^2 \times (0.979)^2 \times 0.043 = 0.01468 + 0.00259 + 0.00264 + 0.00213$$

$$+ 0.00168 + 0.00038 \rightarrow \int_0^{b/2} c^2 y^2 dy \approx 0.02410 \rightarrow$$

$$\left[\left(C_{\tilde{L}Dp}\right)_{ac}\right]_w = -\frac{4(C_{LD\alpha})_w}{Sb^3} \int_0^{b/2} c^2 y^2 dy = -\frac{4 \times (-3.910)}{0.556 \times 2.0^3} \times 0.02410 \rightarrow \left[\left(C_{\tilde{L}Dp}\right)_{ac}\right]_w = 0.0848$$

Choose this value, so that:

$$\left(C_{LDp}\right)_w = -\frac{-0.1288}{2.0} \times (-0.0003) - \frac{-0.1300}{2.0} \times 0.00077 + 0.0848 \rightarrow \underline{\left(C_{LDp}\right)_w = 0.0848}$$

The summary value is:

$$C_{\tilde{L}Dp} = \left(C_{\tilde{L}Dp}\right)_{body} + \left(C_{\tilde{L}Dp}\right)_f + \left(C_{\tilde{L}Dp}\right)_w = 0.0 - 0.0 + 0.0848 \rightarrow \underline{\underline{C_{\tilde{L}Dp} = 0.0848}}$$

In the flight-dynamic equations, this stability derivative appears in conjunction with the non-dimensional rolling moment of inertia, i_{xx}:

$$\hat{d}_{2,2} = i_{xx} - C_{\tilde{L}Dp}, \quad \text{where} \quad i_{xx} = I_{xx} / \left[\rho S (b/2)^3\right]$$

Now, I_{xx} may be estimated from:

$$I_{xx} \approx m\left(r_x b/2\right)^2, \text{ where } r_x \text{ is the non-dimensional rolling radius-of-gyration}$$

As stated earlier, www.eng-tips.com, provides typical radii-of-gyration values for a single-engine high-wing monoplane (reproduced later in this section). In particular, r_x is approximated by:

$$r_x = 0.242 \rightarrow I_{xx} \approx 3.70 \times (0.242 \times 2.0/2)^2 \rightarrow I_{xx} = 0.2167 \text{ kg-m}^2$$

Therefore, at standard sea-level conditions, one obtains that:

$$i_{xx} = \frac{I_{xx}}{\rho S(b/2)^3} = \frac{0.2167}{1.225 \times 0.556 \times (2.0/2)^3} \rightarrow i_{xx} = 0.3181$$

In this instance $C_{\tilde{L}Dp}$ is 26.7% of i_{xx}, and is certainly not negligible. Therefore, one has to conclude that even for "Heavier-Than-Air" aircraft, certain acceleration stability derivatives may be significant if the wing loading is low enough.

Full-scale airplanes with the lowest wing loadings are the human-powered examples, such as the AeroVironment "Gossamer Albatross 2", which was flight-tested by NASA Dryden Flight Research Center:

https://www.nasa.gov/centers/dryden/multimedia/imagegallery/Albatross/ECN-12665.html

This was instrumented in order to characterize its flight-dynamic behavior, including its stability and control. The resulting data were compared with predictions from flight-dynamic analysis, where apparent-mass effects were included. Some comparisons were favorable, but not all. Subsequently, a few of the author's graduate students undertook their own analysis which included both apparent-mass *and* indicial-lift effects, as described in this chapter, and the comparisons were much more favorable, which supports the veracity of this approach.

Examples of lateral stability derivatives are given in the following table:

Lateral stability derivative	PA 24-250 (piper comanche)	Pazmany PL-4 home-built airplane	"Scholar" model airplane	"Wingfoot 2" airship
$C_{Y\beta}$	−0.60	−0.632	−0.4721	−0.7119
C_{Yr}	0.1011	0.396	0.1535	0.4550
C_{Yp}	−0.031	0.122	0.1736	−0.0323
$C_{N\beta}$	0.109	0.0244	0.0118	−0.2258
C_{Nr}	−0.0503	−0.122	−0.0848	−0.1975
C_{Np}	−0.0907	−0.0466	−0.0437	−0.0009
$C_{\tilde{L}\beta}$	−0.057	−0.0876	−0.1277	−0.0507
$C_{\tilde{L}r}$	0.0643	0.0932	0.1701	0.0209
$C_{\tilde{L}p}$	−0.399	−0.55	−0.4655	−0.0074
$C_{YD\beta}$	neglected	neglected	−0.1700	−0.9533
C_{YDr}	"	"	0.0245	−0.0646
C_{YDp}	"	"	0.0052	−0.0827
$C_{ND\beta}$	"	"	0.01210	−0.0323
C_{NDr}	"	"	−0.0156	−0.0898
C_{NDp}	"	"	0.0001	−0.0030
$C_{\tilde{L}D\beta}$	"	"	0.0026	−0.0412
$C_{\tilde{L}Dr}$	"	"	−0.0001	−0.0030
$C_{\tilde{L}Dp}$	"	"	0.0848	−0.0042

A few observations may be made about these derivatives:

1. The $C_{Y\beta}$ values should always be negative, because the sideslip velocity causes an opposing lateral force (visualize a fin in sideslip).
2. The C_{Yr} values are generally positive for a tail-aft design (again, visualize the yaw rate causing a negative β at the fin, thus giving a lateral force in the y-direction).
3. The C_{Yp} values may be positive or negative. For example, if the fin's aerodynamic center is above the x-axis, then roll rate will produce a positive β at the fin. This gives a lateral force in the negative y-direction. However, when the aircraft is at different trim angles, z_f may change sign (the fin's aerodynamic center may go below the x-axis, for example) and give a force in the positive y-direction.

4. $C_{N\beta}$ is generally positive for a conventional airplane configuration, and it is easy to see that an aft-mounted fin in sideslip will produce a positive yawing moment.
5. For C_{Nr}, a positive yaw rate will generally produce a negative yawing moment. Again, visualize the role of a vertical fin. Whether fore or aft mounted, it will oppose the yaw rate.
6. C_{Np} may be either negative or positive. An aft-located fin whose aerodynamic center is above the rolling axis, x, will make a positive contribution to the yawing moment. However, a wing can give rise to a negative contribution in the manner described earlier.
7. For an airplane with a vertical fin whose aerodynamic center is above the rolling axis, x, and a wing with positive dihedral, $C_{\bar{L}\beta}$ will almost always be negative.
8. The $C_{\bar{L}r}$ derivative may be negative or positive. However, observe that an aft-located fin whose aerodynamic center is above the rolling axis, x, will produce a positive rolling moment. This is likewise the case for a lifting surface, when the advancing left-hand panel will gain lift while the retreating right-hand panel loses lift.
9. The $C_{\bar{L}p}$ value should always be negative, because the aerodynamic forces and moments produced by the rolling will oppose the rolling velocity.

RADII-OF-GYRATION VALUES FOR REPRESENTATIVE AIRPLANES

Moments of inertia were estimated for the Scholar example airplane. These were found at www.eng-tips.com. This site gives a table of radii-of-gyration values for several types of airplanes:

Type of airplane	r_x, roll	r_y, pitch	r_z, yaw
Single-engine, low wing	0.248	0.338	0.393
Single-engine, high wing	0.242	0.397	0.393
Light twin-engine design	0.373	0.269	0.461
Light business jet	0.293	0.312	0.420
Heavy business jet	0.370	0.356	0.503
Twin turbo prop	0.235	0.363	0.416
Four-engine jet airliner	0.322	0.339	0.464
Three aft-engine jet airliner	0.249	0.375	0.452
Wing-mounted two-engine jet airliner	0.246	0.382	0.456
Four-engine prop airliner	0.322	0.324	0.456
Two-engine prop airliner	0.308	0.345	0.497
Jet fighter	0.266	0.346	0.400
Single-engine prop fighter	0.268	0.360	0.420
Twin-engine prop fighter	0.330	0.299	0.447
Twin-engine prop bomber	0.270	0.320	0.410
Four-engine prop bomber	0.316	0.320	0.376
Concorde delta-wing airliner	0.253	0.380	0.390

From these values, one may calculate:

$$I_{xx} = \frac{mb^2}{4} r_x^2, \quad I_{yy} = \frac{ml^2}{4} r_y^2, \quad I_{zz} = m\left[r_z \frac{(b+l)}{4} \right]^2, \quad I_{xz} \approx 0$$

where l is the airplane's total length, excluding the propeller.

Note that these values are for initial estimations. A detailed design of the airplane, including its structure, will give more accurate values including the polar moment of inertia I_{xz}.

DEFINITIONS OF STABILITY

A vehicle is considered to be stable if, after a disturbance, it automatically assumes a motion (or non-motion, as the case may be) that is acceptable to its mission. This implies that there is more than one type of stability. In particular, for aircraft, there are generally three levels of stability:

Straight-Line Stability

In this case, the aircraft settles down after a disturbance into a linear path that is generally in a different direction and altitude than that of the original path. This is what is known as controls-fixed stability, and it would be typical for a free-flight model glider (or paper airplane).

Directional Stability

This is the case where the aircraft, after a disturbance, settles down to a linear path that is parallel to the original path, though it may be displaced sideways and vertically. This would be the behavior of an aircraft whose controls are keyed to a compass heading.

Path-Keeping Stability

This is when the aircraft returns, after a disturbance, to its original path. The control system is keyed to the aircraft's exact position and orientation in the ground-fixed reference frame. However, the distance traveled along the path may differ from that if the aircraft had not been disturbed.

The highest level of stability is sought for tethered aerostats where, after a disturbance, the aerostat returns to its same position and orientation. This is sometimes called "photographic stability"

Flight Dynamics

because, after the aerostat settles down, it occupies exactly the same spot in the sky as it did before the disturbance.

With adequately designed non-linear automatic control systems, the above stabilities may be attained for large disturbances. But, in general, if the previously derived small-perturbation equations are used to investigate stability, it is understood that the validity of the results is limited to a region of small disturbances with respect to the aircraft's reference equilibrium position.

In practice, however, this has rarely proven to be a great restriction for the analysis of an aircraft's operational stability, provided that the stability derivatives stay reasonably constant for moderate perturbations from the reference position. Also, this requires that there is only one stable equilibrium position for large perturbations.

Small-perturbation stability analysis is a classical technique for many problems in mathematical physics, and the philosophy of the method is embodied in the question:

"If a system, which is in a reference equilibrium state, is subjected to small perturbations of its coordinates, will these perturbations decrease (stable) or increase (unstable)?"

Two non-aeronautical examples of small-perturbation stability analyses are presented below:
Buckling:

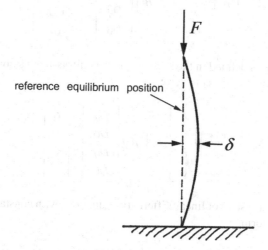

In this case, the concept is that the straight slender column has an axial force, F, applied to it as shown. Then, it is conceptually pulled out by a small displacement, δ. Upon release, the column will either spring back straight or δ will increase till the column collapses. The load at which this divergence occurs is called the buckling force, $F_{buckling}$.

Wave generation:

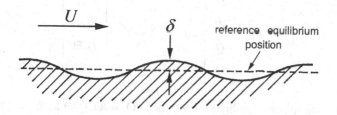

Assume that a body of water has a wind blowing over it. If the water's surface is conceptually perturbed into a steady sinusoidal form, the wind will produce suction loads over the humps and pressure loads over the shallows. The weight of the water resists this perturbation, but there is a wind speed at which the aerodynamic forces and the weight balance. Any speed above this will cause divergence and wave generation.

There are numerous other examples. For instance, there is a wavy-pavement model that shows how potholes are formed by the pressure of tires rolling over it. However, it is aircraft that are of interest here, and it will be shown that their dynamic behavior can include both damped oscillations as well as divergence.

Finally, the next section will concentrate on controls-fixed dynamic stability. Although modern stability-augmentation systems are playing an increasing role in allowing aerodynamically optimized but unstable designs to fly safely, there is still a strong interest in configurations that offer inherent stability.

LONGITUDINAL DYNAMIC STABILITY

In the "Aircraft Longitudinal Small-Perturbation Dynamic Equations" section, the following equations were presented:

$$[\hat{A}]\begin{bmatrix}\hat{u}\\ \alpha\\ \hat{q}\\ \theta\end{bmatrix} + [\hat{B}]\begin{bmatrix}D\hat{u}\\ D\alpha\\ D\hat{q}\\ D\theta\end{bmatrix} = \begin{bmatrix}\Delta C_{X_C}\\ \Delta C_{Z_C}\\ \Delta C_{M_C}\\ 0\end{bmatrix}$$

with the matrix elements as defined in the section. Since controls-fixed stability will be studied, the right-hand control matrix may be set to zero:

$$[\hat{A}]\begin{bmatrix}\hat{u}\\ \alpha\\ \hat{q}\\ \theta\end{bmatrix} + [\hat{B}]\begin{bmatrix}D\hat{u}\\ D\alpha\\ D\hat{q}\\ D\theta\end{bmatrix} = \begin{bmatrix}0\\ 0\\ 0\\ 0\end{bmatrix}$$

Because these represent a system of linear differential equations with constant coefficients, one may assume solutions of the form:

$$\hat{u} = \hat{U}e^{\sigma \hat{t}}, \quad \alpha = \hat{W}e^{\sigma \hat{t}}, \quad \hat{q} = \hat{Q}e^{\sigma \hat{t}}, \quad \theta = \Theta e^{\sigma \hat{t}}$$

where $\hat{U}, \hat{W}, \hat{Q}$ and Θ are constants (possibly complex). When these solutions are substituted into the matrix equations, one obtains:

$$[\hat{A} + \sigma \hat{B}]\begin{bmatrix}\hat{U}\\ \hat{W}\\ \hat{Q}\\ \Theta\end{bmatrix} = 0 \rightarrow [\hat{A} + \sigma \hat{B}]\begin{bmatrix}\hat{U}/\Theta\\ \hat{W}/\Theta\\ \hat{Q}/\Theta\\ 1\end{bmatrix} = 0$$

The σ value is the longitudinal stability root and the $\hat{U}/\Theta, \hat{W}/\Theta, \hat{Q}/\Theta$ terms are the modal vectors, normalized to a Θ of unity. Upon solution, the equations give four longitudinal stability roots of the form:

$$\sigma = \sigma_r + i\sigma_j$$

Flight Dynamics

and each of these is associated with a set of modal vectors. For example, σ_2 is the complex stability root for the motion characterized by $(\hat{U}/\Theta)_2$, $(\hat{W}/\Theta)_2$ and $(\hat{Q}/\Theta)_2$.

Upon returning to the assumed solutions, these may be rewritten as:

$$\hat{u} = \left(\frac{\hat{U}}{\Theta}\right)\Theta e^{\sigma_r \hat{t}} e^{i\sigma_j \hat{t}}, \quad \alpha = \left(\frac{\hat{W}}{\Theta}\right)\Theta e^{\sigma_r \hat{t}} e^{i\sigma_j \hat{t}}, \quad \hat{q} = \left(\frac{\hat{Q}}{\Theta}\right)\Theta e^{\sigma_r \hat{t}} e^{i\sigma_j \hat{t}}, \quad \theta = \Theta e^{\sigma_r \hat{t}} e^{i\sigma_j \hat{t}}$$

Common to all four equations is the term $e^{i\sigma_j \hat{t}}$ which, in complex notation, is a harmonic oscillation of frequency,

$$\omega_j = (2U_0/\bar{c})\sigma_j \text{ (rad/sec)}$$

This gives, as described earlier:

$$\text{Frequency in } Hz = f = \frac{\omega_j}{2\pi}, \quad \text{Period of oscillation} = T = \frac{2\pi}{\omega_j}$$

It sometimes happens that the mode of motion is aperiodic ($\sigma_j = 0$); however, in general, the σ_r term is non-zero. Further, σ_r determines the degree of stability (or instability) of the aircraft. For example, if σ_r is negative, the motion will be stable. However, if σ_r is positive, the motion will increase (unstable). As described earlier in this chapter, one way of characterizing this is with the Time-to-half (or double) amplitude:

$$\text{Time-to-half (or double) amplitude} = T_{1/2(or2)} = \frac{\ln 0.5}{\sigma_r} \frac{\bar{c}}{2U_0} = \frac{-0.69315}{\sigma_r} \frac{\bar{c}}{2U_0}$$

This gives, *for an oscillatory mode*, the number of cycles to half (or double) amplitude:

$$\text{Number of cycles to half (or double) amplitude} = N_{1/2(or2)} = \frac{T_{1/2(or2)}}{T} = \frac{0.11032\sigma_j}{|\sigma_r|}$$

It is seen that a mode may have four types of motion:

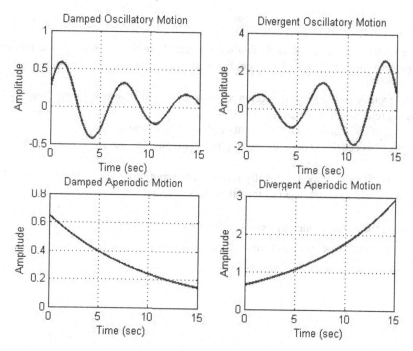

Further, as introduced earlier in this chapter, the modal vectors have a convenient pictorial representation through the "modal vector diagram" (or "Argand diagram"), as shown below for $\hat{U}$, $\hat{W}$ and Θ:

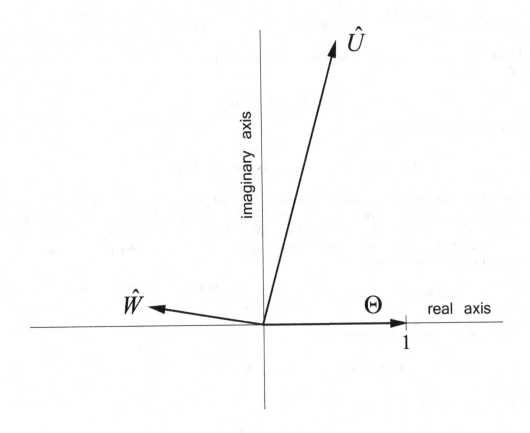

By studying such a diagram, one may obtain considerable information about the motion that the aircraft is performing for a particular mode. For example, the above diagram shows:

1. The magnitude of the surging motion, $\hat{U}$, is much larger than that for pitching, Θ and the plunging relative-angle-of-attack, $\hat{W}$.
2. Plunging, $\hat{W}$, is nearly 180 degrees out of phase with pitching, Θ.
3. Surging, $\hat{U}$, is nearly 90 degrees out of phase with pitching, Θ.

Further, observe that one may graphically represent the motions of *a* mode (say Mode 1) with the projection on the real axis of a triad of modal vectors that are:

1. Changing their magnitudes by $\exp\left(\frac{2U_0}{\bar{c}}\right)(\sigma_r)_1 t$.
2. Rotating counter-clockwise at frequency, ω_1

Flight Dynamics

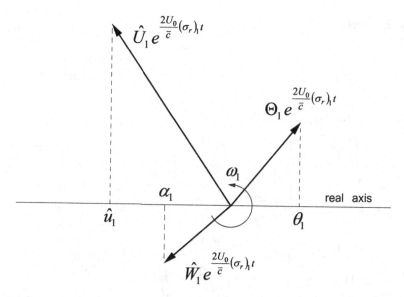

At this point, it is important to remember that the *total* motion is due to that of all four modes, which may be expressed as:

$$\hat{u} = \left(\frac{\hat{U}}{\Theta}\right)_1 \Theta_1 \exp\left(\sigma_1 \hat{t}\right) + \left(\frac{\hat{U}}{\Theta}\right)_2 \Theta_2 \exp\left(\sigma_2 \hat{t}\right) + \left(\frac{\hat{U}}{\Theta}\right)_3 \Theta_3 \exp\left(\sigma_3 \hat{t}\right) + \left(\frac{\hat{U}}{\Theta}\right)_4 \Theta_4 \exp\left(\sigma_4 \hat{t}\right)$$

$$\alpha = \left(\frac{\hat{W}}{\Theta}\right)_1 \Theta_1 \exp\left(\sigma_1 \hat{t}\right) + \left(\frac{\hat{W}}{\Theta}\right)_2 \Theta_2 \exp\left(\sigma_2 \hat{t}\right) + \left(\frac{\hat{W}}{\Theta}\right)_3 \Theta_3 \exp\left(\sigma_3 \hat{t}\right) + \left(\frac{\hat{W}}{\Theta}\right)_4 \Theta_4 \exp\left(\sigma_4 \hat{t}\right)$$

$$\hat{q} = \left(\frac{\hat{Q}}{\Theta}\right)_1 \Theta_1 \exp\left(\sigma_1 \hat{t}\right) + \left(\frac{\hat{Q}}{\Theta}\right)_2 \Theta_2 \exp\left(\sigma_2 \hat{t}\right) + \left(\frac{\hat{Q}}{\Theta}\right)_3 \Theta_3 \exp\left(\sigma_3 \hat{t}\right) + \left(\frac{\hat{Q}}{\Theta}\right)_4 \Theta_4 \exp\left(\sigma_4 \hat{t}\right)$$

$$\theta = \Theta_1 \exp\left(\sigma_1 \hat{t}\right) + \Theta_2 \exp\left(\sigma_2 \hat{t}\right) + \Theta_3 \exp\left(\sigma_3 \hat{t}\right) + \Theta_4 \exp\left(\sigma_4 \hat{t}\right)$$

If one has initial values for the motions: $\hat{u}_{\text{Initial}}$, α_{Initial}, $\hat{q}_{\text{Initial}}$ and θ_{Initial}, then these equations allow one to solve for Θ_1, Θ_2, Θ_3 and Θ_4. For example, if the initial values are taken at $t = 0$, then the equations give:

$$\begin{bmatrix} \left(\hat{U}/\Theta\right)_1 & \left(\hat{U}/\Theta\right)_2 & \left(\hat{U}/\Theta\right)_3 & \left(\hat{U}/\Theta\right)_4 \\ \left(\hat{W}/\Theta\right)_1 & \left(\hat{W}/\Theta\right)_2 & \left(\hat{W}/\Theta\right)_3 & \left(\hat{W}/\Theta\right)_4 \\ \left(\hat{Q}/\Theta\right)_1 & \left(\hat{Q}/\Theta\right)_2 & \left(\hat{Q}/\Theta\right)_3 & \left(\hat{Q}/\Theta\right)_4 \\ \Theta_1 & \Theta_2 & \Theta_3 & \Theta_4 \end{bmatrix} \begin{bmatrix} \Theta_1 \\ \Theta_2 \\ \Theta_3 \\ \Theta_4 \end{bmatrix} = \begin{bmatrix} \hat{u}(0) \\ \alpha(0) \\ \hat{q}(0) \\ \theta(0) \end{bmatrix}$$

This gives four simultaneous equations from which the Θ_i values may be solved.

Numerical Example

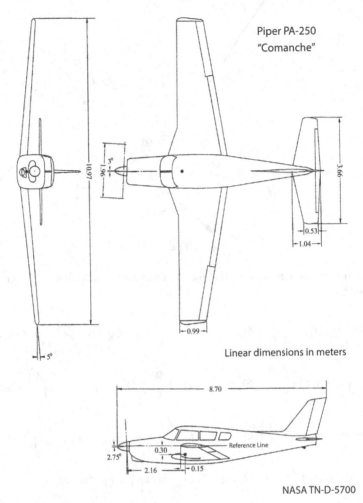

Piper PA-250 "Comanche"

Linear dimensions in meters

NASA TN-D-5700

Now, the longitudinal dynamic-stability characteristics will be calculated for the Piper Comanche whose stability derivatives were previously listed. In addition, the following parameters are required:

mass, $m = 1315$ kg; gross buoyancy, $B \approx 0$;
climb/descent angle, $\Theta_0 = 0$; reference (wing) area, $S = 16.537 \, \text{m}^2$;
mean aerodynamic chord, $\bar{c} = 1.509$ m; wing span, $b = 10.97$ m; flight speed, $U_0 = 76.444$ m/sec;
  altitude is 2840 m $\rightarrow \rho = 0.924 \, \text{kg/m}^3$;
rolling moment of inertia, $I_{xx} = 4648.0 \, \text{kg} - \text{m}^2$;
pitching moment of inertia, $I_{yy} = 8383.0 \, \text{kg} - \text{m}^2$;
yawing moment of inertia, $I_{zz} = 15071.0 \, \text{kg} - \text{m}^2$;
product of inertia, $I_{xz} = 195.0 \, \text{kg} - \text{m}^2$;
coordinates of the aircraft volumetric center from the mass-center, $x_b, z_b = 0$;
coordinates of the propeller hub from the mass-center, $x_t = -2.14$, $z_t \approx 0$.
level flight, $L = mg \rightarrow C_L = 0.2889 \rightarrow C_{Z0} = -0.2889$.

Also, the aircraft's aerodynamic polar equation is given below, from which the C_D may be calculated:

$$C_D = 0.026 + 0.058 \, C_L^2 \rightarrow C_D = 0.0308 \rightarrow C_{X0} = -0.0308$$

Flight Dynamics

Further, the aircraft is assumed to be fully trimmed about the mass-center, so that:

$$C_{M0} = 0$$

Next, the elements of matrices $\left[\hat{A}\right]$ and $\left[\hat{B}\right]$ may be calculated, which are $\hat{a}_{i,j}$ and $\hat{b}_{i,j}$ as defined earlier in the chapter, and the stability roots and modes may be obtained from the MATLAB statement:

```
[modes,stabroots]=eig(a,-b)
```

The stability roots for this example are:

$$\sigma_{1,2} = -0.02355 \pm 0.04304 j, \quad \sigma_{3,4} = -0.000260 \pm 0.001609 j$$

Note that these are two complex-conjugate pairs, which likewise give complex-conjugate modal vectors. However, the motion described by the modal vector of one stability root of the pair is exactly the same as that for its complex conjugate. This is because, even though the signs of the imaginary parts of the modal vectors differ, the direction of the rotation is also reversed (the modal vector diagrams are mirror images about the real axis). Therefore, the motions defined by the vectors' projections on the real axis (as described before) are identical.

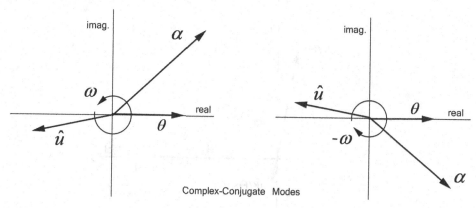

Complex-Conjugate Modes

Now, consider the first mode. The motion is stable (σ_r is negative) and one has that:

$$T = 1.4407 \text{ sec}, \quad T_{1/2} = 0.1904 \text{ sec}, \quad N_{1/2} = 0.2016$$

Further, the normalized modal vectors are:

$$\hat{U}/\Theta = 0.0115 + 0.0126 j, \quad \hat{W}/\Theta = 0.9838 + 0.4584 j, \quad \hat{Q}/\Theta = -0.0236 + 0.0430 j$$

The $\hat{U}/\Theta$ and $\hat{W}/\Theta$ vectors plot out as:

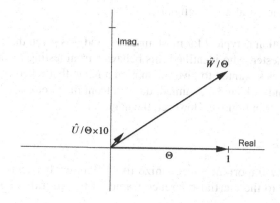

These results show that this mode of motion is:

1. Heavily damped (achieves damping to half amplitude in a fraction of a second),
2. A short period (the period of oscillation is less than two seconds),
3. Primarily an $\alpha - \theta$ motion, which are nearly equal in magnitude and close in phase,
4. The surging motion, $\hat{u}$, is negligible.

This is a mode of motion that is typical for most airplanes. In fact, this is so much so that it has a name: the "Short-Period Mode".

Now, for the second mode, one obtains:

$$T = 38.5440 \text{ sec.}, \quad T_{1/2} = 26.2791 \text{ sec.}, \quad N_{1/2} = 0.6818$$

and the normalized modal vectors are:

$$\hat{U}/\Theta = -0.1448 + 0.7594 j, \quad \hat{W}/\Theta = 0.0041 - 0.0173 j, \quad \hat{Q}/\Theta = -0.0003 + 0.0016 j$$

The $\hat{U}/\Theta$ and $\hat{W}/\Theta$ vectors plot out as:

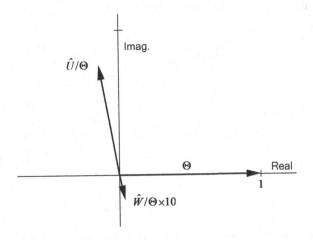

These results show that the mode is:

1. Lightly damped (takes nearly half a minute to reach half amplitude, which is long for this size airplane),
2. Long period (the period is 27 times longer than that for the Short-Period mode),
3. Primarily a surging-pitching motion, with nearly equal magnitudes and pitching lagging surging by approximately 90 degrees,
4. The plunging motion is almost negligible.

Again, this mode of motion is typical for most airplanes and was given the name "Phugoid Mode" by Sir Frederick Lanchester, who identified this behavior from testing model gliders early in the 20th century. Sir Frederick wanted to give the motion a name that reflected classical scholarship; hence he chose "Phugoid" which, he assumed, derived from the Greek word for flight. In fact, the word denotes "fleeing" as in fugitive. However, the name stuck.

COMMENTS ON α AND θ

At this point, it is very important to recognize the difference between θ and α. θ is a pitch-angle change, relative to the inertial-reference frame, of the aircraft's body-fixed coordinates.

Flight Dynamics

However, α is the instantaneous angle of the *perturbation* flow velocity at the mass-center, relative to the aircraft's reference line. More precisely, it is the instantaneous component of the aircraft's perturbation velocity in the z-axis direction, w, divided by the steady-state flight speed, U_0.

For example, in a wind-tunnel test where the model is rigidly supported, the model's pitch angle, θ and angle of attack, α, are identical. The situation is different if the model is allowed to freely move, as is seen from the modal-vector diagrams. For the Short-Period mode, the $\hat{W}/\Theta$ and Θ vectors are of similar magnitude, but have a distinct phase angle between them. Therefore, this is not like the pinned wind-vane example described at the beginning of this chapter. In that case, the two modal vectors would be identical with a zero-phase angle between them.

For the Phugoid mode, it is seen that α is very small compared with the non-dimensional surging, $\hat{u}$ and pitching, θ. In fact, the aircraft may be envisioned as skimming along the flight path, staying more-or-less tangent to the path. Therefore, the flow-velocity angle relative to the aircraft's x-axis (i.e. α) moves much less than the pitching angle.

The distinction is made clearer by the following sketch of two extreme examples:

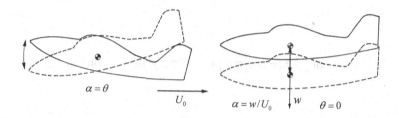

The picture on the left shows the "pinned" example, as discussed before; and the picture on the right shows an example executing a pure plunging motion. Of course, α is generally a function of both plunging motion and the perturbation pitch angle; but in this chapter w has been referred to as a "plunging" velocity, in the sense that it is the flow-velocity component along the body-fixed z-axis. In the same way, u has been referred to as a "surging" velocity.

FLIGHT PATHS

Further insights into both modes may be obtained by calculating their flight paths. For level flight, the small-perturbation forms of the flight-path equations for motion with respect to the inertial reference frame (x', z' coordinates) are obtained from Appendix C:

$$dx'/dt = U_0 + u = U_0(1+\hat{u}) \rightarrow dx'/d\hat{t} = \bar{c}/2(1+\hat{u})$$

$$dz'/dt = -U_0\theta + w = -U_0(\theta - \alpha) \rightarrow dz'/d\hat{t} = -\bar{c}/2(\theta - \alpha)$$

Upon noting recalling that $\hat{u} = \hat{U}e^{\sigma\hat{t}}$, $\alpha = \hat{W}e^{\sigma\hat{t}}$ and $\theta = \Theta e^{\sigma\hat{t}}$, these may be integrated to give:

$$x'(\hat{t}) = \frac{\bar{c}}{2}\int_0^{\hat{t}}(1+\hat{u})d\hat{t} + x'(0) = \frac{\bar{c}}{2}\left[\hat{t} + \frac{\hat{U}}{\sigma}\left(e^{\sigma\hat{t}} - 1\right)\right] + x'(0)$$

$$z'(\hat{t}) = -\frac{\bar{c}}{2}\int_0^{\hat{t}}(\theta - \alpha)d\hat{t} + z'(0) = -\frac{\bar{c}}{2}\left[\frac{\Theta}{\sigma}\left(e^{\sigma\hat{t}} - 1\right) - \frac{\hat{W}}{\sigma}\left(e^{\sigma\hat{t}} - 1\right)\right] + z'(0)$$

The modal-vector terms are generally complex, and therefore $x'(\hat{t})$ and $z'(\hat{t})$ are generally complex and can be divided into real and imaginary parts. It is the real parts that are of interest, which necessitates finding the real parts of these equations. For example, observe that:

$$\frac{\hat{U}}{\sigma}\left(e^{\sigma\hat{t}}-1\right) = \frac{\left(\hat{U}_r + j\hat{U}_j\right)}{\left(\sigma_r + j\sigma_j\right)}\left(\exp\left(\sigma_r\hat{t} + j\sigma_j\hat{t}\right)-1\right) \rightarrow$$

$$\frac{\hat{U}}{\sigma}\left(e^{\sigma\hat{t}}-1\right) = \frac{\left(\hat{U}_r + j\hat{U}_j\right)}{\left(\sigma_r^2 + \sigma_j^2\right)}(\sigma_r - j\sigma_j)\left[\exp\left(\sigma_r\hat{t}\right)\exp\left(j\sigma_j\hat{t}\right)-1\right]$$

Further,

$$\exp\left(j\sigma_j\hat{t}\right) = \cos\left(\sigma_j\hat{t}\right) + j\sin\left(\sigma_j\hat{t}\right), \text{ so one obtains:}$$

$$\text{Re}\left[\frac{\hat{U}}{\sigma}\left(e^{\sigma\hat{t}}-1\right)\right] = \frac{\exp\left(\sigma_r\hat{t}\right)}{\left(\sigma_r^2+\sigma_j^2\right)}\left[\hat{U}_r\left(\sigma_r\cos\left(\sigma_j\hat{t}\right)+\sigma_j\sin\left(\sigma_j\hat{t}\right)\right)\right.$$
$$\left.-\hat{U}_j\left(\sigma_r\sin\left(\sigma_j\hat{t}\right)-\sigma_j\cos\left(\sigma_j\hat{t}\right)\right)\right]-\frac{\left(\hat{U}_r\sigma_r+\hat{U}_j\sigma_j\right)}{\left(\sigma_r^2+\sigma_j^2\right)}$$

Likewise, one obtains:

$$\text{Re}\left[\frac{\Theta}{\sigma}\left(e^{\sigma\hat{t}}-1\right)\right] = \frac{\exp\left(\sigma_r\hat{t}\right)}{\left(\sigma_r^2+\sigma_j^2\right)}\left[\Theta_r\left(\sigma_r\cos\left(\sigma_j\hat{t}\right)+\sigma_j\sin\left(\sigma_j\hat{t}\right)\right)\right.$$
$$\left.-\Theta_j\left(\sigma_r\sin\left(\sigma_j\hat{t}\right)-\sigma_j\cos\left(\sigma_j\hat{t}\right)\right)\right]-\frac{\left(\Theta_r\sigma_r+\Theta_j\sigma_j\right)}{\left(\sigma_r^2+\sigma_j^2\right)}$$

$$\text{Re}\left[\frac{\hat{W}}{\sigma}\left(e^{\sigma\hat{t}}-1\right)\right] = \frac{\exp\left(\sigma_r\hat{t}\right)}{\left(\sigma_r^2+\sigma_j^2\right)}\left[\hat{W}_r\left(\sigma_r\cos\left(\sigma_j\hat{t}\right)+\sigma_j\sin\left(\sigma_j\hat{t}\right)\right)\right.$$
$$\left.-\hat{W}_j\left(\sigma_r\sin\left(\sigma_j\hat{t}\right)-\sigma_j\cos\left(\sigma_j\hat{t}\right)\right)\right]-\frac{\left(\hat{W}_r\sigma_r+\hat{W}_j\sigma_j\right)}{\left(\sigma_r^2+\sigma_j^2\right)}$$

Flight Dynamics

Recall that $\hat{U}_r, \hat{U}_j, \Theta_r, \Theta_j, \hat{W}_r$ and $\hat{W}_j$ are related through the modal vectors $\hat{U}/\Theta$ and $\hat{W}/\Theta$. Therefore, at $\hat{t} = 0$, only two of these six parameters may be independently chosen. For this case, Θ_r and Θ_j are chosen. Thus, one has:

$$\hat{U}_r = \text{Re}\left[\left(\frac{\hat{U}}{\Theta}\right)\Theta\right], \quad \hat{U}_j = \text{Im}\left[\left(\frac{\hat{U}}{\Theta}\right)\Theta\right], \quad \hat{W}_r = \text{Re}\left[\left(\frac{\hat{W}}{\Theta}\right)\Theta\right], \quad \hat{W}_j = \text{Im}\left[\left(\frac{\hat{W}}{\Theta}\right)\Theta\right]$$

For this numerical example, chose $\Theta_r = 10°$ (0.1745 radians) and $\Theta_j = 0$. Also, $x'(0)$ and $z'(0)$ equal zero. So, from above, one has that for the Short-Period Mode,

$$\hat{U}_r = \text{Re}\left[(0.0115 + 0.0126\,j) \times 0.1745\right] = 0.002007$$

likewise,

$$\hat{U}_j = 0.002199, \quad \hat{W}_r = 0.171673, \quad \hat{W}_j = 0.079991$$

When substituted into the equations, the motion plots out as below:

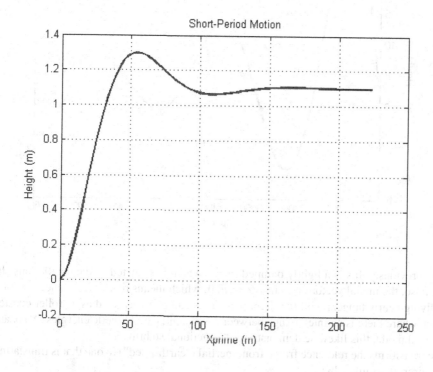

Because z' points downward (what Theodore Von Karman called "the pessimistic coordinate system"), the negative of z' (Height) is plotted. This motion shows a very small height excursion relative to the distance traveled. In fact, the Short-Period mode behaves much like the pinned wind-vane example at the beginning of this chapter, with virtually no vertical movement, and coupled α & θ motions nearly equal in magnitude and relatively close in phase. Also observe that, with reference to the "Definitions of Stability" section, this is "directional stability". After the perturbations the aircraft returns to level flight, but generally at a different altitude.

Next, the flight path for the Phugoid Mode will be calculated for the case where $\Theta_r = 4°$ (0.0698 radians), $\Theta_j = 0$, $x'(0) = 0$, $z'(0) = 0$. Therefore,

$$\hat{U}_r = \text{Re}\left[(-0.1448 + 0.7594\,j) \times 0.0698\right] = -0.010107$$

Likewise,

$$\hat{U}_j = 0.053006, \quad \hat{W}_r = 0.000286, \quad \hat{W}_j = -0.001208$$

When these terms are substituted into the equations, the motion plots out as shown:

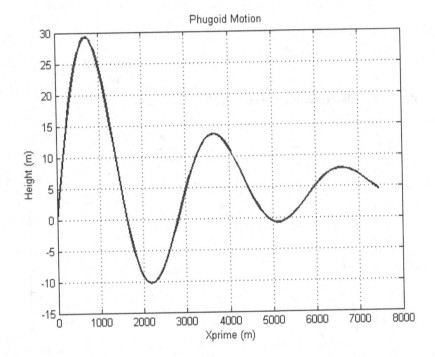

As found previously this is a lightly damped mode, which is reflected in the oscillations shown in the plot. Also, the modal vector gives that $\alpha \ll \hat{u}$, θ, which means that the reference line of the aircraft is flying nearly tangent with the flight path. This can be interpreted as a "roller coaster" type of motion, where there is an interchange between the potential and kinetic energies. Also, as for the Short-Period mode, this likewise demonstrates "directional stability".

If one transforms the reference frame from inertial ("earth-fixed") to one that is translating along with the aircraft, as given by:

$$x'' = x'(t) - U_0 t \rightarrow x'(\hat{t}) - 0.5\bar{c}\,\hat{t}$$

one obtains the "Phugoid Spiral":

Flight Dynamics

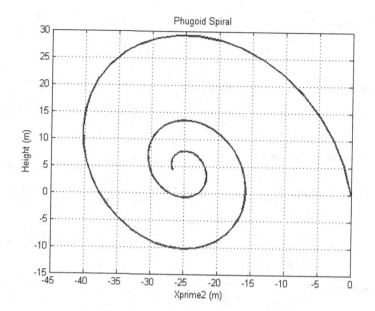

This would be the trajectory traced out by the mass-center of the aircraft if one was looking out the side window of another aircraft flying alongside at a steady speed equal to U_0. This shows that after the perturbation, the aircraft settles down to a position in the transformed reference frame that differs from the position it would have had if the perturbation hadn't occurred. In particular, it is now ≈ 25 m back and ≈ 5 m higher than it would have been otherwise.

Motion in the inertial reference frame may also be assessed by introducing two new modal vectors, $\hat{Z}'/\Theta$ and $\hat{X}''/\Theta$. $\hat{Z}'/\Theta$ is obtained as follows:

$$\frac{dz'}{dt} = -U_0(\alpha - \theta) = -U_0\left(\frac{\hat{W}}{\Theta} - 1\right)\Theta \exp(\sigma\hat{t}) \rightarrow \frac{d\hat{z}'}{d\hat{t}} = -\frac{1}{2}\left(\frac{\hat{W}}{\Theta} - 1\right)\Theta \exp(\sigma\hat{t})$$

This integrates to give:

$$\hat{z}'(\hat{t}) = -\frac{1}{2\sigma}\left(\frac{\hat{W}}{\Theta} - 1\right)\Theta \exp(\sigma\hat{t}) + \hat{z}'(0)$$

One thus obtains the vertical modal flight-path vector:

$$\frac{\hat{Z}'}{\Theta} = -\frac{1}{2\sigma}\left(\frac{\hat{W}}{\Theta} - 1\right)$$

$\hat{X}''/\Theta$ is the modal vector for the transformed horizontal flight path. That is, x'' is the horizontal motion relative to a reference frame translating at a uniform velocity U_0:

$$x''(t) = x'(t) - U_0 t$$

The modal vector is obtained from:

$$\frac{dx''}{dt} = \frac{dx'}{dt} - U_0 = U_0(1 + \hat{u}) - U_0 = U_0\hat{u} \rightarrow \frac{d\hat{x}''}{d\hat{t}} = \frac{1}{2}\frac{\hat{U}}{\Theta}\Theta \exp(\sigma\hat{t})$$

which integrates to give:

$$\hat{x}''(\hat{t}) = \frac{1}{2\sigma}\frac{\hat{U}}{\Theta}\Theta\exp(\sigma\hat{t}) + \hat{x}''(0)$$

This gives the horizontal modal flight-path vector:

$$\frac{\hat{X}''}{\Theta} = \frac{1}{2\sigma}\frac{\hat{U}}{\Theta}$$

For the example airplane, one obtains:

Short-Period Mode: $\left(\frac{\hat{Z}'}{\Theta}\right)_{SP} = -4.1767 + 2.0977\,j$, $\left(\frac{\hat{X}''}{\Theta}\right)_{SP} = 0.0570 - 0.1643\,j$

Phugoid Mode: $\left(\frac{\hat{Z}'}{\Theta}\right)_{Phugoid} = -43.555 - 302.46\,j$, $\left(\frac{\hat{X}''}{\Theta}\right)_{Phugoid} = 237.07 + 6.6486\,j$

which are illustrated below:

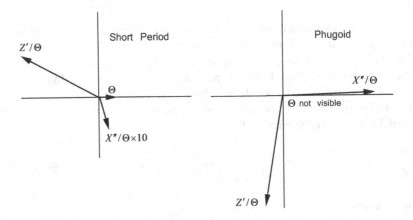

For the Short-Period case, the Θ vector looks small in comparison to the vertical-displacement vector, Z'/Θ. However, in comparison with the Phugoid case, Θ is nearly three orders of magnitude smaller than the Z'/Θ vector. Further, the relative magnitudes and phases of Z'/Θ and X'/Θ relate to the information seen in the Phugoid Spiral.

APPROXIMATE EQUATIONS

The distinct motions of the Short-Period and Phugoid modes allow approximations and simplifications to be made to their dynamic equations. This can give insights into the important parameters driving these motions, as well as offering a means for convenient estimations of an aircraft's stability.

Short-Period Mode

It was seen that surging motion, $\hat{u}$, is very small compared with pitching, θ, and heaving, α. Therefore, the dynamic equations may be approximated by:

$$[\tilde{A}]\begin{bmatrix}\alpha\\\hat{q}\\\theta\end{bmatrix} + [\tilde{B}]\begin{bmatrix}D\alpha\\D\hat{q}\\D\theta\end{bmatrix} = \begin{bmatrix}0\\0\\0\end{bmatrix}$$

Flight Dynamics

where $\left[\tilde{B}\right]$ and $\left[\tilde{B}\right]$ are the truncated versions of matrices $\left[\hat{A}\right]$ and $\left[\hat{B}\right]$.
With the introduction of:

$$\alpha = \hat{W} e^{\sigma \hat{t}}, \quad \hat{q} = \hat{Q} e^{\sigma \hat{t}}, \quad \theta = \Theta e^{\sigma \hat{t}}$$

these equations become:

$$\left[\tilde{A} + \sigma \tilde{B}\right] \begin{bmatrix} \hat{W} \\ \hat{Q} \\ \Theta \end{bmatrix} = \begin{bmatrix} 0 \\ 0 \\ 0 \end{bmatrix} \rightarrow \text{Characteristic Equation} \rightarrow \left|\tilde{A} - \sigma \tilde{B}\right| = 0$$

In the past, before it was possible to easily solve eigenvalue problems or higher-order polynomials, it was a great computational convenience to reduce the characteristic equation from fourth order to third order. Even so, third-order polynomials were enough of a challenge that an entire book was written on solution methodologies ("Table for The Solution of Cubic Equations" by Salzer, Richards and Arsham, McGraw-Hill, 1958). This, of course, is no longer a consideration and there is no particular advantage in time or effort to solving the approximate problem compared with the full equations.

However, some interesting "back-of-the-envelope" insights may be obtained if one notes that for a great majority of aircraft, $C_{XD\alpha}$, $\hat{T}_{X\alpha}$, $\hat{B}$ and the Dq derivatives are negligible. The characteristic equation then becomes:

$$C_3 \sigma^3 + C_2 \sigma^2 + C_1 \sigma = 0$$

where

$$C_3 = \left(2\mu - C_{ZD\alpha}\right) i_{yy}, \quad C_2 = \left[-\left(2\mu - C_{ZD\alpha}\right) C_{Mq} - \left(C_{Z\alpha} + \hat{T}_{Z\alpha}\right) i_{yy} - C_{MD\alpha}\left(2\mu + C_{Zq}\right)\right]$$

$$C_1 = \left[\left(C_{Z\alpha} + \hat{T}_{Z\alpha}\right) C_{Mq} - \left(C_{M\alpha} - \hat{x}_T \hat{T}_{Z\alpha}\right)\left(2\mu + C_{Zq}\right)\right]$$

Also, observe that σ may be factored out, which leaves a quadratic equation that may be expressed as:

$$\sigma^2 + 2\varsigma\omega_n \sigma + \omega_n^2 = 0$$

where

$$\omega_n^2 = \frac{C_1}{C_3}, \quad \varsigma = \frac{C_2}{2\left(C_3 C_1\right)^{1/2}}$$

This form of the equation, as pointed out by Etkin ("Dynamics of Flight", John Wiley & Sons, 1959), is analogous to that for a spring-mass-dashpot system where ω_n^2 is the "spring stiffness", and $2\varsigma\omega_n$ is the viscous damping constant.

For the numerical example, one obtains:

$$C_1 = 716.3521, \quad C_2 = 1.4007 \times 10^4, \quad C_3 = 2.9755 \times 10^5 \rightarrow \omega_n^2 = 0.0024, \quad \varsigma = 0.4797$$

From the solution for a quadratic equation, an approximate solution for σ is found:

$$\sigma_{approx} = -\left(\varsigma\omega_n\right) \pm \left[\left(\varsigma\omega_n\right)^2 - \omega_n^2\right]^{1/2} \rightarrow -0.0235 \pm 0.0431 j$$

This is seen to match the full solution to the fourth decimal place.

Phugoid Mode

For this case, it was seen that the α motion was much less than the surging, $\hat{u}$, and pitching, θ, motion. Therefore, the dynamic equations may be approximated by:

$$[\breve{A}] \begin{bmatrix} \hat{u} \\ \hat{q} \\ \theta \end{bmatrix} + [\breve{B}] \begin{bmatrix} D\hat{u} \\ D\hat{q} \\ D\theta \end{bmatrix} = \begin{bmatrix} 0 \\ 0 \\ 0 \end{bmatrix}$$

where $[\breve{A}]$ and $[\breve{B}]$ are the truncated versions of matrices $[\hat{A}]$ and $[\hat{B}]$.

With the introduction of:

$$\hat{u} = \hat{U}e^{\sigma\hat{t}}, \quad \hat{q} = \hat{Q}e^{\sigma\hat{t}}, \quad \theta = \Theta e^{\sigma\hat{t}}$$

these equations become:

$$[\breve{A} + \sigma\breve{B}] \begin{bmatrix} \hat{W} \\ \hat{Q} \\ \Theta \end{bmatrix} = \begin{bmatrix} 0 \\ 0 \\ 0 \end{bmatrix} \rightarrow \text{Characteristic Equation} \rightarrow |\breve{A} - \sigma\breve{B}| = 0$$

For the vast majority of heavier-than-air aircraft, the Du and Dq derivatives are negligible as well as C_{Xq}, $\hat{T}_{Zu}$ and $\hat{B}$. Therefore, for level flight the characteristic equation becomes:

$$D_2 \sigma^2 + D_1 \sigma + D_0 = 0$$

where

$$D_2 = 2\mu(2\mu + C_{Zq}), \quad D_1 = -(C_{Xu} + 2C_{X0} + \hat{T}_{Xu})(2\mu + C_{Zq}), \quad D_0 = -\hat{m}\hat{g}(C_{Zu} + 2C_{Z0})$$

Just as for the Short-Period case, the quadratic equation may be re-written as:

$$\sigma^2 + 2\breve{\zeta}\breve{\omega}_n \sigma + \breve{\omega}_n^2 = 0$$

where

$$\breve{\omega}_n^2 = \frac{D_0}{D_2}, \quad \breve{\zeta} = \frac{D_1}{2(D_2 D_0)^{1/2}}$$

For the numerical example, one obtains:

$$D_0 = 0.1670, \quad D_1 = 28.2122, \quad D_2 = 5.0722 \times 10^4 \rightarrow \breve{\omega}_n^2 = 3.2919 \times 10^{-6}, \quad \breve{\zeta} = 0.1533$$

From the solution for a quadratic equation, an approximate solution for σ is found:

$$\sigma_{\text{approx}} = -(\breve{\zeta}\breve{\omega}_n) \pm \left[(\breve{\zeta}\breve{\omega}_n)^2 - \breve{\omega}_n^2\right]^{1/2} \rightarrow -0.000278 \pm 0.0018j$$

Recall that the full solution gave:

$$\sigma_{\text{full}} = -0.0003 \pm 0.0016j$$

Flight Dynamics

This is not as good a match as for the Short-Period mode, but it is still reasonably close.

Now, more may be learned if further approximations are made. First of all, note that in the "spring-mass-dashpot" formulation of the quadratic equation, ω_n is referred to as the "natural frequency", which is the oscillation frequency when the damping is zero. For the low-damping situation of the Phugoid mode, $\omega_n \approx \sigma_j$. In this example, for instance, $\omega_n = 0.0018$, which matches σ_j. Further, ς is known as the "damping ratio" which, of course, is a measure of the system's damping. In particular, if the equation for σ_{approx} is re-written as:

$$\sigma_{approx} = -\varsigma \breve{\omega}_n \pm j\omega_n \left(1 - \varsigma^2\right)^{1/2}$$

one sees that if $\breve{\varsigma} = 0$, there is no damping in the system and the motion is purely oscillatory. On the other hand, if $\breve{\varsigma} = 1$, then the system's motion becomes aperiodic. As Etkin points out, $\breve{\varsigma} = 1$ represents the boundary between oscillatory and aperiodic motion and is thus the criterion for *critical damping*. This may occur for the heavily damped Short-Period Mode, but rarely for the lightly damped Phugoid Mode.

Further, for the Phugoid Mode, the following conditions apply to most aircraft:

$$C_{Zq} \ll 2\mu, \quad C_{Zu} \ll C_{Z0}$$

Also, for level flight, one has:

$$C_{X0} = -C_{D0}, \quad C_{Z0} = -C_{L0}, \quad \hat{m}\hat{g} = C_{L0}$$

Therefore,

$$\breve{\omega}_n \approx \frac{C_{L0}}{\sqrt{2}\mu}, \quad \breve{\varsigma} \approx \frac{-C_{Xu} + 2C_{D0} - \hat{T}_{Xu}}{(2)^{3/2} C_{L0}}$$

From the first equation, one may observe the following:

1. The natural frequency, $\breve{\omega}_n$, increases with lift coefficient, C_{L0}, and hence (for level flight) with decreasing flight speed, U_0.
2. $\breve{\omega}_n$ decreases with altitude because the atmospheric density, ρ, decreases, which increases μ.

If the aircraft is in a flight regime where $C_{Xu} \approx 0$ (which is generally the case), then the second equation gives the following information:

a. For propeller-driven aircraft, the damping term is given by:

$$\breve{\varsigma} \approx \frac{2C_{D0} + \hat{T}_0}{(2)^{3/2} C_{L0}} = \frac{3}{(2)^{3/2}} \left(\frac{C_{D0}}{C_{L0}}\right)$$

b. For constant-thrust (jet, rocket) or zero-thrust aircraft (sailplane with small glide slope), the equation for the damping term is:

$$\breve{\varsigma} = \frac{2C_{D0}}{(2)^{3/2} C_{L0}} = \frac{1}{\sqrt{2}} \left(\frac{C_{D0}}{C_{L0}}\right)$$

In both cases the damping is inversely proportional to the equilibrium Lift/Drag ratio. Therefore, one may conclude that well-streamlined aircraft that fly at large values of C_{L0}/C_{D0} suffer a damping penalty in their Phugoid mode. Qualitatively this is true, but one must remember that these conclusions have been obtained from the approximate equations and that the secondary effects of $C_{M\alpha}$, C_{Mq} and $C_{MD\alpha}$ have been ignored. Nonetheless, this is a valuable insight into the role of the Lift/Drag ratio on Phugoid-mode damping.

ROOTS-LOCUS PLOTS

In the course of evaluating an aircraft design, it is often useful to investigate the effects of systematically varying certain parameters. This is conveniently done by plotting, on a complex plane, the variation of the stability roots with the parameter. The resulting curves are called a "roots-locus plot". Note that the independent parameter may be a stability derivative, an inertial property (e.g., μ), or even a static control gain.

For the PA24-250 (Piper Commache) numerical example, $C_{M\alpha}$ is increased from its baseline value of -2.60 in $+0.20$ increments. The first root-locus plot is for the Short-Period mode, taken up to $C_{M\alpha} = 0.2$:

Example Short-Period Mode

[Plot: Imaginary Sigma vs Real Sigma, showing baseline $C_{M\alpha} = -2.60$, $C_{M\alpha} = -0.1196$, $C_{M\alpha} = 0$, $C_{M\alpha} = 0.20$]

From this, the following observations may be made:

a. σ_j initially decreases with increasing $C_{M\alpha}$, with nearly constant σ_r, until $C_{M\alpha} = -0.1196$.
b. At $C_{M\alpha} > -0.1196$, the roots split into two aperiodic modes. The left branch becomes increasingly stable while the right branch approaches instability.
c. At $C_{M\alpha} = 0$ the right branch is still stable.
d. Recall, from the "Complete Aerodynamics" section of Chapter 2, that $C_{M\alpha} < 0$ is the criterion for longitudinal static stability (the mass-center is forward of the neutral point). Clearly, the Short-Period Dynamic Mode is so heavily damped that this mode is still stable for a positive range of $C_{M\alpha}$ values.
e. In truth, the "Short-Period" designation only applies to the oscillatory part of this solution. The two aperiodic modes should have separate designations.

Flight Dynamics

The second roots-locus plot is for the Phugoid Mode, where $C_{M\alpha}$ is increased to 0.60:

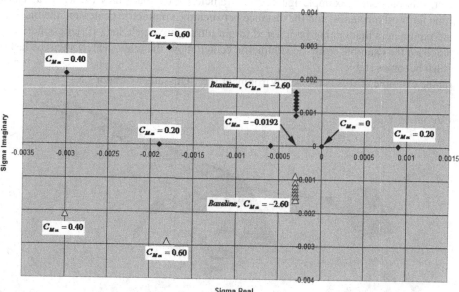

Example Phugoid Mode

From this, one sees the following:

a. σ_j initially decreases with increasing $C_{M\alpha}$, with nearly constant σ_r, until $C_{M\alpha} = -0.0192$. This behavior is similar to that for the Short-Period Mode, except for different magnitudes.
b. At $C_{M\alpha} > -0.0192$ the mode splits into two aperiodic roots. The left branch becomes increasingly stable while the right branch approaches instability, and reaches this when $C_{M\alpha} = 0$.
c. This longitudinal dynamic-instability condition matches that for static instability, and is typical for most airplanes. Therefore, this justifies the static criterion of $C_{M\alpha} < 0$ for airplane stability, as discussed in Chapter 2.
d. If $C_{M\alpha}$ is further increased beyond 0.3059, a new oscillatory root occurs. What happens is that the right aperiodic branch from the Short-Period plot combines with the left aperiodic branch from the Phugoid plot to give the "Third Oscillatory Mode".
e. The Third Oscillatory Mode is mainly of academic interest because most airplanes would have crashed by now.

Recall that the example airplane's mass, m, is 1315.0 kg, and pitching moment of inertia, I_{yy}, 8383.0 kg−m². At the flight conditions previously listed, these give the non-dimensional inertial quantities of:

$$\mu = 114.06, \quad i_{yy} = 1277.3, \quad \hat{mg} = 0.2889$$

The orders of magnitude of these parameters are typical for most airplanes. However, consider an extraordinarily light aircraft, such as a human-powered airplane or an airship (or a neutrally buoyant underwater vehicle) where, for example,

$$\mu = 1.0, \quad i_{yy} = 1.0, \quad \hat{mg} = 0.0025$$

For the very statically-unstable case of $C_{M\alpha} = +0.6$, the stability roots are all aperiodic:

$$(\sigma_r)_1 = -25.6251, \quad (\sigma_r)_2 = -0.7967, \quad (\sigma_r)_3 = -0.0702, \quad (\sigma_r)_4 = 0.0$$

These are modes that cannot be identified as Short-Period or Phugoid, and this speaks to a very different flight-dynamic regime. Most important, despite the strong *static* instability, the vehicle has *dynamic* neutral stability. So, this is an example where the usual static-margin criterion does not apply.

Further, this particular example ignored the vehicle's gross buoyancy (which is a reasonable assumption for most airplanes) and the distance between the volumetric center and the mass-center. These are parameters that cannot be ignored for an ultralight vehicle and, if representative values are included for the example (with the volumetric center above the mass-center), then the last stability root also becomes stable.

A graphic example of this is the AeroVironment Inc. "Gossamer Condor":

The mass-center of this airplane is well behind its neutral point, which would normally be a disaster. However, its extraordinary lightness gave manageably stable stability roots and a controllable and successful design.

Another example is provided by airships, which also have neutral points well ahead of their mass-center.

Flight Dynamics

In Chapter 2, it was described how a body of revolution has a strong upsetting moment because its pressure center is typically well forward of the nose. The relatively small fins on an airship do not come even close to bringing the pressure center aft of the mass-center. However, most airships are controllable. This is especially the case for the blimp in the figure, which is made by the American Blimp Corporation. The author had the privilege of piloting two of these designs and, despite his having a lack of experience, found these to be very controllable and pleasant to fly.

There is a way to visualize this phenomenon. First of all, consider an aircraft that has pitching freedom only. Namely, it is pivoted at the mass-center. If the neutral point is forward of the mass-center, then the lift vector gives an unstable pitching moment. Recall that this is what the static-stability criterion states: that the neutral point must be aft of the mass-center for stability. As mentioned, this criterion has been accepted by airplane designers since the earliest days. Another name for this is "weather-vane stability". As stated before, for most airplanes this criterion is correct.

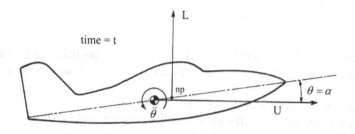

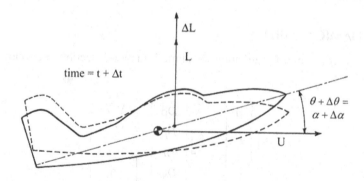

Aircraft Free to Pitch Only ("Pinned")

However, now consider the aircraft to be free to pitch and heave, as is the case in flight. This behavior is now different from that for the pinned case. As the figure below shows, the lift vector not only gives a pitching acceleration, but also heaving (vertical motion) acceleration:

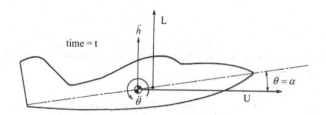

After a time increment Δt, the pitch angle has increased by $\Delta\theta$. However, the angle of attack has decreased by $\Delta\alpha$. This gives rise to a change in the lift vector, ΔL, that is less, or even opposite, the value for the pinned case.

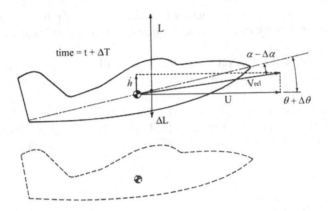

Therefore, the rate of unstable divergence for this case will be less than that for the pinned case, and one sees that heaving tends to stabilize an aircraft where the neutral point is forward of the mass-center.

Now, the mass for most airplanes is such that very little heaving acceleration occurs compared with the pitching acceleration. Thus, the static-stability criterion applies. However, if the aircraft is very light (as for the previous examples) then the heaving acceleration is significant and serves to stabilize the aircraft.

LATERAL DYNAMIC STABILITY

In the "Aircraft Lateral Small-Perturbation Dynamic Equations" section, the following equations were presented:

$$[\hat{C}]\begin{bmatrix}\beta\\\hat{p}\\\hat{r}\\\varphi\end{bmatrix}+[\hat{D}]\begin{bmatrix}D\beta\\D\hat{p}\\D\hat{r}\\D\varphi\end{bmatrix}=\begin{bmatrix}\Delta C_{Yc}\\\Delta C_{\bar{L}c}\\\Delta C_{Nc}\\0\end{bmatrix}$$

with the matrix elements as defined in the section. Since controls-fixed stability will be studied, the right-hand control matrix may be set to zero:

$$[\hat{C}]\begin{bmatrix}\beta\\\hat{p}\\\hat{r}\\\phi\end{bmatrix}+[\hat{D}]\begin{bmatrix}D\beta\\D\hat{p}\\D\hat{r}\\D\phi\end{bmatrix}=\begin{bmatrix}0\\0\\0\\0\end{bmatrix}$$

Because these represent a system of linear differential equations with constant coefficients, one may assume solutions of the form:

$$\beta=\hat{V}e^{\lambda\hat{t}},\quad \hat{p}=\hat{P}e^{\lambda\hat{t}},\quad \hat{r}=\hat{R}e^{\lambda\hat{t}},\quad \varphi=\Phi e^{\lambda\hat{t}}$$

where $\hat{V}$, $\hat{P}$, $\hat{R}$ and Φ are constants (possibly complex). When these solutions are substituted into the matrix equations, one obtains:

$$\left[\hat{C}+\lambda\hat{D}\right]\begin{bmatrix}\hat{V}\\\hat{P}\\\hat{R}\\\Phi\end{bmatrix}=0 \rightarrow \left[\hat{C}+\lambda\hat{D}\right]\begin{bmatrix}\hat{V}/\Phi\\\hat{P}/\Phi\\\hat{R}/\Phi\\1\end{bmatrix}=0$$

The λ value is the lateral stability root, and the $\hat{V}/\Phi$, $\hat{P}/\Phi$ and $\hat{R}/\Phi$ terms are the modal vectors normalized to a Φ of unity. Upon solution, the equations give four lateral stability roots of the form:

$$\lambda = \lambda_r + j\lambda_j$$

and each of these is associated with a set of modal vectors. For example, λ_2 is the complex stability root for the motion characterized by $(\hat{V}/\Phi)_2$, $(\hat{P}/\Phi)_2$ and $(\hat{R}/\Phi)_2$.

Upon returning to the assumed solutions, these may be rewritten as:

$$\beta = \left(\frac{\hat{V}}{\Phi}\right)\Phi e^{\lambda_r \hat{t}}e^{j\lambda_j \hat{t}}, \quad \hat{p}=\left(\frac{\hat{P}}{\Phi}\right)\Phi e^{\lambda_r \hat{t}}e^{j\lambda_j \hat{t}}, \quad \hat{r}=\left(\frac{\hat{R}}{\Phi}\right)\Phi e^{\lambda_r \hat{t}}e^{j\lambda_j \hat{t}}, \quad \phi = \Phi e^{\lambda_r \hat{t}}e^{j\lambda_j \hat{t}}$$

Common to all four equations is the term $e^{j\lambda_j \hat{t}}$ which, in complex notation, is a harmonic oscillation of frequency:

$$\omega_j = (2U_0/b)\lambda_j \quad \text{(rad/sec)}$$

This gives, as described earlier:

$$\text{Frequency in } Hz = f = \frac{\omega_j}{2\pi}, \quad \text{Period of oscillation} = T = \frac{2\pi}{\omega_j}$$

It sometimes happens that the mode of motion is aperiodic ($\lambda_j = 0$); however, in general, the λ_r term is non-zero. Further, λ_r determines the degree of stability (or instability) of the aircraft. For example, if λ_r is negative, the motion will be stable. However, if λ_r is positive, the motion will increase (unstable). As with the longitudinal case, described earlier in this chapter, one way of characterizing this is with the Time to half (or double) amplitude:

$$\text{Time to half (or double) amplitude} = T_{1/2(\text{or }2)} = \frac{\ln 0.5}{\lambda_r}\frac{b}{2U_0} = \frac{-0.69315}{\lambda_r}\frac{b}{2U_0}$$

This gives, *for an oscillatory mode*, the Number of cycles to half (or double) amplitude:

$$\text{Number of cycles to half (or double) amplitude} = N_{1/2(\text{or }2)} = \frac{T_{1/2(\text{or }2)}}{T} = \frac{0.11032\lambda_j}{|\lambda_r|}$$

Regarding the modal vectors, it is useful to introduce another vector, Ψ/Φ, that relates the heading angle, ψ, to the rolling angle, ϕ. The relationship between yaw rate, r, and the time rate of change of heading angle, ψ, is

$$r = \dot{\psi}\cos\Theta_0 \rightarrow \hat{r} = D\psi\cos\Theta_0$$

This then gives:

$$D\psi = \hat{r}\sec\Theta_0 \rightarrow D\psi = \hat{R}e^{\lambda\hat{i}}\sec\Theta_0 \rightarrow \psi = \left(\frac{\hat{R}}{\lambda}e^{\lambda\hat{i}}\right)\sec\Theta_0 + \psi_0$$

The constant of integration, ψ_0, denotes the final heading for a stable aircraft. This shows that this controls-fixed case has "straight-line" stability. That is, after a perturbation, the aircraft generally settles down to a different, but straight-line, heading. For convenience choose $\psi_0 = 0$, so that the new modal vector is:

$$\frac{\Psi}{\Phi} = \frac{1}{\lambda}\frac{\hat{R}}{\Phi}\sec\Theta_0$$

A typical lateral modal-vector diagram would thus look like the following figure.

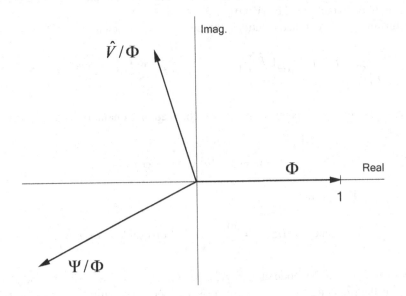

where Φ = *roll angle*, Ψ = *yaw angle*, $\hat{V}$ = *sideslip angle*.

Now, consider the numerical example again. The aircraft is assumed to be in level flight, so that $\Theta_0 = 0$.

Next, the elements of matrices $[\hat{C}]$ and $[\hat{D}]$ may be calculated, which are $\hat{c}_{i,j}$ and $\hat{d}_{i,j}$ as defined earlier in the chapter, and the stability roots and modes may be obtained from the MATLAB statement:

[modes,stabroots]=eig(c,-d)

The stability roots for this example are:

$$\lambda_1 = 0.0007297, \quad \lambda_2 = -0.228337, \quad \lambda_{3,4} = -0.00835 \pm 0.143927j$$

In this case, there are two aperiodic roots, plus one complex-conjugate pair. Further, the first aperiodic root is unstable, and has a time-to-double amplitude of

$$T_2 = 68.156 \text{ sec}$$

Flight Dynamics

This root's normalized modal vectors are:

$$\hat{V}/\Phi = 0.00519, \quad \hat{R}/\Phi = 0.00913, \quad \hat{P}/\Phi = 0.00073, \quad \Psi/\Phi = 12.517$$

and these plots out to give:

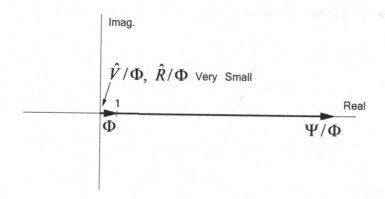

This motion is seen to be a slightly unstable aperiodic divergence involving primarily Ψ. Since Ψ is the aircraft's heading angle, this motion when seen from above describes a tightening spiral (if stable, it would be a straightening spiral). Therefore, this is called the "Spiral Mode".

The second root is also aperiodic, but it is very stable with a time-to-half-amplitude of

$$T_{1/2} = 0.218 \text{ sec}$$

The normalized modal vectors are

$$\hat{V}/\Phi = -0.09660, \quad \hat{R}/\Phi = -0.01081, \quad \hat{P}/\Phi = -0.22834, \quad \Psi/\Phi = 0.04735$$

and these plot out to give:

This motion is a heavily damped aperiodic convergence involving primarily Φ. Because Φ is the aircraft's roll angle, this is called the "Rolling Mode".

The third and fourth roots are complex conjugates which describe an oscillatory mode where

$$T = 3.132 \text{ sec } (Period), \quad T_{1/2} = 5.960 \text{ sec}, \quad N_{1/2} = 1.903$$

The modal vectors are also complex:

$$\hat{V}/\Phi = 0.57598 - 1.03900j, \quad \hat{R}/\Phi = -0.14700 - 0.07208j, \quad \hat{P}/\Phi = -0.00834 + 0.14393j$$

$$\Psi/\Phi = -0.44009 + 1.04690j$$

and the modal-vector diagram for this is:

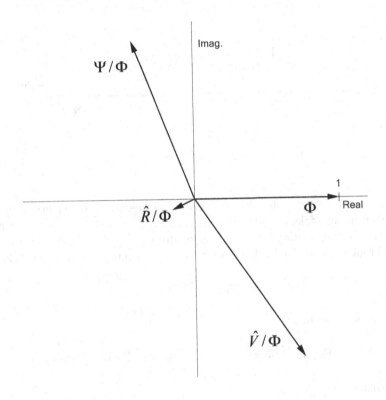

This is a Short-Period moderately damped motion involving approximately equal magnitudes of β, ψ and ϕ. Further, note that $\psi \approx -\beta$, which shows that sideslip angle, β, is due primarily to the yaw angle, ψ. Therefore, the motion is an oscillatory yaw-roll wallow, and is called the "Dutch-Roll Mode".

ASIDE

The origin of the name "Dutch Roll" is usually attributed to a type of ice-skating posture popular in Holland at one time. With arms tucked up behind the skater's body, the skater would bend forward and proceed with a motion that involved oscillatory side-to-side leaning and twisting. Supposedly, some early flight-dynamicist thought this was akin to this lateral-directional mode.

END ASIDE

FLIGHT PATHS

Further insights into Spiral and Dutch-Roll Modes may be obtained by calculating their flight paths. For level flight, the *lateral* small-perturbation form of the flight-path equations for

Flight Dynamics

motion with respect to the inertial reference frame (x', y', z' coordinates) are obtained from Appendix C:

$$dx'/dt = U_0 \rightarrow dx'/d\hat{t} = b/(2U_0)U_0 \rightarrow dx'/d\hat{t} = b/2$$

$$dy'/dt = U_0\psi + v = U_0(\psi + \beta) \rightarrow dy'/d\hat{t} = b/(2U_0)U_0(\psi + \beta) = (b/2)(\psi + \beta)$$

$$dz'/dt = 0 \rightarrow dz'/d\hat{t} = 0$$

The first equation may be integrated to give:

$$x'(t) = U_0 t + x'(0) \rightarrow x'(\hat{t}) = (b/2)\hat{t} + x'(0)$$

and, upon recalling that $\psi = \Psi e^{\lambda \hat{t}}$ and $\beta = \hat{V} e^{\lambda \hat{t}}$, the second equation may be integrated to give:

$$y'_r(\hat{t}) + j y'_j(\hat{t}) = \frac{b}{2}\int_0^{\hat{t}} \left(\Psi e^{\lambda \hat{t}} + \hat{V} e^{\lambda \hat{t}}\right) d\hat{t} + y'(0) = \frac{b}{2}\left[\frac{\Psi}{\lambda}\left(e^{\lambda \hat{t}} - 1\right) + \frac{\hat{V}}{\lambda}\left(e^{\lambda \hat{t}} - 1\right)\right] + y'(0)$$

The modal-vector terms, Ψ/Φ and $\hat{V}/\Phi$ are generally complex, and therefore $y'(\hat{t})$ is generally complex and can be divided into real and imaginary parts. It is the real parts that are of interest, which necessitates finding the real parts of this equation.

$$y'(\hat{t}) \equiv y'_r(\hat{t}) = \frac{b}{2}\left\{\text{Re}\left[\frac{\Psi}{\lambda}\left(e^{\lambda \hat{t}} - 1\right)\right] + \text{Re}\left[\frac{\hat{V}}{\lambda}\left(e^{\lambda \hat{t}} - 1\right)\right]\right\} + y'(0)$$

Observe that

$$\frac{\Psi}{\lambda}\left(e^{\lambda \hat{t}} - 1\right) = \frac{(\Psi_r + j\Psi_j)}{(\lambda_r + j\lambda_j)}\left(\exp(\lambda_r \hat{t} + j\lambda_j \hat{t}) - 1\right) \rightarrow$$

$$\frac{\Psi}{\lambda}\left(e^{\lambda \hat{t}} - 1\right) = \frac{(\Psi_r + j\Psi_j)}{(\lambda_r^2 + \lambda_j^2)}(\lambda_r - j\lambda_j)\left[\exp(\lambda_r \hat{t})\exp(j\lambda_j \hat{t}) - 1\right]$$

Further,

$$\exp(j\lambda_j \hat{t}) = \cos(\lambda_j \hat{t}) + j\sin(\lambda_j \hat{t}), \text{ so one obtains.}$$

$$\text{Re}\left[\frac{\Psi}{\lambda}\left(e^{\lambda \hat{t}} - 1\right)\right] = \frac{\exp(\lambda_r \hat{t})}{(\lambda_r^2 + \lambda_j^2)}\left[\Psi_r\left(\lambda_r \cos(\lambda_j \hat{t}) + \lambda_j \sin(\lambda_j \hat{t})\right) - \Psi_j\left(\lambda_r \sin(\lambda_j \hat{t}) - \lambda_j \cos(\lambda_j \hat{t})\right)\right]$$

$$- \frac{(\Psi_r \lambda_r + \Psi_j \lambda_j)}{(\lambda_r^2 + \lambda_j^2)}$$

Likewise, one obtains:

$$\text{Re}\left[\frac{\hat{V}}{\lambda}\left(e^{\lambda \hat{t}} - 1\right)\right] = \frac{\exp(\lambda_r \hat{t})}{(\lambda_r^2 + \lambda_j^2)}\left[\hat{V}_r\left(\lambda_r \cos(\lambda_j \hat{t}) + \lambda_j \sin(\lambda_j \hat{t})\right) - \hat{V}_j\left(\lambda_r \sin(\lambda_j \hat{t}) - \lambda_j \cos(\lambda_j \hat{t})\right)\right]$$

$$- \frac{(\hat{V}_r \lambda_r + \hat{V}_j \lambda_j)}{(\lambda_r^2 + \lambda_j^2)}$$

Recall that $\hat{V}_r$, $\hat{V}_j$, Φ_r, Φ_j, Ψ_r and Ψ_j are related through the modal vectors $\hat{V}/\Phi$ and Ψ/Φ. Therefore, at $\hat{t} = 0$, only two of these six parameters may be independently chosen. For this case, Φ_r and Φ_j are chosen. Thus, one has:

$$\hat{V}_r = \text{Re}\left[\left(\frac{\hat{V}}{\Phi}\right)\Phi\right], \quad \hat{V}_j = \text{Im}\left[\left(\frac{\hat{V}}{\Phi}\right)\Phi\right], \quad \Psi_r = \text{Re}\left[\left(\frac{\Psi}{\Phi}\right)\Phi\right], \quad \Psi_j = \text{Im}\left[\left(\frac{\Psi}{\Phi}\right)\Phi\right]$$

where,

$$\hat{V} = \left[\left(\frac{\hat{V}}{\Phi}\right)\Phi\right] = \left[\left(\frac{\hat{V}}{\Phi}\right)_r + j\left(\frac{\hat{V}}{\Phi}\right)_j\right](\Phi_r + j\Phi_j) \rightarrow$$

$$\text{Re}\,\hat{V} = \hat{V}_r = \left(\hat{V}/\Phi\right)_r \Phi_r - \left(\hat{V}/\Phi\right)_j \Phi_j, \quad \text{Im}\,\hat{V} = \hat{V}_j = \left(\hat{V}/\Phi\right)_j \Phi_r + \left(\hat{V}/\Phi\right)_r \Phi_j, \text{ etc.}$$

For the *Spiral* numerical example chose $\Psi_r = 10°$ (0.1745 radians) and $\Psi_j = 0$, which means that $\Phi_r = 0.7989°$ and $\Phi_j = 0$. Also, $x'(0)$ equals zero. So, from above, one obtains:

$$x'(\hat{t}) = 5.485\hat{t} \quad (m)$$

$$y'(\hat{t}) = 1309.4\left(1 - \exp(0.000733\hat{t})\right) \quad (m)$$

Also, the roll angle is given by:

$$\varphi(\hat{t}) = \Phi_r \exp(\lambda_r \hat{t}) = 0.7989 \exp(0.000730\hat{t}) \quad (\text{deg})$$

These results are plotted out below, up to $\hat{t} = 2997 \rightarrow 215.04$ sec:

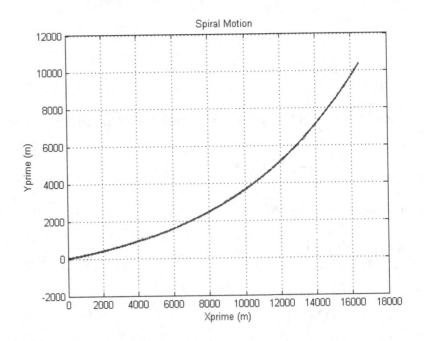

Flight Dynamics

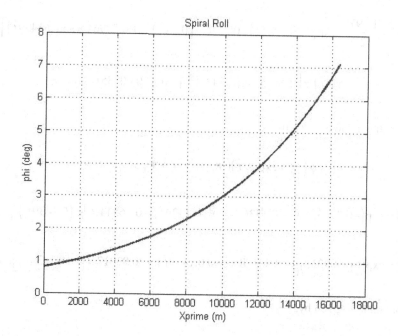

This shows a slowly divergent curve in the horizontal plane, with an increasing roll angle. Taken far enough it would trace a spiral of ever-decreasing radius and increasing bank (roll) angle, eventually developing into a spiral dive. It's an interesting fact that most full-sized airplanes of conventional configuration have mild spiral instability. However, the time-to-double-amplitude is so long that the pilot scarcely notices this.

For the *Dutch-Roll* numerical example, chose $\Phi_r = 20°$ (0.3491 radians) and $\Phi_j = 0$. Also, $x'(0)$ and $y'(0)$ equals zero. So, from above, one obtains:

$$x'(\hat{t}) = 5.485\hat{t} \quad (m)$$

$$\Psi_r = -0.44009 \times 0.3491 - 1.04690 \times 0 \rightarrow \Psi_r = -0.15364$$

$$\Psi_j = 1.04690 \times 0.3491 - 0.44009 \times 0 \rightarrow \Psi_j = 0.36547$$

$$\hat{V}_r = 0.57598 \times 0.3491 + 1.03900 \times 0 \rightarrow \hat{V}_r = 0.20108$$

$$\hat{V}_j = -1.03900 \times 0.3491 + 0.57598 \times 0 \rightarrow \hat{V}_j = -0.36272$$

$$\text{Re}\left[\frac{\Psi}{\lambda}\left(e^{\lambda \hat{t}} - 1\right)\right] = \exp(-0.00835\hat{t})\left[2.59296 \times \cos(0.1439\hat{t}) - 0.917216 \times \sin(0.1439\hat{t})\right]$$
$$-2.59296$$

$$\text{Re}\left[\frac{\hat{V}}{\lambda}\left(e^{\lambda \hat{t}} - 1\right)\right] = \exp(-0.00835\hat{t})\left[-2.59296 \times \cos(0.1439\hat{t}) + 1.246860 \times \sin(0.1439\hat{t})\right]$$
$$+2.59296$$

$$y'(\hat{t}) = \frac{10.97}{2} \times \exp(-0.00835\hat{t})\left[0.0 \times \cos(0.1439\hat{t}) + 0.32964 \times \sin(0.1439\hat{t}) \right] \rightarrow$$

$$y'(\hat{t}) = \left[1.8081 \sin(0.1439\hat{t}) \right] \exp(-0.00835\hat{t}) \quad (m)$$

Also, the heading angle is given by:

$$\psi(\hat{t}) = \exp(\lambda_r \hat{t})(\Psi_r \cos\lambda_j \hat{t} - \Psi_j \sin\lambda_j \hat{t}) \rightarrow$$

$$\psi(\hat{t}) = \exp(-0.00835\hat{t})\left[-0.15362 \times \cos(0.1439\hat{t}) - 0.36544 \times \sin(0.1439\hat{t}) \right], \quad (rad)$$

$$\psi(\hat{t}) = \exp(-0.00835\hat{t})\left[-8.80178 \times \cos(0.1439\hat{t}) - 20.93817 \times \sin(0.1439\hat{t}) \right], \quad (deg)$$

And, the roll angle is given by:

$$\varphi(\hat{t}) = \exp(\lambda_r \hat{t})(\Phi_r \cos\lambda_j \hat{t} - \Phi_j \sin\lambda_j \hat{t}) \rightarrow$$

$$\varphi(\hat{t}) = \exp(-0.00835\hat{t})\left[0.34907 \cos(0.1439\hat{t}) \right], \quad (rad)$$

$$\phi(\hat{t}) = \exp(-0.00835\hat{t})\left[20.0 \cos(0.1439\hat{t}) \right], \quad (deg)$$

These are plotted out below, up to $\hat{t} = 297 \rightarrow t = 21.31 \sec$:

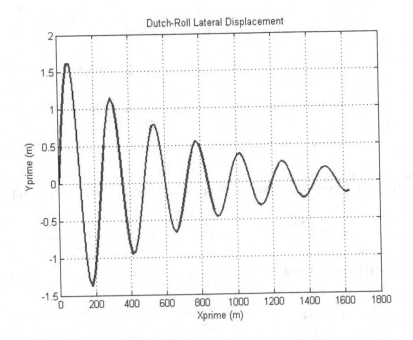

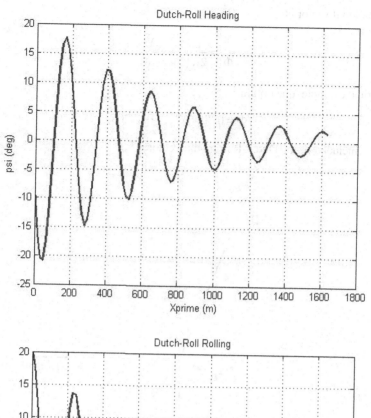

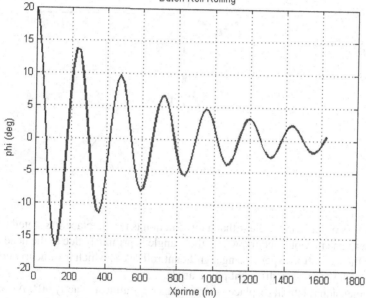

As seen from these plots, the lateral-directional motion is an oscillatory combination of roll, heading, and lateral displacement. These are of equal frequency, but different phases. This motion is a bit difficult to visualize, but it may help to introduce a lateral-displacement modal vector, $\hat{Y}'/\Phi$:

$$\frac{dy'}{dt} = U_0(\psi + \beta) = U_0\left(\frac{\hat{\Psi}}{\Phi} + \frac{\hat{V}}{\Phi}\right)\Phi\exp(\lambda\hat{t}) \rightarrow D\hat{y}' = \frac{1}{2}\left(\frac{\hat{\Psi}}{\Phi} + \frac{\hat{V}}{\Phi}\right)\Phi\exp(\lambda\hat{t}) \rightarrow$$

$$\hat{y}' = \frac{1}{2\lambda}\left(\frac{\hat{\Psi}}{\Phi} + \frac{\hat{V}}{\Phi}\right)\Phi\exp(\lambda\hat{t}) + const.$$

Therefore, the modal vector is:

$$\frac{\hat{Y}'}{\Phi} = \frac{1}{2\lambda}\left(\frac{\hat{\Psi}}{\Phi} + \frac{\hat{V}}{\Phi}\right)$$

For the numerical example, one obtains:

$$\hat{Y}/\Phi = 0.000637 - j\, 0.47208,$$

and this is now included in the Dutch-Roll modal vector:

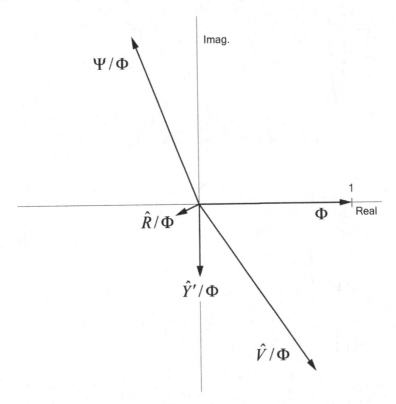

When this mode is observed, it is clear that a lot of action is taking place. As stated before, a type of lateral "weather-vane motion" occurs, where the β angle is primarily due to the heading angle ψ. At the same time the aircraft is experiencing significant rolling, ϕ, which is of nearly equal magnitude to β and ψ, though with a very different phase angle.

The non-dimensional lateral displacement, $\hat{y}'$, lags ϕ by almost exactly 90°. Also, its magnitude is 39.7% relative to β, which would imply that it has similar aerodynamic significance. However, that's not so. As stated before for this case, β is primarily due to heading angle and not sideslip (motion in the y-direction). What this modal-vector diagram does shows is that besides rolling and yawing, the airplane also has side-to side displacement. From a physiological standpoint, this is very uncomfortable and designers try to insure that the Dutch-Roll Mode is well damped. The means for doing this is described in a subsequent section.

APPROXIMATE EQUATIONS

The distinct motions of the Spiral, Rolling and Dutch-Roll Modes allow approximations and simplifications to be made to their dynamic equations. This can give insights into the important

Flight Dynamics

parameters driving these motions, as well as offering a means for convenient estimations of an aircraft's stability.

Spiral Mode

The modal vector for the example showed that β is very small compared with ψ and ϕ. This would lead one to consider dropping the β terms in the dynamic equations. However, this would be wrong because ψ itself produces no force or moment on the aircraft (it is an orientation parameter). Instead, $D\psi$, through $\hat{r}$, produces an aerodynamic force. Because $\hat{r}$ is of the same order of magnitude as β, it is clear that it cannot be neglected.

The same argument may be applied to ϕ, which is also an orientation angle and does not, itself, produce an aerodynamic force or moment. However $D\phi$, through $\hat{p}$, can produce an aerodynamic force. Etkin ("Dynamics of Flight", J. Wiley & Sons, 1959, Page 231) argues that β, $\hat{p}$ and $\hat{r}$ are of the same orders of magnitude, and a simple approximation to the Spiral Mode is not possible. However, D. Caughey ("Introduction to Stability and Control Course Notes", Cornell University, 2011, M & AE 5070, https://courses.cit.cornell.edu/mae5070/Caughey_2011_04.pdf) observes that $\hat{p}$ is generally much smaller than β and $\hat{r}$, and that an approximation based on this has value for observing the mechanism of the Spiral Mode.

For this particular example, upon normalizing with respect to β, the relative magnitudes are:

$$\beta : \hat{p} : \hat{r} = 1 : 0.1407 : 1.7608$$

So, though $\hat{p}$ is not negligible, it is an order of magnitude smaller than β and $\hat{r}$; and an approximate solution will be based on dropping it.

The first step is to assume that the Spiral Mode is driven primarily by the yawing and rolling moments. The dynamic equations for these are:

$$-C_{\tilde{L}\beta}\beta - C_{\tilde{L}p}\hat{p} - C_{\tilde{L}r}\hat{r} - \hat{B}\hat{z}_B \cos\Theta_0 \,\varphi - C_{\tilde{L}D\beta}D\beta + \left(i_{xx} - C_{\tilde{L}Dp}\right)D\hat{p} - \left(C_{\tilde{L}Dr} + i_{xz}\right)D\hat{r} = 0$$

$$-C_{N\beta}\beta - C_{Np}\hat{p} - C_{Nr}\hat{r} + \hat{B}\hat{x}_B \cos\Theta_0 \,\varphi - C_{ND\beta}D\beta - \left(i_{xz} + C_{NDp}\right)D\hat{p} + \left(i_{zz} - C_{NDr}\right)D\hat{r} = 0$$

For most airplanes with conventional configurations and densities, i_{xz}, $\hat{B}$, and the "acceleration" stability derivatives are negligible. In this case, the equations become:

$$-C_{\tilde{L}\beta}\beta - C_{\tilde{L}p}\hat{p} - C_{\tilde{L}r}\hat{r} + i_{xx}D\hat{p} = 0$$

$$-C_{N\beta}\beta - C_{Np}\hat{p} - C_{Nr}\hat{r} + i_{zz}D\hat{r} = 0$$

Next, upon dropping the $\hat{p}$ stability derivatives, the equations further reduce to:

$$-C_{\tilde{L}\beta}\beta - C_{\tilde{L}r}\hat{r} + i_{xx}D\hat{p} = 0, \quad -C_{N\beta}\beta - C_{Nr}\hat{r} + i_{zz}D\hat{r} = 0$$

Further, from

$$D\hat{p} = \lambda \hat{P} e^{\lambda \hat{t}} = \lambda \hat{p},$$

it is seen that, for this example, $D\hat{p}$ is extremely small. Therefore, the first equation becomes:

$$-C_{\tilde{L}\beta}\beta - C_{\tilde{L}r}\hat{r} = 0 \rightarrow \beta = -\frac{C_{\tilde{L}r}}{C_{\tilde{L}\beta}}\hat{r}.$$

Upon substitution into the second equation, one obtains:

$$D\hat{r} - \frac{1}{i_{zz}}\left(C_{Nr} - \frac{C_{N\beta} C_{\tilde{L}r}}{C_{\tilde{L}\beta}}\right)\hat{r} = 0$$

A solution for this may be assumed to be $\hat{r} = \tilde{R}e^{\tilde{\lambda}\hat{t}}$ which, upon substitution into the above equation, gives the expression for $\tilde{\lambda}$:

$$\tilde{\lambda} = \frac{1}{i_{zz}}\left(C_{Nr} - \frac{C_{N\beta} C_{\tilde{L}r}}{C_{\tilde{L}\beta}}\right)$$

For the example airplane, the result is:

$$\tilde{\lambda} = \frac{1}{5.9770}\left(-0.0503 - \frac{0.109 \times 0.0643}{-0.057}\right) = 0.01216$$

This is nowhere close to the exact value of $\lambda_{\text{Spiral}} = 0.0007297$. Therefore, Etkin is correct in stating that a simple approximation does not exist for the Spiral Mode. However, the result does have the right sign, and Caughey points out that this solution offers insights into how certain stability derivatives influence the Spiral mode. First of all, C_{Nr} is always negative (yaw damping) and clearly acts to stabilize the Spiral Mode. For the second term, a positive dihedral angle gives a negative $C_{\tilde{L}\beta}$, and $C_{N\beta}$ & $C_{\tilde{L}r}$ are normally positive. So, according to the equation, the second term acts to destabilize the airplane. As the magnitude of $C_{N\beta}$ increases (the "weather-cock" effect), this approximate mode becomes less stable. However, if the magnitude of $C_{\tilde{L}\beta}$ is large enough (the "dihedral" effect), the approximate mode becomes more stable.

The shortcomings of this simplistic solution become more evident later, when it is clearly shown that a positive dihedral effect (negative $C_{\tilde{L}\beta}$) strongly stabilizes the Spiral Mode. The equation above would predict otherwise.

Although this exercise does not give a close quantitative approximation, the above equation, when rearranged, does give a useful qualitative indication of whether a design is likely to have spiral stability. This criterion is expressed as

$$C_{L\beta} C_{Nr} - C_{N\beta} C_{\tilde{L}r} > 0 \quad \text{for stability}$$

The value for the example airplane is -0.00414, so it is confirmed to be spirally unstable.

Rolling Mode
In this case, it will be assumed that this mode is primarily driven by the rolling moment equation. Drawing from the previous work, the equation is

$$-C_{\tilde{L}\beta}\beta - C_{\tilde{L}p}\hat{p} - C_{\tilde{L}r}\hat{r} + i_{xx} D\hat{p} = 0$$

For the example airplane, the relative magnitudes of the motion variables, normalized to $\hat{p}$, are

$$\beta : \hat{p} : \hat{r} = 0.4231 : 1 : 0.04734$$

It is seen that the $\hat{r}$ motion is very small and can be neglected. However, β and $\hat{p}$ are of similar orders of magnitude. Note, though, that $C_{\tilde{L}p} = -0.399$, whereas $C_{\tilde{L}\beta} = -0.057$. Therefore, the β stability derivative will be dropped and the equation becomes:

$$-C_{\tilde{L}p}\hat{p} + i_{xx} D\hat{p} = 0$$

Flight Dynamics

Choose $\hat{p} = \tilde{P} e^{\tilde{\lambda}\hat{t}}$ and, upon substitution into the equation, one obtains:

$$-C_{\tilde{L}p} + i_{xx}\tilde{\lambda} = 0 \to \tilde{\lambda} = C_{\tilde{L}p}/i_{xx}$$

For the numerical example, the equation gives:

$$\tilde{\lambda} = -0.399/1.8433 = -0.2165$$

This value compares favorably with the exact value: $\lambda_{\text{Roll}} = -0.2284$.

Dutch-Roll Mode

In this case, recall that $\beta \approx -\psi$. Also, because $\hat{r} = D\psi \cos\Theta_0$, one obtains:

$$\hat{r} \approx -D\beta \cos\Theta_0$$

Therefore, one may include this with the previous equations to give:

$$\begin{bmatrix} 0 & 0 & 1 \\ -C_{\tilde{L}\beta} & -C_{\tilde{L}p} & -C_{\tilde{L}r} \\ -C_{N\beta} & -C_{Np} & -C_{Nr} \end{bmatrix} \begin{bmatrix} \beta \\ \hat{p} \\ \hat{r} \end{bmatrix} + \begin{bmatrix} \cos\Theta_0 & 0 & 0 \\ 0 & i_{xx} & 0 \\ 0 & 0 & i_{zz} \end{bmatrix} \begin{bmatrix} D\beta \\ D\hat{p} \\ D\hat{r} \end{bmatrix} = 0$$

If one chooses $\cos\Theta_0 \approx 1$ and introduces:

$$\beta = \tilde{V} e^{\tilde{\lambda}\hat{t}}, \quad \hat{p} = \tilde{P} e^{\tilde{\lambda}\hat{t}}, \quad \hat{r} = \tilde{R} e^{\tilde{\lambda}\hat{t}}$$

these equations become:

$$\left[\tilde{C} + \tilde{\lambda}\tilde{D}\right] \begin{bmatrix} \tilde{V} \\ \tilde{P} \\ \tilde{R} \end{bmatrix} = \begin{bmatrix} 0 \\ 0 \\ 0 \end{bmatrix} \to \text{characteristic equation} \to \left|\tilde{C} + \tilde{\lambda}\tilde{D}\right| = 0$$

Expansion of the characteristic equation gives:

$$E_1 \tilde{\lambda}^3 + E_2 \tilde{\lambda}^2 + E_3 \tilde{\lambda} + E_4 = 0$$

where

$$E_1 = i_{zz} i_{xx}, \quad E_2 = -i_{zz} C_{\tilde{L}p} - i_{xx} C_{Nr}, \quad E_3 = -C_{\tilde{L}r} C_{Np} + C_{Nr} C_{\tilde{L}p} + i_{xx} C_{N\beta}$$

$$E_4 = -C_{N\beta} C_{\tilde{L}p} + C_{L\beta} C_{Np}$$

For the example airplane, one obtains:

$$E_1 = 11.01770, \quad E_2 = 2.47755, \quad E_3 = 0.22683, \quad E_4 = 0.048661$$

from which the roots are:

$$\tilde{\lambda}_1 = -0.22182, \quad \tilde{\lambda}_2 = -0.001525 \pm 0.14110j$$

A consequence of this analysis is that an approximation to the rolling root, $\tilde{\lambda}_1$, is obtained. Again, it closely matches the exact value. However, this cannot be said for the Dutch-Roll approximation. The exact solution is:

$$\lambda_{\text{Dutch-Roll}} = -0.008345 \pm 0.14393j$$

The frequencies are close, but the real parts of the roots differ by a factor of 5. Clearly, these lateral stability approximations (except for roll stability, which is rarely an issue) are not as quantitatively useful as those for the longitudinal stability approximations. These methods are only useful for the most "back-of-envelope" estimations, and their limitations must be understood.

ROOTS-LOCUS PLOTS

Roots-locus plots can be made for the lateral stability case, as was done for the longitudinal stability study. For the PA24-250 (Piper Comanche) numerical example, two stability derivatives will be varied: $C_{N\beta}$ and $C_{\tilde{L}\beta}$. The baseline value of $C_{N\beta}$ is 0.109, and it will vary from 0 to 0.30 in 0.015 increments. From an approximate-design standpoint, one could say that the vertical fin area goes from less than baseline to greater.

The baseline value of $C_{\tilde{L}\beta}$ is -0.057, and it will vary from 0.10 to -1.0 in 0.05 increments. In this case, one could envision that the dihedral angle goes from less than baseline to greater.

The first root-locus plot, where $C_{N\beta}$ is varied, is shown below:

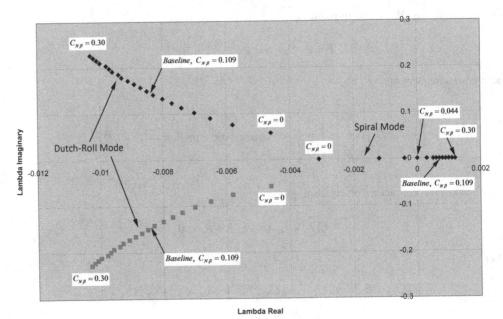

The following observations may be made:

a. The Dutch-Roll Mode is stable for the range of $C_{N\beta}$ values chosen, though it is least stable for $C_{N\beta} = 0$ and increasingly stable as $C_{N\beta}$ becomes larger.
b. The Dutch-Roll frequency of oscillation increases as $C_{N\beta}$ increases.
c. The Spiral Mode is stable when $C_{N\beta} = 0$ and only goes unstable when $C_{N\beta} > 0.044$.
d. The Rolling Mode is not shown, far to the left and off the graph, because of its extreme stability. λ_{Rolling} varies from -0.2321 to -0.2249 when $C_{N\beta}$ varies from 0 to 0.30.

Flight Dynamics

As previously stated, the stability derivative $C_{N\beta}$ may be related to the vertical fin area of the example airplane. Of course, that's not perfectly true because the wing, fuselage and propeller also make contributions to $C_{N\beta}$. However, the dominant effect in this case is due to the vertical fin.

In light of that, one may conclude that no vertical fin would give stability for both Dutch Roll and Spiral Modes. This is not desirable, though, and brings up a very important observation. Note that this small-perturbation analysis predicts stability characteristics "in-the-small". Namely, this deals with the behavior of an aircraft subject to small disturbances to its motion. This does not necessarily give information about the aircraft's behavior if subjected to large perturbations.

As it turns out, if most aircraft are predicted to be stable for small perturbations then they will also have acceptable stability when subjected to large perturbations. The no-fin situation, though, would be an exception. In this case the example airplane would go into an unrecoverable flat spin. What is important to remember is that for dynamic systems, stability in-the-small doesn't necessarily guarantee stability in-the-large (and vice versa).

Finally, regarding the Spiral Mode, it is possible to visualize the role of the vertical fin by considering its effect when the airplane is in a sideslip, β. In this case the resulting aerodynamic force on the fin is, such as to turn the airplane into the direction of the sideslip. If the fin is above a certain size, the airplane will go into an unstable spiral dive (as is the situation for this example).

The second roots-locus plot, where $C_{\tilde{L}\beta}$ is varied, is shown in the figure below.

Example Lateral Modes

The following observations may be made:

a. The Dutch-Roll Mode becomes increasingly unstable as $C_{\tilde{L}\beta}$ becomes increasingly negative, crossing from stable to unstable when $C_{\tilde{L}\beta} = -0.1393$.
b. The Dutch-Roll frequency of oscillation increases as $C_{\tilde{L}\beta}$ decreases.
c. The Spiral Mode becomes increasingly stable as $C_{L\beta}$ decreases and crosses from unstable to stable when $C_{\tilde{L}\beta} = -0.1393$.
d. Again, the Rolling mode is not shown because it would be far off the graph to the left. λ_{Rolling} varies from -0.1851 to -0.3406 when $C_{\tilde{L}\beta}$ varies from 0.10 to -1.0.

STABILITY-BOUNDARY PLOT

It is clear that the stability derivatives, $C_{N\beta}$ and $C_{\tilde{L}\beta}$, are significant parameters for determining the example airplane's lateral stability. The combinations of these derivatives that give stability or instability may be graphically represented by plotting stability boundaries on a $C_{\tilde{L}\beta}$ vs. $C_{N\beta}$ field. This is accomplished by a systematic sequence of stability-root calculations over prescribed ranges of $C_{\tilde{L}\beta}$ and $C_{N\beta}$, from which one may obtain a typical plot as shown below:

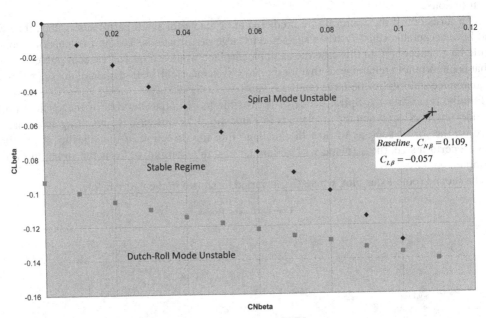

This shows that there are distinct ranges of $C_{\tilde{L}\beta}$ and $C_{N\beta}$ for which the airplane will have complete lateral dynamic stability. However, this particular airplane example lies outside of the stable region. As mentioned before, there may be other important design constraints that have nothing to do with small-perturbation dynamic stability. In any case, as shown before, the time-to-double amplitude, T_2, of the instability is so large that the Spiral Mode is easily controlled.

Finally, recall that $C_{\tilde{L}\beta}$ is primarily given by the wing dihedral angle for this example. With increasing dihedral angle, $C_{\tilde{L}\beta}$ becomes increasingly negative. When the airplane is in a sideslip, β, a negative value of $C_{\tilde{L}\beta}$ produces a rolling moment that acts to level the airplane. Normally, this would be a stabilizing effect. However, the plot above shows that this can be too much of a good thing if $C_{\tilde{L}\beta}$ becomes too negative (too much dihedral angle). In that case, Dutch-Roll instability occurs.

ASIDE

The author has been involved with the design and testing of ornithopters (flapping-wing aircraft). One such test involved a 3-m-span engine-powered design. During the flight, the engine quit and left the wings at a very high dihedral angle. At that point, during its glide, Dutch-Roll instability began building up. The author's first reaction was "wow, that's a classic Dutch-Roll Mode", followed by the realization that "oh no, it's going to crash". Fortunately, Eric Edwards, who had remote control of the aircraft, was able to successfully damp the Dutch-Roll motion using rudder control (no ailerons on this aircraft) and bring it in to a safe landing.

END ASIDE

Flight Dynamics

ADDENDUM

A considerable effort went into the estimation of the stability derivatives for the "Scholar" high-wing monoplane. So, it's worth the extra effort to calculate its dynamic stability characteristics for its given flight conditions. Recall that, at sea level, the inertial properties were estimated to be:

$$m = 3.70 \, kg, \quad I_{xx} = 0.2167 \, kg-m^2, \quad I_{yy} = 0.3116 \, kg-m^2, \quad I_{zz} = 0.4281 \, kg-m^2, \quad I_{xz} \approx 0$$

The longitudinal dynamic stability roots are:

$$\sigma_{1,2} = -0.11580 \pm 0.07859 j, \quad \sigma_{3,4} = -0.001597 \pm 0.007597 j$$

The first complex-conjugate pair represents the Short-Period Mode, for which

$$T = 0.8295 \, sec, \quad T_{1/2} = 0.0621 \, sec, \quad N_{1/2} = 0.0749$$

With an initial pitch angle of 10°, the Short-Period trajectory is as shown:

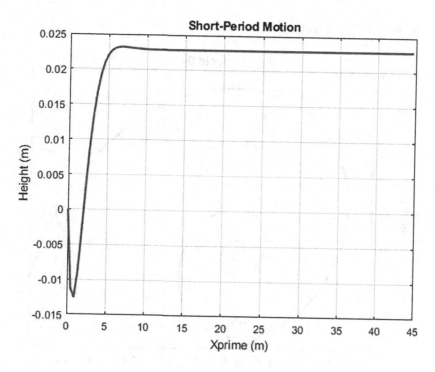

The second complex-conjugate pair represents the Phugoid Mode, for which

$$T = 8.5807 \, sec, \quad T_{1/2} = 4.65038 \, sec, \quad N_{1/2} = 0.54249$$

With an initial pitch angle of 4°, the Phugoid trajectory is as shown:

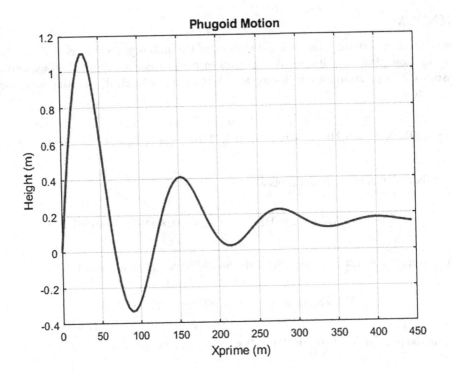

From this, the Phugoid Spiral is obtained:

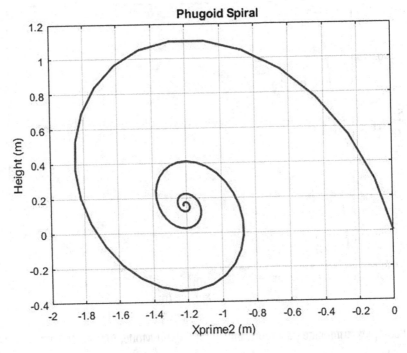

It is seen that, for this flight condition, the Scholar model airplane has very acceptable inherent longitudinal dynamic stability.

As for the lateral case, the stability roots are:

$$\lambda_1 = -0.02738, \quad \lambda_2 = -1.98082, \quad \lambda_{3,4} = -0.08961 \pm 0.20520 j$$

Flight Dynamics

The first root, which is aperiodic, represents the Spiral Mode, for which:

$$(T_{1/2})_{Spiral} = 1.7556 \text{ sec}$$

Its flight-path trajectory, for an initial perturbation heading angle of 10° ($\Psi_r = 10°$, $\Psi_j = 0$) is shown below:

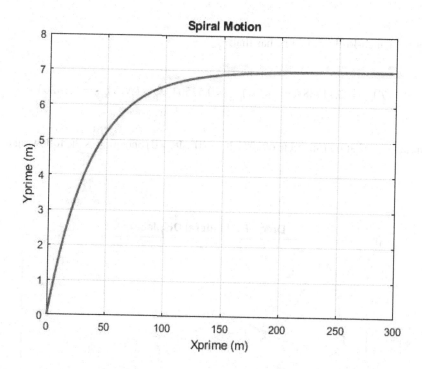

Also, the aircraft's corresponding roll angle is as shown in the figure given below.

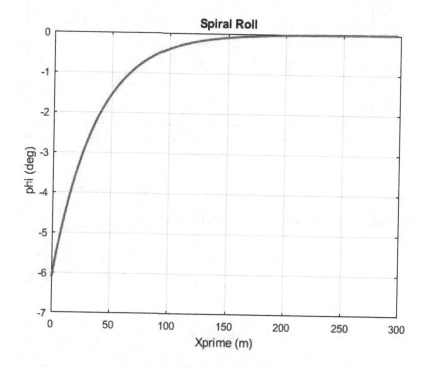

This mode is seen to be solidly stable, so that the Scholar is not susceptible to a spiral dive. Likewise, the second root, which represents the Rolling Mode, is even more stable with:

$$(T_{1/2})_{\text{Roll}} = 0.02420 \text{ sec}$$

Finally, for the Dutch-Roll Mode, one has that:

$$(T)_{\text{D-R}} = 2.11788 \text{ sec}, \quad (T_{1/2})_{\text{D-R}} = 0.53504 \text{ sec}, \quad (N_{1/2})_{\text{D-R}} = 0.25263$$

With an initial perturbation roll angle of 20° $(\Phi_r = 20°, \Phi_j = 0)$ one obtains the following flight-path trajectory:

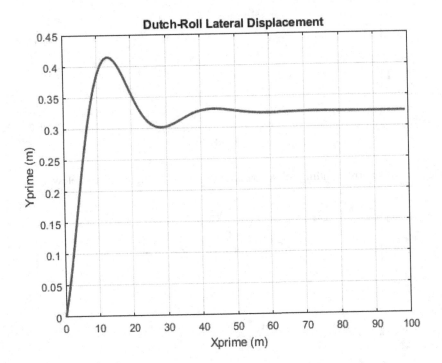

Likewise, the heading angle is shown by the following figure.

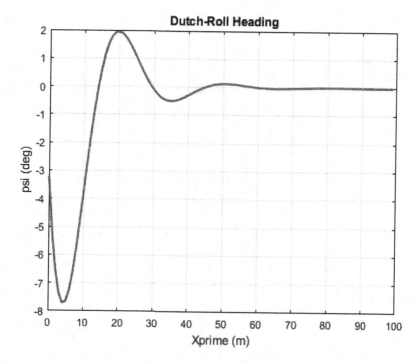

And finally, the Dutch-Roll rolling motion is shown below.

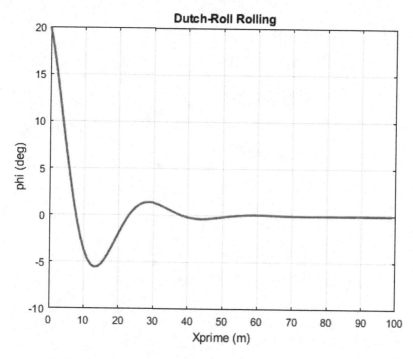

It is seen that the Scholar is inherently controls-fixed stable in all modes. That would make it an easy-to-fly model airplane and a suitable trainer to practice radio-controlled flying (hence the name).

7 Performance

GLIDE TESTS

If the aircraft is a model, such as was typically constructed in the Aircraft Design course the author taught, it is possible to obtain a few experimental points to compare with the predicted C_L vs. C_D polar curve.

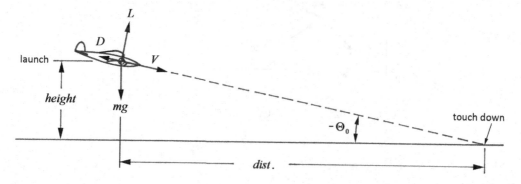

First, the propeller should be removed and substituted with a small weight equal to that of the propeller. Then the procedure is to carefully launch the model from a given height, and measure the time it takes to cover a distance until touchdown. In trimmed equilibrium flight, one has that:

$$L\cos\Theta_0 + D\sin\Theta_0 = mg$$

If the model is reasonably aerodynamically efficient, say $L/D \geq 6$, then

$$L \approx mg$$

Further, the velocity, V, of the model is approximately:

$$V \approx \frac{dist.}{t_{\text{touch-down}} - t_{\text{launch}}}$$

Therefore, upon knowing the weight of the model and its characteristic area, one may find the lift coefficient:

$$C_L \approx \frac{2mg}{\rho V^2 S}$$

Next, the drag coefficient may be obtained from:

$$C_D \approx -C_L \tan\Theta_0 = C_L \frac{height}{dist.}$$

The test should be run several times to obtain averaged results. It takes a certain amount of skill to launch the model at the angle and speed that best corresponds to the flight conditions. In other words, the launch transients should be minimal. Also, the moment of touchdown requires judgment because the model will tend to flare out due to the ground effect.

The graph below shows a data point obtained by a University of Toronto team, in 2004, consisting of Andrew Lam, Ryan McCoubrey, Graham Murray and Devi Soondarsingh:

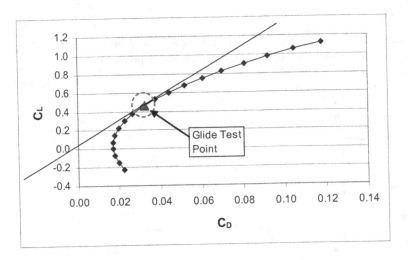

Their design was a 1.5-m-span flying wing of extraordinary performance:

To obtain several averaged points on the C_L vs. C_D polar graph, it is necessary to repeat the test for different trim states. Therefore, the mass-center is moved and the controls are adjusted (and fixed) to give another configuration for stable equilibrium flight. Obviously, there are limits to this. The mass-center can only be moved forward so much before the controls become saturated and can no longer give a trim condition. Also if the mass-center is moved aft of the neutral point, this is a condition for static instability. Nonetheless, this exercise gives a good "reality check" on the predicted aerodynamic performance.

The graph below shows such results for a 1 m-span low aspect-ratio flying-wing design by a 2004 University of Toronto team consisting of Tahoura Soltani, Benjamin Avdicevic, Amy Bilton and Hadi Farashahi:

Performance

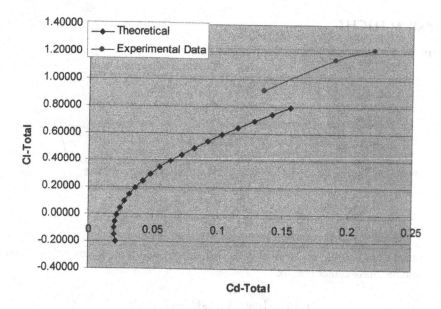

In this case, three data points were obtained. Also it was clear that their design was more aerodynamically efficient than they had calculated (which was a pleasant surprise). This was borne out by its fine flight performance. A top view of the design is shown below:

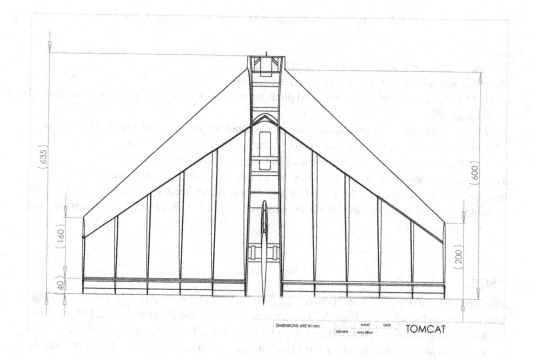

EQUILIBRIUM FLIGHT

TRIM STATE

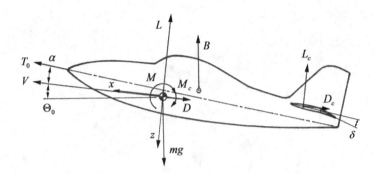

The condition for an equilibrium trim state is that the lift and drag forces balance and the net moment about the mass-center is zero:

$$L + L_c - (mg - B)\cos\Theta_0 - (T_0)_z = 0 \tag{7.1}$$

$$D + D_c + (mg - B)\sin\Theta_0 - (T_0)_x = 0 \tag{7.2}$$

$$M + M_c + x_B B\cos\Theta_0 + z_B B\sin\Theta_0 - x_T(T_0)_z + z_T(T_0)_x = 0 \tag{7.3}$$

where:
- B = the aircraft's gross buoyancy, $B = \rho g (Vol)_{moulded}$,
- Θ_0 = the aircraft's climb angle (negative for descent),
- L, D, M = the aerodynamic lift, drag and pitching moment (about the mass-center) as defined in Chapter 2,
- L_c, D_c, M_c = the lift, drag and pitching moment (about the mass-center) due to the control,
- $(T_0)_z$ = the component of the thrust vector in the z (negative lift) direction,
- $(T_0)_x$ = the component of the thrust vector in the x (negative drag) direction,
- x_B, z_B = the coordinates of the buoyancy center relative to the aircraft's mass-center,
- x_T, z_T = the coordinates of the origin of the thrust vector relative to the aircraft's mass-center (notice the upper-case subscript, which distinguishes this from "tail" terms that use a lower-case "t").

ASIDE

The aircraft's "molded volume", $(Vol)_{moulded}$, is the displacement volume of a solid body in the exact shape and size of the aircraft. The term comes from naval architecture. For most airplanes, the magnitude of the resulting buoyancy force is negligible compared with the gravity force. This is not necessarily so for ultra-light human-powered airplanes and certainly not for airships. In fact, for such aircraft, the mass of the air and gas enclosed inside the structure must be included in the calculation for mass "m", as well as the other inertial terms.

END ASIDE

Performance

As defined in Chapter 2, the aerodynamic forces and pitching moment may be expressed in coefficient form as:

$$L = C_L\, qS, \quad D = C_D\, qS, \quad M = C_M\, q\bar{c}\, S$$

where:
- q = the "dynamic pressure", $q = 0.5\rho V^2$,
- S = the aircraft's reference area,
- $\bar{c}$ = the aircraft's characteristic length.

Upon following the methods described in the chapter, one may obtain equations for C_L, C_D and C_M as functions of α:

$$C_L = (C_L)_0 + C_{L\alpha}\alpha \tag{7.4}$$

$$C_D = (C_D)_0 + C_1\alpha + C_2\alpha^2 \tag{7.5}$$

$$C_M \approx (C_{M0})_{np} - C_L(l_{np} - l_{cm})/\bar{c} \rightarrow C_M = (C_{M0})_{np} - (C_L)_0(l_{np} - l_{cm})/\bar{c}$$
$$- \left[C_{L\alpha}(l_{np} - l_{cm})/\bar{c}\right]\alpha \tag{7.6}$$

Further, the control terms may be expressed as:

$$C_{Lc} = (C_{L\delta})_c\, \delta, \quad C_{Dc} = (C_{D\delta})_c\, \delta, \quad (C_{Mc})_{cm} = (C_{M\delta})_c\, \delta$$

where, for the aircraft illustrated, δ is the deflection of the control surface on the tail. Also,

$$(C_{M\delta})_c = (C_{Mac\delta})_c - (C_{L\delta})_c(l_t - l_{cm})/\bar{c}$$

where $(C_{Mac\delta})_c$ is the derivative with respect to δ of the pitching-moment coefficient about the stabilizer's aerodynamic center.

Further for the airplane drawn above, which has the thrust vector parallel to its reference line, one has:

$$(T_0)_x = T_0\cos\alpha \approx T_0, \quad (T_0)_z = -T_0\sin\alpha \approx -(T_0)\alpha$$

Therefore, the thrust variable is simply T_0.

Now, consider an aircraft where the thrust vector is not parallel with the reference axis:

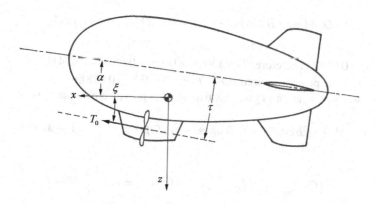

One has that:

$$(T_0)_x = T_0 \cos\xi \rightarrow (T_0)_x = T_0 \cos(\tau+\alpha) = T_0(\cos\tau\cos\alpha - \sin\tau\sin\alpha),$$

$$(T_0)_z = -T_0 \sin\xi \rightarrow (T_0)_z = -T_0 \sin(\tau+\alpha) = -T_0(\sin\tau\cos\alpha + \cos\tau\sin\alpha)$$

where τ is the angle of the thrust vector to the aircraft's reference line.

The figure above speaks to the author's career-long interest in Lighter-Than-Air (LTA) technology. Hence, the inclusion of buoyancy terms in these equations, which explains the inclusion of buoyancy terms in these equations.

Now, these equations may be used to find the *controls-neutral* ($\delta = 0$) equilibrium states for different mass-center locations. The use of these will be illustrated by a numerical example where it is desired to trim the example airplane from Chapter 2 for level flight at a lift coefficient of 0.5. The airplane's properties (propeller sideways effects are ignored) are:

$$S = 0.556\,\text{m}^2,\ \bar{c} = 0.30\,\text{m},\ B \approx 0,\ l_{\text{buoy}} = 0.6\,\text{m},\ h_{\text{buoy}} = 0.07\,\text{m},\ l_T = 0.0,\ h_T = 0.0,$$

$$(\tau)_{\text{thrust}} = 0,\ l_{np} = 0.603\,\text{m},\ (C_L)_0 = 0.418,\ C_{L\alpha} = 4.788,\ (C_D)_0 = 0.0509,$$

$$C_{D1} = 0.2001,\ C_{D2} = 1.3439,\ (C_{M0})_{np} = 0.1770$$

$$C_D = 0.0437 - 0.0072\, C_L + 0.0586\, C_L^2$$

And the mass is chosen to be $m = 3.7\,\text{kg}$.

Equation (7.1) gives:

$$L + L_c - (mg - B)\cos\Theta_0 - (T_0)_z = 0 \rightarrow L - mg = 0 \rightarrow qSC_L = mg \rightarrow$$

$$q = \frac{mg}{SC_L} = \frac{3.7 \times 9.81}{0.556 \times 0.5} = 130.56\,\text{N/m}^2$$

At sea-level flight ($\rho = 1.225\,\text{kg/m}^3$), the speed is thus $V = \sqrt{2q/\rho} = 14.6\,\text{m/s}$

Next, equation (7.2) gives that:

$$D + D_c + (mg - B)\sin\Theta_0 - (T_0)_x = 0 \rightarrow D - T_0 = 0 \rightarrow qSC_D = T_0$$

At $C_L = 0.5$, $C_D = 0.0548$. Therefore, $T_0 = 130.56 \times 0.556 \times 0.0548 = 3.974\,\text{N}$.

Also, this gives that the power required is $P = T_0 V = 58.03\,\text{W}\ (0.078\,\text{hp})$.

Finally, the mass-center location to give this trim state is provided by equations (7.3) and (7.6):

$$M + M_c + x_B B\cos\Theta_0 + z_B B\sin\Theta_0 - x_T(T_0)_z + z_T(T_0)_x = 0 \rightarrow M = 0 \rightarrow$$

$$C_M \approx (C_{M0})_{np} - C_L(l_{np} - l_{cm})/\bar{c} = 0 \rightarrow l_{cm} = -\bar{c}\frac{(C_{M0})_{np}}{C_L} + l_{np} \rightarrow$$

$$l_{cm} = -0.30 \times \frac{0.1770}{0.5} + 0.603 = 0.497\,\text{m}$$

The static margin for this case is:

$$(l_{np} - l_{cm})/\bar{c} = (0.603 - 0.497)/0.30 = 0.353 \to 35.3\%$$

This is rather "stiff", and speaks to the "longitudinal dihedral $(i_w - i_t)$" being excessive at 3°. The angle of attack for this flight condition is:

$$\alpha = (C_L - C_{L0})/C_{L\alpha} = (0.5 - 0.418)/4.788 = 0.0171\,\text{rad} = 0.981°$$

From Chapter 2, the lift coefficient for the tail, based on its own area, is:

$$(C_L)_t = -0.141 + 2.815\alpha = -0.093$$

This shows that the tail has to exert a down-force in order to achieve trimmed level flight. Because this force is "pointing in the wrong direction", most designers seek a trim condition where the tail's lift coefficient is nearly zero. This will generally give the best overall aerodynamic efficiency for the chosen C_L flight condition, as well as the widest range of effective $\pm\delta$ angles for the control surface.

Now consider the case where the airplane is trimmed to fly level at its maximum lift/drag ratio. As found in Chapter 2,

$$(C_L)_{\text{Max L/D}} = 0.864,\ (C_D)_{\text{Max L/D}} = 0.0812$$

Therefore, upon following the previous procedures, one obtains:

$$q = 75.56\,\text{N/m}^2 \to V = 11.11\,\text{m/s},\ T_0 = 3.41\,\text{N} \to P = 37.89\,\text{W}\,(0.051\,\text{hp})$$

Further, the mass-center location to achieve this flight condition is given by:

$$l_{cm} = -\bar{c}\frac{(C_{M0})_{np}}{C_L} + l_{np} = -0.30 \times \frac{0.1770}{0.864} + 0.603 = 0.542\,\text{m}$$

this gives a static margin of,

$$(l_{np} - l_{cm})/\bar{c} = (0.603 - 0.542)/0.30 = 0.203 \to 20.3\%$$

which again falls outside the recommended range of 5–15% as stated in the "Complete-Aircraft Aerodynamics" section in Chapter 2. However, this is much closer than that for the $C_L = 0.5$ case. Further,

$$\alpha = (C_L - C_{L0})/C_{L\alpha} = (0.864 - 0.418)/4.788 = 0.0932\,\text{rad} = 5.34°$$

and the tail lift coefficient (referenced to its own area) is:

$$(C_L)_t = -0.141 + 2.815\alpha = 0.121$$

In this case, the stabilizer is now carrying a positive aerodynamic load.

> **ASIDE**
>
> Later in this chapter it will be shown that the most desirable operational lift coefficient might not necessarily be that for maximum L/D. Mission requirements might require higher cruising speeds or minimum power. Also, as seen by the high value of α, the airplane's wing might be operating too close to the stall angle to be comfortable in gusty conditions.
>
> **END ASIDE**

Finally for level flight with $(C_L)_t = 0$, one obtains that:

$$\alpha = 0.141/2.815 = 0.0501 \, \text{rad} = 2.87° \rightarrow$$

$$C_L = 0.418 + 4.788 \times 0.0501 = 0.658, \quad C_D = 0.0643 \rightarrow C_L/C_D = 10.23$$

Further, the level-flight conditions are:

$$q = 99.21 \, \text{N/m}^2 \rightarrow V = 12.73 \, \text{m/s} \rightarrow T_0 = 3.55 \, \text{N} \rightarrow P = 45.15 \, \text{W} \; (0.0605 \, \text{hp})$$

Also,

$$l_{cm} = 0.522 \, \text{m} \rightarrow \text{Static Margin} = 27\%$$

For this flight condition, the wing's lift coefficient is found from previous work in Chapter 2:

$$(C_L)_w = 0.454 + 4.066\alpha = 0.454 + 4.066 \times 0.0501 = 0.658$$

This is very comfortably lower than the stall value of $(C_L)_{stall} \approx 1.3$, and the C_L/C_D value is only 4% less than the maximum value of 10.64.

FULL SOLUTION

The author has often worked with curve-fitted wind-tunnel data for the aerodynamic lift, drag and moment coefficients. The forces and moments are assumed to act at a fixed point on the model, and the nose (furthest forward point of the reference line) is often chosen as an easily identifiable reference point:

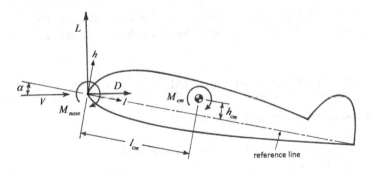

$$C_L = C_{L0} + C_{L1}\alpha + C_{L2}\alpha^2 + C_{L3}\alpha^3$$

$$C_D = C_{D0} + C_{D1}\alpha + C_{D2}\alpha^2 + C_{D3}\alpha^3$$

> **ASIDE**
>
> Note the difference between C_{D0}, which is the drag coefficient at $\alpha = 0$ and $(C_D)_0$, which is the drag coefficient at $C_L = 0$ (which is also written in this book as $(C_D)_{C_L=0}$).
>
> **END ASIDE**

$$C_M = (C_M)_{\text{nose}} + C_L \left(\hat{l}_{\text{cm}} \cos\alpha + \hat{h}_{\text{cm}} \sin\alpha \right) + C_D \left(\hat{l}_{\text{cm}} \sin\alpha - \hat{h}_{\text{cm}} \cos\alpha \right)$$

where:

$$\hat{l}_{\text{cm}} \equiv l_{\text{cm}}/\bar{c}, \quad \hat{h}_{\text{cm}} \equiv h_{\text{cm}}/\bar{c}$$

From the definitions in Chapter 2, l and h are length and height coordinates based at the nose point of the aircraft, with l along the aircraft's reference line and h perpendicular upwards. Further,

$$(C_M)_{\text{nose}} = C_{M0} + C_{M1}\alpha + C_{M2}\alpha^2 + C_{M3}\alpha^3$$

With a bit of algebra, one may directly relate these expanded, generalized, equations to the simpler ones from equations (7.4–7.6). For example, if, for the moment equation, the drag and $\hat{h}_{\text{cm}}$ contributions are neglected, then the equation becomes:

$$C_M \approx (C_M)_{\text{nose}} + C_L \hat{l}_{\text{cm}} \cos\alpha$$

Further, if small α values are assumed, and the square and cubed terms of α are neglected, then

$$C_M \approx C_{M0} + C_{M1}\alpha + C_{L0}\hat{l}_{\text{cm}} + C_{L1}\hat{l}_{\text{cm}}\alpha \rightarrow C_M = \left(C_{M0} + C_{L0}\hat{l}_{\text{cm}} \right) + \left(C_{M1} + C_{L1}\hat{l}_{\text{cm}} \right)\alpha$$

In comparison with equation (7.6), one obtains:

$$\left(C_{M0} + C_{L0}\hat{l}_{\text{cm}} \right) = \left(C_{M0} \right)_{\text{np}} - (C_L)_0 \left(\hat{l}_{\text{np}} - \hat{l}_{\text{cm}} \right)$$

$$\left(C_{M1} + C_{L1}\hat{l}_{\text{cm}} \right)\alpha = -C_{L\alpha}\left(\hat{l}_{\text{np}} - \hat{l}_{\text{cm}} \right)\alpha$$

Since $C_{L0} \equiv (C_L)_0$ and $C_{L1} \equiv C_{L\alpha}$, then one finds that:

$$C_{M0} = \left(C_{M0} \right)_{\text{np}} - (C_L)_0 \hat{l}_{\text{np}}, \quad C_{M1} = -C_{L\alpha}\hat{l}_{\text{np}}$$

The reason for this exercise is to convert the theoretical equations derived in Chapter 2 into a form for use in the more generalized equilibrium equations. In this case the square and cubic terms for C_L and C_M are zero, and the cubic term for C_D is zero. However, wind-tunnel data may extend α into a range where these terms are important. Therefore, a great deal of generality is gained by the use of the cubic curve-fit equations. Equilibrium states may then be calculated for angles of attack beyond the veracity of the linear equations for lift and moment coefficients and the quadratic equation for the drag coefficient.

When these cubic equations and the thrust components are introduced into equations (7.1) and (7.2), one obtains:

$$qS\left[C_{L0} + (C_{L\delta})_c \delta\right] + qSC_{L1}\alpha + qSC_{L2}\alpha^2 + qSC_{L3}\alpha^3 - (mg - B)\cos\Theta_0$$
$$+ T_0 \sin\tau \cos\alpha + T_0 \cos\tau \sin\alpha = 0 \tag{7.7}$$

$$qS\left[C_{D0} + (C_{D\delta})_c \delta\right] + qSC_{D1}\alpha + qSC_{D2}\alpha^2 + qSC_{D3}\alpha^3 + (mg - B)\sin\Theta_0$$
$$- T_0 \cos\tau \cos\alpha + T_0 \sin\tau \sin\alpha = 0 \tag{7.8}$$

Also, the full expression of equation (7.3) is:

$$qS\bar{c}\left[(C_M)_{\text{nose}} + C_L\left(\hat{l}_{cm}\cos\alpha + \hat{h}_{cm}\sin\alpha\right) + C_D\left(\hat{l}_{cm}\sin\alpha - \hat{h}_{cm}\cos\alpha\right)\right]$$
$$+ qS\bar{c}\left[(C_{Mc})_{\text{nose}} + C_{Lc}\left(\hat{l}_{cm}\cos\alpha + \hat{h}_{cm}\sin\alpha\right) + C_{Dc}\left(\hat{l}_{cm}\sin\alpha - \hat{h}_{cm}\cos\alpha\right)\right]$$
$$+ x_B B \cos\Theta_0 + z_B B \sin\Theta_0 + x_T T_0 \sin\tau\cos\alpha + x_T T_0 \cos\tau\sin\alpha$$
$$+ z_T T_0 \cos\tau\cos\alpha - z_T T_0 \sin\tau\sin\alpha = 0 \tag{7.9}$$

where:

$$x_B = (l_{cm} - l_{\text{buoy}})\cos\alpha + (h_{cm} - h_{\text{buoy}})\sin\alpha, \quad z_B = (h_{cm} - h_{\text{buoy}})\cos\alpha - (l_{cm} - l_{\text{buoy}})\sin\alpha$$

$$x_T = (l_{cm} - l_T)\cos\alpha + (h_{cm} - h_T)\sin\alpha, \quad z_T = (h_{cm} - h_T)\cos\alpha - (l_{cm} - l_T)\sin\alpha$$

Also, C_L, C_D and $(C_M)_{\text{nose}}$ are represented by the cubic equations presented earlier.

Further, control terms are seen: C_{Lc}, C_{Dc} and $(C_{Mc})_{\text{nose}}$, which are likewise referenced to the aircraft's fixed nose point. These are obtained from:

$$C_{Lc} = (C_{L\delta})_c \delta, \quad C_{Dc} = (C_{D\delta})_c \delta, \quad (C_{Mc})_{\text{nose}} = (C_{M\delta})_c^{\text{nose}} \delta$$

where δ is the magnitude of the control input. For movable control surfaces, this is usually the deflection angle (positive down).

It is seen that equations (7.7), (7.8), and (7.9) are three equations with five variables: q, α, Θ_0, δ and T_0. The author has found it most useful to choose δ and T_0 as the independent variables and

Performance

solve for the remaining dependent variables. This method of solution will be illustrated with the previous numerical example, namely the example airplane from Chapter 2. Again, its characteristics are:

$$S = 0.556\,\text{m}^2,\ \bar{c} = 0.30\,\text{m}, B \approx 0,\ l_{\text{buoy}} = 0.6\,\text{m},\ h_{\text{buoy}} = 0.07\,\text{m},\ m = 3.7\,\text{kg},\ l_T = 0.0,$$

$$h_T = 0.0,\ (\tau)_{\text{thrust}} = 0$$

Also, the calculated aerodynamic characteristics are:

$$C_{L0} = 0.418,\ C_{L1} = 4.788,\ C_{L2} = 0,\ C_{L3} = 0$$

$$C_{D0} = 0.0509,\ C_{D1} = 0.2001,\ C_{D2} = 1.3439,\ C_{D3} = 0$$

$$C_{M0} = (C_{M0})_{np} - (C_L)_0\,\hat{l}_{np} = 0.1770 - 0.418 \times \frac{0.603}{0.30} = -0.663$$

$$C_{M1} = -C_{L\alpha}\,\hat{l}_{np} = -4.788 \times \frac{0.603}{0.30} = -9.624,\ C_{M2} = 0,\ C_{M3} = 0$$

The longitudinal control consists of a movable surface ("elevator") in the horizontal portion of the tail ("stabiliser").

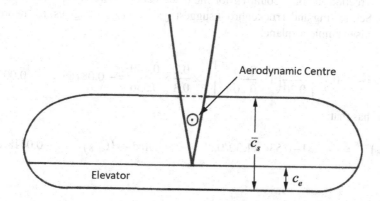

From "Airplane Performance Stability and Control" by C. Perkins and R. Hage (J. Wiley and Sons, 1953, p. 250), one may obtain $(C_{L\delta})_c$ from:

$$(C_{L\delta})_c = (C^*_{L\alpha})_t \frac{d\alpha_t}{d\delta} \frac{S_t}{S} \eta_t \eta_q$$

where, from Chapter 2, the isolated tail's lift-curve slope is $(C^*_{L\alpha})_t = 4.244/\text{rad}$, the area ratio is $S_t/S = 0.140/0.556 = 0.252$, the tail-body efficiency factor is $\eta_t = 0.960$, the dynamic-pressure ratio is $\eta_q \approx 1$, and the $d\alpha_t/d\delta$ parameter is given by:

$$\frac{d\alpha_t}{d\delta} = 3.3602\frac{S_e}{S_t} - 9.3204\left(\frac{S_e}{S_t}\right)^2 + 17.100\left(\frac{S_e}{S_t}\right)^3 - 16.722\left(\frac{S_e}{S_t}\right)^4 + 6.7308\left(\frac{S_e}{S_t}\right)^5$$

The ratio of the elevator area to the total stabilizer area is $S_e/S_t = 0.0433/0.140 = 0.31$.

Therefore, $d\alpha_t/d\delta = 0.520$. One thus obtains:

$$(C_{L\delta})_c = 4.244 \times 0.520 \times 0.252 \times 0.960 \times 1.0 = 0.534/\text{rad} \rightarrow (C_{L\delta})_c = 0.0093/\text{deg}$$

For most airplanes with the relative elevator size of the example, $(C_{D\delta})_c \approx 0$; and the pitching-moment coefficient about the nose due to the elevator deflection is thus given by:

$$(C_{M\delta})_c^{\text{nose}} = (C_{Mac\delta})_c - (C_{L\delta})_c \hat{l}_t$$

where $(C_{Mac\delta})_c$ is the derivative with respect to δ of the pitching-moment coefficient about the stabilizer's aerodynamic center. From "Aerodynamics of the Airplane" by H. Schlichting and E. Truckenbrodt (McGraw-Hill, 1979, p. 486), and referring to the figure above, this is given by:

$$(C_{Mac\delta})_c = -2\chi \left[\frac{c_e}{c_s}\left(1 - \frac{c_e}{c_s}\right)^3 \right]^{1/2} \times \frac{\bar{c}_s}{\bar{c}} \frac{S_s}{S}$$

where χ is a correction factor accounting for the discrepancy between the theoretical and experimental values. Schlichting and Truckenbrodt suggest $\chi = 0.75$ ($\chi = 1.0$ gives the theoretical value). Therefore, for this example airplane,

$$(C_{Mac\delta})_c = -2 \times 0.75 \left[\frac{0.06}{0.20}\left(1 - \frac{0.06}{0.20}\right)^3 \right]^{1/2} \times \frac{0.2}{0.3} \times \frac{0.140}{0.556} = -0.081/\text{rad} \rightarrow -0.0014/\text{deg}$$

Therefore, one has that:

$$(C_{M\delta})_c^{\text{nose}} = -0.081 - 0.534 \times 1.30/0.30 = -2.395/\text{rad} \rightarrow (C_{M\delta})_c^{\text{nose}} = -0.0418/\text{deg}$$

ASIDE

1. There are numerous more-sophisticated methods for estimating the control derivatives, such as described in the USAF DATCOM, Chapter 6.1.4. These can account for detailed differences in the nose shape of the elevator, the gap between the elevator and the stabilizer, and the thickness ratio of the stabilizer's airfoil. The equations for this example assume a relatively-thin stabilizer airfoil, a simple hinge between the elevator and the stabilizer, and negligible gap between the two. This is typical for most small-scale airplanes.
2. Recall, from Chapter 2, that the subscript "t" was chosen to refer to the tail properties in order to accommodate a "V" tail that incorporates both fin and stabilizer functions. However, for a cruciform tail consisting of a distinct stabilizer and fin, the subscript "t" is chosen to be interchangeable with that for the stabilizer, "s". There is no confusion in this case because the fin has its own distinct subscript "f" as seen in Chapter 6.

END ASIDE

Performance

The equilibrium equations were coded in a MATLAB program, which is applied to this numerical example. As stated before, the independent variables are the elevator angle, δ, and the propulsive thrust, T_0; and the program gives the dynamic pressure, q (and hence the flight speed, V), climb-or-descent angle, Θ_0, and angle of attack of the aircraft's reference line to the velocity vector, α. The specific numerical example is the $C_L = 0.5$, $\delta = 0$ case discussed earlier. The previously calculated values of $T_0 = 3.974$ N and $l_{cm} = 0.497$ m are used:

```
clc
% Program to find an aircraft's equilibrium flight conditions
clear all:   % delete all the variables from the workspace to make sure they
% don't interfere
% "Scholar" example airplane
S=0.556; % reference area (m^2)
c=0.30; % reference length, (m)

% Inputs
g=9.81; % gravitational constant (m/sec^2)
ro=1.225; % atmospheric density (kg/m^3)
B=0.0; % gross buoyancy (N)
Lbuoy=0.6; % axial gross-buoyancy centre (m)
Hbuoy=0.07; % vertical gross-buoyancy centre (m)
m=3.70; % vehicle total mass (kg)
Lcm=0.497; % axial mass centre (m)
Hcm=0.03; % vertical mass centre (m)
T0=3.974; % thrust magnitude (N)
Lthrust=0.0; % axial thruster location
Hthrust=0.0; % vertical thruster location
taudeg=0.0; % thrust angle wrt the ref. line (deg., pos, up)
tau=taudeg*pi/180;
deltacontrol=0.0; % elevator angle in degrees (pos. down)

% Longitudinal-control characteristics
CLdelta=0.0093; % per degree
CDdelta=0.0;
CMdeltanose=-0.0418; % per degree

% Overall CL, CD, and CM Polynomial Coefficients
CL0=0.418;
CL1=4.788;
CL2=0.0;
CL3=0.0;
CD0=0.0509;
CD1=0.2001;
CD2=1.3439;
CD3=0.0;
CM0=-0.663;
CM1=-9.624;
CM2=-0.0;
CM3=0.0;

% Defined coefficients
A1=S*(CD0+CDdelta*deltacontrol);
A2=S*CD1;
A3=S*CD2;
A4=S*CD3;
```

```
A5=(m*g)-B;
A6=-T0*cos(tau);
A7=T0*sin(tau);
B1=S*(CL0+CLdelta*deltacontrol);
B2=S*CL1;
B3=S*CL2;
B4=S*CL3;
B5=B-(m*g);
B6=A7;
B7=-A6;
C1=S*c*CM0;
C2=S*c*CM1;
C3=S*c*CM2;
C4=S*c*CM3;
C5=S*c*CMdeltanose*deltacontrol;
C6=Lcm*B1;
C7=Lcm*B2;
C8=Lcm*B3;
C9=Lcm*B4;
C10=Hcm*B1;
C11=Hcm*B2;
C12=Hcm*B3;
C13=Hcm*B4;
C14=Lcm*A1;
C15=Lcm*A2;
C16=Lcm*A3;
C17=Lcm*A4;
C18=-Hcm*A1;
C19=-Hcm*A2;
C20=-Hcm*A3;
C21=-Hcm*A4;
C22=B*(Lcm-Lbuoy);
C23=B*(Hcm-Hbuoy);
C24=B*(Hcm-Hbuoy);
C25=-B*(Lcm-Lbuoy);
C26=T0*((Lcm-Lthrust)*sin(tau)+(Hcm-Hthrust)*cos(tau));
D1=C6+C18;
D2=C7+C19;
D3=C8+C20;
D4=C9+C21;
D5=C10+C14;
D6=C11+C15;
D7=C12+C16;
D8=C13+C17;
x0=[100,0,0];
options=optimset('Display','iter');
[x]=fsolve('equilib',x0,options,A1,A2,A3,A4,A5,A6,A7,B1,B2,B3,B4,B5,...

B6,B7,C1,C2,C3,C4,C5,C22,C23,C24,C25,C26,D1,D2,D3,D4,D5,D6,D7,D8)
V=sqrt(2*x(1)/ro)
AngleOfAttack=x(2)*180/pi
FlightpathAngle=x(3)*180/pi
```

And, within the same folder, the following function program is required:

```
function F=equilib(X,A1,A2,A3,A4,A5,A6,A7,B1,B2,B3,B4,B5,B6,B7,...
    C1,C2,C3,C4,C5,C22,C23,C24,C25,C26,D1,D2,D3,D4,D5,D6,D7,D8)
```

```
q=X(1);
a=X(2);
Theta=X(3);

F(1)=A1*q+A2*q*a+A3*q*a^2+A4*q*a^3+A5*sin(Theta)+A6*cos(a)+A7*sin(a);
F(2)=B1*q+B2*q*a+B3*q*a^2+B4*q*a^3+B5*cos(Theta)+B6*cos(a)+B7*sin(a);
F(3)=q*(C1+C2*a+C3*a^2+C4*a^3+C5)+q*cos(a)*(D1+D2*a+D3*a^2+...
    D4*a^3)+q*sin(a)*(D5+D6*a+D7*a^2+D8*a^3)+(C22*cos(Theta)+...

C24*sin(Theta))*cos(a)+(C23*cos(Theta)+C25*sin(Theta))*sin(a)+...
    C26;
```

The results are:

$$V = 14.4578 \text{ m/s}, \ \alpha = 1.0871°, \ \Theta_0 = 0.0715°$$

These values are very close to the hand-calculated ones, and the slight differences may be due to round-off errors in the hand calculations and the inclusion of the vertical displacement of the mass-center to the reference line.

What is interesting to look at is the effect of increased thrust on the performance. If the thrust is increased from the level-flight value of 3.974 N to 6.0 N, one obtains the following results:

$$V = 14.3067 \text{ m/s}, \ \alpha = 1.1972°, \ \Theta_0 = 3.3459°$$

So, one has a climb rate of $V \tan \Theta_0 = 0.836$ m/s.

Now, if the mass-center is moved aft to that for $(L/D)_{\text{Max}}$ in level flight, as previously calculated ($l_{\text{cm}} = 0.542$ m), then one obtains for the 6.0 N thrust case:

$$V = 10.1768 \text{ m/s}, \ \alpha = 7.0327°, \ \Theta_0 = 4.0749°$$

The climb rate is now 0.725 m/s, which is 13% less than that for the previous case. However, the power to achieve this is $T_0 V$, which is 29% less than that of the previous case. Clearly, there are opportunities for optimization, depending on the mission goals.

Performance Parameters

Upon returning to equations (7.1) and (7.2), for an airplane in level flight with zero control deflection and negligible buoyancy, so that trim is controlled with l_{cm}, one obtains:

$$mg = qSC_L \rightarrow C_L = \frac{mg}{qS}, \ T_0 = D \rightarrow T_0 = qSC_D = qS\left[(C_D)_{C_L=0} + AC_L + BC_L^2\right] =$$

$$T_0 = qS(C_D)_{C_L=0} + qSAC_L + qSBC_L^2 \equiv Term1 + Term2 + Term3$$

From the analysis presented in Chapter 2, one has that:

*Term*1 is the zero-lift drag, which consists of viscous, "form", interference and excrescence drag.
*Term*2 is the drag contribution from the airfoils in the lifting surfaces (wing, tail, etc.), and is mostly due to leading-edge suction efficiency that is less than 100%.

*Term*3 also has a contribution from the airfoil characteristics, but is usually dominated by the induced drags of the lifting surfaces. This is especially true for higher Reynolds numbers.

Upon returning to the numerical-example airplane, where:

$$(C_D)_{C_L=0} = 0.0437, \ A = -0.0072, \ B = 0.0586, \ S = 0.556 \text{ m}^2, \ m = 3.7 \text{ kg},$$

one may use the above equations to obtain plots of T_0, *Term*1, *Term*2 and *Term*3 as functions of flight speed, $V = \sqrt{2q/\rho}$:

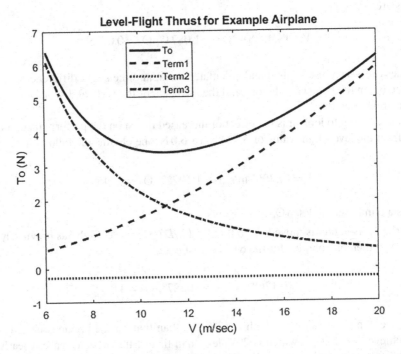

It is seen that at low speeds (and correspondingly high C_L values) that *Term*3, and hence the induced drag, dominates. However, at high speeds (and correspondingly low C_L values), the zero-lift drag dominates. *Term*2 is seen to be constant and small.

In practice, this relates to the design speed for the airplane. The wing area is largely determined by the need to generate sufficient lift for take-off and landing at reasonable speeds and distances. This means that an airplane is generally carrying more wing area at higher speeds than what would be required for flying at the highest-efficiency condition of maximum *L/D*. Extreme examples are racing airplanes that operate at high speeds. In this case, their operational C_L values are small and the drag is dominated by *Term*1. As this relates to the example graph, the thrust required is mainly to overcome the viscous, form, interference and excrescence drag of the airplane; and the induced-drag contribution is relatively small.

Recall, from Chapter 2, that the induced-drag coefficient of a wing is inversely proportional to its aspect ratio. If, as for the racing-plane example, the operational C_L is small, then a high aspect ratio is not required for the wing design (as compared with making it sturdy and compact). An example of this is the racing plane shown in the figure below:

D. Ramey Logan, Wikipedia Commons

At the other extreme are sailplane designs that generally operate near their maximum L/D values. In this case, induced drag is as important as the zero-lift drag, and high aspect ratios are sought. The figure below shows the dramatic difference in wing design as compared to the racing plane:

Arpingstone, Wikipedia Commons

If the landing was not a consideration for the racing plane, then its high-speed performance would be improved by replacing the broad, low aspect-ratio wing with a higher aspect-ratio wing operating at maximum L/D. Such a wing would have a smaller area because it is operating at a higher C_L. Therefore, the $Term1$ (zero-lift) drag would be less. Also, if carefully designed, the $Term3$ (induced) drag could be the same (or less) as for the broader wing. Therefore, improved performance is obtained. Of course, the plane would now have to land at approximately its racing speed, which would generally be a disaster.

An example of such an airplane, for which safe landings are not a consideration, is a cruise missile:

The skinny high aspect-ratio wing is evident and no undercarriage is in sight!

The persistent goal of airplane designers is to somehow morph the wing of an airplane from the optimum skinny shape to a broader high-lift design for landings and take-offs. The most popular approach is to use retractable flaps and slats, as was shown in the "Airfoils" section of Chapter 2:

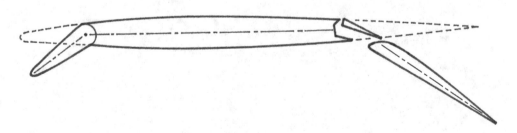

This not only increases the wing area, but also transforms the airfoil into a cambered shape capable of a higher C_L. Anyone looking out the window of a commercial airliner, observing the wing, will see this concept in action. It is rather amazing how much the chord of the wing extends for landing and take-off.

Another approach is to vary the span of the wing. The earliest application of that concept was the Ivan Makhonine designs built in France in the 1930s. It can be seen, in the figure, how the narrower outer portion of the wing can retract within the wider inner portion.

More recently, a very innovative remotely-piloted telescoping "flying-wing" aircraft was designed and built by Patrick Zdunich while he was at Advanced Subsonics, Inc. (he is now at the Canadian National Research Council):

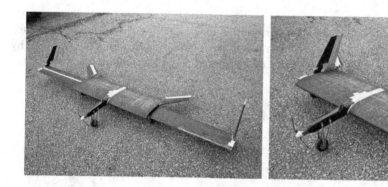

The calculated C_L vs. C_D polar plots for both extended and contracted configurations are given below:

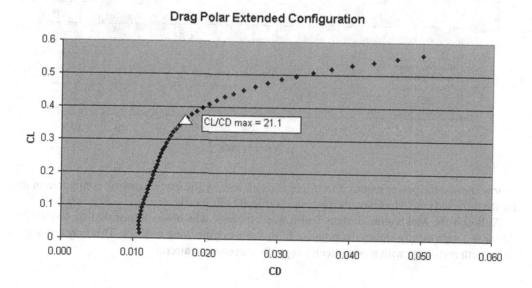

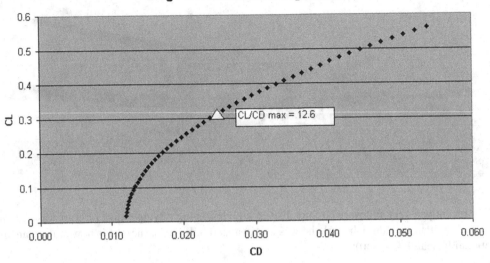

Drag Polar Retracted Configuration

Based on the coefficients alone, the extended-wing design looks superior. However, the advantage of the retractable-wing concept is best seen from the power-required plots, which are obtained by calculating the thrust required for level-flight (as demonstrated before) and multiplying this by the speed:

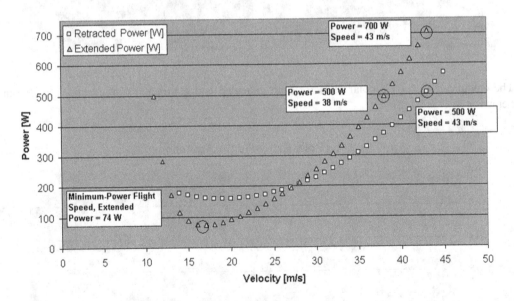

Power Required as a Function of Velocity

At low speeds, the power required for flight is much less for the extended-wing configuration than for the retracted case. However, the opposite is true at high speeds.

Although the Makhonine designs had limited success, which was terminated by the war, the Zdunich design clearly shows the promise of this telescoping-wing concept. This concept may be well worth revisiting, with modern technology, for larger-scale aircraft.

Performance

Upon returning to the earlier example airplane (which has a non-morphing wing), one may obtain the sea-level power-required plot shown below. As stated before, this is calculated by multiplying the thrust required for level flight, T_0, by the flight speed V.

$$P_{req} = T_0 V = q S C_D V = 0.5 \rho V^3 S C_D$$

Also,

$$q = \frac{mg}{C_L S} \rightarrow V = \left(\frac{2}{\rho}\right)^{1/2} \frac{(mg)^{1/2}}{S^{1/2}} \frac{1}{C_L^{1/2}} \rightarrow V^3 = \left(\frac{2}{\rho S}\right)^{3/2} (mg)^{3/2} \frac{1}{C_L^{3/2}}$$

Therefore,

$$P_{req} = \left(\frac{2}{\rho S}\right)^{1/2} (mg)^{3/2} \frac{C_D}{C_L^{3/2}}$$

This plot shows that there is a distinct speed at which the power required is a minimum value. From the equation, this occurs when $C_D/C_L^{3/2}$ is minimum.

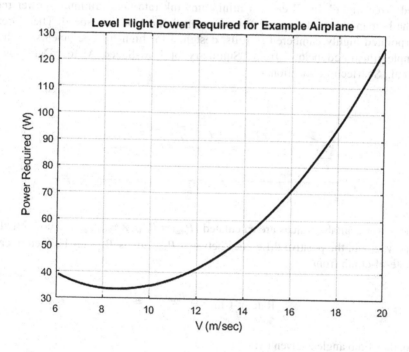

It is accepted terminology to refer to the inverse of this, $C_L^{3/2}/C_D$, as the "endurance parameter" because this would be the condition for flying at minimum energy consumption and, therefore, maximum endurance. For this example,

$$\left(C_L^{3/2}/C_D\right)_{max} = 11.16 \text{ at } C_L = 1.441 \rightarrow (V)_{Max\,Edur.\,Param.} = 8.60 \text{ m/s}$$

By comparison with the maximum L/D case,

$$(C_L/C_D)_{max} = 10.64 \text{ at } C_L = 0.864 \rightarrow (V)_{Max\,L/D} = 11.11 \text{ m/s}$$

There are several points to notice from this:

1. The maximum endurance parameter is achieved at a high lift coefficient. In fact, from the C_L vs. α plot for the Clark-Y airfoil used in the wing of the example airplane (seen in Chapter 2, the lift coefficient becomes nonlinear at $C_L \approx 1.2$ and will probably be stalled at $C_L \approx 1.35$. Therefore, the example airplane is unlikely to achieve the maximum theoretical endurance parameter.
2. The maximum achievable lift coefficient increases with increasing Reynolds number. For example, at $RN = 10^6$, $(C_L)_{max} \approx 1.7$. Therefore, a full-scale version of the example airplane should be able to realize its maximum endurance parameter. It would, however, still be close to stalling. As discussed in the "Airfoils" section of Chapter 2, an airfoil is selected based on the operational requirement of the airplane. For a given wing area, payload and desired speed (cruising speed for most airplanes and maximum speed for racers), a lift coefficient is calculated. Then, an airfoil is sought whose "ideal angle of attack" corresponds to that lift coefficient. This identifies the camber of the airfoil. Most airplanes are not designed to operate at a maximum endurance parameter (they would be too slow to go anywhere!). Exceptions are sailplanes in the early 20th century, human-powered airplanes and a class of model airplanes called "free flight". In all cases a minimum sink rate (and minimum-power required for the human-powered airplanes) was/is more important than speed. Therefore, these incorporated highly cambered airfoils, designed for high lift coefficients, such as the example illustrated below (from "Summary of Low-Speed Airfoil Data" by Selig. M, et al., Soartech Publications):

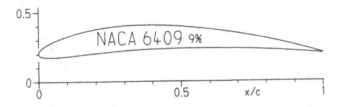

3. If the power-available values are calculated ($P_{avail} = P_{motor} \times \eta_{propeller}$) and co-plotted on the P_{req} vs. V graph, the positive difference between P_{avail} minus P_{req} may be used to calculate the rate-of-climb from:

$$\text{Rate-of-Climb} = \frac{P_{avail} - P_{req}}{mg}$$

Also, the climb angle is given by:

$$\Theta_0 = \sin^{-1}(\text{Rate-of-Climb})/V$$

For the numerical example where $T_0 = 6.0\,\text{N}$ at a velocity of $V = 14.307\,\text{m/s}$, one obtains that the power available for this flight condition is $P_{avail} = 85.840\,\text{W}$. Also from the calculations for P_{req} vs. V, the power required for level flight at this speed is $P_{req} = 55.687\,\text{W}$. Therefore, the rate of climb is given by:

$$\text{Rate-of-Climb} = \frac{85.840 - 55.687}{3.7 \times 9.81} = 0.831\,\text{m/s}$$

This is very close to the 0.836 m/sec value calculated from the equilibrium program. Further, the climb angle is:

$$\Theta_0 = \sin^{-1}\left(\frac{0.831}{14.307}\right) = 3.33°$$

This compares favorably with the 3.35° obtained from the equilibrium program.

4. As mentioned before, operating at $(P_{req})_{min}$ allows sustained flight at the lowest energy consumption. However, this does not give the greatest range, which is achieved (in still air) by flying at $(L/D)_{max}$. Just as most airplanes do not operate at their maximum endurance parameter, this is also true for their maximum lift-to-drag ratio. Modern sailplanes and cruise missiles are the exceptions. It would be thought that airliners would try to maximize their $(L/D)_{max}$. However, the profitability equation also includes the time spent in the air. Once passengers have paid their fare, the airline wants the quickest turn-around time possible. Therefore, the profitability equation requires faster speeds and a lower L/D. It was interesting that, during the energy crisis of the 1970s, the average airliner speed diminished because the higher cost of fuel compelled a speed with lower energy consumption. This quest for energy conservation continues with aerodynamic, propulsion and combustion research.

ASIDE

As a historical note, the famous aviator Charles Lindbergh was sent to the Pacific in World War II to advise on achieving maximum range for the fighter airplanes operating over large areas of the ocean ("The Wartime Journals of Charles A. Lindbergh", Harcourt, Brace, Jovanovich, 1970). The particular airplane in question was the Lockheed P-38:

United States Air Force

> The pilots had intuitively assumed that the greatest range was obtained by flying at the design cruising speed, which was well above that for $(L/D)_{max}$. However, Lindbergh demonstrated that throttling back and flying at the $(L/D)_{max}$ condition gave an extraordinary increase in range. Further, the engine is not straining under that condition and the fuel-air ratio can be reduced ("leaning the mixture") to provide further fuel economy. The only "problem" was that the flight now became long and boring, with some physiological discomfort. At least, though, the pilots were able to achieve some extraordinary missions.

> **END ASIDE**

TAKE-OFF RUN

The complete analysis of an aircraft's take-off dynamics can be very complex, with variable forces requiring a step-by-step calculation for the acceleration and subsequent integration for the take-off distance and time. Also, the pilot's individual technique can affect the results. However, there are various useful approximate techniques, one of which is presented below.

Wood's Methodology

In Karl Wood's book, "Technical Aerodynamics" (McGraw-Hill, 1935, page 183), he makes linear approximations between the static thrust, T_{Static} and the propulsive thrust at take-off, T_{TO}. Also, the net thrust at the beginning of the take-off run is:

$$T_{Initial} = T_{Static} - \mu mg$$

where μ is the rolling-force coefficient. Next, the net thrust at the moment of take-off is:

$$T_{Final} = T_{TO} - D_{TO} = T_{TO} - mg/(L/D)_{TO}$$

Next, the variation of the net force on the aircraft with speed is assumed to be linear:

$$F_{net} = T_{Initial} - \frac{(T_{Initial} - T_{Final})}{V_{TO}} V$$

where V_{TO} is the aircraft's speed at the moment of take-off. From Newton's 2nd law, one may obtain the aircraft's acceleration:

$$a = \frac{F_{net}}{m} = \frac{T_{Initial}}{m} \left(1 - \frac{(T_{Initial} - T_{Final})}{T_{Initial}} \frac{V}{V_{TO}} \right)$$

It is seen that the acceleration is likewise assumed to be linear. Wood now introduces a "K" term, so that the acceleration equation becomes:

$$a = \frac{T_{Initial}}{m} \left(1 - K \frac{V}{V_{TO}} \right)$$

Where:

$$K \equiv \frac{T_{\text{Initial}} - T_{\text{Final}}}{T_{\text{Initial}}} = 1 - \frac{T_{\text{Final}}}{T_{\text{Initial}}} = 1 - \frac{T_{\text{TO}} - mg/(L/D)_{\text{TO}}}{T_{\text{Static}} - \mu mg} \equiv 1 - \text{Thrust Ratio}$$

Now, the time for take-off t_{TO}, and the distance for take-off s_{TO}, may be found by integrating:

$$dt = dV/a \text{ and } ds = V \, dV/a$$

from $V = 0$ to V_{TO}. Doing so results in the following equations:

$$t_{\text{TO}} = \frac{-m}{T_{\text{Initial}}} V_{\text{TO}} \frac{\ln(1-K)}{K}, \quad s_{\text{TO}} = \frac{m(V_{\text{TO}})^2}{T_{\text{Initial}}} \frac{1}{K}\left[-1 - \frac{\ln(1-K)}{K}\right]$$

Now, the use of this methodology is best demonstrated with the numerical example:

1. The first step is to calculate V_{TO}. Obviously, for this case, a large value of C_L is sought. However, this must not be too close to that for stalling. Usually, controlling the aircraft to give the C_L for maximum L/D will suffice. For the numerical example this is $(C_L)_{\text{MaxL/D}} = 0.864$. Therefore, $V_{\text{TO}} = 11.11 \text{ m/sec}$ (note that "image-plane" ground effects are neglected for this analysis, which makes it conservative).
2. The propulsor's static thrust, T_{Static} must be found. One means would be to simply measure this using a spring scale or digital load cell. Also, for a propeller, if the geometry is known along with the motor characteristics, the static thrust may be calculated from the program in Chapter 3. Note that a small forward velocity is required to avoid division by zero. Also, the blade angle-of-attack plot should be checked for stalling values, which would compromise the results.

   The internet can also be of help. A guide to electric flight may be found at the website: http://adamone.rchomepage.com/guide5.htm. From this, a rule of thumb is that for non-aerobatic flying, the airplane should have 110W/kg. This means that for the example airplane, a $(P_{\text{Motor}})_{\text{Max}} \approx 500 \text{ W}$. Also, a 0.38 m diameter is selected for the propeller (15 inches). From a list of commercial designs, a candidate propeller is the "APC E 15x8". From a calculator provided on the site, the resulting static thrust is $T_{\text{Static}} = 8.26 \text{ N}$, which will be chosen for this numerical example.
3. The rolling-friction coefficient must be obtained next. Woods gives the following example values for μ:

Smooth hard surface: 0.02
Hard turf: 0.04
Average field, short grass: 0.05
Average field, long grass: 0.10
Soft ground, sand: 0.10 to 0.30

These values are for full-scale airplanes. The coefficients would differ for smaller-scale aircraft. However, in this case, pulling the aircraft with a spring scale over the surface would give an accurate value.

For this numerical example assume a smooth hard surface and choose $\mu = 0.02$, which gives:

$$T_{\text{Initial}} = T_{\text{Static}} - \mu mg = 8.26 - 0.02 \times 3.7 \times 9.81 = 7.534 \text{ N}$$

4. The next required value is the thrust at take-off, T_{TO}, for which a plot of the thrust variation with speed is needed. Again, this may be obtained from the propeller program, or from wind-tunnel tests. For this example, it will be assumed that $T_{TO} = 6.0\,\text{N}$. Therefore,

$$T_{\text{Final}} = T_{TO} - mg/(L/D)_{TO} = 6.0 - 3.7 \times 9.81/10.64 = 2.589$$

5. Thus:

$$\text{Thrust Ratio} = 2.589/7.534 = 0.344 \rightarrow K = 0.656$$

6. From the equations, the results are:

$$t_{TO} = \frac{-m}{T_{\text{Initial}}} V_{TO} \frac{\ln(1-K)}{K} = \frac{-3.7}{7.534} \times 11.11 \times \frac{\ln(1-0.656)}{0.656} = 8.876\,\text{sec}$$

$$s_{TO} = \frac{(V_{TO})^2 m}{T_{\text{Initial}}} \frac{1}{K}\left[-1 - \frac{1}{K}\ln(1-K)\right] = \frac{(11.11)^2}{7.534} \frac{3.7}{0.656}\left(-1 - \frac{\ln(1-0.656)}{0.656}\right) = 57.91\,\text{m}$$

It is important to note that there are various ways in which an airplane can take off. If the design is a "tail dragger", like the numerical example, it accelerates at its ground-orientation angle until the speed is sufficient for the tail to lift up. The acceleration continues until take-off speed is reached, at which point the airplane is given a nose-up control to "rotate" and commences its climb-out. For an airplane with a tricycle undercarriage, the tail lift portion of the take-off is avoided which, in principle, would shorten the take-off run if it wasn't for the additional weight and drag of the nose gear.

Wood suggests that for smooth low-friction runway surfaces the C_L should be kept low, thus also keeping the C_D low and allowing a rapid acceleration. On the other hand, for higher-friction surfaces (grass, etc.), one should take off with a high value of C_L so as to minimize the resistance of the ground when the aerodynamic lift builds up.

What is interesting, according to Wood, is that these different take-off techniques do not give great variations in time and distance. Comparisons of this analytical methodology with experiments supposedly show variations within 10%.

Finally, it is obvious that a headwind during take-off can greatly reduce t_{TO} and s_{TO}. Therefore, take-off (and landing) should always be in a headwind, if possible.

FINAL COMMENTS

This chapter is intended to introduce certain important methodologies for aircraft performance predictions. However, it does not, by any means, cover the full scope of aircraft performance analysis. There are many other considerations such as range, time-to-climb, landing over an obstacle, engine characteristics (specific fuel consumption, power variation with altitude), power required during a turn, operating envelope, etc. Some excellent references that address these topics are listed below:

"Airplane Performance Stability and Control" by Courtland D. Perkins and Robert E. Hage, John Wiley and Sons, 1960.
"Aircraft Design: A Conceptual Approach" by Daniel P. Raymer, AIAA Education Series, 1992.
"Synthesis of Subsonic Airplane Design" by Egbert Torenbeek, Delft University Press, Kluwer Academic Publishers, 1982.

8 Balloons and Airships

FREE BALLOONS

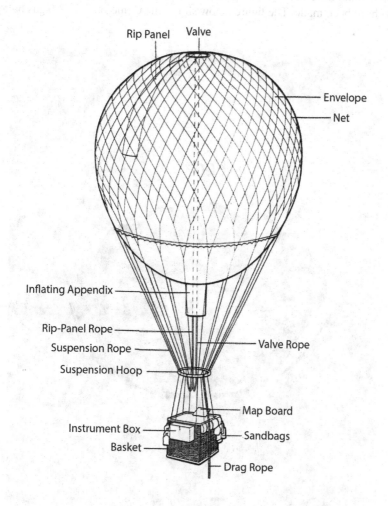

The basic gas-filled free balloon (non-tethered) is the oldest human-carrying aircraft, and it has changed little since its invention in 1783 by Jacques Alexandre Charles. The envelope (the bag) is filled with a "lifting gas", such as helium or hydrogen, whose internal pressure produces the round shape shown. To rise, ballast (from the sand bags) is dropped. As altitude is gained, the lower atmospheric pressure causes the lifting gas to expand and vent through the inflating appendix. To descend, lifting gas is released through the valve at the top of the balloon. The flight ends when there is no longer sufficient lifting gas or ballast for altitude control.

The passenger basket is traditionally woven from wicker because of its light weight and durable construction. Landings typically consist of impact and dragging, and the wicker construction flexes to withstand these loads. The weight of the basket and its contents are supported by a net of ropes, distributing the load over the surface of the envelope. Note that if the envelope bursts and deflates,

DOI: 10.1201/9781315228167-8

it sometimes collapses into the net to form a crude parachute (this can't necessarily be counted on as a safety feature!).

The rip panel, as shown, is a segment of fabric near the top of the balloon that can be torn away from the envelope. Purposeful deflation is accomplished by pulling the rip-panel rope, which is done after landing to eliminate all buoyancy and bring the partially deflated envelope under control as well as allowing the envelope to be folded and packed.

These fundamental features remained essentially invariant over the decades, and a gas balloon from 1860 would have been very similar to one from 1960. However, some important recent improvements have been made. The figure below shows the Cameron GB1000 gas balloon.

The GB1000 design, as well as most previous sport balloons, uses hydrogen for the lifting gas. The cost of helium is typically prohibitive for such personal use. Therefore, great care is exercised in the management of this inflammable gas. The GB1000 is unique in that the pure hydrogen is sealed within the envelope, isolating it from the air that would make a combustible mixture. In an over-pressure situation, the gas is vented from a top-mounted metal valve instead of through the bottom fabric appendix, which is the only occasion when hydrogen and air mix. In the unlikely event of ignition, the valve incorporates a chimney extension to isolate the combustion until the valve is closed.

Another type of free balloon, most often used for high-altitude research, is the "zero-pressure" design. Because the envelope is made from thin flexible plastic, easily torn, it is open at the bottom to minimize the difference between internal and external pressures. When the balloon is launched, there is a small "bubble" of lifting gas at the top of the envelope. As the balloon rises, this bubble expands until the envelope is fully inflated at operational altitude (usually 30 km). The "beet-like" fully inflated shape is designed to produce only vertical stresses and not circumferential stresses ("hoop stresses").

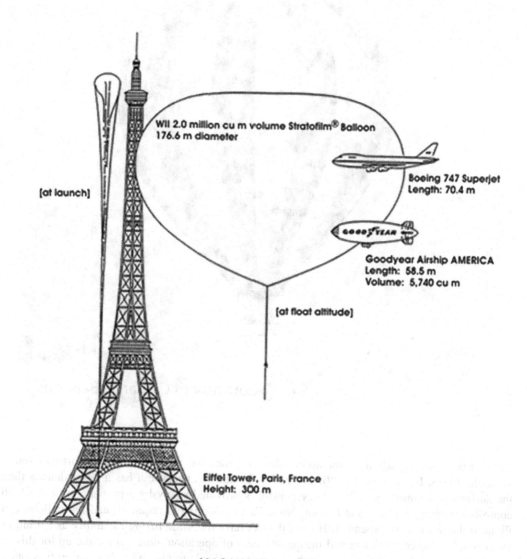

NASA balloon size

These types of free balloons typically have an operational life of several days and circumnavigate the earth. In daylight, the lifting gas expands and is vented from the bottom opening. At night, the gas contracts and the balloon loses altitude. Eventually, after several such cycles, sufficient gas is lost and the mission must be terminated. At that point the payload is released and parachuted to the earth. The envelope is lost, but this is rarely a problem because the material is so unsubstantial (like plastic dry-cleaner bags) that it breaks up and scatters. Note that these balloons are typically huge, as illustrated.

Courtesy of Cameron Balloons

Another type of free balloon is the hot-air design, which has been popular for sport ballooning since the 1970s. In this case the "lifting gas" is heated air, because it has a lower density than the surrounding atmosphere. Buoyancy control is provided solely by a propane burner, which controls the temperature of the hot air. No ballast or venting is required, and the balloon can fly until the propane is expended. Hot air doesn't have the same lifting capability as helium or hydrogen, but the convenience and inexpensiveness of operation more than make up for this. It should be observed that aerodynamics has a very small role to play with these aircraft because a free balloon moves with the wind. This allows hot-air balloons to be shaped into a variety of unique configurations.

As a historical note, it should be mentioned that the earliest free balloons were sustained by heated air and were invented by the Montgolfier brothers in the late 18th century. Their work culminated with the first human-carrying free balloon, flown in 1783 with Jean-Francois Pilatre de Rozier and Marquis Francois d'Arlandes in the gondola. A model of this from the Science Museum (London) is shown.

Mike Young, Wikipedia Commons

The source of heat was damp burning straw in a brazier. This produced considerable smoke, which the Montgolfier brothers wrongly assumed caused the buoyancy. The envelope was made from paper, and floating embers from the fire burned holes. This was a very hazardous and lucky flight. Just a month later Jacques Charles and Nicolas-Louis Robert successfully flew in a hydrogen-filled free balloon of their own design, and free ballooning for nearly the next two centuries consisted solely of gas balloons. Hot-air balloons only became practical after the development of modern fabrics (usually rip-stop nylon) and propane burner units (which produce heat with no embers).

Pilatre de Rozier continued with ballooning and had an idea for combining the greater buoyancy of a gas balloon with the simple buoyancy control from a hot-air balloon. The illustration shows his design, in which hydrogen gas is contained within the spherical portion of the envelope and heated air is contained in the cylindrical portion with the brazier at its base. This is a brilliant idea, except for the fact that placing an open fire anywhere near hydrogen is asking for disaster. That is exactly what happened, and the brave and ingenious M. de Rozier along with his passenger Pierre Romain lost their lives while on an attempt to cross the English Channel in 1785.

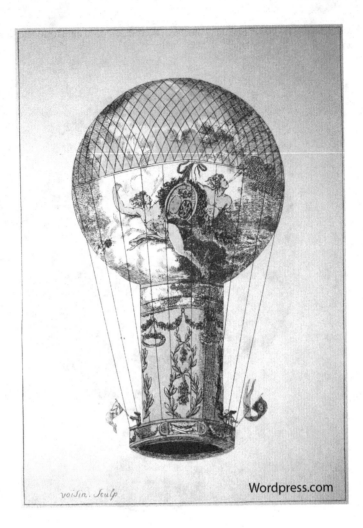

However, the idea was recently revisited for attempts on circumnavigation of the earth with human-carrying free balloons. In this case, helium is used and the heat source comes from propane burners. The picture shows the successful example designed and built by Cameron Balloons in the U.K., and piloted by Betrand Piccard and Brian Jones for their history-making flight in 1999.

Courtesy of Cameron Balloons

As with de Rozier's original design the upper portion is the gas balloon, with a smaller gas balloon on top which gives an upper conical shape from the external cover draped over it. The lower cone is the heated-air portion, with the pressurized gondola and propane tanks below. The overall inflated height of the balloon was 55 m (180 ft), and the duration of the circumnavigation was over 19 days. During that time the balloon reached an altitude of 11,373 m (37,313 ft) and the total distance traveled was 40,814 km (25,361 miles).

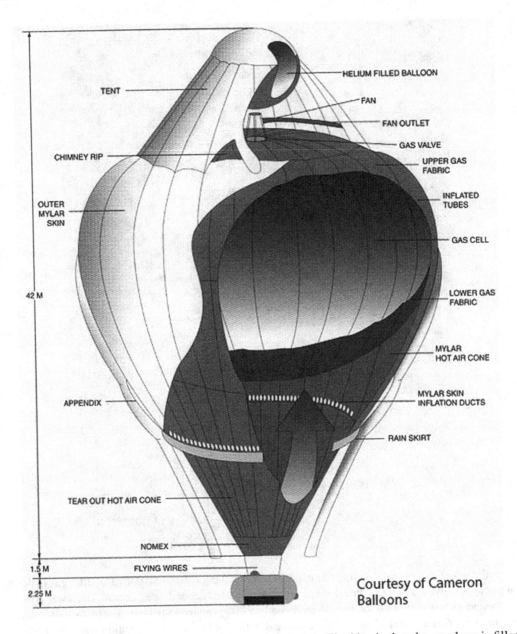

Another type of free balloon is the "super-pressure" design. The idea is that the envelope is filled with lifting gas under pressure and then completely sealed. The envelope has to be extraordinarily strong to withstand the stresses caused by rising to altitude because the lower atmospheric pressure causes the trapped gas to exert increasing internal pressure.

NASA/Columbia Scientific Balloon Facility

Solar heating likewise increases pressure, and the envelope's strength comes from its space-age material as well as the "pumpkin" gores of its shape. These billows reduce the circumferential stresses ("hoop stresses") as explained later. Because no gas is vented or ballast dropped, very long flights are possible (over 100 days), subject only to molecular gas migration through the material and environmental deterioration. It is envisioned that such balloons, well at the upper edge of the atmosphere, could provide relatively inexpensive platforms for telescopes and other scientific instruments. There is also consideration of these being used to explore planets with an atmosphere, like Mars.

THE PHYSICS OF BUOYANCY

Buoyancy is due to the variation of atmospheric pressure with altitude. As the figure shows, the atmospheric pressure integrates over the surface of the balloon to give the gross buoyancy, B_{gross}

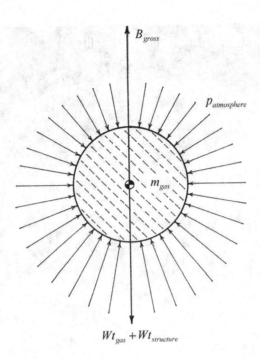

It is important to recognize that B_{gross} is independent of the balloon's weight and would be the same if the balloon was made from foam or lead. If one performed the math, it would be found that the integrated pressure force becomes the familiar Archimedean buoyancy equation:

$$B_{gross} = \rho_{atmosphere} \, g \, Vol_{moulded}$$

where $\rho_{atmosphere}$ is the atmospheric density, g is the gravitational constant and $Vol_{moulded}$ is the volume of air that the balloon displaces (the "Molded Volume", a name taken from naval architecture). The net buoyancy, B_{net}, is the upward force that is actually available to lift the payload, and is given by:

$$B_{net} = B_{gross} - Wt_{gas} - Wt_{structure}$$

where Wt_{gas} is the weight of the lifting gas, given by:

$$Wt_{gas} = \rho_{gas} \, g \, Vol_{gas}$$

and $Wt_{structure}$ is the weight of the envelope, ropes, etc. For normal thin-skinned balloons whose internal volume is completely filled with gas,

$$Vol_{gas} = Vol_{moulded}$$

The fact that the lifting gas density is much less than the atmospheric density is what allows a positive net buoyancy to be attained.

Note that this analysis assumes that $\rho_{atmosphere}$ and ρ_{gas} are constant over the height of the balloon. This is a fair assumption for most balloon sizes, even though the variation of atmospheric pressure with height is what produces buoyancy. However, the density variation might have to be considered in the buoyancy calculations for huge balloons.

ASIDE

It is interesting that gravitational force produces the vertical atmospheric pressure gradient that gives buoyancy. Theoretically, buoyancy cannot exist in a weightless condition, as on the International Space Station. However, in this case objects can "float" about without the benefit of buoyancy.

END ASIDE

TETHERED BALLOONS

One of the first practical uses for balloons were as elevated observation platforms for military surveillance. At the battle of Fleurus in 1794 a tethered gas balloon gave the French army valuable information about the Coalition Army led by Great Britain, which helped France achieve a crucial victory. However, this didn't become a regular component of the French army. The concept was tried again by the American Union Army during the Civil War, using a tethered-balloon system developed by Thaddeus Lowe. As was the case for the Fleurus balloon, it was tethered to the ground. This worked well in calm conditions; but any significant winds gave rise to large random lateral oscillations, which inhibited the utility of this as a stable elevated observation platform. Lowe tried a tripod of lines to help steady the system, but the oscillations persisted. This, plus the complex logistics of operating the balloon, caused the disbandment of the Union Balloon Corps despite occasionally producing useful information.

As mentioned before, a free balloon experiences negligible aerodynamic forces because it travels with the wind. The exceptional case is a wind-shear condition, which is a major factor when launching huge, fragile, scientific balloons that are most vulnerable in their tall partially inflated state. Otherwise, when the author took a flight in a free-flying hot-air balloon, the calmness in the basket was almost eerie compared with the way the balloon was drifting over the ground below. This is not the case, though, for a tethered balloon, which must be shaped to be stable in a wind and sufficiently streamlined to reduce downwind displacement.

An early attempt to achieve this is shown by the French Caquot observation balloon from World War I, where the envelope is elongated to reduce drag and fins are added for stabilization.

A forward-facing scoop was placed on the bottom of the lower fin to provide ram air from the wind to inflate all three fins (when the wind was calm, the fins drooped down). A telephone connection was run up the tether line so that the observer in the basket could communicate with the ground personnel. The mission was usually to provide aiming corrections for artillery bombardment, and the effectiveness of this was demonstrated by these balloons being prime targets for airplane attack. In fact, some airplanes were equipped with crude rockets for igniting the hydrogen lifting gas. In this case the observer had to make a quick exit, and parachutes were developed for this purpose. The German and British armies also had such balloons, and the British used a very similar design for anti-aircraft purposes in World War II.

Balloons and Airships

British Ministry of Defence

The idea was that the steel tether cable would damage low-flying enemy airplanes, thus inhibiting dive bombing and strafing. The picture shows a cluster of these balloons. The tether cables are not visible, but that was the idea. As to effectiveness, that is still being debated. However, they did provide a certain assurance to the public and even gave rise to the expression: "the balloon is up", meaning that a situation is about to become very serious or unpleasant (Cambridge Dictionaries Online).

In the 1970's the idea of tethered balloons was revisited for the purpose of telecommunications relay and surveillance.

Courtesy of TCOM, L.P.

In this case a fresh look was given to the technology, such as new materials as well as advanced aerodynamic and flight dynamic analyses. For example, the materials now used are typically Dacron/

Mylar/Tedlar laminates, as compared with cotton/Neoprene/silver paint laminates used for earlier tethered balloons. Also, an active blower-fan system, for both the hull and fins, gives consistent higher pressures allowing a much-improved aerodynamic shape to be attained (the fins keep their shape for all wind speeds). Therefore, a modern tethered balloon has much lower drag and better stability than earlier designs.

> **ASIDE**
>
> It is now common practice to refer to modern streamlined tethered balloons as "aerostats". Henceforth, this will be the terminology used. Also, all modern aerostats use non-flammable helium for a lifting gas. Historically, helium had only been available in the United States and was expensive. Therefore, European balloons and airships used hydrogen, with sometimes dire consequences.
>
> **END ASIDE**

A modern aerostat is pressurized through a ballonet, which is an envelope of air within the envelope of the hull.

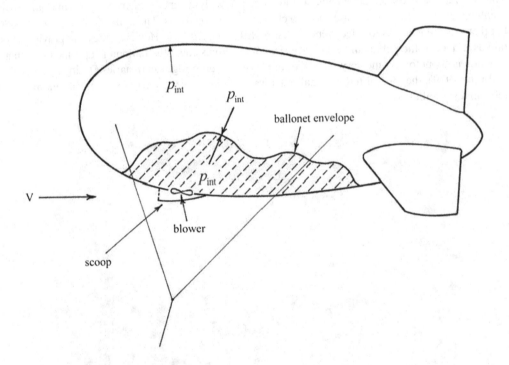

The ballonet is pressurized, from the atmosphere, with ram air and a blower fan as illustrated. This pressure is carried across the ballonet curtain to the main envelope. Note that because the pressures on both sides are equal, the curtain itself can be slack (un-stretched). This can give rise to an interesting phenomenon called ballonet slosh, where the ballonet air can move to-and-fro like water in the bottom of a rowboat. This is because of the different densities between the ballonet air and the lifting gas. Ballonet slosh can sometimes couple with the aerostat system's overall dynamic stability, and it is important to account for this.

Balloons and Airships

The idea behind a ballonet is that it can vary its volume to compensate for the expansion and contraction of the lifting gas while maintaining a consistent inner pressure, thus allowing the aerostat's shape to be retained.

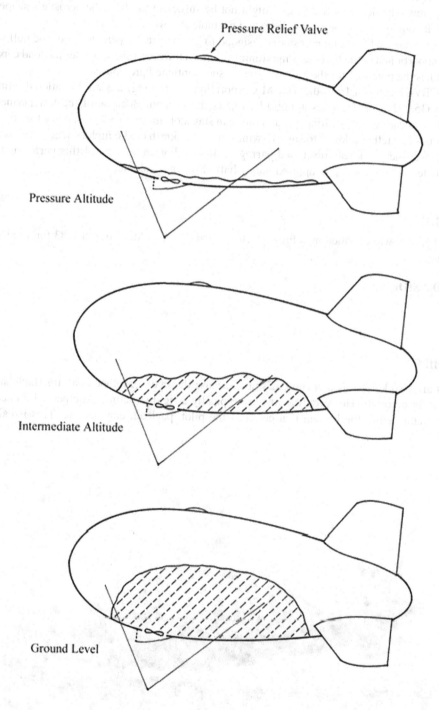

The illustration shows how the ballonet is nearly full at ground level. With increasing altitude, the lifting gas expands and the ballonet curtain collapses to compensate, pushing air out of the ram scoop. At "pressure altitude", the ballonet is fully collapsed and the curtain lies along the bottom of the main envelope. If the aerostat is allowed to rise above pressure altitude, the lifting gas continues to expand and must be vented through a relief valve to prevent the hull from bursting. However, if possible, flying above pressure altitude is undesirable because when the aerostat is brought back to ground level the depleted lifting gas might not be sufficient to allow the aerostat's shape to be retained. It could go slack, becoming limp and unstable at lower altitudes.

It is seen that the higher an aerostat is designed to go, the greater percentage of the hull volume must be apportioned to the ballonet's maximum volume. However, this decreases payload capability and also has the potential for the ballonet-slosh issue mentioned previously.

The CBV-71 aerostat from the TCOM Corporation (71 m long) has an operational altitude of 4,600 m (15,092 ft) and carries a 1,600 kg (3,527 lb) payload. Because of the low permeability of its fabric, it can retain its lifting gas and thus can stay aloft for up to 30 days. Also, the strength of its materials and tether allow it to survive winds up to 167 km/h (103.6 mph). Clearly, this is a great advance over the previously discussed barrage balloons. For an aerostat of this performance, the fully inflated ballonet occupies approximately half of the hull volume.

ASIDE

The ballonet was invented by a French officer, Jean Baptiste Meusnier in 1783 for an airship design.

END ASIDE

AIRSHIPS

Blimps are also known as non-rigid airships because the main envelope is an inflatable structure like that for aerostats. However, the fins are typically a rigid structure. Another difference is the structure attached to the bottom that contains the pilot, payload, engines, etc. This is called the gondola.

Allan Judd, Wikipedia Commons

Balloons and Airships

Blimps also use ballonets to retain their internal pressure with altitude. Typically, two are used to give the capability for static trim, as shown below in the figure adapted from "Blimp!" by George Hall, Baron Wolman and George Larson (Van Nostrand Reinhold, 1981, now owned by Wiley).

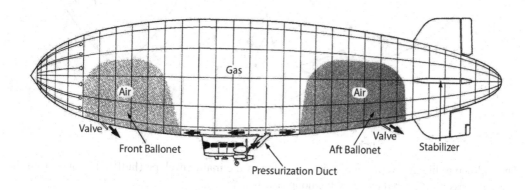

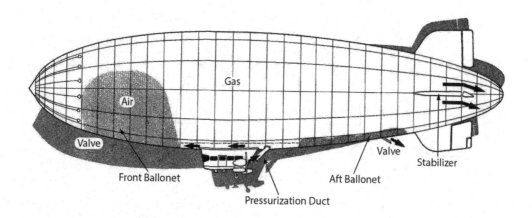

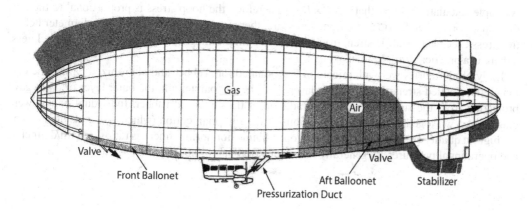

Observe that the blimp's propellers do double duty as pressurization blowers.

At this point it is useful to describe hoop stress, S, which is a crucial loading for inflatable structures like aerostats and blimps, and which can limit their size.

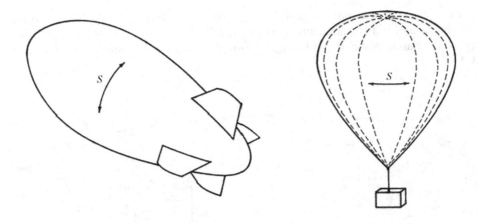

The following diagram represents a cross section of the main envelope (hull). The force from the internal pressure, P, is balanced by the hoop stresses, S.

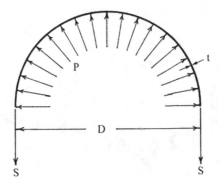

A simple calculation shows that $S = P \times D/2$. Therefore, the hoop stress is proportional to the hull diameter. For a given fabric and internal pressure, there is a clear upper limit to the diameter before the stress exceeds the fabric strength. Although one might imagine increasing the fabric thickness, t, there is a practical limit to this.

Early blimp fabrics were a laminate of Egyptian cotton, rubber and aluminized paint. The cotton cloth provided strength, the rubber layer gave the gas barrier and the outer layer of paint gave protection from the sun's ultraviolet degradation of the cloth and rubber. Individual panels were glued and sewn together. Strength limitations of the Egyptian cotton (which was considered to be the highest quality cotton) limited the envelope volume, for normal fineness ratios (length/diameter) and operational pressures, to generally less than 4,000 m³ (140,000 ft³).

In the late 1940s blimp fabric was improved to use Dacron instead of cotton, with a Neoprene gas barrier. The Dacron's added strength allowed larger volumes to be attained, such as seen with the U.S. Navy's ZPG-3W blimp with a 41,300 m³ hull volume (1,465,000 ft³). An added benefit was that Dacron, being a non-organic material, was less susceptible to mold from operation or storage in humid conditions (a common problem with blimps, as well as bagpipes). Neoprene was also more durable than the early natural rubber layers. Numerous successful blimps have been made from this fabric.

Another and early, solution to the blimp's hull-volume limitation was provided by the invention of "rigid" airships.

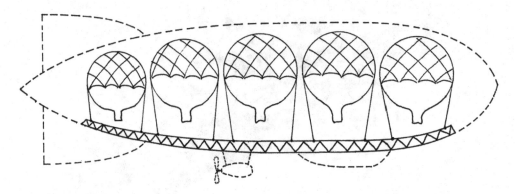

The drawing illustrates the idea behind this. Namely, a series of free balloons are tethered to the inside of a lightweight, rigid, streamlined structure. Because these balloons have very little pressure and are protected from the environment, their fabric can be very lightweight. The structure provides support for the balloons, gondola, engines and fins, as well as giving a low-drag and stable configuration. As previously described for free balloons, buoyancy control comes from venting and ballast.

A more realistic diagram of a rigid airship is shown below, for the 1931 U.S. Navy ZRS-4 "Akron" (credit, U.S. Navy archives):

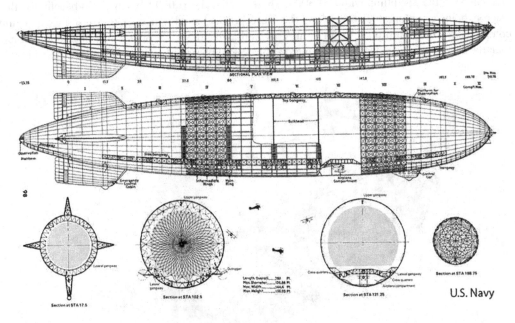

It can be seen that the structure consists of rings and girders, which were fabricated from a special light and strong alloy of aluminum called "duralumin". The gas cells were located between the deep rings and took on a cylindrical shape when fully inflated. Also, the gas cells were separated by heavy netting to prevent rubbing and wear between them. Care was also taken, in the design, to prevent the cells from chaffing on the frames.

The gas volume of the Akron was 184,000 m³ (6,500,000 ft³) and its design was typical for rigid airships of its time: duralumin frame covered with aircraft linen and painted silver (aluminized paint) for solar reflectance.

European airships used hydrogen, which was inexpensive but flammable. U.S. airships used helium, which was expensive but inert. European airships could afford to vent hydrogen for buoyancy control as the fuel weight was consumed, but the Akron (and its sister ship "Macon") used condensers along the sides (dark squares in the photo) to recover moisture from the engines' exhaust. This water recovery would, to a certain extent, offset the weight of the fuel being consumed. This reduced the requirement for venting the costly and rare helium gas.

Although the great rigid airships looked massive and substantial, it has to be remembered that they primarily consisted of a relatively fragile light-metal open-frame structure. These airships could be destroyed by strong winds and mishandling. A picture is shown, that is also presented in "Airshipwreck" by Len Deighton and Arnold Schwartzman (Jonathan Cape Ltd., 1978), of the 1911 Zeppelin LZ 8 wrapped around its hangar. That a whole book could be devoted to airship wrecks speaks to the troubled life spans of these huge aircraft.

A 1908 British airship named "Mayfly" was destroyed before its first flight, thus living up to its unfortunate name in more ways than one. Military airships were even more vulnerable than commercial craft because they were expected to be operational in all conditions.

The greatest success was enjoyed by the Zeppelin Corporation, who maintained flexible schedules and assiduously avoided bad weather. The most successful historic rigid airship was the 1928 LZ 127 "Graf Zeppelin", which flew safely for over 1,600,000 km (1,000,000 miles) in its career, before being scrapped for its metal in World War II.

Grombo, Wikipedia Commons

However, the story would not be complete without mentioning the 1936 LZ 129 "Hindenburg". The flaming destruction of this great airship is now an iconic finale to the end of the age of rigid airships. The sad story is that the Hindenburg had actually been designed to use helium. An arrangement had been made with the United States to purchase this; but because of concern with the Nazi regime, the Secretary of The Interior, Harold Ickes, cancelled the order. However, because the Zeppelin Corporation had experience with using hydrogen (all previous Zeppelin airships had used hydrogen, including the highly successful Graf Zeppelin), they decided that their usual strict protocol would be acceptable for the Hindenburg. Perhaps that would have been true if something extraordinary and tragic hadn't happened during its mooring at Lakehurst New Jersey on 6 May 1937. The cause is still a mystery as to whether it was sabotage or some sort of natural phenomenon like a static-electric spark.

Another approach to overcoming the hoop-stress limitation was the "semi-rigid" design. This concept consists of a rigid-structure keel, to which an inflated envelope is attached.

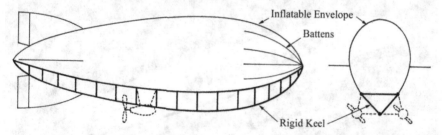

The idea is that the keel takes up a major portion of the imposed loads, so that the envelope does not have to be inflated to as high a pressure as that required to stiffen non-rigid airships. Thus, hoop stresses are reduced. However, the nose has to be reinforced with battens to prevent it from being caved in by air pressure at the airship's higher speeds.

One of the most successful designers of semi-rigid airships was Italian General Umberto Nobile, and the picture shows the "Norge", which was the first aircraft to fly over the North Pole (1926). The Norge had a volume of 19,000 m^3 (670,980 ft^3), which was much larger than that for contemporary blimps. However, this is still much less than that for the great rigid airships. Even though hoop stresses were reduced, they were still a limiting factor for the inflated envelope.

Aeronautica Militare, Wikipedia Commons

General Nobile designed several semi-rigid airships including the 1921 "Roma", which crashed into high-voltage lines igniting its hydrogen lifting gas, and the 1928 "Italia", which crashed in icy conditions during another polar flight. The ordeal of the survivors is the subject of an excellent 1969 movie called "The Red Tent".

A modern, and very successful, form of semi-rigid airship has been developed by Zeppelin Luftschifftechnik, a corporate descendent of the original Zeppelin company. This is their first airship since the 1938 LZ 130 "Graf Zeppelin II".

AngMoKio, Wikipedia Commons

The diagram below shows that the internal structure is a triangular cross-section beam, which is attached along three lengthwise seams of the envelope. This arrangement gives rise to a slightly three-lobed configuration when inflated. Also, the structure gives support to the fins, thrusters and gondola.

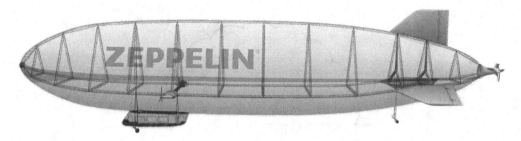

Courtesy of ZLT Zeppelin Luftschifftechnik

This airship design has a maximum lifting-gas volume of 8,225 m^3 (290,463 ft^3) and has been flying since 1998. Several models have been built and sold; and the customers include the Goodyear Corporation.

Another approach for dealing with hoop-stress limitations was the 1929 all-metal ZMC-2 airship. The envelope consisted of 2 mm Duralumin sheets attached together with a type of metal wire "stitching". A bituminous compound was used to provide gas sealing where the sheets overlapped to form a seam. Ballonets were used to provide pressure, just like in a blimp. However, because of internal metal stringers, the metal shell could hold its shape even when deflated. The envelope was constructed on two wooden mandrels for the fore and aft sections, and then joined together. Although the fineness ratio (length/maximum diameter) was low, the eight fins reportedly gave excellent stability and control. This innovative airship had an internal envelope volume of 5,663 m^3 (200,000 ft^3), and this metal-skin concept had the potential for much larger airships because the strength/weight ratio of Duralumin was better than that for the blimp fabrics of that time.

U.S. Navy

The ZMC-2 was a "proof-of-concept" model, which role it certainly fulfilled because it flew successfully for 12 years. It was finally scrapped in 1941 to recover its metal for wartime airplane construction.

Note that a fabric envelope is constructed from longitudinal "gores" attached to each other. With a sufficient number of gores, when the envelope is inflated the compliance of the fabric gives a smooth hull shape. This offers another way to deal with hoop stress, where the longitudinal compliance might be constrained along the gore seams with longitudinal tension lines, so that the fabric is free to billow in-between these lines.

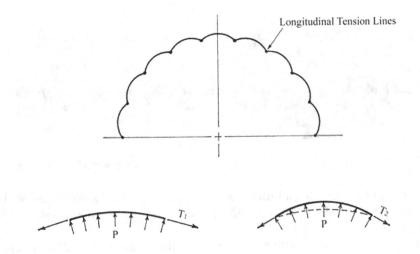

This was seen in the previously discussed super-pressure free balloons (where the tension lines are vertical in that case), and the same idea could be used for non-rigid airships. As can be seen from the diagrams, the curvature of the billows decreases the local hoop stress from T_1 to T_2. The longitudinal tension lines are crucial, supporting the hoop stresses, T, along their length. For the natural-shape free balloon, the tension lines form a type of catenary shape. However, in practice, this feature has not often been used for blimp designs. The most the author has seen is three-lobe (three-billow) configurations. It could be that the fabric is typically too stiff to produce the sort of multiple billows desired for the streamlined non-catenary shape of a typical airship.

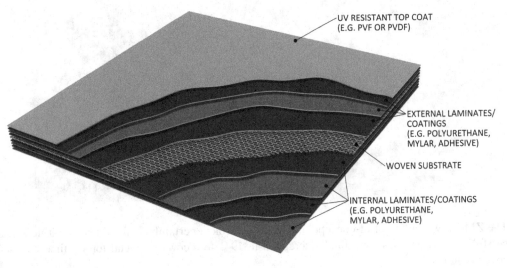

Courtesy of Steve Wallace

Modern laminated fabrics now have sufficient strength that non-rigid airships can be made as large as the great rigid airships of the past. As the figure shows, the load-carrying portion of the fabric is Dacron or Kevlar; and the gas barrier is Mylar film. Also, protection against environmental degradation (e.g. ultraviolet radiation from the sun) is provided by an outer layer of

Tedlar. There are variations of this concept for modern aerostats and airships, but this example is typical.

Courtesy of Cargolifter, GmbH & Co.

A major German project of the late 20th century was the proposed CargoLifter airship. This was a non-rigid design with a maximum lifting-gas volume of 560,000 m³ (19,800,000 ft³), which would have been three times the volume of the LZ-129 "Hindenburg". The project concluded without the airship being built. There were a variety of reasons for this, but from an engineering standpoint the design was running into the issue of "gigantism". What that means is that a design solution that makes sense at one scale might not work well when the scale is greatly increased. One of these challenges involved altitude control. Most airships do not fly with neutral buoyancy, but instead fly a bit "heavy". This is so that, with engine failure, the airship can "glide" to earth without having to valve lifting gas. The extra lift needed to maintain altitude is provided by flying the airship at a small positive angle of attack. As will be seen in the "Aerodynamics of Finned Axisymmetric Bodies" section, this lift force is a direct function of the airship's characteristic area, S. Since S is proportional to the square of the airship's length, L, one may say that the aerodynamic lift force at a given speed and atmospheric density is proportional to L^2. However, the buoyant lift is a direct function of the volume, Vol, which is proportional to the cube of the length, L^3. For normal-sized airships, variations in net buoyancy due to solar heating, fuel burn, etc. are readily compensated for by adjusting the airship's aerodynamic lift with the control surfaces. However, these same-percentage buoyancy variations for huge airships can swamp the available aerodynamic control. This is because, with an increasing value of L, the buoyancy force perturbation varies at a much greater magnitude (L^3) than the available aerodynamic control can manage (L^2), at a given speed and atmospheric density. The solution is to fly faster to increase the aerodynamic control force. There is definitely a limit on this for an airship. Also, it might not be economical or safe to always fly at top speed.

A solution to this might be to design an envelope that augments the aerodynamic force. Such a possible configuration was designed by the author and test flown in 2010. As seen, the envelope is "flattened" into a buoyant flying-wing. As well as flight tests, wind-tunnel and theoretical studies have been conducted that show promise for this idea as a large cargo-carrying aircraft. The photograph shows a scaled proof-of-concept single-seat design.

Another issue with gigantism pertains to the ballonets. As the height of the ballonet increases, increasing blower power is needed to lift the weight of the air within the ballonet. Recall that the ballonet air is surrounded by the lighter lifting gas, so this is analogous to pumping up a columnar bag of water surrounded by the atmosphere. The higher the bag, the more pressure (and power) is needed. The buoyant flying-wing design, though, uses tubular ballonets running span-wise within the envelope. Therefore height (and blower power) is less of an issue.

As mentioned before, the dramatic 1937 destruction of the Hindenburg ended interest in the great rigid airships for virtually everyone everywhere. However, U.S. Navy blimps performed valuable service for anti-submarine patrol during World War II.

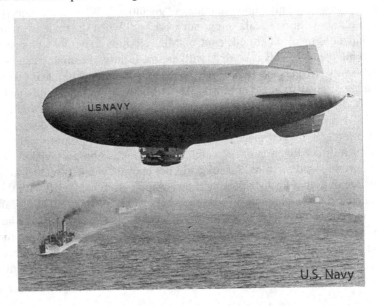

Their slow pace allowed them to escort convoys of ships, and they had durations that lasted for days. This is a unique feature of airships: because lift is primarily provided by buoyancy, engine power is thus primarily used for forward motion over a wide range of speeds (including close to zero). This is in contrast to airplanes where the engine's power is continuously applied to provide the speed required for aerodynamic lift to sustain the aircraft. These are the reasons why an airship has the potential for very long durations as well as loitering capability. Such desirable features, as well as energy efficiency, continue to motivate considerations of modern blimps for surveillance missions.

One consistent application, though, has been the presence of the blimp itself. These aircraft are unique enough that they have considerable "visibility" to the general public. This makes them a natural billboard for advertising purposes. The photo shows the "Snoopy Two" blimp advertising life insurance.

Matthew Field, Wikipedia Commons

These commercial blimps are typically small, with a volume of 6,000 m³ (approx. 200,000 ft³), though the larger Zeppelin NT has been used by the Goodyear Corporation for advertisement at sporting events.

This is a profitable business, and several manufacturers have been involved with building such aircraft. One of the most notable is The American Blimp Corporation in Hillsboro Oregon. Their designs can have a feature where a powerful internal light illuminates the envelope at night, giving an attention-getting "Japanese-lantern" effect.

Courtesy of Hokan Colting

Another noticeable design is the spherical shape from 21st Century Airships in Newmarket Ontario, Canada. As can be seen, both propulsion and altitude control are provided by vectored thrust. This has a very eye-catching futuristic design (which is the idea), and one vehicle was decorated as a baseball for the World Series. However, such a configuration is not exactly low drag, and operation is limited to relatively low-wind conditions.

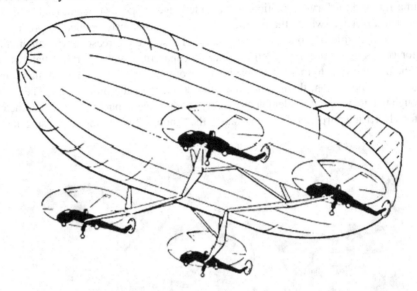

Another persistent interest of modern airship design is the Heavy-Lift Airship (HLA) concept. The idea (and promise) is simple to understand. Consider that for helicopters, the rotor lift is proportional to the square of the rotor diameter whereas the structural weight of a helicopter is approximately proportional to the cube of the rotor's diameter. This means that, as the helicopter gets larger, an increasing proportion of the rotor's lift is devoted to just lifting the helicopter's structure itself. Therefore, a smaller percentage is available for payload. This can be referred to as the "square-cube law", and is illustrated by the first graph below.

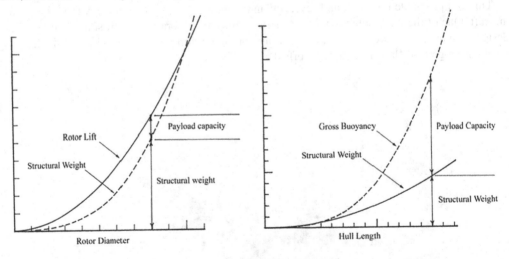

The structural-weight curve follows the cubic law, and the rotor lift follows the square law. At the mid ranges of rotor diameter there is an optimum payload capability. However, as size increases a crossover point is reached where the aircraft can only stay aloft with no payload capability at all. The Sikorsky "Skycrane" is near the optimum point. It has a rotor diameter of 22 m (72 ft) and a payload capacity of 10,382 kg (22,840 lb).

By comparison, an airship is subject to the "cube-square law" in that the volume varies as the cube of the hull length and the structural weight varies approximately as the square of the length.

Balloons and Airships

As the second graph shows, the payload capacity increases dramatically with the airship's size. The idea of an HLA is to use buoyancy to augment the lifting capability of the rotors. In other words, the buoyant lift will support the structural weight of the airship and helicopter units, thus freeing the rotor lift for the payload. The exciting promise is that an HLA could carry payloads an order of magnitude greater than those of any possible heavy-lift helicopter.

However, despite this appealing theoretical possibility for an HLA, there are several important design issues to be addressed. The first is altitude control while flying, as had been discussed before. This may limit the ultimate size for these types of aircraft. Another issue is the power required for huge ballonets, as mentioned previously (the engineers at CargoLifter calculated that ballonet-blower power would be equivalent to propulsion power). However, a particularly important issue is controllability while hovering. Many proponents of the HLA envision the aircraft hovering over a specific location while lowering or raising the payload. This always looks great on artists' conceptions, but the reality is that such huge aircraft would be very difficult to precisely control. This was pointed out by the late Professor Howard "Pat" Curtis of Princeton when he identified a measure of controllability as the ratio of available control forces to the inertial-reaction forces. By this measure, the controllability of a helicopter is very good. However, if you add the inertia (and apparent inertia) of the airship portion, then the hover capability becomes much less, even with dedicated control thrusters. In fact, for the huge HLA's envisioned, their ability to station-keep would be akin to that for a super tanker. In a more colorful way, this author has also stated that station-keeping control of a huge HLA would be akin to controlling a cow by poking it with your finger. Until this challenge is solved, the dream of a precisely hovering HLA is unlikely to be fulfilled.

Despite these caveats, it is clear from this chapter that modern airships, with their new materials and computer-optimized designs, have considerable promise for applications beyond being displayed for their novelty (i.e., advertising). Heavy-Lift Airships have the potential for transporting outsized payloads an order of magnitude greater than any other type of aircraft. If some form of reliable station-keeping control can be developed (e.g., external mooring lines), then these payloads can be picked up and lowered from site to site. Also, modern blimps continue to be studied for long-range long-endurance surveillance patrols. Balloons, aerostats, blimps and airships have a long history and are by no means of obsolete.

AERODYNAMICS OF FINNED AXISYMMETRIC BODIES

Since the great majority of aerostat and airship designs consist of finned axisymmetric bodies, it is appropriate to discuss the aerodynamic predictions for such configurations. The following two illustrations come from "Aerodynamic Considerations of LTA Vehicles" by William F. Putman (AIAA Paper No. 77-1176), which show the lift distribution on a slender axisymmetric body at a positive angle of attack.

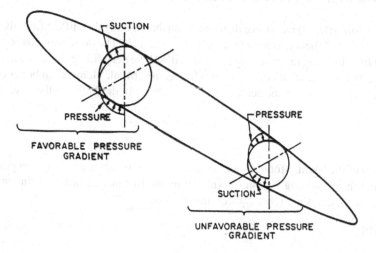

The first picture shows circumferential pressure distributions on the fore-body and aft-body portions of the hull, where there are suction pressures on the top of the fore-body and positive pressures on the underside. This integrates to give a positive lifting force on the fore-body, and the reverse is true on the aft-body. The distribution of this force along the axial (x) coordinate of the body, dF/dx, is illustrated in the next figure for a body at an angle of attack. The term "q" is the dynamic pressure, $0.5\rho V^2$. This figure also illustrates the results from two theoretical methods for calculating this, along with experimental results.

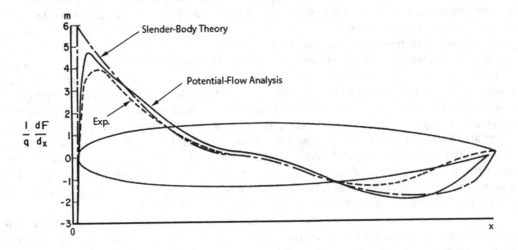

The first theoretical method is known as "Slender-Body Theory", which gives that:

$$\frac{1}{q}\frac{dF}{dx} = \frac{dS}{dx}(k_3 - k_1)\sin(2\alpha)$$

where S is the cross-sectional area, k_1, k_3 are apparent-mass coefficients defined later in this chapter, and α is the angle of attack. Notice that when the cross-sectional area is increasing with axial distance from the nose, the force is positive; and the reverse is true for the aft-body.

The next theoretical method involves a more complete potential-flow analysis using source-sink and doublet distributions along the axis. Since both theoretical methods assume potential flow, no viscous effects are accounted for and thus there is no flow separation. If dF is integrated along the length, it will be found that F is zero and the body experiences a pure couple (moment).

In reality, viscous effects give a force distribution on the aft-body that differs from that predicted by cross-flow and potential theory, as seen in the experimental plot. The body will thus produce a certain amount of net lift. However, the upsetting moment will still be very strong. In the author's experience, most aerostat and airship analysts refer to that effect as the "Munk Moment" in honor of Max Munk, who did pioneering research on the aerodynamics of airships in the 1920's at the NACA.

ASIDE

The strength of the Munk Moment can be demonstrated by filling an elongated party balloon with air, and then dropping it point down. Within the distance of one balloon length, it will turn sideways, which is its stable equilibrium position.

END ASIDE

Fins are added to airships to counter the Munk Moment. However, fin sizes are constrained by weight and balance considerations, as well as by the simple fact that the airship has to fit into the available hangar space. Therefore, for all conventional airships, the fins cannot completely compensate for the strong Munk Moment. This means that the center of pressure is forward of the mass center, and such a condition is called "static instability". This condition can be so extreme that it results in "dynamic instability", which means that the airship has to be continually controlled to stay stable. This was very tiring for the helmsmen of the great airships of the past, who had to be relieved on a regular basis. More advanced fin designs on the U.S. Navy blimps made them more manageable, but they were still unstable aircraft. In truth, all airships that this author is aware of are statically unstable. The picture of the TCOM aerostat shows the huge fin sizes required to achieve static stability. The dynamic-stability requirements for a tethered aerostat are much more exacting than those for a freely flying airship, thus requiring the large fins shown. Compare these with the fins in the previous pictures of blimps and airships.

Courtesy of TCOM, L.P.

However, as previously discussed in the "Longitudinal Stability" section of Chapter 6, an aircraft whose average density is of the same order of magnitude as that of the air it is flying through can be *dynamically* stable even though it is *statically* unstable, provided the fins are adequately sized. This also applies to lateral stability.

Luftschiffeseiten.de, Wikipedia Commons

The photo shows the Airship Industries "Skyship 600". The author had the opportunity to be at the controls of two of these aircraft. It was laterally dynamically unstable and required constant control to follow a linear flight path. Because response times are so long, this was not a problem; and many examples of this aircraft have performed well over thousands of miles for many years. Of course, being at the controls, this author experimented with this airship's maneuverability and stability. It was found that if the rudder control was held neutral, the airship would eventually diverge left or right and go into a stable circular flight path. This has been referred to as the "model-train mode", for obvious reasons. By comparison, the author also had time at the controls of the American Blimp Corporation's A-60 design (previously illustrated), and found this to be just on the positive side of dynamically stable: if the rudder control was held neutral, the airship would continue on a linear flight path for a significant length of time.

A Method for Calculating the Longitudinal Static Aerodynamic Coefficients

This is a methodology developed by Samuel P. Jones of TCOM, LP and the author ("Aerodynamic Estimation Techniques for Aerostats and Airships", AIAA Journal of Aircraft, Vol. 20, No. 2, February 1983), and it has been successfully applied to a large variety of airships and tethered aerostats. The theoretical foundation is based on slender-body cross-flow aerodynamics, which assumes that the local aerodynamic force on any cross section is due only to that segment's geometry and the local incident flow. This concept was first formulated by the previously mentioned Max Munk, and was elaborated upon by H.J. Allen and E.W. Perkins in NACA Report 1048, "A Study of Viscosity on Flow over Slender Inclined Bodies of Revolution" (1951). Further, the cross-flow assumption can apply not only to the body, but to the finned portion if the fins are of very low aspect ratio (e.g. slender delta planforms). This concept is exactly correct in the limit of infinite slenderness, but it is still accurate for a typical missile configuration.

Obviously, an airship is far from slender and it might be assumed that slender-body theory is of limited usefulness for such shapes. However, it turns out that the concept is, indeed, applicable if appropriate mutual-interference factors are included. Comparisons with experimental data have been very favorable, even for such low fineness-ratio shapes as tethered aerostats. This has made the aerodynamic estimations for airships with axisymmetric hulls a very straightforward procedure. Another benefit is that the cross-flow model lends itself to easy calculation of atmospheric turbulence-profile effects and structural loading.

The assumed configuration is an axisymmetric body with cruciform fins or fins configured in a "Y" (or "Inverted Y") geometry. The previous photo of a TCOM aerostat shows an example of an Inverted-Y configuration, whereas the earlier photo of the Akron airship shows the cruciform arrangement.

Also, note that the word "fins" is generally used to describe the empennage on an airship. However, from this point on, the term "fin" will refer to the vertical component of the empennage, and "stabilizer" will refer to the component of the empennage that gives a vertical force. In other words, this is exactly the terminology used to describe the tail components of an airplane, as used in Chapter 6.

The figure below illustrates the longitudinal force components on an airship at an angle-of-attack, α:

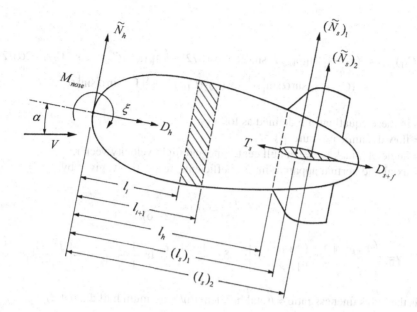

Observe that the forces are in the body-axis system: Normal Forces and Axial Forces. The Normal-Force equation is:

$$\tilde{N} = q C_{\tilde{N}} S$$

where:

$$C_{\tilde{N}} = (k_3 - k_1)(\eta_k)_{\text{long}} \hat{I}_1 \sin(2\alpha)\cos(\alpha/2) + (C_{Dc})_h \sin\alpha \sin|\alpha| \hat{J}_1$$
$$+ \left[(C^*_{\tilde{N}\alpha})_s \eta_s \cos^2 \Gamma_s \sin(2\alpha)/2 + (C_{Dc})_s \cos^3|\Gamma_s| \sin\alpha \sin|\alpha| \right] \hat{S}_s$$

A tilde is placed over the normal force term, $\tilde{N}$, to distinguish it from the yawing-moment term, N, used in the lateral equations described later.

The axial force equation is:

$$D_{\text{axial}} = q(C_D)_{\text{axial}} S$$

where:

$$(C_D)_{\text{axial}} = \left[(C_{Dh})_0 \hat{S}_h + (C_{Ds})_0 \hat{S}_s + (C_{Df})_0 \hat{S}_f \right] \cos^2\alpha - (k_3 - k_1)(\eta_k)_{\text{long}} \hat{I}_1 \sin(2\alpha)\sin(\alpha/2) - C_{Ts} \hat{S}_s$$

The pitching moment is defined to be about the nose (the "soft nose" in airship terminology, i.e., the nose-cone mooring structure is not included). The equation is:

$$M_{\text{nose}} = q C_{M\text{nose}} S \bar{c}$$

where:

$$(C_M)_{nose} = -(k_3 - k_1)(\eta_k)_{long} \hat{I}_3 \sin(2\alpha)\cos(\alpha/2) - \hat{S}_s \eta_s (\hat{l}_s)_1 (C^*_{N\alpha})_s \cos^2 \Gamma_s \sin(2\alpha)/2$$
$$- (C_{Dc})_h \hat{J}_2 \sin\alpha \sin|\alpha| - \hat{S}_s (\hat{l}_s)_2 (C_{Dc})_s \cos^3|\Gamma_s||\sin\alpha \sin|\alpha|$$

The terms in these equations are defined as follows:
  q = flow dynamic pressure, $\rho V^2/2$
  α = angle of attack between hull centerline and flight velocity vector,
  k_1, k_3 = axial and normal apparent-mass coefficients, respectively, given by:

$$k_1 = \frac{\gamma}{(2-\gamma)}; \quad k_3 = \frac{\delta}{(2-\delta)}$$

$$\gamma = 2\left(\frac{1-e^2}{e^3}\right)\left[\frac{1}{2}\ln\left(\frac{1+e}{1-e}\right)-e\right]; \quad \delta = \frac{1}{e^2} - \frac{1-e^2}{2e^3}\ln\left(\frac{1+e}{1-e}\right); \quad e = \left(\frac{f^2-1}{f^2}\right)^{1/2}$$

where f is the hull's fineness ratio = (total hull length)/(maximum hull diameter).

ASIDE

These apparent-mass equations are exact potential-flow solutions for an ellipsoid of revolution, as derived in Chapter 7 of "Aerodynamic Theory" edited by W.F. Durand (Julius Springer, publisher, 1934). Airship hulls are rarely exact ellipsoids of revolution, but these equations have sufficed for this application.

END ASIDE

$(\eta_k)_{long}$ is the longitudinal hull-efficiency factor accounting for the aerodynamic effect of the stabilizer on the hull (a correction to the slender-body model). This is a function of the ratio of the total stabilizer area, S_s, to the hull's complete horizontal projected area, $(J_1)_{total}$ (which may also be interpreted as the maximum horizontal *shadow area* of the hull).

S_s is defined in the following figure. Observe that this is the "flattened" area. The effect of dihedral (or inverted dihedral, "anhedral"), Γ_s, is accounted for in the equations.

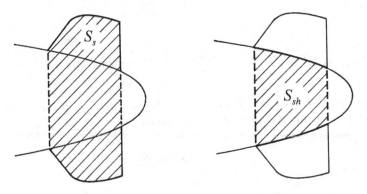

And the relationship between $(\eta_k)_{long}$ and $S_s/(J_1)_{total}$ is given by the following graph:

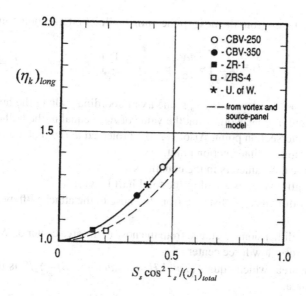

Two curves are shown: one based on wind-tunnel data and one derived from a vortex and source-panel model of airship configurations ("An Application of Source-Panel and Vortex Methods for Aerodynamic Solution of Airship Configurations" by K.Y. Wong, L. Zhiyung and J. DeLaurier, AIAA Technical Paper 85-0874). Both curves are fairly close and one can approximate the function by:

$$(\eta_k)_{long} = 1.0 + 1.40 \left[S_s \cos^2 \Gamma_s / (J_1)_{total} \right]^2$$

where, as stated before, $(J_1)_{total}$ is the hull's complete horizontal projected "shadow area", and Γ_s is the stabilizer's dihedral angle.

In a similar fashion η_s is the stabilizer's efficiency factor, which accounts for the aerodynamic effect of the hull on the stabilizer. It has been found that this can be simply represented as a function of S_{sh}/S_s (see earlier Figure for the definition of S_{sh}):

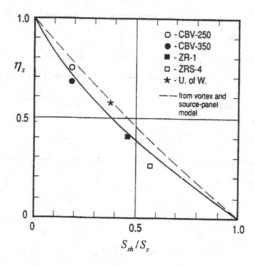

Again, a reasonable approximate equation for this is:

$$\eta_s = 1 - 1.4(S_{sh}/S_s) + 0.4\left(S_{sh}/S_s\right)^2$$

The $\hat{I}$ and $\hat{J}$ integrals, taken from the nose to the hull/stabilizer intersection point (0 to l_h), are:

$$\hat{I}_1 = \frac{1}{S}\int_0^{l_h}\frac{dA}{d\xi}d\xi = \frac{A_h}{S}; \quad \hat{I}_3 = \frac{1}{S\bar{c}}\int_0^{l_h}\xi\frac{dA}{d\xi}d\xi; \quad \hat{J}_1 = \frac{1}{S}\int_0^{l_h}2r\,d\xi = \frac{J_1}{S}; \quad \hat{J}_2 = \frac{1}{S\bar{c}}\int_0^{l_h}2r\xi\,d\xi$$

where A is the hull cross-sectional area, ξ is the axial coordinate along the hull's centerline from the nose and r is the hull radius. Notice that the value of A_h is equal to the hull cross-sectional area at the hull/stabilizer intersection point. Also, J_1 is the projected horizontal area of the hull (*shadow area*) up to the hull/stabilizer intersection point.

The other geometrical parameters in the equations are:
$\bar{c}$ is the reference length, which is equal to the total hull length.

$(\hat{l}_s)_1 = (l_s)_1/\bar{c}$, non-dimensional distance from the nose to the attached-flow aerodynamic center of the stabilizer.

$(\hat{l}_s)_2 = (l_s)_2/\bar{c}$, non-dimensional distance from the nose to the area center of $(S_s - S_{sh})$, which is the exposed stabilizer's cross-flow force center.

S is the reference area, which equals $(Total\ Hull\ Volume)^{2/3}$; $\hat{S}_s = S_s/S$ is the non-dimensional stabilizer reference area.

$\hat{S}_f = S_f/S$ is the non-dimensional vertical-fin area.

$\hat{S}_h = S_h/S$ is the non-dimensional hull reference area (depends on the reference area chosen for $(C_{Dh})_0$).

Γ_s is the stabilizer's dihedral angle (this is negative for anhedral angles, such as for the two bottom fins on the TCOM aerostat).

The aerodynamic parameters are:
$(C_{Dc})_h$ = hull cross-flow drag coefficient, referenced to J_1. This value is estimated from:

$$(C_{Dc})_h = (C_{Dc})_{cylinder}\,\gamma_{fineness}$$

where $(C_{Dc})_{cylinder}$ is the cross-flow drag of an infinitely long circular cylinder, which is ≈ 0.29 for a cross-flow Reynolds number greater than $\approx 3\times 10^5$. The correction factor for a finite hull fineness ratio, $\gamma_{fineness}$, is obtained from A.B. Wardlaw in "High-Angle-of-Attack Missile Aerodynamics", AGARD Lecture Series 98, Feb. 1979, pp. (5-1) to (5-53):

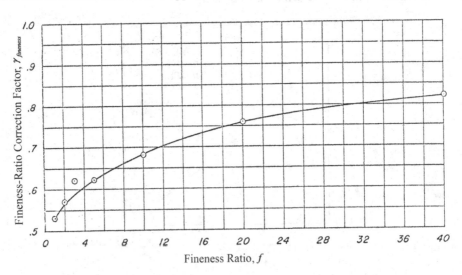

For a fineness ratio up to 10, $\gamma_{fineness}$ may be approximated by the equation:

$$\gamma_{fineness} = 0.5088 + 0.02750\,f - 0.001011\,f^2, f \leq 10$$

$(C_{Dc})_s$ = stabilizer cross-flow drag coefficient, referenced to S_s. From Wardlaw, this value is approximately equal to 2.0 for most typical airship and aerostat fins. However, a more exact value may be found from analytical curve fits of Wardlaw's cross-flow plots for straight-tapered missile fins (limited to $0 \leq AR \leq 3$):

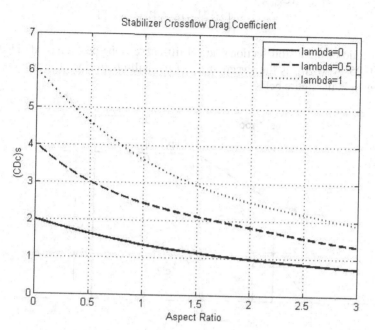

$$(C_{Dc})_s = A_0 + A_1 \times AR + A_2 \times (AR)^2 + A_3 \times (AR)^3 + A_4 \times (AR)^4$$

where:

$$A_0 = 2.0001 + 4.0002\lambda; \quad A_1 = -0.8667 - 4.5383\lambda + 2.177\lambda^2$$

$$A_2 = 0.2373 + 4.1957\lambda - 3.4854\lambda^2; \quad A_3 = -0.0339 - 1.6913\lambda + 1.6298\lambda^2$$

$$A_4 = 0.0012 + 0.2316\lambda - 0.2376\lambda^2$$

AR is the stabilizer's aspect ratio, defined by $AR = b_s^2/S_s$, and λ is the taper ratio, defined by $\lambda = c_{tip}/c_{root}$. It is important to note that, for fins with a dihedral (or anhedral) angle, b_s is twice the semi-span length from the hull's center line to the stabilizer's tip. In other words, it is *not* the projected length on the horizontal plane.

Also observe that the stabilizer's defined planform geometry, if it has a dihedral angle, is *not* the projected shadow area. Instead, it is the "flattened" (no dihedral) area, as discussed in the "Wings" section of Chapter 2. Further, the stabilizer's effective planform, shown by the previous illustration of S_s (which represents the *flattened* planform), does not generally have a simple straight-tapered geometry. Therefore, as illustrated in the following drawing, it has been found adequate to define an equivalent straight-tapered planform with a taper ratio given by:

$$\lambda = c_{tip}/c'_{root}$$

$(C^*_{N\alpha})_s$ is the derivative of the isolated stabilizer's (no hull) normal-force coefficient with respect to α, referenced to S_s. For the low aspect ratios typical of airship fins, the appropriate equation is from Lowry and Polhamus (NACA TN 3911, 1957), as given in the "Wings" section of Chapter 2:

$$(C^*_{N\alpha})_s = \frac{2\pi(AR)}{2 + \left[\frac{(AR)^2}{\kappa^2}(1 + \tan^2\Lambda_{c/2}) + 4\right]^{1/2}}$$

where κ is the ratio of the airfoil section's actual lift-curve slope per radian, divided by 2π. This is typically ≈ 0.95. Also $\Lambda_{c/2}$ is the sweep angle of the half-chord line. The way in which this is obtained is shown in the figure below:

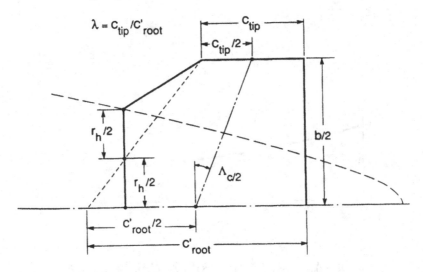

Note how an equivalent root chord, c'_{root} is defined so that the equations for a straight-tapered wing, presented in the "Wings" section of Chapter 2, may be applied to the fin or stabilizer's planform.

$(C_{Dh})_0$ is the hull zero-angle drag coefficient, referenced to S_h. This value includes skin friction, base drag and excrescence drag (gondola, engine pods, load patches, etc.). The flat-plate friction-drag coefficient is estimated from the turbulent boundary-layer equation presented in the "Airfoils" section of Chapter 2, where $(RN)_h$ is the Reynolds Number based on the hull's length.

$$(C_{fric.})_h = \frac{0.455}{[\log_{10}(RN)_h]^{2.58}}$$

When applied to a body of revolution, this value is increased by a function of the body's fineness ratio (given by S. Hoerner, "Fluid-Dynamic Drag, 1965, page 6–17):

$$(C_{Dwet})_h = (C_{fric.})_h\left[1 + 1.5\left(\frac{1}{f}\right)^{3/2} + 7\left(\frac{1}{f}\right)^3\right]$$

where $(C_{Dwet})_h$ is the "wetted area" drag coefficient. With respect to the hull's defined reference area, S_h, this becomes:

$$(C_{Dh})_0^{fric} = (C_{Dwet})_h (S_{wet})_h / S_h$$

Additionally, certain inflated shapes have an abruptly closed aft portion called a "boat tail", from which the flow may be assumed to be separated. This gives a base drag, $(C_{Dh})_0^{base}$.

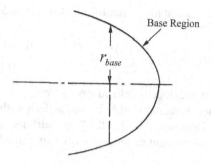

An equation to estimate this is obtained from Hoerner (Fluid-Dynamic Drag, 1965, p. 3–19) which, when adjusted for the hull's reference area, S_h, is given by:

$$(C_{Dh})_0^{base} = 0.029 \frac{(S_{base}/S_h)}{\left[(C_{fric.})_h (S_{wet})_h / S_{base}\right]^{1/2}}$$

Therefore, the hull's zero-angle drag coefficient, with no excrescence drag, is given by:

$$(C_{Dh})_0' = (C_{Dh})_0^{fric} + (C_{Dh})_0^{base}$$

For an airship, the excrescence drag might only give an increase to $(C_{Dh})_0'$ of 5%, whereas for a tethered aerostat the load patches, handing lines, lightning protection, huge payload enclosure, etc., might give an increase of 15%.

$(C_{Ds})_0$ and $(C_{Df})_0$ are the axial drag coefficients of the stabilizer and fin, referenced to S_s and S_f respectively. These values include the profile drag of the exposed surfaces as well as the interference drag where the roots of the stabilizer and fin panels intersect the hull. If the stabilizer has a smooth-surfaced airfoil with a non-blunt trailing edge, its profile drag may be approximated by calculating the total exposed "wetted" area of the stabilizer's exposed panels, $(S_{wet})_s$, multiplying this by skin-friction drag coefficient, $C_{fric.}$, dividing by its reference area S_s and then multiplying by a correction factor accounting for the airfoil's thickness ratio (S. Hoerner, "Fluid-Dynamic Drag" 1965, page 6–7):

$$(C_{Ds})_0 = (C_{fric.})_s \frac{(S_{wet})_s}{S_s} \left[1 + 2\left(\frac{t}{c}\right)_s + 60\left(\frac{t}{c}\right)_s^4\right]$$

where t/c is the airfoil's maximum thickness divided by its chord length.

The same procedure applies to the fin's drag coefficient:

$$(C_{Df})_0 = (C_{fric.})_f \frac{(S_{wet})_f}{S_f} \left[1 + 2\left(\frac{t}{c}\right)_f + 60\left(\frac{t}{c}\right)_f^4\right]$$

Again, upon assuming that the boundary layer is totally turbulent (which is usually the case), the previous skin-friction drag coefficient equation may be applied:

$$C_{fric.} = \frac{0.455}{(\log_{10} RN)^{2.58}}$$

The Reynolds Number, RN, is now based on the mean chord length of the *exposed* panels. For very low-altitude flight in the Earth's atmosphere, one has:

$$RN = 6.85 VL \times 10^4 \text{ (mks)}, \quad RN = 6.36 VL \times 10^3 \text{ (ft–lb–sec)}$$

$(C_T)_s$ = leading-edge suction coefficient for the stabilizer, referenced to S_s. This may be calculated from

$$(C_T)_s = (C^*_{\tilde{N}\alpha})_s \eta_s \eta_t \cos^2 \Gamma_s \frac{\sin(2\alpha)}{2} \tan\alpha - \frac{\left[(C^*_{\tilde{N}\alpha})_s \eta_s \cos^2 \Gamma_s \sin(2\alpha)\right]^2 \eta_t}{4\pi(AR)_s}$$

All of these terms have been previously defined except for the leading-edge suction efficiency parameter η_t. For fins that incorporate smoothly curved airfoils with rounded leading edges (such as the NACA 0012), $\eta_t \approx 1.0$. However, for thin "flat-plate" airfoils, $\eta_t \approx 0$.

If $\alpha \leq 30°$, then the force and moment equations may be simplified:

$$\tilde{N} = C_{\tilde{N}} q S \quad \text{where}$$

$$C_{\tilde{N}} = \left[(k_3 - k_1)(\eta_k)_{\text{long}} \hat{I}_1 + 0.5(C^*_{\tilde{N}\alpha})_s (\cos^2 \Gamma_s)\eta_s \hat{S}_s\right] \sin(2\alpha)$$

$$+ \left[(C_{Dc})_h \hat{J}_1 + (C_{Dc})_s \cos^3|\Gamma_s|\hat{S}_s\right] \sin\alpha \sin|\alpha|$$

$$\tilde{D} = \tilde{C}_D q S \quad \text{where}$$

$$(C_D)_{\text{axial}} = \left[(C_{Dh})_0 \hat{S}_h + (C_{Ds})_0 \hat{S}_s + (C_{Df})_0 \hat{S}_f\right] \cos^2\alpha$$

$$- (k_3 - k_1)(\eta_k)_{\text{long}} \hat{I}_1 \sin(2\alpha)\sin(\alpha/2) - (C_T)_s \hat{S}_s$$

$$M_{\text{nose}} = C_{M\text{nose}} q S \bar{c} \quad \text{where}$$

$$(C_M)_{\text{nose}} = -\left[(k_3 - k_1)(\eta_k)_{\text{long}} \hat{I}_3 + 0.5(\hat{l}_s)_1 (C^*_{\tilde{N}\alpha})_s (\cos^2\Gamma_s)\eta_s \hat{S}_s\right] \sin(2\alpha)$$

$$- \left[(C_{Dc})_h \hat{J}_2 + (C_{Dc})_s \cos^3|\Gamma_s|(\hat{l}_s)_2 \hat{S}_s\right] \sin\alpha \sin|\alpha|$$

EXAMPLE 1: SMALL AEROSTAT

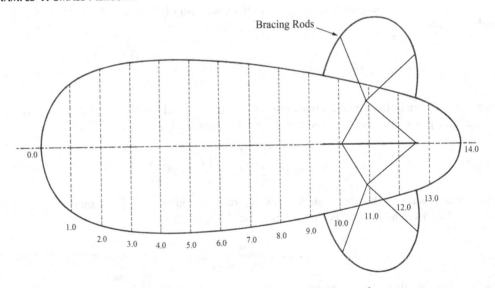

The drawing shows the configuration of a 14-foot-long aerostat used for advertising. This has externally braced "flat-plate" fins made by stretching fabric over tensioned slender fiberglass rods. The hull ordinates, in Imperial units, are:

ξ (ft)	r (ft)	ξ (ft)	r (ft)	ξ (ft)	r (ft)
0.0	0.0	3.0	2.686	11.0	1.860
0.1	0.730	3.5	2.732	11.5	1.740
0.2	1.020	4.0	2.750	12.0	1.610
0.3	1.210	4.5	2.752	12.3	1.527
0.4	1.375	5.0	2.741	12.6	1.430
0.5	1.515	5.5	2.718	12.8	1.351
0.6	1.638	6.0	2.687	13.0	1.257
0.7	1.752	6.5	2.644	13.2	1.148
0.8	1.855	7.0	2.597	13.3	1.086
0.9	1.942	7.5	2.535	13.4	1.020
1.00	2.023	8.0	2.462	13.5	0.938
1.2	2.157	8.5	2.382	13.6	0.848
1.4	2.265	9.0	2.286	13.7	0.739
1.7	2.394	9.5	2.190	13.8	0.608
2.0	2.493	10.0	2.086	13.9	0.430
2.5	2.612	10.5	1.978	14.0	0

From this one may calculate that the hull volume is 221.115 cubic feet. Thus,

$$S = S_h = (total\ hull\ volume)^{2/3} = 36.566 ft^2, \quad \bar{c} = 14 ft$$

The fineness ratio, f, is 2.544, from which one obtains:

$$k_1 = 0.1527; \quad k_3 = 0.7661; \quad (k_3 - k_1) = 0.6134; \quad (C_{Dc})_h = 0.1659$$

Also, $l_h = 9.42 ft$, so the hull integrals are:

$$\hat{I}_1 = 0.41797; \quad \hat{I}_3 = -0.07359; \quad \hat{J}_1 = 1.25694; \quad \hat{J}_2 = 0.46067$$

These were obtained from trapezoidal numerical integration (with the exception of $\hat{I}_1$, which is directly obtained from the cross-sectional area at l_h). Also, numerical integration gives the total surface area of the hull ("wetted area") as $S_{wet} = 203.31 ft^2$.

The assumed wind speed is $V = 15.3 ft/sec$. Therefore, the hull Reynolds number at standard sea-level conditions is $RN = 6.36 \times 10^3\ V\bar{c} = 13.62 \times 10^5$ for which the skin-friction coefficient is $(C_{fric.})_h = 0.004223$.

From the previously presented wetted-area drag equation for a body of revolution, one obtains:

$$(C_{Dwet})_h = (C_{fric.})_h \left[1 + 1.5 \left(\frac{1}{f}\right)^{3/2} + 7 \left(\frac{1}{f}\right)^3 \right] = 0.004223 \times 1.7948 = 0.00758$$

Thus, the hull friction drag, based on the hull reference area, S_h, is:

$$(C_{Dh})_0^{fric} = (C_{Dwet})_h (S_{wet}/S_h) = 0.00758 \times 203.31/36.566 = 0.0421$$

Additionally, the hull has a significantly-blunt aft portion ("boat tail"). Upon assuming that the hull radius at the flow-separation point is $r_{base} = 1.24 ft$, the cross-sectional area at that point is $S_{base} = 4.8305 \text{ ft}^2$. Thus, from the previously presented base-drag coefficient equation, one obtains:

$$(C_{Dh})_0^{base} = 0.029 \frac{S_{base}/S_h}{\left[(C_{fric.})_h S_{wet}/S_{base}\right]^{1/2}} = 0.029 \times \frac{4.8305/36.566}{(0.004223 \times 203.31/4.8305)^{1/2}} =$$

$$(C_{Dh})_0^{base} = 0.0091$$

Therefore, the hull drag coefficient, without accounting for excrescence-drag effects, is:

$$(C_{Dh})_0' = (C_{Dh})_0^{fric} + (C_{Dh})_0^{base} = 0.0512$$

The effect of load patches, external lines and dilation panel (an area tensioned by bungee cords to compensate for gas expansion and contraction) is to add approximately 15% to the $(C_{Dh})_0'$ value. Therefore, the final axial drag coefficient for the hull is estimated to be:

$$(C_{Dh})_0 \approx 0.059$$

For the hull cross-flow drag coefficient, $(C_{Dc})_h$, first note that, as stated before, the super-critical cross-flow drag of an infinite cylinder is $(C_{Dc})_{cylinder} \approx 0.29$. This is corrected for the fineness ratio of the example, $f = 2.544$, by the equation presented earlier:

$$\gamma_{fineness} = 0.5088 + 0.02750 \times 2.544 - 0.001011 \times (2.544)^2 = 0.5722 \rightarrow$$

$$(C_{Dc})_h = 0.1659$$

Regarding the empennage, calculations are now performed for the stabilizer where an equivalent trapezoidal panel is defined to represent the curved planform:

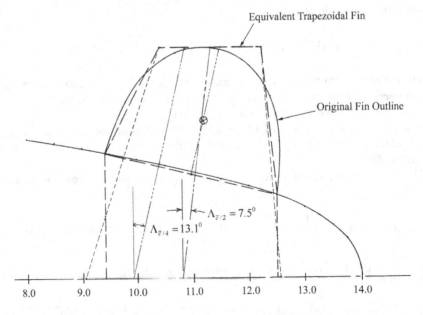

From the drawing and the earlier definitions, one may obtain the following terms:

$$c_{tip} = 1.823 \text{ ft}, \ c'_{root} = 3.51 \text{ ft}, \ taper \ ratio, \ \lambda = 0.519, \ b_s = 8.12 \text{ ft}, \ \Lambda_{c/2} = 7.5^0$$

And S_s is the total area of the projected equivalent trapezoidal planform (accounting for both panels on either side of the hull's center line). From the definitions in the illustration below, this value is given by:

$$S_s = 2c_r y_{te} + (c_r + c_0) \times (y_0 - y_{te}) + (c_t + c_0) \times (b_s/2 - y_0) = 22.489 \, \text{ft}^2$$

Therefore, the aspect ratio is $AR = b_s^2/S_s = 2.93$.

Recall that b_s and S_s refer to the complete "flattened" stabilizer. That is, these parameters are defined for the fin shown above in conjunction with its mirror image.

Because the airfoil is a thin flat plate, one may estimate from Figure B.1.1 of Appendix B1 in Bernard Etkin's "Dynamics of Flight" (Wiley, 1959) that the sectional lift-curve slope is 5.65/rad. Thus, $\kappa = 0.90$. One may now obtain the derivative of the normal-force coefficient with respect to α:

$$(C^*_{N\alpha})_s = \frac{2\pi \times 2.93}{2 + \left[\frac{2.93^2}{0.90^2}(1 + \tan^2(7.5°)) + 4\right]^{1/2}} = 3.150 \, /\text{rad}$$

Next, from the cross-flow drag equations, one obtains:

$$(C_{Dc})_s = 4.0762 - 2.6357 \times AR + 1.4760 \times AR^2 - 0.4727 \times AR^3 + 0.0574 \times AR^4 = 1.365$$

Now, consider the geometry of the equivalent trapezoidal configuration (illustrated above) for the stabilizer's planform. The dimensions are:

$c_t \equiv c_{tip} = 1.823 \, \text{ft}$
$c_r = 3.08 \, \text{ft}$
$c_0 = 3.007 \, \text{ft}$
$y_0 = r_h = 2.205 \, \text{ft}$
$y_{te} = 1.46 \, \text{ft}$
$\Lambda_{le} = 28.264°$
$\Lambda_{te} = 5.643°$
$b = 8.12 \, \text{ft}$

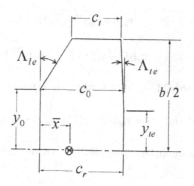

One may obtain the aerodynamic center, $\bar{x}$, (stabilizer center of pressure) from:

$$\bar{x} = \frac{2}{S_s}\left[\frac{c_r}{8}(2c_r y_0 - y_0^2 \tan \Lambda_{te}) + AC\left(\frac{b}{2} - y_0\right) + \frac{BD}{3}\left(\frac{b^3}{8} - y_0^3\right) + \frac{(BC+DA)}{2}\left(\frac{b^2}{4} - y_0^2\right)\right]$$

where:

$$A = c_r + y_0 \tan \Lambda_{le} = 3.08 + 2.205 \times \tan(28.264°) = 4.2655$$

$$B = -\tan \Lambda_{le} - \tan \Lambda_{te} = -0.6365$$

$$C = \frac{c_r}{4} - \frac{3}{4} y_0 \tan \Lambda_{le} = -0.1191$$

$$D = \frac{3}{4} \tan \Lambda_{le} - \frac{1}{4} \tan \Lambda_{te} = 0.3785$$

Thus, one has that:

$$\bar{x} = 0.08893 \times (5.0444 - 0.9424 - 4.5134 + 9.8220) = 0.8369 \approx 0.84 \text{ ft}$$

From this, one may obtain:

$$(l_s)_1 = l_h + \bar{x} = 9.42 + 0.84 = 10.26 \text{ ft} \rightarrow (\hat{l}_s)_1 = 0.7329$$

ASIDE

The mean aerodynamic chord of this planform is given by:

$$\bar{c}_s = (2/S_s)(Int_1 + Int_2 + Int_3)$$

where,

$$Int_1 = c_r^2 \, y_{te}, \quad Int_2 = A' - B' + C' + D' - E' + F', \quad Int_3 = G' - H' - J' + K' + L' + M'$$

$$A' = c_r^2(y_0 - y_{te}), \quad B' = c_r(y_0^2 - y_{te}^2)\tan\Lambda_{te}, \quad C' = 2c_r(y_0 - y_{te})\tan\Lambda_{te},$$

$$D' = \tan\Lambda_{te}(y_0^3 - y_{te}^3)/3, \quad E' = \tan\Lambda_{te}(y_0^2 - y_{te}^2)y_{te}, \quad F' = \tan\Lambda_{te}(y_0 - y_{te})y_{te}^2,$$

$$G' = c_0^2(b/2 - y_0), \quad H' = c_0(b^2/4 - y_0 b + y_0^2)\tan\Lambda_{le},$$

$$J' = c_0(b^2/4 - y_{te}b - y_0^2 + 2y_{te}y_0)\tan\Lambda_{te},$$

$$K' = 2\left[b^3/24 - (y_{te} + y_0)b^2/8 + y_0 y_{te} b/2 - y_0^3/3 + (y_{te} + y_0)y_0^2/2 - y_0^2 y_{te}\right]\tan\Lambda_{le}\tan\Lambda_{te}$$

$$L' = (b^3/24 - y_0 b^2/4 + y_0^2 b/2 - y_0^3/3)\tan\Lambda_{le}$$

$$M' = (b^3/24 - y_{te}b^2/4 + y_{te}^2 b/2 - y_0^3/3 + y_{te}y_0^2 - y_{te}^2 y_0)\tan\Lambda_{te}$$

From a computer solution of these equations, the stabilizer's mean aerodynamic chord for this numerical example is:

$$\bar{c}_s = 2.845 \text{ ft}$$

END ASIDE

From measurement on the geometry of the exposed stabilizer panel, one has:

$$(l_s)_2 = l_h + 1.76 = 9.42 + 1.76 = 11.18 \text{ ft} \rightarrow (\hat{l}_s)_2 = 0.7986$$

The average chord of the exposed stabilizer panel is $\bar{c}_{exposed} \approx 2.42$ ft, so that for $V = 15.3$ ft/s, the Reynolds number of the stabilizer's exposed portion is:

$$(RN)_s = 6.36 \, V \, \bar{c}_{exposed} \times 10^3 = 6.36 \times 15.3 \times 2.42 \times 10^3 = 2.35 \times 10^5$$

This gives a skin-friction coefficient of $(C_{fric.})_s = 0.00595$.

Balloons and Airships

Because the fins are thin flat plates, it will be assumed that thickness effects are negligible. That is, the thickness correction factor discussed earlier is unity. Thus,

$$(C_{Dwet})_s = (C_{fric.})_s$$

The single-side area of an exposed stabilizer panel is measured to be 5.591 ft², so the *total wetted area* of the exposed stabilizer is:

$$(S_{wet})_{fins} = 4 \times 5.591 = 22.36 \, \text{ft}^2$$

Therefore, the drag coefficient of the stabilizer due to skin friction, relative to the stabilizer's reference area, S_s, is:

$$(C_{Ds})_0^{fric} = (C_{Dwet})_s (S_{wet})_s / S_s = 0.00595 \times 22.36/22.489 = 0.0059$$

Note that the fins, for this particular example, are externally braced by "V-struts" of fiberglass rods. Normally aerostat fins are supported by guy lines, and the drag of these is accounted for in the excrescence increment. However, in this case, the rod drag will be included in the stabilizer and fin drag. The drag equation used is that for an inclined cylinder, derived from Hoerner, page 3–11:

$$(C_{Ds})_0^{bracing} = Cd'_{cylinder} (\sin^3 \theta) d \times L/S_s$$

where:

$Cd'_{cylinder} \equiv$ cross-flow drag coefficient based on diameter (equals 1.1 for sub-critical flow),
$\theta \equiv$ inclination of the cylinder to the flow (for this case $\theta \approx 45°$),
$d \equiv$ cylinder diameter (measured to be 0.184 inches = 0.01533 ft),
$L \equiv$ total rod length to support the stabilizer = 22 ft.

This gives:

$$(C_{Ds})_0^{bracing} = 1.1 \times \sin^3(45°) \times 0.01533 \times 22/22.489 = 0.0058$$

Therefore, the stabilizer's total drag coefficient, with respect to the stabilizer's reference area, S_s, is:

$$(C_{Ds})_0 = (C_{Ds})_0^{fric} + (C_{Ds})_0^{bracing} = 0.0059 + 0.0058 = 0.0117$$

Also, because the vertical fins and bracing are geometrically identical, one immediately obtains that its total drag coefficient, with respect to the fin's reference area, S_f, is:

$$(C_{Df})_0 = 0.0117$$

Because of the fact that the fins are thin flat plates, the leading-edge suction efficiency is assumed to be zero:

$$\eta_t \approx 0.0$$

Now, the calculated value of $(J_1)_{total}$ is 60.519 ft², so $S_s/(J_1)_{total} = 0.3716$. From the curve-fit equation for η_k one obtains:

$$(\eta_k)_{long} = 1.0 + 1.40 \times 0.3716^2 = 1.193$$

Also, S_{sh} is given by the portion of horizontal hull area between the leading-edge and trailing-edge hull-fin intersection points:

$$S_{sh} = (J_1)_{12.5} - (J_1)_{9.42} = 57.268 - 45.961 = 11.307 \, \text{ft}^2$$

Therefore, $S_{sh}/S_s = 11.307/22.489 = 0.5028$, which gives:

$$\eta_s = 1 - 1.4 \times 0.5028 + 0.4 \times 0.5028^2 = 0.397$$

This provides all of the inputs required to obtain the $C_{\tilde{N}}$, $(C_D)_{\text{axial}}$ and $(C_M)_{\text{nose}}$ coefficients. Also note that the wind-axis coefficients may then be obtained from:

$$C_L = C_{\tilde{N}} \cos\alpha - (C_D)_{\text{axial}} \sin\alpha$$
$$C_D = (C_D)_{\text{axial}} \cos\alpha + C_{\tilde{N}} \sin\alpha$$

With these parameters, the results for C_L, C_D and $(C_M)_{\text{nose}}$ at angles of attack from -30^0 to 30^0 were calculated and are shown in the following plot:

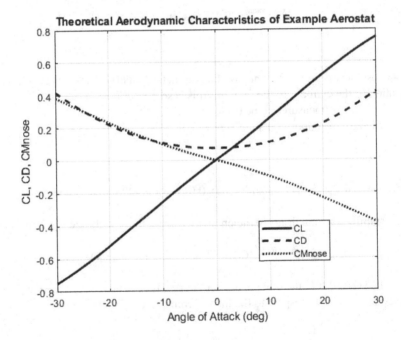

It is seen that the C_L curve doesn't show the abrupt stalling behavior at the higher magnitudes of α, as is typically the case for airplanes. This is confirmed by experiment, as will be shown in subsequent examples. Therefore, higher-order polynomial curve-fit equations are required to match these results. In *radian* measure, these equations are:

$$C_L = 1.42645\alpha + 0.96762\alpha^3 - 3.48993\alpha^5$$
$$C_D = 0.07276 + 1.12420\alpha^2 + 1.12233\alpha^4 - 2.29847\alpha^6$$
$$(C_M)_{\text{nose}} = -0.53869\alpha - 1.35937\alpha^3 + 2.41861\alpha^5$$

In *degree* measure, these equations are:

$$C_L = 2.48962 \times 10^{-2} \alpha + 5.14440 \times 10^{-6} \alpha^3 - 5.65202 \times 10^{-9} \alpha^5$$

$$C_D = 0.07276 + 3.42450 \times 10^{-4} \alpha^2 + 1.04143 \times 10^{-7} \alpha^4 - 6.49690 \times 10^{-11} \alpha^6$$

$$(C_M)_{nose} = -9.40200 \times 10^{-3} \alpha - 7.22722 \times 10^{-6} \alpha^3 + 3.91699 \times 10^{-9} \alpha^5$$

EXAMPLE 2: AIRSHIP ZRS-4 "AKRON" 5.98M WIND-TUNNEL MODEL

The U.S. Navy airship ZRS-4 "Akron", from 1933, was mentioned earlier in this chapter. As part of its development, a 5.98m long model was wind-tunnel tested by NACA (predecessor agency to NASA). The results are presented in a report by Hugh B. Freeman: "Force Measurements on a 1/40-Scale Model of the U.S. Airship "Akron", NACA TR 432, 1933. The cross-flow methodology will now be applied to this model, and comparisons will be made between the predictions and the experimental results. First, the model's hull ordinates are presented below:

Station No.	ξ (m)	r (m)	Station No.	ξ (m)	r (m)
1	0	0	12	2.9901	0.5029
2	0.1196	0.1257	13	3.2891	0.4976
3	0.2990	0.2530	14	3.5881	0.4856
4	0.5980	0.3607	15	3.8871	0.4689
5	0.8970	0.4229	16	4.1861	0.4445
6	1.1960	0.4671	17	4.4852	0.4102
7	1.4951	0.4856	18	4.7842	0.3668
8	1.7941	0.4981	19	5.0832	0.3122
9	2.0931	0.5042	20	5.3822	0.2441
10	2.3921	0.5055	21	5.6812	0.1656
11	2.6911	0.5055	22	5.9802	0.0

From this, trapezoidal integration gives that the hull's volume is 3.2657m³. Therefore,

$$S = S_h = (total\ hull\ volume)^{2/3} = 2.2012\,m^2, \quad \bar{c} = 5.9802\,m$$

The fineness ratio, f, is 5.9151, from which one obtains:

$$k_1 = 0.0462, \ k_3 = 0.9155, \ (k_3 - k_1) = 0.8693,$$

Also, $l_h = 4.6174\,\text{m}$, so the hull integrals are:

$$\hat{I}_1 = 0.21887, \ \hat{I}_3 = -0.05570, \ \hat{J}_1 = 1.84505, \ \hat{J}_2 = 0.80585$$

Again, from numerical integration, the total wetted surface area is $S_{\text{wet}} = 15.0777\,\text{m}^2$. Further, since the wind speed was 44.2 m/sec ($\approx$ 100 mph), the test Reynolds Number (based on length, $\bar{c}$, and sea-level atmospheric conditions) was $RN = 1.8105 \times 10^7$. This gives a skin-friction drag coefficient of $(C_{\text{fric.}})_h = 0.00274$. Upon correcting for the hull's "fatness factor", the wetted-area drag coefficient is:

$$(C_D)_{\text{wet}} = 0.00274 \times 1.1381 = 0.00311$$

So, the hull friction drag, based on the hull's reference area, S_h, is:

$$(C_{Dh})_0^{\text{fric}} = (C_D)_{\text{wet}} (S_{\text{wet}}/S_h) = 0.00311 \times 15.0777/2.2012 = 0.02130$$

Also, the hull has a relatively small blunted aft portion ("boat tail"). Upon assuming that the hull radius at the flow-separation point is $r_{\text{base}} = 0.166\,\text{m}$, the cross-sectional area at that point is $S_{\text{base}} = 0.0866\,\text{m}^2$. Thus, from the previously presented base-drag coefficient equation, one obtains:

$$(C_{Dh})_0^{\text{base}} = 0.029 \frac{S_{\text{base}}/S_h}{\left[(C_{\text{fric.}})_h S_{\text{wet}}/S_{\text{base}}\right]^{1/2}} = 0.029 \times \frac{0.0866/2.2012}{(0.00274 \times 15.0777 / 0.0866)^{1/2}} =$$

$$(C_{Dh})_0^{\text{base}} = 0.00165$$

Therefore, the hull drag coefficient, without accounting for excrescence-drag effects, is:

$$(C_{Dh})_0' = (C_{Dh})_0^{\text{fric}} + (C_{Dh})_0^{\text{base}} = 0.0230$$

This example is aerodynamically much "cleaner" than the "Small Aerostat" example. Nonetheless there is some excrescence drag from the engine pods and gondola. It is estimated that these add approximately 5% to the $(C_{Dh})_0'$ value. Therefore, the final axial drag coefficient for the hull is estimated to be:

$$(C_{Dh})_0 \approx 0.0241$$

For the hull cross-flow drag coefficient, $(C_{Dc})_h$, the wind-tunnel model is large enough that the cross-flow may be assumed to be super-critical. So, one may again assume a cylindrical cross-flow drag coefficient of $(C_{Dc})_{\text{cylinder}} \approx 0.29$. This is corrected for the fineness ratio of the example, $f = 5.9151$, by the equation discussed earlier:

$$\gamma_{\text{fineness}} = 0.5088 + 0.02750 \times 5.9151 - 0.001011 \times (5.9151)^2 = 0.6361 \rightarrow$$

$$(C_{Dc})_h = 0.1845$$

Regarding the empennage, calculations are now performed for the stabilizer, where its geometry is shown below:

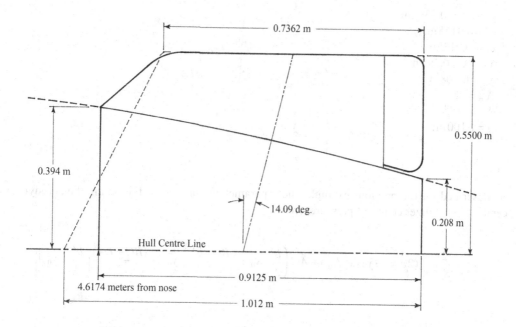

From this drawing, one may obtain the following values for the equivalent "straight-tapered" planform:

$c_{tip} = 0.7362$ m, $c'_{root} = 1.012$ m, *taper ratio*, $\lambda = 0.7275$, $b_s = 1.1000$ m, $\Lambda_{c/2} = 14.09°$

Further, for the defined "trapezoidal" planform, one obtains that:

$$S_s = 0.9757 \text{m}^2, \quad AR = b_s^2/S_s = 1.2401$$

Recall that b_s and S_s refer to the complete "flattened" stabilizer. That is, these parameters are defined for the configuration shown above in conjunction with its mirror image.

Now, because the airfoil is a smooth symmetrical shape, with a rounded leading edge, it is assumed that $\kappa = 1.0$. Therefore, the derivative of the stabilizer's normal-force coefficient with respect to α is given by:

$$(C^*_{\tilde{N}\alpha})_s = \frac{2\pi \times 1.2401}{2 + \left[\frac{1.2401^2}{1.0^2}(1 + \tan^2(14.09°)) + 4\right]^{1/2}} = 1.7815 \text{/rad}$$

Next, from the cross-flow drag equations, one obtains:

$$(C_{Dc})_s = 4.9110 - 3.0161 \times AR + 1.4450 \times AR^2 - 0.4018 \times AR^3 + 0.0439 \times AR^4 = 2.7307$$

From the dimensioned drawing of the stabilizer panel, the summary geometric parameters of the equivalent trapezoidal configuration are:

$c_t \equiv c_{tip} = 0.7361\,\text{m}$,
$c_r = 0.9125\,\text{m}$,
$c_0 = 0.9125\,\text{m}$,
$y_0 = r_h = 0.394\,\text{m}$,
$y_{te} = 0.208\,\text{m}$,
$\Lambda_{le} = 48.5^0$,
$\Lambda_{te} = 0.0^0$,
$b = 1.100\,\text{m}$.

As described for the previous example, these parameters may be used to obtain the aerodynamic center, $\bar{x}$, (stabilizer center of pressure) from:

$$\bar{x} = \frac{2}{S_s}\left[\frac{c_r}{8}(2c_r y_0 - y_0^2 \tan\Lambda_{te}) + AC\left(\frac{b}{2} - y_0\right) + \frac{BD}{3}\left(\frac{b^3}{8} - y_0^3\right) + \frac{(BC+DA)}{2}\left(\frac{b^2}{4} - y_0^2\right)\right]$$

where:

$$A = c_r + y_0 \tan\Lambda_{le} = 1.3545,\quad B = -\tan\Lambda_{le} - \tan\Lambda_{te} = -1.1303,$$
$$C = \frac{c_r}{4} - \frac{3}{4} y_0 \tan\Lambda_{le} = -0.1033,\quad D = \frac{3}{4}\tan\Lambda_{le} - \frac{1}{4}\tan\Lambda_{te} = 0.8477$$

Thus, one has that:

$$\bar{x} = 2.0498 \times (0.0814 - 0.0223 - 0.0340 + 0.0946) = 0.2454\,\text{m}$$

From this, one may obtain:

$$(l_s)_1 = l_h + \bar{x} = 4.6174 + 0.2454 = 4.8628\,\text{m} \rightarrow (\hat{l}_s)_1 = 0.8131$$

Also, from the equations for the stabilizer's mean aerodynamic chord, given in the "Example 1: Small Aerostat" subsection, one may obtain:

$$\bar{c}_s = 0.8884\,\text{m}$$

Next, from measurements on the geometry of the exposed stabilizer panel, one has:

$$(l_s)_2 = l_h + 0.5424 = 4.6174 + 0.5424 = 5.1598\,\text{m} \rightarrow (\hat{l}_s)_2 = 0.8628$$

The average chord of the exposed stabilizer panel is $\bar{c}_{exposed} \approx (c_0 + c_t)/2 = 0.824\,\text{m}$, so that for $V = 44.2\,\text{m/sec}$, the Reynolds number of the fins is,

$$(RN)_s = 6.85\, V\, \bar{c}_{exposed} \times 10^4 = 6.85 \times 44.2 \times 0.824 \times 10^4 = 2.50 \times 10^6$$

This gives a skin-friction coefficient of $(C_{fric.})_s = 0.00379$.

The average thickness/chord ratio, t/c, equals 0.09. Therefore,

$$(C_{Dwet})_s = (C_{fric.})_s \left[1 + 2\left(\frac{t}{c}\right)_s + 60\left(\frac{t}{c}\right)_s^4\right] = 0.00379 \times (1 + 0.180 + 0.0039) = 0.00449$$

The single-side area of an exposed stabilizer panel is measured to be $0.2146\,\text{m}^2$, so the *total wetted area* of the exposed stabilizer is:

$$(S_{wet})_s = 4 \times 0.2146 = 0.8584\,\text{m}^2$$

So, the zero-angle drag coefficient of the stabilizer with respect to its reference area, is:

$$(C_{Ds})_0 = (C_{Dwet})_s (S_{wet})_s / S_s = 0.00449 \times 0.8584 / 0.9757 = 0.00395$$

The vertical fin is virtually identical (except for a slight reconfiguring of the lower panel's tip to accommodate angular ground clearance for take-off). Therefore, one has:

$$(C_{Df})_0 = 0.00395$$

As previously mentioned, the fins incorporate airfoils with rounded, though somewhat sharp, leading edges. Therefore, the leading-edge suction efficiency may be assumed to be 0.9:

$$\eta_t \approx 0.90$$

Now, the calculated value of $(J_1)_{total}$ is $4.7324\,\text{m}^2$, so $S_s/(J_1)_{total} = 0.2062$. From the curve-fit equation for η_k one obtains:

$$(\eta_k)_{long} \approx 1.0 + 1.40 \times 0.2062^2 = 1.0595$$

Also, S_{sh} is given by the portion of horizontal hull area between the leading-edge and trailing-edge hull-fin intersection points:

$$S_{sh} = (J_1)_{5.5299} - (J_1)_{4.6174} = 4.6210 - 4.0614 = 0.5596\,\text{m}^2$$

Therefore $S_{sh}/S_s = 0.5596/0.9757 = 0.5735$, which gives:

$$\eta_s = 1 - 1.4 \times 0.5735 + 0.4 \times 0.5735^2 = 0.3287$$

This provides all of the inputs required to obtain the $C_{\tilde{N}}$, $(C_D)_{axial}$ and $(C_M)_{nose}$ coefficients. Also note that the wind-axis coefficients may then be obtained from:

$$C_L = C_{\tilde{N}} \cos\alpha - (C_D)_{axial} \sin\alpha$$

$$C_D = (C_D)_{axial} \cos\alpha + C_{\tilde{N}} \sin\alpha$$

With these parameters, the results for C_L, C_D and $(C_M)_{nose}$ at angles of attack from $-30°$ to $30°$ were calculated and are shown in the following plot:

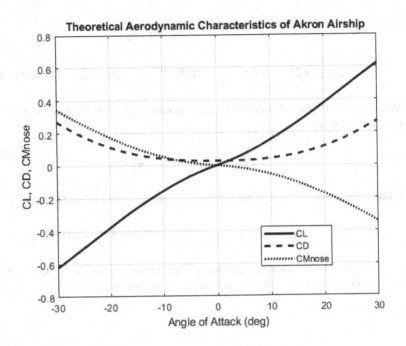

Polynomial curve-fit equations of these results, with α in *radians*, are:

$$C_L = 0.81003\alpha + 2.89158\alpha^3 - 5.63873\alpha^5$$
$$C_D = 0.02679 + 0.45571\alpha^2 + 2.56606\alpha^4 - 3.59228\alpha^6$$
$$(C_M)_{nose} = -0.24022\alpha - 2.58331\alpha^3 + 3.94087\alpha^5$$

And, the curve-fit equations in *degrees* are:

$$C_L = 1.41337 \times 10^{-2}\alpha + 1.53733 \times 10^{-5}\alpha^3 - 9.13204 \times 10^{-9}\alpha^5$$
$$C_D = 0.02679 + 1.38817 \times 10^{-4}\alpha^2 + 2.38110 \times 10^{-7}\alpha^4 - 1.01539 \times 10^{-10}\alpha^6$$
$$(C_M)_{nose} = -4.19260 \times 10^{-3}\alpha - 1.37344 \times 10^{-5}\alpha^3 + 6.38232 \times 10^{-9}\alpha^5$$

Also, comparisons were made between the theoretical predictions and the experimental data from the NACA tests, from $\alpha = -3°$ to $20°$:

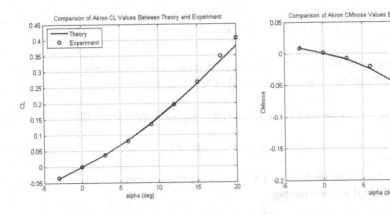

It is seen that theory and experiment agree well for C_L and $(C_M)_{nose}$, only significantly diverging for α values above 15°. However, a larger divergence is seen for the C_D comparison.

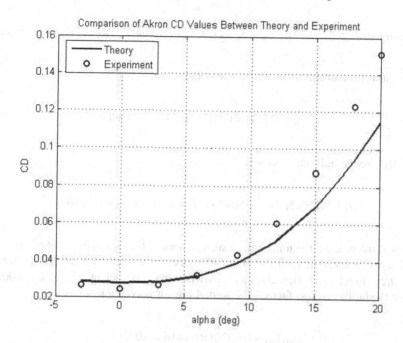

Although the theoretical values initially track fairly well with the experimental points below 6°, above this value the theory increasingly under-predicts the drag coefficient. The reason for this is not clear, although it's worth noting that, at higher angles, the cross-flow drag coefficients, $(C_{Dc})_h$ and $(C_{Dc})_s$ become increasingly important. It may be that an adjustment in the way these values are calculated would be appropriate. In any case, the comparison is still good within the airship's normal operational angles of attack.

EXAMPLE 3: GOODYEAR "WINGFOOT2" AIRSHIP (ZEPPELIN LZ N07)

This is a modern airship, designed and built by the Zeppelin Corporation for Goodyear Airship Operations. The author was involved, in 1996, with the initial aerodynamic estimations and wind-tunnel testing of this design, and again in 2017, with further analysis and testing of the Wingfoot 2 version of this design. That experience is drawn upon for this example:

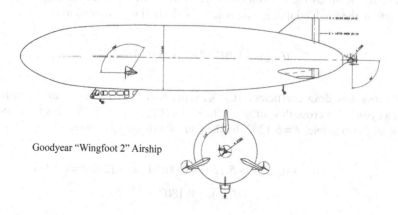

Goodyear "Wingfoot 2" Airship

From hull ordinates supplied to the author, the following parameters were obtained by using trapezoidal numerical integration:

$$\text{Total hull volume} = 8225.12\,\text{m}^3,\ S = S_\text{h} = (\text{Total hull volume})^{2/3} = 407.47\,\text{m}^2,\ \bar{c} = 72.5\,\text{m}$$

The fineness ratio, f, is 5.1202, from which one obtains:

$$k_1 = 0.0571,\ k_3 = 0.8975,\ (k_3 - k_1) = 0.8404,$$

Also, $l_\text{h} = 57.0\,\text{m}$, so the hull integrals are:

$$\hat{I}_1 = 0.25665,\ \hat{I}_3 = -0.04958,\ \hat{J}_1 = 1.76793,\ \hat{J}_2 = 0.75801$$

From numerical integration, the total wetted surface area of the hull is $S_\text{wet} = 2680.26\,\text{m}^2$. Further, at the nominal cruise speed of $V = 15.433\,\text{m/s}$ (30 knots), the sea-level Reynolds Number (based on length, $\bar{c}$) is $RN = 7.664 \times 10^7$. This gives a skin-friction drag coefficient of $(C_\text{fric.})_\text{h} = 0.00221$. Upon correcting for the hull's "fatness factor", the wetted-area drag coefficient is:

$$(C_\text{D})_\text{wet} = 0.00221 \times 1.1816 = 0.00261$$

So, the hull friction drag, based on the reference area, S_h, is:

$$(C_\text{Dh})_0^\text{fric} = (C_\text{D})_\text{wet}(S_\text{wet}/S_\text{h}) = 0.00261 \times 2680.26/407.47 = 0.01717$$

Because of the streamlined aft-body shape and in-flow from the aft propeller, the base drag is assumed to be zero. Therefore, the hull drag coefficient, without accounting for excrescence-drag effects, is:

$$(C_\text{Dh})_0' = (C_\text{Dh})_0^\text{fric} + (C_\text{Dh})_0^\text{base} = 0.01717 + 0 = 0.01717$$

This design is as "clean" as the "Akron" airship, so the same excrescence drag increase of 5% is chosen. Therefore, the final axial drag coefficient for the hull is estimated to be:

$$(C_\text{Dh})_0 = 0.01717 \times 1.05 = 0.01803$$

For the hull cross-flow drag coefficient, $(C_\text{Dc})_\text{h}$, super-critical cross-flow may again be assumed, which gives a cylindrical cross-flow drag coefficient of $(C_\text{Dc})_\text{cylinder} \approx 0.29$. This is corrected for the fineness ratio of the example, $f = 5.1202$, by the equation discussed earlier:

$$\gamma_\text{fineness} = 0.5088 + 0.02750 \times 5.1202 - 0.001011 \times (5.1202)^2 = 0.6231 \rightarrow$$

$$(C_\text{Dc})_\text{h} = 0.1807$$

Regarding the empennage, calculations are now performed for the stabilizer, where its geometry is shown below:

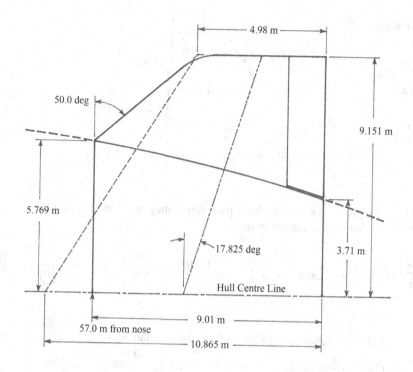

From this drawing, one may obtain the following terms for the equivalent "straight-tapered" planform:

$$c_{tip} = 4.98 \, \text{m}, \quad c'_{root} = 10.865 \, \text{m}, \quad \text{taper ratio}, \; \lambda = 0.4584, \quad b_s = 18.302 \, \text{m}, \quad \Lambda_{c/2} = 17.825°$$

Further, for the defined "trapezoidal" planform, one obtains that:

$$S_s = 151.272 \, \text{m}^2, \quad AR = b_s^2/S_s = 2.2143$$

Recall that b_s and S_s refer to the complete "flattened" stabilizer. That is, these parameters are defined for the planform shown above in conjunction with its mirror image.

Because the airfoil is a smooth symmetrical shape, with a rounded and blended leading edge, it is assumed that $\kappa = 1.0$. Therefore, the derivative of the stabilizer's normal-force coefficient with respect to α is given by:

$$(C^*_{\tilde{N}\alpha})_s = \frac{2\pi \times 2.2143}{2 + \left[\dfrac{2.2143^2}{1.0^2}(1 + \tan^2(17.825°)) + 4\right]^{1/2}} = 2.7455 \, /\text{rad}$$

Next, from the cross-flow drag equations, one obtains:

$$(C_{Dc})_s = 3.8345 - 2.4895 \times AR + 1.4282 \times AR^2 - 0.4667 \times AR^3 + 0.0574 \times AR^4 = 1.6383$$

From the dimensioned drawing of the stabilizer panel, the summary geometric parameters for the equivalent trapezoidal configuration are:

$c_t \equiv c_{tip} = 4.98\,\text{m},$
$c_r = 9.01\,\text{m},$
$c_0 = 9.01\,\text{m},$
$y_0 = r_h = 5.769\,\text{m},$
$y_{te} = 3.71\,\text{m},$
$\Lambda_{le} = 50.0°,$
$\Lambda_{te} = 0.0°,$
$b = 18.302\,\text{m}$

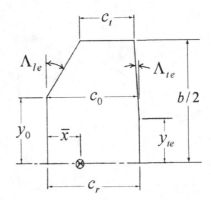

As described for the previous example, these parameters may be used to obtain the aerodynamic center, $\bar{x}$, (stabilizer center of pressure) from:

$$\bar{x} = \frac{2}{S_s}\left[\frac{c_r}{8}(2c_r y_0 - y_0^2 \tan\Lambda_{te}) + AC\left(\frac{b}{2} - y_0\right) + \frac{BD}{3}\left(\frac{b^3}{8} - y_0^3\right) + \frac{(BC+DA)}{2}\left(\frac{b^2}{4} - y_0^2\right)\right]$$

where,

$A = c_r + y_0 \tan\Lambda_{le} = 15.8852,\ B = -\tan\Lambda_{le} - \tan\Lambda_{te} = -1.1918,$

$C = \dfrac{c_r}{4} - \dfrac{3}{4} y_0 \tan\Lambda_{le} = -2.9039,\ D = \dfrac{3}{4}\tan\Lambda_{le} - \dfrac{1}{4}\tan\Lambda_{te} = 0.8938$

From this, one may obtain:

$$(l_s)_1 = l_h + \bar{x} = 57.0 + 2.6798 = 59.6798\,\text{m} \rightarrow (\hat{l}_s)_1 = 0.8232$$

Also, from the equations for the stabilizer's mean aerodynamic chord, given in the "Example 1: Small Aerostat" subsection, one obtains:

$$\bar{c}_s = 8.4012\,\text{m}$$

Next, from measurements on the geometry of the exposed stabilizer panel, one has:

$$(l_s)_2 = l_h + 5.5824 = 57.0 + 5.5824 = 62.5824\,\text{m} \rightarrow (\hat{l}_s)_2 = 0.8632$$

The average chord of the exposed stabilizer panel is $\bar{c}_{exposed} \approx (c_0 + c_t)/2 = 6.995\,\text{m}$, so that for $V = 15.433\,\text{m/s}$, the Reynolds number on the fins is:

$$(RN)_s = 6.85\,V\,\bar{c}_{exposed} \times 10^4 = 6.85 \times 15.433 \times 6.995 \times 10^4 = 7.395 \times 10^6$$

This gives a skin-friction coefficient of $(C_{fric.})_s = 0.00315$.
The average thickness/chord ratio, t/c, equals 0.114. Therefore,

$$(C_{Dwet})_s = (C_{fric.})_s\left[1 + 2\left(\frac{t}{c}\right)_s + 60\left(\frac{t}{c}\right)_s^4\right] = 0.00315 \times (1 + 0.228 + 0.0101) = 0.00390$$

The single-side area of an exposed stabilizer panel is measured to be $32.9329\,\text{m}^2$, so the *total wetted area* of the exposed stabilizer is:

$$(S_{\text{wet}})_s = 4 \times 32.9329 = 131.732\,\text{m}^2$$

Therefore, the zero-angle drag coefficient of the stabilizer, with respect to its reference area, is:

$$(C_{Ds})_0 = (C_{D\text{wet}})_s (S_{\text{wet}})_s / S_s = 0.00390 \times 131.732/151.2715 = 0.00340$$

The single vertical fin is identical to the individual stabilizer fin panels. Therefore, with respect to S_f, one has:

$$(C_{Df})_0 = (C_{D\text{wet}})_f (S_{\text{wet}})_f / S_f = 0.00390 \times 65.8666/75.6358 \rightarrow (C_{Df})_0 = 0.00340$$

As previously mentioned, the fins incorporate airfoils with rounded and blended leading edges. Therefore, the leading-edge suction efficiency may be assumed to be 0.90:

$$\eta_t \approx 0.90$$

Now, the calculated value of $(J_1)_{\text{total}}$ is $836.970\,\text{m}^2$. Also, $\Gamma_s = -25^0$, so $S_s \cos^2 \Gamma_s / (J_1)_{\text{total}} = 0.1485$. From the curve-fit equation for η_k one obtains:

$$(\eta_k)_{\text{long}} = 1.0 + 1.40 \left[S_s \cos^2 \Gamma_s / (J_1)_{\text{total}} \right]^2 = 1.0 + 1.40 \times 0.1485^2 = 1.0309$$

Also, S_{sh} is given by the portion of horizontal hull area between the leading-edge and trailing-edge hull-fin intersection points:

$$S_{sh} = (J_1)_{66.01} - (J_1)_{57.0} = 807.523 - 720.377 = 87.146\,\text{m}^2$$

Therefore $S_{sh}/S_s = 87.146/151.272 = 0.5761$, which gives:

$$\eta_s = 1 - 1.4 \times 0.5761 + 0.4 \times 0.5761^2 = 0.3262$$

This provides all of the inputs required to obtain the $C'_{\tilde{N}}$, $(C_D)_{\text{axial}}$ and $(C_M)_{\text{nose}}$ coefficients. Also note that the wind-axis coefficients may then be obtained from:

$$C_L = C_{\tilde{N}} \cos\alpha - (C_D)_{\text{axial}} \sin\alpha$$
$$C_D = (C_D)_{\text{axial}} \cos\alpha + C_{\tilde{N}} \sin\alpha$$

With these parameters, the results for C_L, C_D and $(C_M)_{\text{nose}}$ at angles of attack from -30^0 to 30^0 were calculated and are shown in the following plot:

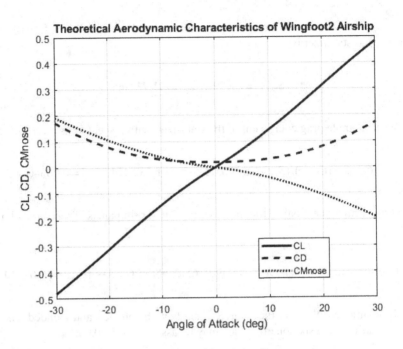

Polynomial curve-fit equations of these results, with α in *radians*, are:

$$C_L = 0.78486\alpha + 1.23768\alpha^3 - 2.77274\alpha^5$$

$$C_D = 0.01951 + 0.36249\alpha^2 + 1.19972\alpha^4 - 1.78078\alpha^6$$

$$(C_M)_{nose} = -0.19713\alpha - 1.08236\alpha^3 + 1.73186\alpha^5$$

Also, these polynomial-fit equations in *degrees* are:

$$C_L = 1.36983 \times 10^{-2}\alpha + 6.58024 \times 10^{-6}\alpha^3 - 4.49052 \times 10^{-9}\alpha^5$$

$$C_D = 0.01951 + 1.10421 \times 10^{-4}\alpha^2 + 1.11324 \times 10^{-7}\alpha^4 - 5.03354 \times 10^{-11}\alpha^6$$

$$(C_M)_{nose} = -3.44051 \times 10^{-3}\alpha - 5.75447 \times 10^{-6}\alpha^3 + 2.80478 \times 10^{-9}\alpha^5$$

As mentioned previously, wind-tunnel tests were performed in 1996 on a model of this configuration. The scale was 1/100 (much smaller than the NACA Akron model), and the tests were performed in the author's wind-tunnel laboratory at the University of Toronto Institute for Aerospace Studies. A comparison between the experimental results and the theoretical predictions are shown below:

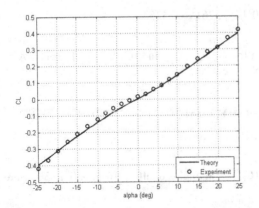
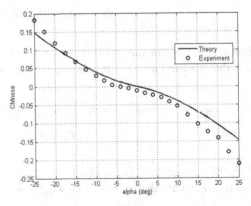

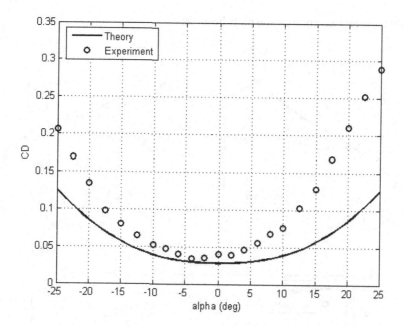

Although the C_L values compare very favorably, the $(C_M)_{nose}$ values are under-predicted as are the C_D values. However, it should be noted that the full-scale cylinder cross-flow drag coefficient, $(C_{Dc})_{cylinder} = 0.29$, was assumed. This model, however, at a wind speed of $V = 16.57$ m/s, is at a Reynolds number (based on model length of 0.725m) of 8.229×10^5. For the fineness ratio of 5.1202, this equates to a maximum cross-flow RN of 1.607×10^5. The figure shows cross-flow drag-coefficient data for an infinite circular cylinder as a function of cross-flow Reynolds number based on diameter (adapted from Figure 12 on Page 3-9 in "Fluid-Dynamic Drag" by Sighard Hoerner, Hoerner Fluid Dynamics, 1965).

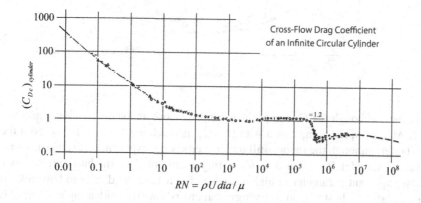

It is seen that the flow becomes super-critical in the region just short of $RN \approx 10$. In light of the wind-tunnel's small-scale turbulence, the assumption of the full-scale $(C_{Dc})_h$ might possibly be justified for angles-of-attack approaching 90^0. For the lower angles of interest, though, the effective cross-flow RN would be in the transition region between sub- and super-critical flows. If one assumes that $(C_{Dc})_{cylinder} = 0.7$, then the model's effective cross-flow drag coefficient is given by:

$$(C_{Dc})_{hull} = \gamma_{fineness} \times (C_{Dc})_{cylinder} = 0.6231 \times 0.7 = 0.4362$$

With this value, the revised comparison plots become:

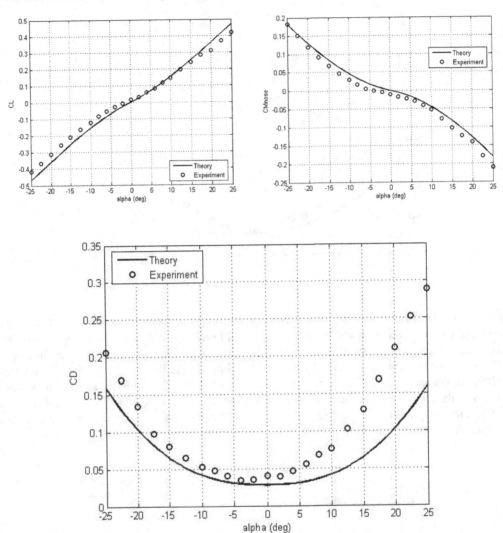

In this instance, although the C_L comparison is not as close as for the previous case, it still tracks fairly well. Moreover, the comparison with $(C_M)_{nose}$ is much improved. Further, even the C_D comparison is better; although the theory still under-predicts the experimental values (as was also seen for the Akron example). It is clear that, for testing airship models, the hull's cross-flow drag coefficient is an important parameter at higher angles of attack. If the data is to have relevance to the full-sized aircraft, it is best to insure super-critical cross-flow from either higher Reynolds numbers or sufficient surface roughness to trip the flow.

Example 4: TCOM CBV-71 Aerostat

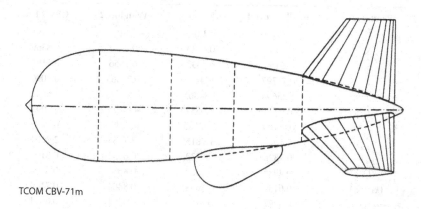

TCOM CBV-71m

The cross-flow analysis was applied to the 71 m-long aerostats designed and built by the TCOM Corporation, and the input parameters are listed in the summary table below.

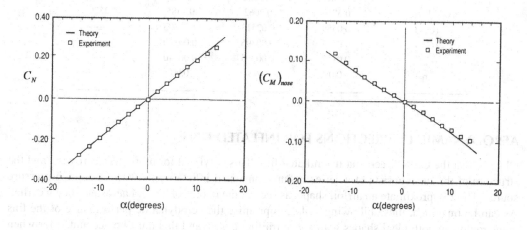

Wind-tunnel tests were performed on a model of this aerostat, and the resulting comparison between the analytical and experimental results for the normal-force coefficient, $C_{\tilde{N}}$ and pitching-moment coefficient, $(C_M)_{nose}$, is very satisfactory (no axial-force information is available). This, plus the close comparison between theory and experiment for the other examples, gives confidence in the cross-flow model for other applications such as calculations of stability derivatives, gust loading and estimations of force and moment distributions along the length of the hull.

Summary of Example Parameters

A summary table of the parameters for the four examples is given below:

Parameters	Small aerostat	ZRS-4	Wingfoot2	CBV-71 m
$\hat{S}_h$	1.0	1.0	1.0	1.0
$\hat{S}_s$	0.6150	0.4433	0.3712	0.8369
$\hat{S}_f$	0.6150	0.4433	0.1856	0.4185
$\hat{I}_1$	0.41797	0.21887	0.25665	0.2090
$\hat{J}_1$	1.25694	1.84505	1.76793	1.5542
$\hat{I}_3$	−0.07359	−0.05570	−0.04958	−0.1691
$\hat{J}_2$	0.46067	0.80585	0.75801	0.5498
$(C^*_{\tilde{N}\alpha})_s$	3.1508	1.7815	2.7455	2.7101
$(\hat{l}_s)_1$	0.7326	0.8132	0.8232	0.8108
$(\hat{l}_s)_2$	0.7985	0.8628	0.8632	0.8798
$(k_3 - k_1)$	0.6134	0.8693	0.8404	0.7169
η_s	0.3972	0.3288	0.3261	0.6684
$(\eta_k)_{long}$	1.1933	1.0595	1.0309	1.0034
$(C_{Dc})_s$	1.3667	2.7307	1.6383	1.1145
$(C_{Dc})_h$	0.1659	0.1845	0.18070	0.1200
S_{sh}/S_s	0.5028	0.5735	0.5761	0.2851
Γ_s	0.0 deg	0.0 deg	−25 deg	−45 deg
$S_s \cos^2 \Gamma_s /(J_1)_{total}$	0.3716	0.2062	0.1485	0.2329
$(C_{Dh})_0$	0.0589	0.02413	0.01803	----------
$(C_{Ds})_0$	0.0117	0.00395	0.00340	----------
$(C_{Df})_0$	0.0117	0.00395	0.00340	----------
η_t	0.0	0.90	0.90	----------

AERODYNAMIC CORRECTIONS FOR INFLATED FINS

Observe that the CBV-71 aerostat has inflated fins. This is typical for most large aerostats, and the structure of the fins consists of a series of interconnected inflated cells with a common pressure source. These approximate an airfoil shape, as seen in the photos of TCOM aerostats shown earlier. As can be imagined, these billowing cells compromise the aerodynamic performance of the fins compared to smooth ideal shapes with sharp trailing edges, and this must be accounted for when calculating $(C^*_{\tilde{N}\alpha})_s$, $(C_{Ds})_0$ and $(C_{Df})_0$. To quantitatively assess these effects, TCOM engaged the services of the author, in 1980, to perform a series of wind-tunnel tests on wing models fabricated to simulate various numbers of inflated cells. The photo below shows three of these models.

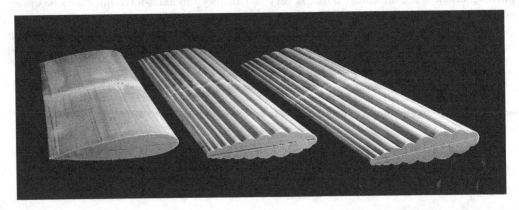

The reference wing incorporated a NACA 0018 symmetrical airfoil, to be compared with the performances of a 9-cell, 11-cell and 15-cell airfoil. The test Reynolds number was 2×10^5 and the models were hot-wire cut from blue insulating foam, giving a textured surface that increased the effective RN. The results are summarized in the following graphs, the first one giving the profile drag-coefficient variation as a function of the number of cells:

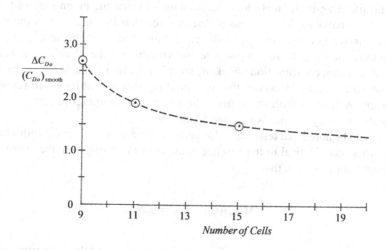

$(C_{Do})_{smooth}$ is the profile drag coefficient of the reference smooth wing with a sharp trailing edge, and ΔC_{Do} is the drag-coefficient increase above that value. For example, if a smooth-surfaced fin had a $(C_D)_0 = 0.02$, then the drag-coefficient increase for the 15-cell equivalent of this is $\Delta C_D \approx 1.50 \times 0.02 = 0.03$, which gives a total profile-drag coefficient for the 15-cell fin of $(C_D)_{15-Cell} = 0.05$. This shows the huge drag penalty caused by such an inflated-fin design. However, this is not actually a significant operational problem for tethered aerostats, in that any additional downwind-displacement caused by fin drag may be readily compensated for by a relatively small increase in net buoyancy. For aerostats, the convenience of inflated fins for storage, transportation and impact resistance are of greater practical importance than low drag.

The next graph shows the variation of lift-curve slope, between $-4^\circ \leq \alpha \leq 4^\circ$, as a function of the number of cells:

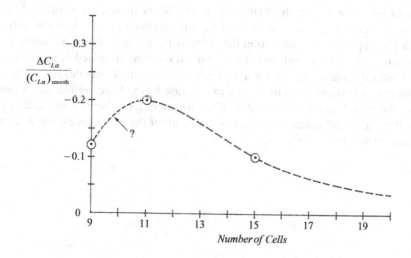

$(C_{L\alpha})_{\text{smooth}}$ is the lift-coefficient slope for the smooth-surfaced reference wing and $\Delta C_{L\alpha}$ is the increase in lift-coefficient slope above that value. For example, if a smooth fin has a $(C_{L\alpha})_{\text{smooth}} = 3.50/\text{rad}$, then the lift-slope decrease for the 15-cell equivalent of this is $\Delta C_{L\alpha} \approx -0.10 \times 3.50 = -0.35$. This gives a corrected lift-coefficient slope of $(C_{L\alpha})_{\text{15-Cell}} = 3.15/\text{rad}$. The effect is not nearly as dramatic as for the drag coefficient, but this does decrease the stability and must be compensated for with a relatively larger fin size.

Now, the profile-drag-increment plot "makes sense" in that the magnitude of the increment decreases as the number of cells increase. One can see that in the limit, as the number goes to infinity, the surface becomes the "smooth" case. This is also so for the lift-slope-increment plot, with the exception that there is a possible "maximum" for the 11-cell case. The tests were repeated and the balance calibration checked, so this puzzling behavior must be accepted as valid for these experiments. However, this is something that would be worth revisiting, either with more extensive tests or with an accurate fluid-dynamic computer program. It may well be that these results are very Reynolds number dependent.

Finally, it should be noted that tests were also performed on a smooth airfoil with a blunt rounded trailing edge. This was identical to the baseline smooth-surface wing with the chord cut down by 10%. It was found, for this case, that

$$\Delta C_{Do}/(C_{Do})_{\text{smooth}} = 0.20,$$

which is not an extreme drag increase compared with that from the billows. Further, the lift-curve slopes were virtually identical:

$$(C_{L\alpha})_{\text{blunt}} \approx (C_{L\alpha})_{\text{smooth}}$$

However, it must also be pointed out that reference length (chord) for the blunted wing was 10% less than that for the baseline smooth wing, so the aspect ratio of the blunted wing was 4.67 compared with 4.2 for the baseline wing. This increase in aspect ratio serves to increase the lift-coefficient vs. angle-of-attack slope, which compensates for a reduction of this value caused by the blunt trailing edge. In any case, the differences are slight.

ADDITIONAL OBSERVATIONS ABOUT AEROSTATS

A drawing of an aerostat from the 1970's, the CBV-250, is shown. As mentioned before, what characterizes tethered aerostats is the size of the fins relative to the body because they have to overcome a large upsetting moment from the elongated hull (the "Munk" moment). This applies to both the lateral moment as well as the longitudinal moment discussed in this chapter. In fact, achieving lateral dynamic stability for a *tethered* aircraft is more challenging than that for its longitudinal dynamic stability, as illustrated in this case by the huge vertical fins located further aft than the horizontal fins. A "rule-of-thumb" requirement for tethered aerostat lateral dynamic stability is that the lateral center of pressure must be aft of the confluence point (the tether point where the confluence lines join).

Balloons and Airships

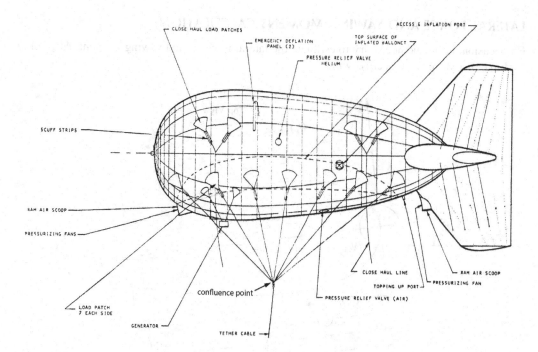

By comparison, the ZRS-4 "Akron" and "Wingfoot2" show relatively small fins compared with the hull. There are several reasons for this. First, fin size is constrained by the requirement to achieve static balance. Fins that are too large and heavy would require an unachievable forward-located mass center. Second, fin sizes are constrained by the mundane need for the airship to fit in a hangar, and to allow ground access to the gondola when moored in a level orientation. Finally, the requirement for dynamic stability is less constrained for an airship than it is for a tethered aerostat. An interesting fact that is well known to designers of airships and underwater vehicles is that even if the center of pressure is forward of the mass center (statically unstable), the vehicle can still have dynamic stability (discussed in detail at the end of the "Longitudinal Dynamic Stability" section of Chapter 6). This is a consequence of the fact that the mass of the vehicle is of the same order of magnitude as the mass of the displaced fluid (air or water). This is not the case for most airplanes, where static instability virtually guarantees dynamic instability.

ASIDE

Observe that the CBV-250 has a cruciform-fin configuration. The CBV-250 was designed in the early 1970's by the Schjeldahl Corporation of Northfield Minnesota, and it was the beginning of a new family of large aerostats using modern materials and computerized analysis for the aerodynamic and flight-dynamic performance predictions. The author was part of this project, which was his introduction to a career in Lighter-Than-Air (LTA) technology.

Because the aerostat was intended for use in the Florida Keys, snow and ice loads on the fins were not a consideration. However, as it became clear that these new aerostats would have useful applications as telecommunication-relay platforms in other climes, the huge inflated horizontal fins would be vulnerable to collapse from ice and snow accumulation. Therefore, the fin configuration was changed to an Inverted-Y arrangement, as shown in the previous pictures of the TCOM aerostats.

As for the cruciform fins on airships, their relatively small size evidently poses less of a weather problem.

END ASIDE

LATERAL FORCE AND YAWING MOMENT CALCULATION

On occasion, it may be necessary to estimate the lateral force, Y, and yawing moment, N_{nose}, of an airship either yawed or in a cross-wind.

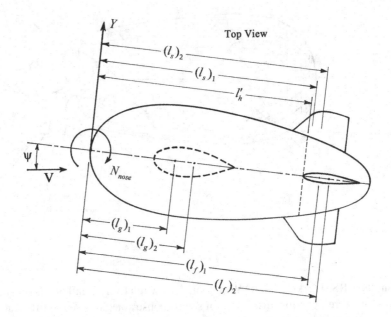

Top View

If it may be assumed that the airship is at a small angle of attack, α, the equations for these values are given by:

$$Y = C_Y \rho U^2 S/2$$

where:

$$C_Y = (k_2 - k_1)(\eta_k)_{lat} \tilde{I}_1 \sin(2\psi)\cos(\psi/2) + (C_{Dc})_h \tilde{J}_1 \sin\psi \sin|\psi|$$
$$+ \left[(C^*_{\tilde{N}\alpha})_f \eta_f \sin(2\psi)/2 \right] \hat{S}_f + \left[(C^*_{\tilde{N}\alpha})_s \eta_s \tilde{K} \sin^2\Gamma_s \sin(2\psi)/2 \right] \hat{S}_s$$
$$+ \left[(C^*_{\tilde{N}\alpha})_g \eta_g \sin(2\psi)/2 \right] \hat{S}_g$$
$$+(C_{Dc})_f \hat{S}_f \sin\psi \sin|\psi| + (C_{Dc})_s \hat{S}_s |\sin^3\Gamma_s| \sin\psi \sin|\psi| + (C_{Dc})_g \hat{S}_g \sin\psi \sin|\psi|$$

$$N_{nose} = (C_N)_{nose} \rho U^2 S b/2$$

where:

$$(C_N)_{nose} = -(k_2 - k_1)(\eta_k)_{lat} \tilde{I}_3 \sin(2\psi)\cos(\psi/2) - (C_{Dc})_h \tilde{J}_2 \sin\psi \sin|\psi|$$
$$- \left[(C^*_{\tilde{N}\alpha})_f \eta_f (\tilde{l}_f)_1 \sin(2\psi)/2 \right] \hat{S}_f - \left[(C^*_{\tilde{N}\alpha})_s \eta_s \tilde{K} \sin^2\Gamma_s (\tilde{l}_s)_1 \sin(2\psi)/2 \right] \hat{S}_s$$
$$- \left[(C^*_{\tilde{N}\alpha})_g \eta_g (\tilde{l}_g)_1 \sin(2\psi)/2 \right] \hat{S}_g$$
$$-(C_{Dc})_f \hat{S}_f (\tilde{l}_f)_2 \sin\psi \sin|\psi| - (C_{Dc})_s \hat{S}_s (\tilde{l}_s)_2 |\sin^3\Gamma_s| \sin\psi \sin|\psi|$$
$$-(C_{Dc})_g \hat{S}_g (\tilde{l}_g)_2 \sin\psi \sin|\psi|$$

Balloons and Airships

The k_1 and k_2 apparent-mass coefficients are the same as defined before where, for a body of revolution, $k_2 = k_3$.

> **ASIDE**
>
> The notational distinction between k_2 and k_3 implies that the aerodynamic equations might be applicable to a hull with elliptical cross sections. This would be a useful possibility, but the author has never experimentally confirmed this.
>
> **END ASIDE**

The integrals are also the same as defined before, except that "b" is used as the reference length (actually equal to $\bar{c}$ for the typical axisymmetric airship and aerostat) and the limits of integration are now from the hull's nose to the intersection point of the *vertical* fin's leading-edge on the hull, l'_h:

$$\tilde{I}_1 = \frac{1}{S}\int_0^{l'_h}\frac{dA}{d\xi}d\xi = \frac{A_h}{S}; \quad \tilde{I}_3 = \frac{1}{Sb}\int_0^{l'_h}\xi\frac{dA}{d\xi}d\xi; \quad \tilde{J}_1 = \frac{1}{S}\int_0^{l'_h}2r d\xi = \frac{J_1}{S}; \quad \tilde{J}_2 = \frac{1}{Sb}\int_0^{l'_h}2r\xi d\xi$$

Regarding the empennage, if the fins are arranged in a cruciform configuration, as for the "Small Aerostat" or "Akron" examples, $(\eta_k)_{lat}$, η_f, $(C^*_{N\alpha})_f$ and $(C_{Dc})_f$ are calculated in the same way as for a zero-dihedral stabilizer. For example:

$$(\eta_k)_{lat} = 1.0 + 1.40\left[S_f/(J_1)_{total}\right]^2, \quad \eta_f = 1 - 1.4(S_{fh}/S_f) + 0.4(S_{fh}/S_f)^2$$

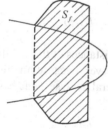
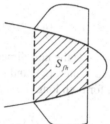

From this S_f planform, one may calculate the aspect ratio $(AR)_f$ and mid-chord sweep angle $(\Lambda_{c/2})_f$ (from the equivalent straight-tapered planform) from which $(C^*_{N\alpha})_f$ may be found, in the same way as for the stabilizer:

$$(C^*_{N\alpha})_t = \frac{2\pi(AR)_f}{2+\left[\frac{(AR)_f^2}{\kappa^2}(1+\tan^2(\Lambda_{c/2})_f)+4\right]^{1/2}}$$

> **ASIDE**
>
> Note that the "α" in $(C^*_{N\alpha})_f$ is now understood to represent a "lateral angle of attack".
>
> **END ASIDE**

Likewise, $(AR)_f$ and the equivalent straight-tapered planform's taper ratio, λ_f, give the parameters required for the calculation of $(C_{Dc})_f$, in the same way as for the stabilizer.

If the empennage has a "Y" or "Inverted-Y" configuration, as for the "Wingfoot2" or CBV-71, one has to define a "virtual planform" that consists of the vertical fin plus its mirror image:

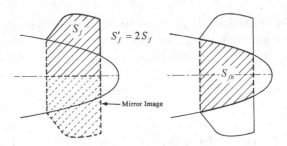

The area of this virtual planform is $S'_f \equiv 2S_f$. Therefore, one obtains:

$$(\eta_k)_{lat} = 1.0 + 1.40\left[S'_f/(J_1)_{total} \right]^2, \quad \eta_f = 1 - 1.4(S_{fh}/S'_f) + 0.4\left(S_{fh}/S'_f \right)^2$$

Further, this virtual planform has an aspect ratio $(AR)_f$ and mid-chord sweep angle $(\Lambda_{c/2})_f$, from which $(C^*_{N\alpha})_f$ may be found using the previous equation. Also, $(C_{Dc})_f$ is calculated in the same way.

Now, a factor $\tilde{K}$ appears in the yawing-moment equation. This accounts for the mutual interference between both panels of the dihedralled stabilizer, as discussed for an airplane's wing in the "Estimation of the Lateral Stability Derivatives" section of Chapter 6. The main effect occurs near the root area of a dihedralled wing. However, in this case, that area is typically subsumed within the hull. Therefore, $\tilde{K} \approx 1.0$.

The parameters $(\tilde{l}_f)_1$ and $(\tilde{l}_f)_2$ are the non-dimensional distances from the hull's nose to the fin's aerodynamic center and exposed-panel area center, respectively:

$$(\tilde{l}_f)_1 = (l_f)_1/b, \quad (\tilde{l}_f)_2 = (l_f)_2/b$$

Likewise, the equations include the non-dimensional distances of the stabilizer's aerodynamic center and exposed-panel area center, $(\tilde{l}_s)_1$ and $(\tilde{l}_s)_2$. However, because airships and aerostats use the same reference length, $\bar{c}$, for both longitudinal and lateral analyses, one has:

$$(\tilde{l}_s)_1 = (\hat{l}_s)_1, \quad (\tilde{l}_s)_2 = (\hat{l}_s)_2$$

Note that airships typically have a gondola beneath the hull, which serves as a control car as well as passenger space.

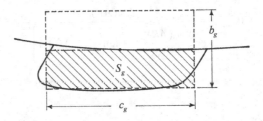

This component may also contribute to the lateral force and yawing moment. An approximate aerodynamic model is to assume that the gondola acts as half of a rectangular fin, as illustrated above. The aspect ratio is simply $(AR)_g = b_g/c_g$. Therefore, its "lateral lift-curve slope" is estimated from:

$$(C^*_{\tilde{N}\alpha})_g \approx \frac{2\pi(AR)_g}{2+\left[(AR)_g^2+4\right]^{1/2}}$$

For the gondola's cross-flow drag coefficient, $(C_{Dc})_g$, the previously mentioned Wardlaw curves would give a value on the order of 5. However, Wardlaw's cross-flow model is meant to be applied to flattened surfaces with small-radius edges (e.g., typical wings and fins). Therefore, $(C_{Dc})_g$ would certainly be significantly attenuated by the generously rounded edges of a typical gondola. In fact, from the previously discussed wind-tunnel tests of the Zeppelin model, it was found that the comparison between lateral theoretical predictions and wind-tunnel data was best when $(C_{Dc})_g \approx 0$. Therefore, barring further research on this effect, one may assume that $(C_{Dc})_g \approx 0$. The gondola cross-flow drag coefficient term is included in the equations for completeness, in the case of an atypical configuration where the inclusion of a finite value for (C_{Dc}) is required.

Finally, the parameters $(\tilde{l}_g)_1$ and $(\tilde{l}_g)_2$ are the non-dimensional distances from the hull's nose to the quarter-chord and half-chord locations, respectively, of the gondola's equivalent rectangular planform:

$$(\tilde{l}_g)_1 = (l_g)_1/b, \quad (\tilde{l}_g)_2 = (l_g)_2/b$$

Also, $\hat{S}_g$ is the gondola's non-dimensional reference area:

$$\hat{S}_g = S_g/S$$

NUMERICAL EXAMPLE:

The Zeppelin model will be chosen for analysis because lateral wind-tunnel data exist for comparison. As noted before, this design incorporates an Inverted-Y empennage, with a stabilizer dihedral angle of $\Gamma_s = -25^0$. Also, all three fins are identical and occupy the same longitudinal position on the hull. Therefore, the parameters are:

$$k_1 = 0.0571, \quad k_2 = 0.8975, \quad \tilde{I}_1 = 0.25665, \quad \tilde{I}_3 = -0.04958, \quad \tilde{J}_1 = 1.76793, \quad \tilde{J}_2 = 0.75801$$

$$(\eta_k)_{lat} = 1.0457, \quad \eta_f = 0.3261, \quad \eta_s = 0.3261, \quad (C^*_{\tilde{N}\alpha})_f = 2.7455, \quad (C^*_{\tilde{N}\alpha})_s = 2.7455$$

$$\hat{S}_f = S_f/S = 0.18562, \quad \hat{S}_s = S_s/S = 0.37125, \quad (\tilde{l}_f)_1 = (l_f)_1/b = 0.82317$$

$$(\tilde{l}_f)_2 = (l_f)_2/b = 0.86321, \quad (\tilde{l}_s)_1 = (l_s)_1/b = 0.82317, \quad (\tilde{l}_s)_2 = (l_s)_2/b = 0.86321$$

$$(C_{Dc})_f = 1.6383, \quad (C_{Dc})_s = 1.6383, \quad (C_{Dc})_h = 0.4362$$

Note that the previously discussed wind-tunnel model value of $(C_{Dc})_h$ has been chosen.

From measurements on the gondola and its location on the hull, one obtains that:

$$\hat{S}_g = 0.0483, \quad (AR)_g = 0.4699 \rightarrow (C^*_{\tilde{N}\alpha})_g = 0.728, \quad (\tilde{l}_g)_1 = 0.2548, \quad (\tilde{l}_g)_2 = 0.2865$$

Also, η_g is assumed to be unity: $\eta_g \approx 1$. It might, in fact, be higher because of the faster cross-flow at the top and bottom of the hull. In any case, though, the gondola's contribution for this example is small.

The resulting theoretical predictions and co-plotted experimental data are shown below:

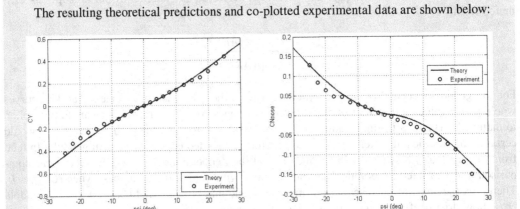

Overall, the comparison is very favorable. So, for this case at least, the veracity of the aerodynamic model is supported.

AIRSHIP AERODYNAMIC MYSTERY

While on the subject of aerodynamic behavior, observe the longitudinal pressure distributions on three airship-hull models (no fins) from wind-tunnel tests by the Zeppelin Corporation in the early 1940's in the figure below (the wind flow is from right to left):

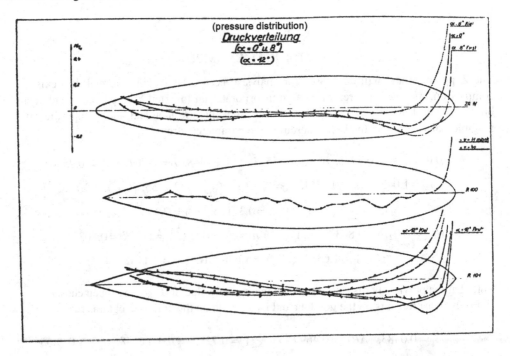

This is taken from "Aerodynamic Model Tests with German and Foreign Airship Designs in the Wind Tunnel of the Zeppelin Airship Works at Friedrichshafen", ZWB Report FB. No. 1647, April, 1942 (TPA3/TIB Translation No.GDC.1Q/1381 T, Issued by the Ministry of Supply); and the author is listed as Eng. M. Schirmer.

The first airship hull model is that for the U.S. Navy ZRS-4, "Akron" and the pressure-coefficient distributions are for:

1. Zero angle of attack
2. Along the top ("keel") and side ("equator") for eight degrees angle of attack

The shapes of the distributions are smoothly continuous and of the form expected from potential-flow theory applied to such a streamlined configuration.

The second hull model is that for the British airship R100, and the distribution is measured for zero angle of attack. It is seen that this distribution has a surprisingly wavy shape. In fact, Mr. Schirmer wrote the following paragraph in his report:

> The wavy course of the R100 pressure distribution measurements over a longitudinal frame at $\alpha = 0^0$ as well as at angles of incidence in the equator and on the lee-side is especially striking. It has to be noted that those measurements have been carried out under exactly the same conditions as for the rest of the models. The author was allowed to see film shots taken from a plane accompanying the R100 in flight. They showed a strangely marked, radial swelling of the ship's envelope, distributed over the whole length of the body, and following the pressure curves of the model body in even closer sequences. As the pressure in the interior of the ship is lower than that of the open air, mainly as a result of the ventilation appliances, a pressure which changes its sign may make its influence felt on the envelope of one field, producing such swellings. It must be noted that this system has not been observed on the form R101 which is very similar to R100, differing mainly only in the pointed bow.

The third hull model is that for the R101 (a British airship that was competitive to the R100) and, as Mr. Schirmer states, its pressure distributions give no surprises.

The wind-tunnel models were large (over 2m long) and built with a steel frame covered with sheet metal (tin) approximately 10mm shy of the final shape, which was then built up with plaster. The surface was carefully shaped, smoothed and finished with a polished waxed surface and perforated with flush pressure holes. So, the wavy pressure distribution is undoubtedly a true aerodynamic effect and not caused by an irregular surface. It would be interesting to apply a modern CFD analysis to the R100 shape to see if this strange behavior can be replicated.

MATLAB PROGRAM

The calculations for both the longitudinal and lateral aerodynamic coefficients were coded in MATLAB and presented below. Note that this program is particularized for the Wingfoot2 airship with its Inverted-Y empennage. Application to other empennage configurations, such as cruciform, would require modification of statements for the vertical fin, such as area as well as aerodynamic and area center locations.

IMPORTANT ASIDE

The following listing differs from the original MATLAB listing in an important way. The original MATLAB font size is smaller than the readable font size given in this book. Therefore, some long statements that originally fitted on one line are now broken into two or more lines. These do not (and generally cannot) incorporate the ellipsis (...) required by MATLAB at the end of the first line for the continuation of a statement in the next line. When this is encountered in transcribing this listing into a MATLAB program, the two (or more) lines must be brought back into a contiguous form (all on one line). Otherwise, an error message will result. Anyone familiar with MATLAB programming will readily recognize this.

END ASIDE

```
clc
fprintf('Aerodynamics for Wingfoot2 Airship')
fprintf('\n')
fprintf('\n')
fprintf('16 March, 2021')
fprintf('\n')
fprintf('\n')
clear all
% Calculation of Hull Integrals for Aerostats and Airships
% Listing of Hull Ordinates (all dimensions in mks system)

NS=99; %number of stations
Lh=57.0; %distance from nose to the hull/fin intersection point

S=407.4692; %hull reference area
c=72.5; %hull reference length

x(1)=0; r(1)=0;
x(2)=0.264; r(2)=0.968;
x(3)=0.892; r(3)=1.8154;
x(4)=1.198; r(4)=2.114;
x(5)=1.463; r(5)=2.3424;
x(6)=1.653; r(6)=2.4944;
x(7)=2.48; r(7)=3.0736;
x(8)=2.5; r(8)=3.0865;
x(9)=2.574; r(9)=3.1331;
x(10)=3.35; r(10)=3.5845;
x(11)=4.037; r(11)=3.94;
x(12)=4.247; r(12)=4.0426;
x(13)=5.164; r(13)=4.461;
x(14)=6.098; r(14)=4.8438;
x(15)=6.248; r(15)=4.9018;
x(16)=6.6; r(16)=5.0339;
x(17)=7.044; r(17)=5.193;
x(18)=8.004; r(18)=5.5049;
x(19)=8.822; r(19)=5.7343;
x(20)=8.975; r(20)=5.7732;
x(21)=9.957; r(21)=5.9951;
x(22)=10.948; r(22)=6.1758;
x(23)=11.947; r(23)=6.3218;
x(24)=12.0; r(24)=6.3287;
x(25)=12.95; r(25)=6.4398;
x(26)=13.957; r(26)=6.5362;
x(27)=14.718; r(27)=6.5987;
x(28)=14.965; r(28)=6.6176;
x(29)=15.972; r(29)=6.6887;
x(30)=16.98; r(30)=6.7508;
x(31)=17.8; r(31)=6.7953;
x(32)=17.987; r(32)=6.8047;
x(33)=18.993; r(33)=6.8512;
x(34)=20.0; r(34)=6.8912;
x(35)=21.007; r(35)=6.9255;
x(36)=22.014; r(36)=6.955;
x(37)=23.021; r(37)=6.9804;
x(38)=23.54; r(38)=6.9922;
x(39)=24.028; r(39)=7.0026;
x(40)=24.3; r(40)=7.0081;
x(41)=25.036; r(41)=7.0221;
```

```
x(42)=26.044;  r(42)=7.0388;
x(43)=27.052;  r(43)=7.0527;
x(44)=28.06;   r(44)=7.0639;
x(45)=28.7;    r(45)=7.0696;
x(46)=29.068;  r(46)=7.0724;
x(47)=30.076;  r(47)=7.0778;
x(48)=31.085;  r(48)=7.0798;  %Maximum Diameter
x(49)=31.1;    r(49)=7.0798;
x(50)=37.6;    r(50)=7.0798;
x(51)=38.593;  r(51)=7.0778;
x(52)=39.601;  r(52)=7.0712;
x(53)=40.609;  r(53)=7.0595;
x(54)=41.617;  r(54)=7.0421;
x(55)=42.624;  r(55)=7.0186;
x(56)=43.632;  r(56)=6.9883;
x(57)=43.658;  r(57)=6.9874;
x(58)=44.5;    r(58)=6.9564;
x(59)=44.639;  r(59)=6.9507;
x(60)=45.646;  r(60)=6.9057;
x(61)=46.653;  r(61)=6.853;
x(62)=47.659;  r(62)=6.7923;
x(63)=48.665;  r(63)=6.7236;
x(64)=49.671;  r(64)=6.6465;
x(65)=50.5;    r(65)=6.5765;
x(66)=50.675;  r(66)=6.5609;
x(67)=51.4;    r(67)=6.4937;
x(68)=51.679;  r(68)=6.4664;
x(69)=52.682;  r(69)=6.3619;
x(70)=53.684;  r(70)=6.2465;
x(71)=54.684;  r(71)=6.119;
x(72)=55.682;  r(72)=5.9783;
x(73)=56.0;    r(73)=5.9305;
x(74)=56.678;  r(74)=5.8236;
x(75)=57.672;  r(75)=5.655;
x(76)=58.2;    r(76)=5.5597;
x(77)=58.663;  r(77)=5.4731;
x(78)=59.653;  r(78)=5.2788;
x(79)=60.64;   r(79)=5.0728;
x(80)=61.5;    r(80)=4.884;
x(81)=61.625;  r(81)=4.856;
x(82)=62.607;  r(82)=4.6284;
x(83)=62.8;    r(83)=4.5823;
x(84)=63.587;  r(84)=4.3885;
x(85)=64.562;  r(85)=4.1346;
x(86)=64.8;    r(86)=4.07;
x(87)=65.532;  r(87)=3.8643;
x(88)=66.496;  r(88)=3.5709;
x(89)=67.0;    r(89)=3.4045;
x(90)=67.454;  r(90)=3.2454;
x(91)=67.5;    r(91)=3.2286;
x(92)=68.397;  r(92)=2.876;
x(93)=69.313;  r(93)=2.4475;
x(94)=70.186;  r(94)=1.9448;
x(95)=70.3;    r(95)=1.87;
x(96)=71.004;  r(96)=1.3569;
x(97)=71.2;    r(97)=1.1979;
x(98)=71.771;  r(98)=0.7;
x(99)=72.5;    r(99)=0;
```

```
% Plot Hull Shape
I=[1:1:NS];
plot(x(I),r(I),'linewidth',2);
grid on
xlabel('Axial Length (m)')
ylabel('Radius (m)')
title('Hull Profile of Wingfoot Airship')

% Calculation of Cross-Sectional Areas at the Longitudinal Hull Stations
for i=1:1:NS
    A(i)=pi*r(i)^2;
    Ihat1(i)=A(i)/S;
end

Volume=trapz(x,A);
Sh=(Volume)^(2/3);

% Calculation of I2 Hull Integral
Ihat2(1)=0;
for j=2:1:NS
    Ihat2(j)=(0.5*(A(j)+A(j-1))*(x(j)-x(j-1)))/(S*c)+Ihat2(j-1);
end

% Calculation of I3 Hull Integral
sum3=0.0;
Ihat3(1)=0.0;
for j=2:1:NS
    sum3=(A(j)+A(j-1))*(x(j)-x(j-1))/2+sum3;
    Ihat3(j)=(x(j)*A(j)-sum3)/(S*c);
end

% Calculation of I4 Hull Integral
Ihat4(1)=0.0;
for j=2:1:NS

Ihat4(j)=0.5*(A(j)+A(j-1))*(x(j)-x(j-1))*(x(j-1)+0.5*(x(j)-x(j-1)))...
        /(S*c^2)+Ihat4(j-1);
end

% Calculation of I5 Hull Integral (superceded)
sum5(1)=0.0;
Ihat5(1)=0.0;
for j=2:1:NS

sum5(j)=0.5*(A(j)+A(j-1))*(x(j)-x(j-1))*(x(j-1)+0.5*(x(j)-x(j-1))^2)...
        +sum5(j-1);
    Ihat5(j)=sum5(j)/(S*c^3)-2*Ihat4(j);
end

% Calculation of I5 Hull Integral
for j=1:1:NS
    Ihat5(j)=Ihat1(j)*(x(j)/c)^2-2*Ihat4(j);
end

% Calculation of J1 and J2 Integrals
Jhat1(1)=0.0;
Jhat2(1)=0.0;
```

```
for m=2:1:NS
    Jhat1(m)=((r(m)+r(m-1))*(x(m)-x(m-1)))/S+Jhat1(m-1);
    Jhat2(m)=((r(m)+r(m-1))*(x(m)-x(m-1)))/(S*c)*x(m)+Jhat2(m-1);
end

fprintf('Total Length=%8.4f, Reference Area=%8.4f, Total
Volume=%9.4f\n'...
    ,c,S,Volume)
fprintf('\n')
fprintf('Summary of Hull Integrals\n')
fprintf('\n')
fprintf('  Axial     Radius    Ihat1     Ihat2     Ihat3     Ihat4
Ihat5     Jhat1     Jhat2\n')
for k=1:1:NS
fprintf('%7.5f   %7.5f   %7.5f   %7.5f   %7.5f   %7.5f   %7.5f   %7.5f
%7.5f\n',...
x(k),r(k),Ihat1(k),Ihat2(k),Ihat3(k),Ihat4(k),Ihat5(k),Jhat1(k),Jhat2(k))
end

fprintf('\n')

% Calculation of apparent-mass coefficients
%fineness ratio:
f=max(x)/(2*max(r));
e=sqrt((f^2-1)/f^2);
gamma=2*((1-e^2)/e^3)*(0.5*log((1+e)/(1-e))-e);
k1=gamma/(2-gamma);
delta=1/e^2-(1-e^2)/(2*e^3)*log((1+e)/(1-e));
k3=delta/(2-delta);
kprime=e^4*(delta-gamma)/((2-e^2)*(2*e^2-(2-e^2)*(delta-gamma)));
fprintf('Fineness Ratio=%6.4f, k1=%5.4f, k3=%5.4f, kprime=%5.4f\n'...
    ,f,k1,k3,kprime)
fprintf('\n')

% Calculate the total wetted area and axial location of the area centre
Swet=0.0;
MomSwet=0.0;
for m=1:1:NS-1
    DeltaSwet=pi*(r(m)+r(m+1))*sqrt((x(m+1)-x(m))^2+(r(m+1)-r(m))^2);
    Swet=Swet+DeltaSwet;
    DeltaMomSwet=DeltaSwet*0.5*(x(m)+x(m+1));
    MomSwet=MomSwet+DeltaMomSwet;
end
LSwet=MomSwet/Swet;
fprintf('Total Wetted Area=%9.4f, Axial Area-Centre Location=%9.4f\n'...
    ,Swet,LSwet)
fprintf('\n')

fprintf('Axial Location of Hull/Stabilizer Intersection
Point=%9.4f\n',Lh)
fprintf('\n')

% Calculation of hull integrals at hull/fin intersection point
for ij=1:1:NS-1
    if x(ij)<Lh
        if x(ij+1)>=Lh
            IHAT1=(Ihat1(ij+1)-Ihat1(ij))/(x(ij+1)-x(ij))*(Lh-x(ij))...
                +Ihat1(ij);
```

```
                IHAT2=(Ihat2(ij+1)-Ihat2(ij))/(x(ij+1)-x(ij))*(Lh-x(ij))...
                    +Ihat2(ij);
                IHAT3=(Ihat3(ij+1)-Ihat3(ij))/(x(ij+1)-x(ij))*(Lh-x(ij))...
                    +Ihat3(ij);
                IHAT4=(Ihat4(ij+1)-Ihat4(ij))/(x(ij+1)-x(ij))*(Lh-x(ij))...
                    +Ihat4(ij);
                IHAT5=(Ihat5(ij+1)-Ihat5(ij))/(x(ij+1)-x(ij))*(Lh-x(ij))...
                    +Ihat5(ij);
                JHAT1=(Jhat1(ij+1)-Jhat1(ij))/(x(ij+1)-x(ij))*(Lh-x(ij))...
                    +Jhat1(ij);
                JHAT2=(Jhat2(ij+1)-Jhat2(ij))/(x(ij+1)-x(ij))*(Lh-x(ij))...
                    +Jhat2(ij);
                rh=(r(ij+1)-r(ij))/(x(ij+1)-x(ij))*(Lh-x(ij))+r(ij);
            end
        end
end
fprintf('Ihat1=%8.5f,  Ihat3=%8.5f,  Jhat1=%8.5f,  Jhat2=%8.5f\n',IHAT1,...
    IHAT3,JHAT1,JHAT2)
fprintf('\n')

fprintf('Ihat2=%8.5f,  Ihat4=%8.5f,  Ihat5=%8.5f\n',IHAT2,IHAT4,IHAT5)
fprintf('\n')

% Location of hull volumetric center (from nose)
Lcv=trapz(x,x.*A)/Volume;
fprintf('Total Hull Volume=%8.4f,  Location of Hull Volumetric 
Center=%8.4f\n'...
    ,Volume,Lcv)
fprintf('\n')

% Calculation of Hull Skin-Friction Drag
%nominal flow speed (30 knots):
U=15.433;
%hull nominal Reynolds number (mks):
RNhull=6.85*10^4*U*c;
%turbulent BL skin friction:
Cfhull=0.455/(log10(RNhull))^2.58;
Cdhullwet=Cfhull*(1+1.5*(1/f)^1.5+7*(1/f)^3);
Cdhullfric=Cdhullwet*(Swet/S);

% Calculation of Base Drag
%radius when the hull flow separates:
rbase=0;
Sbase=pi*rbase^2;
Cdhullbase=0.029*(Sbase/S)/(Cfhull*(Swet/Sbase))^0.5;

% Total zero-angle hull drag
Cdhullbare=Cdhullfric+Cdhullbase;
% Correction for excrescences such at external lines, load patches, etc.
ExcressFactor=1.05;
Cdhull=ExcressFactor*Cdhullbare;
fprintf('Total Zero-Angle Hull Drag Coeff.=%7.5f\n',Cdhull)
fprintf('\n')

% Hull crossflow drag coefficient
gammafineness=0.5088+0.0275*f-0.001011*f^2;
Cdcrosshull=0.29*gammafineness;
```

```
fprintf('Hull Cross-Flow Drag Coeff.=%7.5f\n',Cdcrosshull)
fprintf('\n')
fprintf('\n')

% Empennage Properties
fprintf('Empennage Properties\n')
fprintf('\n')

%horizontal fins dihedral angle (deg):
Gammadeg=-25.0;
Gamma=Gammadeg*pi/180;

% These are the parameters for the "equivalent trapezoidal stabilizer"
cr=9.01;
c0=9.01;
ctip=4.98;
%total flattened span of horizontal fins:
bs=18.302;
y0=rh;
yte=3.71;
%leading-edge sweep in degrees:
lamle=50.0;
%leading-edge sweep in radians
lamlerad=lamle*pi/180;
%trailing-edge sweep in degrees
lamte=0.0;
%trailing-edge sweep in radians
lamterad=lamte*pi/180;
%stabilizer flattened area:
Ss=2*cr*yte+(cr+c0)*(y0-yte)+(ctip+c0)*(bs/2-y0);

% This calculates the axial pressure-center location from leading edge
A=cr+y0*tan(lamlerad);
B=-tan(lamlerad)-tan(lamterad);
C=0.25*cr-0.75*y0*tan(lamlerad);
D=0.75*tan(lamlerad)-0.25*tan(lamterad);
Atilda=0.125*cr*(2*cr*y0-y0^2*tan(lamterad));
Btilda=(A*C)*(0.5*bs-y0);
Ctilda=(B*D/3)*(0.125*bs^3-y0^3);
Dtilda=0.5*(B*C+D*A)*(0.25*bs^2-y0^2);
xbar=(2/Ss)*(Atilda+Btilda+Ctilda+Dtilda);

% This estimates the area and axial centre of the exposed fin area

% Calculations for the trapezoidal portion:
%area of trapezoidal portion:
Sexposed1=0.5*(c0+ctip)*(bs/2-y0);
%height of trapezoidal portion
H1=bs/2-y0;
%vertical location of trapezoidal area centre from hull/fin intersection
%point:
yarea1prime=H1*(c0+2*ctip)/(3*(c0+ctip));
%vertical location of area centre of trapezoidal portion from hull axis:
yarea1=H1*(c0+2*ctip)/(3*(c0+ctip))+y0;
tlamprime=(H1*tan(lamlerad)+0.5*(ctip-c0))/H1;
%axial location of area centre of trapezoidal portion from hull/fin
%intersection point:
```

```
xarea1=0.5*c0+yarea1prime*tlamprime;

% Calculations for the triangular portion:
%base of triangular portion:
lbase=sqrt(cr^2+(y0-yte)^2);
delta=acos(cr/lbase);
%height of triangular portion
hheight=cr*sin(delta);
bbar=hheight*tan(delta+lamterad);
abar=(lbase+bbar)/3;
xprimearea2=lbase-abar;
yprimearea2=hheight/3;
%axial location of area centre of triangular portion from hull/fin
%intersection point:
xarea2=xprimearea2*cos(delta)+yprimearea2*sin(delta);
%vertical location of area centre of triangular portion from hull axis:
yarea2=y0-xprimearea2*sin(delta)+yprimearea2*cos(delta);
%area of triangular portion:
Sexposed2=0.5*lbase*hheight;

% Total exposed-fin properties:
%exposed area of one fin on one side
Sexposed=Sexposed1+Sexposed2;
%axial area centre of exposed fin from hull/fin intersection point
xarea=(xarea1*Sexposed1+xarea2*Sexposed2)/Sexposed;
%vertical area centre of exposed fin from hull axis
yarea=(yarea1*Sexposed1+yarea2*Sexposed2)/Sexposed;
fprintf('xarea=%6.4f, yarea=%6.4f, (Sexposed)
approx=%7.4f\n',xarea,yarea,Sexposed)
fprintf('\n')

% These are the parameters for the "equivalent straight-tapered fin"
croot=10.865;
%half-chord sweep angle in degrees:
lambdahalf=17.825;
%half-chord sweep angle in radians
lambdahalfrad=lambdahalf*pi/180;
taper=ctip/croot;
AR=bs^2/Ss;
%the ratio of the actual section lift-curve slope over 2*pi:
kappa=1.0;

% This is the normal-force slope of the flattened, isolated, fins:
Cnastar=2*pi*AR/(2+sqrt((AR/kappa)^2*(1+(tan(lambdahalfrad))^2)+4));

% This is the cross-flow drag coefficient for the flattened isolated
fins:
A0=2.001+4.0002*taper;
A1=-0.8667-4.5383*taper+2.177*taper^2;
A2=0.2373+4.1957*taper-3.4854*taper^2;
A3=-0.0339-1.6913*taper+1.6298*taper^2;
A4=0.0012+0.2316*taper-0.2376*taper^2;
Cdcrossstar=A0+A1*AR+A2*AR.^2+A3*AR.^3+A4*AR.^4;

% This is the fin's mean aerodynamic chord:
Tte=tan(lamterad);
Tle=tan(lamlerad);
```

```
Int1=cr^2*yte;
Aint=cr^2*(y0-yte);
Bint=cr*Tte*(y0^2-yte^2);
Cint=2*cr*Tte*(y0-yte);
Dint=Tte*(y0^3-yte^3)/3;
Eint=Tte*yte*(y0^2-yte^2);
Fint=Tte*yte^2*(y0-yte);
Int2=Aint-Bint+Cint+Dint-Eint+Fint;
Gint=c0^2*(bs/2-y0);
Hint=c0*Tle*(bs^2/4-y0*bs+y0^2);
Jint=c0*Tte*(bs^2/4-yte*bs-y0^2+2*yte*y0);
Kint=2*Tle*Tte*(bs^3/24-(yte+y0)*bs^2/8+y0*yte*bs/2-...
y0^3/3+(yte+y0)*y0^2/2-y0^2*yte);
Lint=Tle*(bs^3/24-y0*bs^2/4+y0^2*bs/2-y0^3/3);
Mint=Tte*(bs^3/24-yte*bs^2/4+yte^2*bs/2-y0^3/3+yte*y0^2-yte^2*y0);
Int3=Gint-Hint-Jint+Kint+Lint+Mint;
cf=(2/Ss)*(Int1+Int2+Int3);
fprintf('Stab. characteristic chord, c=%6.4f   Stab. characteristic area,
Ss=%6.4f\n'...
    ,cf,Ss)
fprintf('\n')
fprintf('Flattened isolated stabilizer lift-curve slope,(Cna)
star=%6.4f\n'...
    ,Cnastar)
fprintf('\n')
fprintf('Flattened isolated stabilizer cross-flow drag coeff.,(Cdcross)
star=%6.4f\n'...
    ,Cdcrossstar)
fprintf('\n')
fprintf('Lift-curve slope of isolated dihedralled stabilizer,(CL1)
s=%6.4f\n'...
    ,Cnastar*(cos(Gamma))^2)
fprintf('\n')
fprintf('Stabilizer flattened span, bs=%8.4f,   Aspect
Ratio=%7.4f\n',bs,AR)
fprintf('\n')

% Fin-Moment Lengths
%distance from blimp nose to the aerodynamic center of horiz. fins:
Ls1=Lh+xbar;
%distance from blimp nose to area center of exposed horiz. fins:
Ls2=Lh+xarea;
fprintf('Stabilizer Flattened Area, Ss=%6.4f,   Ls1=%6.4f,
Ls2=%6.4f\n',Ss,Ls1,Ls2)
fprintf('\n')
fprintf('Axial distance to stabilizer aerodynamic centre,
Lotab=%6.4f\n',Ls1)
fprintf('\n')
fprintf('Axial distance to stabilizer area centre,
Lareastab=%6.4f\n',Ls2)
fprintf('\n')
fprintf('Vertical distance to fin area centre, Haf=%6.4f\n',yarea)
fprintf('\n')

% Calculation of the spanwise location of the aerodynamic center:
Tte=tan(lamterad);
Tle=tan(lamlerad);
```

```
J1int=bs/(3*pi);
J2int=cr*yte^2/(2*Ss);
J3int=(1/Ss)*((cr+yte*Tte)*(y0^2-yte^2)/2-Tte*(y0^3-yte^3)/3);
J4int=(1/Ss)*((c0+y0*Tle)*((bs/2)^2-y0^2)/2-Tle*((bs/2)^3-y0^3)/3-Tte*...
    ((bs/2)^3-y0^3)/3+Tte*yte*((bs/2)^2-y0^2)/2);
d=J1int+J2int+J3int+J4int;
fprintf('Spanwise distance to fin aerodynamic centre, Hfin=%6.4f\n',d)
fprintf('\n')
fprintf('Spanwise distance to stabilizer aerodynamic centre,
Hstab=%6.4f\n'...
    ,d*sin(Gamma))
fprintf('\n')

% Calculation of zero-angle drag for stabilizer
%the approximate average chord of the exposed stabilizer:
csexposed=(c0+ctip)/2;
%the average thickness-to-chord ratio:
thick=0.114;
%exposed side of single fin panel:
Sspanelexposed=Sexposed;
Swets=4*Sspanelexposed;
%fin nominal Reynolds number (mks):
RNfin=6.85*10^4*U*csexposed;
%turbulent BL skin friction:
Cfstab=0.455/(log10(RNfin))^2.58;
Cdstabwet=Cfstab*(1+2.0*thick+60*thick^4);
Cdstabfric=Cdstabwet*(Swets/Ss);
%correction factor for inflated fins:
CorrectFactor=1.0;
Cdstabfric=Cdstabfric*CorrectFactor;
Cdstabbracing=0.0;

%Zero-Angle Drag Coefficient of Stabilizer
Cdstab=Cdstabfric+Cdstabbracing;
fprintf('Total Zero-Angle Stabilizer Drag Coeff.=%7.5f\n',Cdstab)
fprintf('\n')

%**********************************************************************
%For this tri-fin case, with identical fin panels:
Cdfin=Cdstab;
Sf=Ss/2;
fprintf('Total Zero-Angle Vertical-Fin Drag Coeff.=%7.5f\n',Cdfin)
fprintf('\n')
%**********************************************************************

% Hull Efficiency Factor
J1total=Jhat1(NS)*S;
HullFinParameter=Ss*(cos(Gamma))^2/J1total;
etaklong=1.0+1.40*(HullFinParameter)^2;
etaklat=1.0+1.40*(Ss/J1total)^2;

% Fin Efficiency Factor
%distance from nose to rearward hull/fin intersection point:
Lte=66.01;
for ij=1:1:NS-1
    if x(ij)<Lte
        if x(ij+1)>=Lte
```

```
                    JHAT1te=(Jhat1(ij+1)-Jhat1(ij))/
(x(ij+1)-x(ij))*(Lte-x(ij))+Jhat1(ij);
                    rte=(r(ij+1)-r(ij))/(x(ij+1)-x(ij))*(Lte-x(ij))+r(ij);
            end
        end
end
Sfh=(JHAT1te-JHAT1)*S;
Sprime=(Ss-Sfh)/2;
fprintf('Actual exposed area of individual fin=%8.4f\n',Sprime)
fprintf('\n')
etas=1.0-1.4*(Sfh/Ss)+0.4*(Sfh/Ss)^2;

etat=0.9; %leading-edge suction efficiency

fprintf('Hull Efficiency Factors, etaklong=%6.4f,
etaklat=%6.4f\n',etaklong,etaklat)
fprintf('\n')
fprintf('Stabilizer Efficiency Factor, etas=%6.4f\n',etas)
fprintf('\n')
fprintf('Leading-Edge Suction Efficiency, etat=%6.4f\n',etat)
fprintf('\n')
fprintf('Induced drag coefficient of isolated dihedralled stabilizer,
(CD2)s=%6.4f\n',(Cnastar*(cos(Gamma))^2)^2/(pi*AR))
fprintf('\n')
fprintf('\n')

% Calculation of Static Longitudinal Aerodynamic Coefficients
Anglesweep=60;
for ii=1:1:Anglesweep+1
    adeg(ii)=ii-(Anglesweep/2+1);
    alpha=adeg(ii)*pi/180;
    % Leading-Edge Suction Coefficient
    Cts=0.5*Cnastar*(cos(Gamma))^2*etas*etat*sin(2*alpha)*tan(alpha)-...
        0.25*(Cnastar*(cos(Gamma))^2*etas*sin(2*alpha))^2*etat/(pi*AR);
    Cn=((k3-k1)*etaklong*IHAT1+0.5*Cnastar*(cos(Gamma))^2*etas*Ss/S)...
        *sin(2*alpha)+(Cdcrosshull*JHAT1+Cdcrossstar*(cos(Gamma))^3*
        Ss/S)...
        *sin(alpha)*abs(sin(alpha));
    Cdaxial=(Cdhull*Sh/S+Cdstab*Ss/S+Cdfin*Sf/S)*(cos(alpha))^2-...
        (k3-k1)*etaklong*IHAT1*sin(2*alpha)*sin(alpha/2)-Cts*Ss/S;
    CMnose(ii)=-((k3-k1)*etaklong*IHAT3+...
        0.5*(Ls1/c)*Cnastar*(cos(Gamma))^2*etas*Ss/S)*sin(2*alpha)-...
        (Cdcrosshull*JHAT2+...

Cdcrossstar*(cos(Gamma))^3*(Ls2/c)*Ss/S)*sin(alpha)*abs(sin(alpha));
    CL(ii)=Cn*cos(alpha)-Cdaxial*sin(alpha);
    CD(ii)=Cdaxial*cos(alpha)+Cn*sin(alpha);
    CMcv(ii)=CMnose(ii)+Cn*Lcv/c;
    arad(ii)=adeg(ii)*pi/180;
    Lhatnp(ii)=-CMnose(ii)/Cn;
end

fprintf('Summary of Aerodynamic Properties of Wingfoot2 Airship\n')
fprintf('\n')
fprintf(' Angle      CL          CD         CMnose      CMcv       (Lhat)
np\n')
for k=1:1:Anglesweep+1
```

```
fprintf('%4.2f    %8.5f    %8.5f    %8.5f    %8.5f    %8.5f\n',...
adeg(k),CL(k),CD(k),CMnose(k),CMcv(k),Lhatnp(k))
end

% Plot Aerodynamic Results
figure
I=[1:1:Anglesweep+1];
plot(adeg(I),CL(I),'-k',adeg(I),CD(I),'--k',adeg(I),CMnose(I),':k','linew
idth',2);
grid on
xlabel('Angle of Attack (deg)')
ylabel('CL, CD, CMnose')
title('Theoretical Aerodynamic Characteristics of Wingfoot2 Airship')
legend('CL','CD','CMnose','Location','NW')

% Plot Non-Dimensional Longitudinal Pressure Centre
figure
I=[Anglesweep/2:1:Anglesweep];
plot(adeg(I),Lhatnp(I),'linewidth',2);
grid on
xlabel('Angle of Attack (deg)')
ylabel('(Lhat)np')
title('Non-Dimensional Longitudinal Neutral Point for Wingfoot2 Airship')

% Curve Fits of Aerodynamic Results
fprintf('\n')
fprintf('Lift Coefficient Polynomial Terms (alpha in radians)')
format long

CLterms=polyfit(arad,CL,5);
fprintf('\n')
fprintf('CL0=%7.5f, CL1=%7.5f, CL2=%7.5f, CL3=%7.5f, CL4=%7.5f,
CL5=%7.5f\n',...
    CLterms(6),CLterms(5),CLterms(4),CLterms(3),CLterms(2), CLterms(1))
fprintf('\n')

fprintf('Drag Coefficient Polynomial Terms (alpha in radians)')
CDterms=polyfit(arad,CD,6);
fprintf('\n')
fprintf('CD0=%7.5f, CD1=%7.5f, CD2=%7.5f, CD3=%7.5f, CD4=%7.5f,
CD5=%7.5f, CD6=%7.5f\n',...
    CDterms(7),CDterms(6),CDterms(5),CDterms(4),CDterms(3),CDterms(2),
    CDterms(1))
fprintf('\n')

fprintf('Moment Coefficient Polynomial Terms (about nose, alpha in
radians)')
CMnoseterms=polyfit(arad,CMnose,5);
fprintf('\n')
fprintf('CM0=%7.5f, CM1=%7.5f, CM2=%7.5f, CM3=%7.5f, CM4=%7.5f,
CM5=%7.5f\n',...
    CMnoseterms(6),CMnoseterms(5),CMnoseterms(4),CMnoseterms(3),
    CMnoseterms(2),...
    CMnoseterms(1))
fprintf('\n')

fprintf('Lift Coefficient Polynomial Terms (alpha in degrees)')
```

Balloons and Airships

```
CLterms=polyfit(adeg,CL,5);
fprintf('\n')
fprintf('CL0=%9.5e, CL1=%9.5e, CL2=%9.5e, CL3=%9.5e, CL4=%9.5e,
CL5=%9.5e\n',...
    CLterms(6),CLterms(5),CLterms(4),CLterms(3),CLterms(2), CLterms(1))
fprintf('\n')

fprintf('Drag Coefficient Polynomial Terms (alpha in degrees)')
CDterms=polyfit(adeg,CD,6);
fprintf('\n')
fprintf('CD0=%9.5e, CD1=%9.5e, CD2=%9.5e, CD3=%9.5e, CD4=%9.5e,
CD5=%9.5e, CD6=%9.5e\n',...
    CDterms(7),CDterms(6),CDterms(5),CDterms(4),CDterms(3),CDterms(2),...
    CDterms(1))
fprintf('\n')

fprintf('Moment Coefficient Polynomial Trems (about nose, alpha in
degrees)')
CMnoseterms=polyfit(adeg,CMnose,5);
fprintf('\n')
fprintf('CM0=%9.5e, CM1=%9.5e, CM2=%9.5e, CM3=%9.5e, CM4=%9.5e,
CM5=%9.5e\n',...
    CMnoseterms(6),CMnoseterms(5),CMnoseterms(4),CMnoseterms(3),
    CMnoseterms(2),...
    CMnoseterms(1))
fprintf('\n')

% Lateral Characteristics at Alpha=0
k2=k3;
etaf=etas;
Sf=Ss/2;
LateralHullFinParameter=2*Sf/(J1total);
etaklat=1.0+1.40*(LateralHullFinParameter)^2;
CYbetaf=-Cnastar*etaf*Sf/S;
Ktilda=1.0;
CYbetas=-Ktilda*Cnastar*etaf*(sin(Gamma)^2)*Ss/S;
CYbetahull=-2*(k2-k1)*etaklat*IHAT1;
CYbeta=CYbetaf+CYbetas+CYbetahull;
fprintf('These are the contributions of the fin, stabilizer, and hull to
\n')
fprintf('the total lateral-force derivative, all non-dimensionalized to
S\n')
fprintf('CYbetaf=%6.4f,   CYbetas=%6.4f,   CYbetahull=%6.4f\n',CYbetaf,...
    CYbetas,CYbetahull)
fprintf('\n')
Lf1=Ls1;
CNbetaf=-CYbetaf*Lf1/c;
CNbetas=-CYbetas*Ls1/c;
CNbetanosehull=2*(k2-k1)*etaklat*IHAT3;
CNbetanose=CNbetaf+CNbetas+CNbetanosehull;
fprintf('These are the contributions of the fin, stabilizer, and hull to
\n')
fprintf('the total lateral-moment derivative about the nose, all non-
dimensionalized to S\n')
fprintf('CNbetaf=%6.4f,   CNbetas=%6.4f,   CNbetanosehull=%6.4f\n',...
    CNbetaf,CNbetas,CNbetanosehull)
fprintf('\n')
```

```
Lhatcplat=-CNbetanose/CYbeta;
fprintf('Lateral Parameters\n')
fprintf('CYbeta=%6.4f,  CNbeta=%6.4f, (Lhatcp)lat=%6.4f\n',CYbeta,...
    CNbetanose,Lhatcplat)
fprintf('\n')
fprintf('\n')
fprintf('Inputs for Lateral Stability Derivative Calculations\n')
fprintf('\n')
fprintf('CCYVf=%6.4f, CCYVs=%6.4f\n',-Cnastar,-Cnastar*(sin(Gamma))^2)
fprintf('CLV0s=%6.4f\n',-Cnastar*(d/bs)*sin(Gamma)*(cos(Gamma)^2))
fprintf('CNR0f=%6.4f\n',-pi/4)
fprintf('CNDR0b=%7.5f\n',-2*kprime*(c^2+(c/f)^2)*Volume/(5*S*c^3))
```

Appendix A: Multhropp Body-Moment Equation

From NACA 1036, the body-moment equation is:

$$M_{body} = \frac{\pi q}{2}\int_0^l w^2 \beta \, dx = \frac{\pi q}{2}\int_0^l w^2 (\alpha + \varepsilon_u)\, dx$$

where q is the dynamic pressure, $\rho V^2/2$, l is the body's total length, w is the local body width, α is the body's angle of attack and ε_u is the local up-wash angle due to the presence of the wing. In coefficient form, this equation becomes:

$$(C_M)_{body} = \frac{\pi}{2\bar{c}_b S_b}\int_0^l w^2 (\alpha + \varepsilon_u)\, dx$$

Therefore, the derivative with respect to α is:

$$(C_{M\alpha})_{body} = \frac{\pi}{2\bar{c}_b S_b}\int_0^l w^2 \left(1 + \frac{\partial \varepsilon_u}{\partial \alpha}\right) dx = \frac{\pi}{2\bar{c}_b S_b}\int_0^l w^2 \frac{\partial \beta}{\partial \alpha}\, dx$$

The variation of up-wash rate, $\partial \beta / \partial \alpha$, along the length of the body is illustrated below:

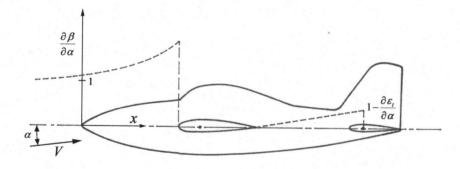

It is seen that from the nose to the wing's leading edge, the up-wash rate increases rapidly. However, at the wing's trailing edge, the up-wash rate is zero because the wake leaves the wing at a constant angle to the body's reference line, for all α values. From this point, it is assumed that the up-wash rate increases linearly to the tail's aerodynamic center, where the value is $1-(\partial \varepsilon/\partial \alpha)_{tail}$.

Appendix B: Alternative Swept-Wing Analysis

Franklin Diederich at the NACA expanded upon the Schrenk methodology for the purposes of refining this for spanwise load estimations of swept wings, as well as obtaining aerodynamic influence coefficients for aeroelastic analysis. Diederich's methodology was not extended into calculating aerodynamic centers or pitching-moment coefficients, so the author thought this would be informative for comparison with his own methodology.

Mr. Diederich worked on this topic for years, and the culmination of his research is presented in NACA Technical Note 2751, "A Simple Approximate Method for Calculating Spanwise Lift Distributions and Aerodynamic Influence Coefficients at Subsonic Speeds", August 1952. For the following equations, Diederich's terminology is used, except for those terms that coincide with the author's. In those cases, the nomenclature from Chapter 4 will be used.

Diederich introduces a spanwise function, $\gamma_a(y)$, which is equivalent to the formerly-defined Basic Lift Distribution divided by C_L:

$$\gamma_a(y) = BDL/C_L$$

Further, γ_a is assumed to be of the form:

$$\gamma_a(y) = C_1 \frac{c_w(y)}{c_{avg}} + C_2 \frac{4}{\pi}\left\{1 - \left(\frac{y}{b/2}\right)^2\right\}^{1/2} + C_3 f(y)$$

where,

$$C_1 + C_2 + C_3 = 1$$

It can be seen, in comparison with equations in the chapter, that $\gamma_a(y)$ is equivalent to $fn(y)c_0/c_{avg}$. So, the alternative basic lift distribution, BLD', is given by:

$$BLD' = \gamma_a(y)C_L$$

Also, note that:

$$S = 2c_{avg} \int_0^{b/2} \gamma_a(y)\,dy$$

Now, in order to proceed with the calculation, C_1, C_2, C_3 and f have to be obtained. Diederich presents graphs for these:

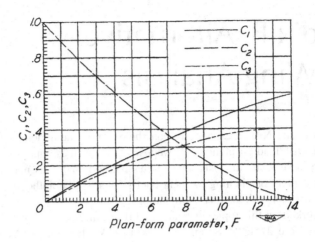

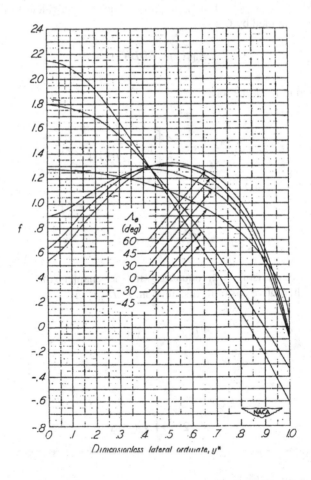

Appendix B: Alternative Swept-Wing Analysis

In order to apply these to the "Example 1, Straight-Tapered Linear-Twisted Wing" in Chapter 4 linearly swept and tapered wing, the "planform parameter" F was calculated:

$$F = \frac{AR}{\kappa \cos \Lambda_{c/4}} = \frac{4.444}{1 \times \cos(22.3°)} = 4.803$$

ASIDE

Recall that κ is the ratio of airfoil lift-curve slope (per radian) to the ideal lift-curve slope of 2π. For the Example 1 wing, κ is assumed to be 1.

END ASIDE

From the C_i vs. F graph, $C_1 = 0.250$, $C_2 = 0.537$, $C_3 = 0.213$. Obtaining $f(y)$, however, is more of a challenge. In this case the curve for $\Lambda_{c/4} = 22.3°$ was approximated by interpolating values between the $\Lambda_{c/4} = 0$ and $\Lambda_{c/4} = 30°$ curves, and then curve-fitting a fifth-order polynomial. The result is shown below:

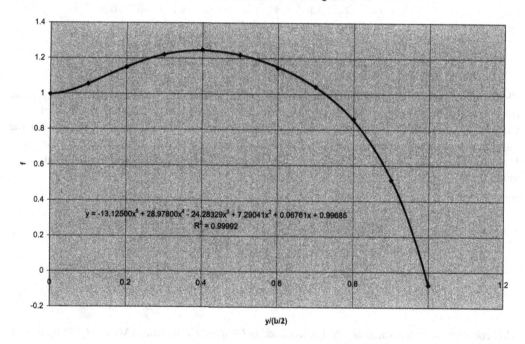

This is an adequate fit, but the author wishes that Diederich had given the equations, or the reference, from which the $f(y)$ curves were obtained. The evaluation of these might have been tedious in 1952 (hence the graphs), but this could have been easily programmed now.

With these values, the basic lift distribution was calculated and plotted:

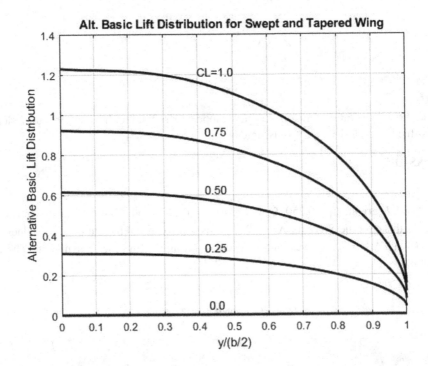

There are some differences in comparison with results from the author's methodology (note the corresponding figure in the "Example 1, Straight-Tapered Linear-Twisted Wing" section in Chapter 4). Diederich's results show less outer loading on the wing's span. Therefore, for the purposes of structural analysis, this example wing would experience smaller bending moments. However, when the Diederich loading is used to calculate the axial aerodynamic center, the value is:

$$\bar{l}_{ac} = \frac{\int_0^{b/2} \gamma_a(y) l_{ac}(y) dy}{\int_0^{b/2} \gamma_a(y) dy} \rightarrow \bar{l}_{ac} = 1.638, \text{ (Diederich Equations)}$$

This compares with the value of 1.661, calculated in Chapter 2 using the NACA TR 572 methodology. The difference between these two methodologies is only 1.38%.

Next, for the twisted lift distribution, Diederich introduces another spanwise function, $\gamma_b(y)$, which is equivalent to the formerly defined Twisted Lift Distribution.

$$\gamma_b = k_1 C_{L\alpha} \Delta\delta(y) \gamma_a = TDL'$$

where k_1 is a function of F, as shown in the following graph:

Appendix B: Alternative Swept-Wing Analysis

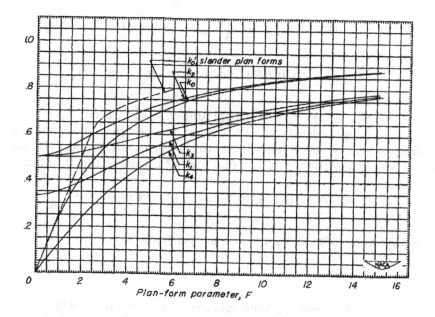

ASIDE

The other k values in the graph are for additional analyses performed by Diederich for rolling wings, etc.

END ASIDE

Further,

$$\Delta\delta(y) = \delta(y) - \overline{\delta}'_{ZLL}$$

where $\overline{\delta}'_{ZLL}$ is the aerodynamic zero-lift line given by:

$$\overline{\delta}'_{ZLL} = \frac{2c_{avg}}{S} \int_0^{b/2} \gamma_a(y)\delta(y)\,dy = -1.2992°$$

Now, when $\left(C'_{Mac}\right)_{loading}$ is calculated for the example wing from:

$$\left(C'_{Mac}\right)_{loading} = \frac{2}{S} \int_0^{b/2} \gamma_b \left[l_{ac}(y) - \overline{l}_{ac} \right] dy$$

The result is $\left(C'_{Mac}\right)_{loading} = 0.0071$. This is far off from that value calculated for the example wing in Chapter 2. However, if the $C_{L\alpha}$ term is replaced with the section lift-curve slope coefficient, $C_{l\alpha} = 0.10966/\deg\,(2\pi/\text{rad})$, then the result is:

$$\left(C'_{Mac}\right)_{loading} = 0.0113 \rightarrow \left(C_{Mac}\right)_{loading} = \left(C'_{Mac}\right)_{loading} c_{avg}/\bar{c} = 0.0113 \times 2.25/2.333 =$$

$$\left(C_{Mac}\right)_{loading} = 0.01090, \text{(Modified Diederich equations)}$$

This compares favorably to the Chapter 2 value of 0.01019, being 7% higher. Therefore, the author can only conclude that the original Diederich equation was either in error or a misprint.

In any case, this exercise demonstrates that the author's simpler methodology also gives reasonable values in comparison to both those from the Chapter 2 example and the Diederich equations:

$$\bar{l}_{ac} = 1.6624, \quad \left(C_{Mac}\right)_{loading} = 0.0116, \text{(Author's Equations)}$$

The Modified Diederich twisted-lift distribution curve is shown below:

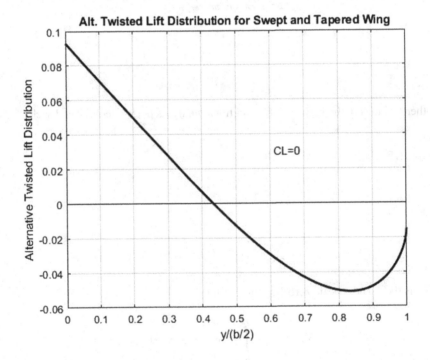

This looks more realistic, especially in comparison with the Multhopp curve shown in the "Example 1, Straight-Tapered Linear-Twisted Wing" section in Chapter 4. The values go to zero at the wing's tip because of the inclusion of γ_a in the TDL' equation. Also, the k_1 factor is less than 1 and, in fact, is 0.53 for the example wing. This provides some veracity for Schrenk's assumption of 1/2.

Further, the Schrenk and Diederich curves are co-plotted below, which shows the results are rather close for this particular example:

Appendix B: Alternative Swept-Wing Analysis

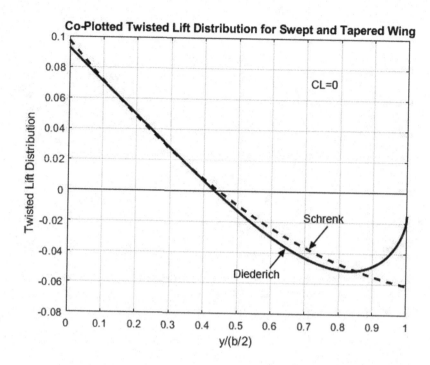

This again supports the replacement of $C_{L\alpha}$ in the original Diederich equation with the sectional lift-curve slope coefficient, $C_{l\alpha}$.

Finally, it would be informative to explore an extension of the Diederich method to wings with multiple sweep angles, such as described in the "Example 2, Double-Swept and Double-Tapered Wing" section of Chapter 4. In Diederich's earlier work, "A Simple Approximate Method for Obtaining Spanwise Lift Distributions over Swept Wings", NACA RM L7107, 1948, he states that this methodology may be applied to such cases by assuming an average sweep angle defined as follows:

$$\overline{\Lambda}_{c/4} = \tan^{-1}\left[\frac{2}{S}\int_0^{b/2} \tan \Lambda_{c/4}\, c_w(y)\, dy\right]$$

This might be a promising direction, though the author would welcome learning more about the origins of the f graphs.

Appendix C: Rigid-Body Equations of Motion

KINEMATICS

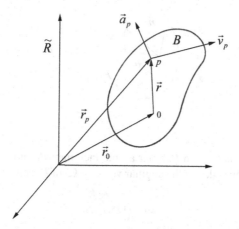

The system shown above is a rigid body, B, in motion relative to an inertial reference frame, $\tilde{R}$. For most aircraft, and all subsonic aircraft, an earth-fixed reference frame may be considered to be "inertial". The body's kinematics relative to the inertial reference frame is fully described if one knows the following information for every point p in B:

1. Location, $\vec{r}_p$, relative to $\tilde{R}$.
2. Velocity, $\vec{v}_p$, relative to $\tilde{R}$.
3. Acceleration, $\vec{a}_p$, relative to $\tilde{R}$.

It will be shown that the above requirements are met if one knows the following information for the body, B:

1. Location, $\vec{r}_0$, of a fixed point, o, in B, relative to $\tilde{R}$.
2. Velocity, $\vec{v}_0$, of point o relative to $\tilde{R}$.
3. Angular velocity, $\vec{\Omega}$, of B relative to $\tilde{R}$.
4. Acceleration, $\vec{a}_0$, of point o relative to $\tilde{R}$.
5. Angular acceleration, $\vec{\alpha}$, of B relative to $\tilde{R}$.

First, note that if $\vec{c}$ is an arbitrary vector, then:

$$\frac{^{\tilde{R}}d\vec{c}}{dt} = \frac{^{B}d\vec{c}}{dt} + \vec{\Omega} \times \vec{c}$$

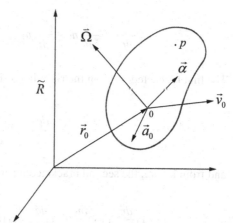

where:

$^{\tilde{R}}d\vec{c}/dt \equiv$ the time rate of change of $\vec{c}$ relative to $\tilde{R}$.
$^{B}d\vec{c}/dt \equiv$ the time rate of change of $\vec{c}$ relative to B.

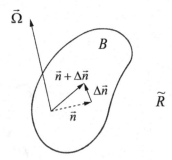

Proof:
Consider a vector, $\vec{n}$, which is *fixed* in B. After a time increment Δt, it rotates (relative to $\tilde{R}$) to a new position $\vec{n} + \Delta\vec{n}$. This rotation is due to the angular velocity, $\vec{\Omega}$, of B relative to $\tilde{R}$, so that $\Delta\vec{n}$ is given by:

$$\Delta\vec{n} = (\vec{\Omega} \times \vec{n})\Delta t$$

From this it is easy to see that the time derivative of $\vec{n}$ relative to $\tilde{R}$ is given by:

$$\left(\frac{\Delta\vec{n}}{\Delta t}\right)_{\Delta t \to 0} \to \frac{^{\tilde{R}}d\vec{n}}{dt} = \vec{\Omega} \times \vec{n}$$

Now, consider a triad of unit vectors, $\vec{n}_1, \vec{n}_2$ and $\vec{n}_3$ fixed in B. Vector $\vec{c}$ may then be expressed as:

$$\vec{c} = c_1\vec{n}_1 + c_2\vec{n}_2 + c_3\vec{n}_3$$

and the time derivative of this is:

$$\frac{d\vec{c}}{dt} = \left(\frac{dc_1}{dt}\vec{n}_1 + \frac{dc_2}{dt}\vec{n}_2 + \frac{dc_3}{dt}\vec{n}_3\right) + \left(c_1\frac{d\vec{n}_1}{dt} + c_2\frac{d\vec{n}_2}{dt} + c_3\frac{d\vec{n}_3}{dt}\right)$$

The first bracketed term on the right-hand side is simply the time-rate-of-change of $\vec{c}$ relative to B:

$$\frac{dc_1}{dt}\vec{n}_1 + \frac{dc_2}{dt}\vec{n}_2 + \frac{dc_3}{dt}\vec{n}_3 = \frac{^{B}d\vec{c}}{dt}$$

and from before, the second bracketed term on the right-hand side may be expressed as:

$$c_1\frac{d\vec{n}_1}{dt} + c_2\frac{d\vec{n}_2}{dt} + c_3\frac{d\vec{n}_3}{dt} = c_1(\vec{\Omega} \times \vec{n}_1) + c_2(\vec{\Omega} \times \vec{n}_2) + c_3(\vec{\Omega} \times \vec{n}_3) = \vec{\Omega} \times \vec{c}$$

Appendix C: Rigid-Body Equations of Motion

Therefore, together, one has:

$$\frac{^{\tilde{R}}d\vec{c}}{dt} = \frac{^{B}d\vec{c}}{dt} + \vec{\Omega} \times \vec{c}$$

which completes the proof.

Now, observe from the first figure that a point, p, in the body, B, may be located relative to a fixed point in an inertial reference frame, $\tilde{R}$, by the vector $\vec{r}_p$, where:

$$\vec{r}_p = \vec{r}_o + \vec{r}$$

and the velocity of p in $\tilde{R}$ is given by:

$$\vec{v}_p = \frac{^{\tilde{R}}d\vec{r}_p}{dt} = \frac{^{\tilde{R}}d\vec{r}_o}{dt} + \frac{^{\tilde{R}}d\vec{r}}{dt}$$

Also, $^{\tilde{R}}d\vec{r}_o/dt = \vec{v}_o$, which is the velocity of point o in the inertial reference frame, and from before:

$$\frac{^{\tilde{R}}d\vec{r}}{dt} = \frac{^{B}d\vec{r}}{dt} + \vec{\Omega} \times \vec{r}$$

So, if $\vec{r}$ is fixed in B, then one obtains that the velocity of point p in $\tilde{R}$ is:

$$\vec{v}_p = \vec{v}_o + \vec{\Omega} \times \vec{r}$$

Further, the acceleration of point p relative to $\tilde{R}$ is given by:

$$\vec{a}_p = \frac{^{\tilde{R}}d\vec{v}_p}{dt} = \frac{^{\tilde{R}}d\vec{v}_o}{dt} + \frac{^{\tilde{R}}d\vec{\Omega}}{dt} \times \vec{r} + \vec{\Omega} \times \frac{^{\tilde{R}}d\vec{r}}{dt}$$

Note the following:

$^{\tilde{R}}d\vec{v}_o/dt = \vec{a}_o$, the acceleration of point o, relative to $\tilde{R}$.
$^{\tilde{R}}d\vec{r}/dt = \vec{\Omega} \times \vec{r}$, from before for $\vec{r}$, fixed in B.
$^{\tilde{R}}d\vec{\Omega}/dt = \vec{\alpha}$, the angular acceleration of B relative to $\tilde{R}$.

So, one obtains that:

$$\vec{a}_p = \vec{a}_o + \vec{\alpha} \times \vec{r} + \vec{\Omega} \times (\vec{\Omega} \times \vec{r})$$

Observe that, in general, $\vec{v}_o, \vec{a}_o, \vec{\Omega}$ and $\vec{\alpha}$ are not constant with respect to B, but may be varying with time.

ASIDE

The example rigid body certainly doesn't look like an aircraft, but the author has found that the derivation is clearer when it isn't obscured by drawing some sort of airplane shape. When the author was an undergraduate, his friends and he would refer to this as "blob theory".

END ASIDE

DYNAMIC EQUATIONS

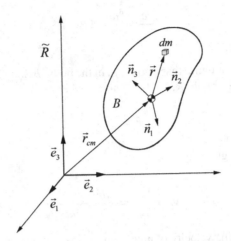

Consider that the inertial reference frame, $\tilde{R}$, is defined by an orthogonal triad of unit vectors, $\vec{e}_1$, $\vec{e}_2$ and $\vec{e}_3$. Also, a triad of orthogonal unit vectors, $\vec{n}_1$, $\vec{n}_2$ and $\vec{n}_3$, are fixed in the body, B, and have their origins at the body's mass center. If the body's total mass is given by m, Newton's force-acceleration equation for the body is:

$$\vec{F} = m\vec{a}_{cm}$$

where $\vec{a}_{cm}$ is the acceleration of the body's mass center in the inertial reference frame. Therefore, $\vec{a}_{cm}$ may also be written as the time derivative of the mass center's velocity in the inertial reference frame:

$$\vec{a}_{cm} = \frac{{}^{\tilde{R}}d\vec{v}_{cm}}{dt}$$

Further, upon using the previously derived kinematic equations, this may be written in terms of the body-fixed reference frame as:

$$\vec{a}_{cm} = \frac{{}^{\tilde{R}}d\vec{v}_{cm}}{dt} = \frac{{}^{B}d\vec{v}_{cm}}{dt} + \vec{\Omega} \times \vec{v}_{cm}$$

Now, $\vec{v}_{cm}$ and $\vec{\Omega}$ may be written in terms of the body-fixed unit vectors as:

$$\vec{v}_{cm} = U\vec{n}_1 + V\vec{n}_2 + W\vec{n}_3, \quad \vec{\Omega} = P\vec{n}_1 + Q\vec{n}_2 + R\vec{n}_3$$

So, the body's mass-center acceleration becomes:

$$\vec{a}_{cm} = \left(\dot{U} + QW - RV\right)\vec{n}_1 + \left(\dot{V} + RU - PW\right)\vec{n}_2 + \left(\dot{W} + PV - QU\right)\vec{n}_3$$

Also, the imposed force on the body may be expressed as:

$$\vec{F} = X\vec{n}_1 + Y\vec{n}_2 + Z\vec{n}_3$$

Therefore, the force-acceleration equations, *in terms of the body-fixed coordinate values*, are:

$$X = m\left(\dot{U} + QW - RV\right)$$
$$Y = m\left(\dot{V} + RU - PW\right)$$
$$Z = m\left(\dot{W} + PV - QU\right)$$

Appendix C: Rigid-Body Equations of Motion

Now, consider a force, $d\vec{F}$ on a mass element, dm. This imparts an acceleration to the element given by:

$$dF = dm\,\vec{a}$$

Likewise, an elemental moment about the mass center, $d\vec{\Gamma}$, is given by:

$$d\vec{\Gamma} = \vec{r}\times d\vec{F} = \vec{r}\times dm\,\vec{a}$$

Note that within the body, equal and opposite elemental moments cancel; so, the reaction of the body to external moments is given by:

$$\vec{\Gamma} = \int_B \vec{r}\times \vec{a}\,dm$$

From the "Kinematics" section it was shown that, for a rigid body, $\vec{a}$ may be expressed as:

$$\vec{a} = \vec{a}_{cm} + \vec{\alpha}\times\vec{r} + \vec{\Omega}\times\left(\vec{\Omega}\times\vec{r}\right)$$

where $\vec{\alpha}$ is the body's angular acceleration. So, the moment equation becomes:

$$\vec{\Gamma} = \int_B \vec{r}\times\left[\vec{a}_{cm} + \vec{\alpha}\times\vec{r} + \vec{\Omega}\times\left(\vec{\Omega}\times\vec{r}\right)\right]dm$$

The first term on the right-hand side becomes, upon noting that the origin of $\vec{r}$ is the mass center:

$$\int_B \vec{r}\times\vec{a}_{cm}\,dm = -\vec{a}_{cm}\times\int_B \vec{r}\,dm = -\vec{a}_{cm}\times 0 = 0$$

For the second term, first define:

$$\vec{r} = x\,\vec{n}_1 + y\,\vec{n}_2 + z\,\vec{n}_3$$

Also, observe that:

$$\vec{\alpha} = \frac{{}^R d\vec{\Omega}}{dt} = \frac{{}^B d\vec{\Omega}}{dt} + \vec{\Omega}\times\vec{\Omega} = \frac{{}^B d\vec{\Omega}}{dt} = \dot{P}\,\vec{n}_1 + \dot{Q}\,\vec{n}_2 + \dot{R}\,\vec{n}_3$$

Therefore, one has that:

$$\vec{r}\times\left(\vec{\alpha}\times\vec{r}\right) = \vec{r}\times\begin{vmatrix}\vec{n}_1 & \vec{n}_2 & \vec{n}_3 \\ \dot{P} & \dot{Q} & \dot{R} \\ x & y & z\end{vmatrix} = \left[(y^2+z^2)\dot{P} - xy\dot{Q} - xz\dot{R}\right]\vec{n}_1$$

$$+ \left[-xy\dot{P} + (x^2+z^2)\dot{Q} - yz\dot{R}\right]\vec{n}_2 + \left[-xz\dot{P} - yz\dot{Q} + (x^2+y^2)\dot{R}\right]\vec{n}_3$$

Further, define the following *moment-of-inertia* terms:

$$I_{xx} \equiv \int_B (y^2+z^2)\,dm,\quad I_{yy} \equiv \int_B (x^2+z^2)\,dm,\quad I_{zz} \equiv \int_B (x^2+y^2)\,dm$$

and the *product-of-inertia* terms:

$$I_{xy} \equiv \int_B xy\,dm, \quad I_{xz} \equiv \int_B xz\,dm, \quad I_{yz} \equiv \int_B yz\,dm$$

From this one obtains:

$$\int_B \vec{r}\times(\vec{\alpha}\times\vec{r})\,dm = \left[I_{xx}\dot{P} - I_{xy}\dot{Q} - I_{xz}\dot{R}\right]\vec{n}_1 + \left[I_{yy}\dot{Q} - I_{xy}\dot{P} - I_{yz}\dot{R}\right]\vec{n}_2 + \left[I_{zz}\dot{R} - I_{xz}\dot{P} - I_{yz}\dot{Q}\right]\vec{n}_3$$

For the third term, the cross-products expand to give:

$$\vec{r}\times\left[\vec{\Omega}\times\left(\vec{\Omega}\times\vec{r}\right)\right] = \left[xy\,PR + \left(y^2 - z^2\right)QR + yz\left(R^2 - Q^2\right) - xz\,PQ\right]\vec{n}_1$$

$$+ \left[-xy\,QR + \left(z^2 - x^2\right)PR + xz\left(P^2 - R^2\right) + yz\,PQ\right]\vec{n}_2$$

$$+ \left[-yz\,PR + \left(x^2 - y^2\right)PQ - xy\left(P^2 - Q^2\right) + xz\,QR\right]\vec{n}_3$$

Upon introducing the inertia definitions, one thus obtains:

$$\int_B \vec{r}\times\left[\vec{\Omega}\times\left(\vec{\Omega}\times\vec{r}\right)\right]dm = \left[I_{xy}PR + \left(I_{zz} - I_{yy}\right)QR + I_{yz}\left(R^2 - Q^2\right) - I_{xz}PQ\right]\vec{n}_1$$

$$+ \left[-I_{xy}QR + \left(I_{xx} - I_{zz}\right)PR + I_{xz}\left(P^2 - R^2\right) + I_{yz}PQ\right]\vec{n}_2$$

$$+ \left[-I_{yz}PR + \left(I_{yy} - I_{xx}\right)PQ - I_{xy}\left(P^2 - Q^2\right) + I_{xz}QR\right]\vec{n}_3$$

ASIDE
Note that:

$$\int_B \left(y^2 - z^2\right)dm = \int_B \left(y^2 + x^2\right)dm - \int_B \left(x^2 + z^2\right)dm = I_{zz} - I_{yy}$$

Similarly,

$$\int_B \left(z^2 - x^2\right)dm = I_{xx} - I_{zz}, \quad \int_B \left(x^2 - y^2\right)dm = I_{yy} - I_{xx}$$

END ASIDE

Now, define the moment term as:

$$\vec{\Gamma} = \tilde{L}\vec{n}_1 + M\vec{n}_2 + N\vec{n}_3$$

Appendix C: Rigid-Body Equations of Motion

ASIDE

The tilde is placed over L, so that it will not be confused with the symbol for lift.

END ASIDE

Therefore, upon gathering terms, the complete moment-dynamic equations are:

$$\tilde{L} = I_{xx}\dot{P} - I_{xy}\dot{Q} - I_{xz}\dot{R} + I_{xy}PR + (I_{zz} - I_{yy})QR + I_{yz}(R^2 - Q^2) - I_{xz}PQ$$
$$M = I_{yy}\dot{Q} - I_{xy}\dot{P} - I_{yz}\dot{R} - I_{xy}QR + (I_{xx} - I_{zz})PR + I_{xz}(P^2 - R^2) + I_{yz}PQ$$
$$N = I_{zz}\dot{R} - I_{xz}\dot{P} - I_{yz}\dot{Q} - I_{yz}PR - (I_{xx} - I_{yy})PQ - I_{xy}(P^2 - Q^2) + I_{xz}QR$$

For the study of aircraft flight dynamics, the convention is to position the $\vec{n}_1$ and $\vec{n}_3$ unit vectors in the vertical plane of symmetry. Therefore, $I_{xy}, I_{yz} = 0$ and the equations become:

$$\tilde{L} = I_{xx}\dot{P} - I_{xz}(\dot{R} + PQ) + (I_{zz} - I_{yy})QR$$
$$M = I_{yy}\dot{Q} + I_{xz}(P^2 - R^2) + (I_{xx} - I_{zz})PR$$
$$N = I_{zz}\dot{R} - I_{xz}(\dot{P} - QR) + (I_{yy} - I_{xx})PQ$$

ASIDE

It must be stated that although the vast number of aircraft have been symmetrical about the vertical plane, there have been a few exceptions. One notable example is the NASA AD-1 oblique-wing design from 1979.

This concept, designed by R. T. Jones, was intended to have efficient flight over a large speed range by positioning the wing with no sweep at low speeds and sweeping (as shown) for high-speed flight.

END ASIDE

Interesting Torque-Free Example

Consider a case where there are no external moments applied to the body ($L, M, N = 0$), and the $\vec{n}_1$, $\vec{n}_2$ and $\vec{n}_3$ unit vectors are aligned with the body's principal axes of inertia ($I_{xz} = 0$). The moment-dynamic equations become:

$$\dot{P} = -\frac{(I_{zz} - I_{yy})}{I_{xx}} QR, \quad \dot{Q} = -\frac{(I_{xx} - I_{zz})}{I_{yy}} PR, \quad \dot{R} = -\frac{(I_{yy} - I_{xx})}{I_{zz}} PQ$$

Next, assume the rotation to be primarily P (rotation about the x-axis), so that $Q, R \ll P$. Thus, the first equation gives that $\dot{P} \ll P$; so that it may be said that $P \approx Const. \equiv P_0$ and the other equations may be rewritten as:

$$\dot{Q} = -\frac{(I_{xx} - I_{zz})}{I_{yy}} P_0 R, \quad \dot{R} = -\frac{(I_{yy} - I_{xx})}{I_{zz}} P_0 Q$$

The time derivative of the first equation combined with the second equation gives the following 2nd-order differential equation:

$$\ddot{Q} - \left[\frac{P_0^2 (I_{xx} - I_{zz})(I_{yy} - I_{xx})}{I_{yy} I_{zz}} \right] Q = 0$$

A solution for this is $Q = Q_0 e^{at}$. So, the characteristic equation is:

$$a = \sqrt{\frac{P_0^2 (I_{xx} - I_{zz})(I_{yy} - I_{xx})}{I_{yy} I_{zz}}}$$

It is seen that if $(I_{xx} - I_{zz})(I_{yy} - I_{xx}) < 0$, then a is imaginary, and thus Q is oscillatory and bounded.

However, if $(I_{xx} - I_{zz})(I_{yy} - I_{xx}) > 0$, then a is real and positive so that Q diverges (unstable).

This solution says that rotation about an intermediate axis is unstable. For example, consider the two cases:

Case 1, $I_{xx} < I_{yy} < I_{zz}$:

$$(I_{xx} - I_{zz})(I_{yy} - I_{xx}) = negative \times positive = negative < 0, \; stable$$

Case 2, $I_{yy} < I_{xx} < I_{zz}$:

$$(I_{xx} - I_{zz})(I_{yy} - I_{xx}) = negative \times negative = positive > 0, \; unstable$$

Appendix C: Rigid-Body Equations of Motion

ASIDE

A simple demonstration of the stability (and instability) of torque-free bodies was performed by Prof. Max Anliker of Stanford University, when the author was a student in his class. This consists of imparting a rotation to a semi-expendable book while tossing it into the air.

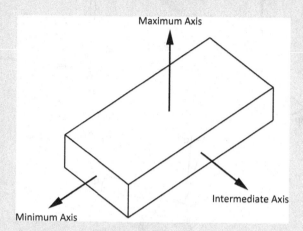

If the rotation is imparted about the minimum or maximum axes, the book will continue rotating about these axes while in the air. However, if the rotation is about the intermediate axis, then the book will tumble. Hint: put a rubber band around the book to keep it from flying open.

Further, this explains why a football, when properly passed, continues rotating about its minimum axis (the axis of symmetry).

END ASIDE

As a final comment it must be noted that, although the full moment-dynamic equations were derived with respect to the body's mass center, this isn't necessarily required. In fact, moment-dynamic equations may be derived relative to *any point fixed* relative to the body. Of course, the resulting equations will have extra terms, but the author found these more-complex equations to be crucial for analyzing the flight dynamics of airships. The reason is that the mass center of an airship can vary depending on the flight condition and even during its motions. Therefore, a fixed point was chosen, at the nose, for the moment-dynamic equations.

However, most other aircraft, airplanes in particular, have relatively invariant mass centers for the period of time that their flight dynamics are being studied. Therefore, most texts use the moment-dynamic equations presented in this Appendix.

ORIENTATION

It is important to describe an aircraft's orientation with respect to the inertial reference frame because:

1. Buoyant and gravity forces acting on the aircraft are fixed in the inertial reference frame and are, therefore, functions of orientation in a body-fixed reference frame.
2. Descriptions of an aircraft's dynamic performance in an inertial reference frame require a means of quantifying its orientation.

One way to accomplish this is with Euler Angles.

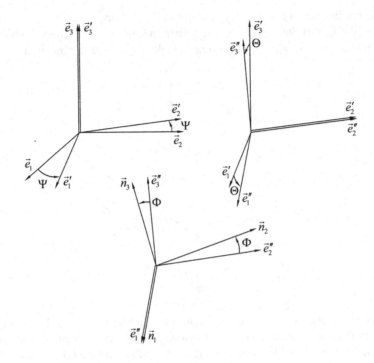

An aircraft is oriented by Euler Angles in the following way:

1. Coordinate system $\vec{e}_1', \vec{e}_2', \vec{e}_3'$ is originally aligned with the inertial coordinate system $\vec{e}_1, \vec{e}_2, \vec{e}_3$.
2. The coordinate system $\vec{e}_1', \vec{e}_2', \vec{e}_3'$ is then rotated about $\vec{e}_3'$ by an angle Ψ.
3. System $\vec{e}_1'', \vec{e}_2'', \vec{e}_3''$ is then aligned with $\vec{e}_1', \vec{e}_2', \vec{e}_3'$.
4. One then rotates system $\vec{e}_1'', \vec{e}_2'', \vec{e}_3''$ about $\vec{e}_2''$ through angle Θ.
5. The body-fixed system $\vec{n}_1, \vec{n}_2, \vec{n}_3$ is now aligned with $\vec{e}_1'', \vec{e}_2'', \vec{e}_3''$.
6. A rotation about $\vec{e}_1''$ through angle Φ brings $\vec{n}_1, \vec{n}_2, \vec{n}_3$ into its final alignment.

Therefore, the orientation of the aircraft is described by three angles: $\Psi, \Theta,$ and Φ. By vector algebra, one can obtain the following relationships between the inertial coordinate system and the body-fixed coordinate system:

$$\vec{e}_1 = \cos\Theta\cos\Psi\,\vec{n}_1 + (\sin\Phi\sin\Theta\cos\Psi - \cos\Phi\sin\Psi)\vec{n}_2 + (\cos\Phi\sin\Theta\cos\Psi + \sin\Phi\sin\Psi)\vec{n}_3$$

$$\vec{e}_2 = \cos\Theta\sin\Psi\,\vec{n}_1 + (\sin\Phi\sin\Theta\sin\Psi + \cos\Phi\cos\Psi)\vec{n}_2 + (\cos\Phi\sin\Theta\sin\Psi - \sin\Phi\cos\Psi)\vec{n}_3$$

$$\vec{e}_3 = -\sin\Theta\,\vec{n}_1 + \sin\Phi\cos\Theta\,\vec{n}_2 + \cos\Phi\cos\Theta\,\vec{n}_3$$

If one defines $\vec{V}_{cm}$ with respect to the inertial reference frame as:

$$\vec{v}_{cm} = \frac{dx'}{dt}\vec{e}_1 + \frac{dy'}{dt}\vec{e}_2 + \frac{dz'}{dt}\vec{e}_2 = U\vec{n}_1 + V\vec{n}_2 + W\vec{n}_3,$$

Appendix C: Rigid-Body Equations of Motion

then the previous equations may be used to obtain the velocity components of the aircraft's mass center with respect to the inertial reference frame in terms of rotations of – and velocities with respect to – the body-fixed reference frame:

$$\frac{dx'}{dt} = U\cos\Theta\cos\Psi + V(\sin\Phi\sin\Theta\cos\Psi - \cos\Phi\sin\Psi) + W(\cos\Phi\sin\Theta\cos\Psi + \sin\Phi\sin\Psi)$$

$$\frac{dy'}{dt} = U\cos\Theta\sin\Psi + V(\sin\Phi\sin\Theta\sin\Psi + \cos\Phi\cos\Psi) + W(\cos\Phi\sin\Theta\sin\Psi - \sin\Phi\cos\Psi)$$

$$\frac{dz'}{dt} = -U\sin\Theta + V\sin\Phi\cos\Theta + W\cos\Phi\cos\Theta$$

The above equations allow one to calculate the flight-path trajectory of an aircraft in the inertial reference frame if the on-board instrumentation gives:

$$U(t), \quad V(t), \quad W(t), \quad \Psi(t), \quad \Theta(t), \quad \Phi(t)$$

It is also possible to obtain a set of equations relating the Euler Angles and the angular velocity components P, Q, R. Observe from the previous figure that each reference frame has an angular velocity relative to the other:

$$\vec{\Omega}_1 = \dot{\Psi}\vec{e}_3, \quad \vec{\Omega}_2 = \dot{\Theta}\vec{e}_2', \quad \vec{\Omega}_3 = \dot{\Phi}\vec{n}_1$$

The sum of these gives the total angular velocity of the aircraft with respect to the inertial reference system:

$$\vec{\Omega} = \vec{\Omega}_1 + \vec{\Omega}_2 + \vec{\Omega}_3 = \dot{\Psi}\vec{e}_3 + \dot{\Theta}\vec{e}_2' + \dot{\Phi}\vec{n}_1 = P\vec{n}_1 + Q\vec{n}_2 + R\vec{n}_3$$

From vector algebra, one obtains:

$$P = \dot{\Phi} - \dot{\Psi}\sin\Theta, \quad Q = \dot{\Theta}\cos\Phi + \dot{\Psi}\sin\Phi\cos\Theta, \quad R = -\dot{\Theta}\sin\Phi + \dot{\Psi}\cos\Phi\cos\Theta$$

The inverse of these equations is:

$$\dot{\Phi} = P + Q\sin\Phi\tan\Theta + R\cos\Phi\tan\Theta, \quad \dot{\Theta} = Q\cos\Phi - R\sin\Phi,$$

$$\dot{\Psi} = Q\sin\Phi\sec\Theta + R\cos\Phi\sec\Theta$$

If one has on-board gyros that give $P(t), Q(t), R(t)$, then the integration of these equations allow the determination of the aircraft's orientation with respect to time in the inertial reference frame.

LINEARIZED VERSIONS OF THE EQUATIONS

For use in an aircraft's dynamic-stability analysis, the equations take on a linearized form. In this case, the body-fixed velocity components become:

$$U = U_0 + u, \quad V = v, \quad W = w \quad \text{where } u/U_0, \quad v/U_0, \quad w/U_0 = O[\varepsilon], \quad \varepsilon \ll 1$$

Also, if l is defined to be a characteristic length of the aircraft (body length, wing span, mean wing chord, etc.), the angular-velocity components become:

$$P = p, \quad Q = q, \quad R = r \quad \text{where } p/(2U_0/l),\ q/(2U_0/l),\ r/(2U_0/l) = O[\varepsilon]$$

Further,

$$\Theta = \Theta_0 + \theta, \quad \Phi = \phi, \quad \Psi = \psi \quad \text{where } \theta, \phi, \psi = O[\varepsilon]$$

Therefore, the linearized dynamic equations become:

$$\Delta X = m(\dot{u} + qw - rv) \rightarrow \underline{\Delta X = m\dot{u}}$$
$$\Delta Y = m[\dot{v} + r(U_0 + u) - pw] \rightarrow \underline{\Delta Y = m(\dot{v} + U_0 r)}$$
$$\Delta Z = m[\dot{w} + pv - q(U_0 + u)] \rightarrow \underline{\Delta Z = m(\dot{w} - U_0 q)}$$
$$\Delta \tilde{L} = I_{xx}\dot{p} - I_{xz}(\dot{r} + pq) + (I_{zz} - I_{yy})qr \rightarrow \underline{\Delta \tilde{L} = I_{xx}\dot{p} - I_{xz}\dot{r}}$$
$$\Delta M = I_{yy}\dot{q} + I_{xz}(p^2 - r^2) + (I_{xx} - I_{zz})pr \rightarrow \underline{\Delta M = I_{yy}\dot{q}}$$
$$\Delta N = I_{zz}\dot{r} - I_{xz}(\dot{p} - qr) + (I_{yy} - I_{xx})pq \rightarrow \underline{\Delta N = I_{zz}\dot{r} - I_{xz}\dot{p}}$$

In the same fashion, the linearized trajectory equations become:

$$dx'/dt = (U_0 + u)\cos\Theta_0 - \theta U_0 \sin\Theta_0 + w\sin\Theta_0$$
$$dy'/dt = v + \psi U_0 \cos\Theta_0$$
$$dz'/dt = -(U_0 + u)\sin\Theta_0 - \theta U_0 \cos\Theta_0 + w\cos\Theta_0$$

And finally, the linearized orientation equations become:

$$p = \dot{\phi} - \dot{\psi}\sin\Theta_0, \quad q = \dot{\theta}, \quad r = \dot{\psi}\cos\Theta_0$$
$$\dot{\phi} = p + r\tan\Theta_0, \quad \dot{\theta} = q, \quad \dot{\psi} = r\sec\Theta_0$$

Appendix D: Apparent-Mass Effects

The aerodynamic concept known as "apparent mass" (or sometimes, "virtual mass" or "added mass") is based on the fact that an accelerating body in a fluid field experiences a reaction force (or moment) that, mathematically, appears as an added mass to the body. This is best explained by referring to the figure below, which shows a body moving in an infinite field of potential fluid.

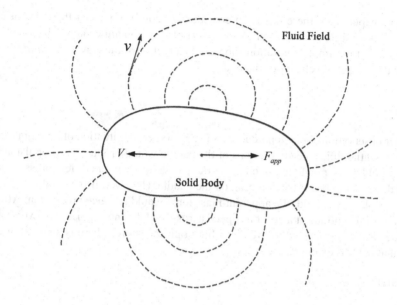

ASIDE

Observe that this is the streamline pattern produced by the body when moving through a stationary fluid field, which is known as the field-fixed viewpoint. A body-fixed viewpoint gives the more usual picture of streamlines.

END ASIDE

If the body has a steady-state velocity, V, the potential-fluid field will exert no aerodynamic force on the body. This is, of course, because potential fluid is non-viscous, which means that the streamlines will close perfectly behind the body (no base drag) and there is no skin-friction drag. However, the body's motion produces kinetic energy throughout the fluid field. This can be seen by observing that the presence of the body, and its velocity, produces proportional velocity perturbations, v, at every point in the fluid field. From this, one may obtain the kinetic energy in the fluid field due to the body's velocity:

$$\tilde{T} = \frac{1}{2} \int_{\text{field}} \rho v^2 d(vol) \equiv \frac{Const}{2} \times V^2$$

Now, if the body is accelerated from a velocity V_1 to a velocity V_2, the field perturbations likewise increase proportionally from v_1 to v_2. The field kinetic energy also increases from $\tilde{T}_1$ to $\tilde{T}_2$. Clearly,

539

in order to put this energy into the field, work must have been done by the body on the fluid. This manifests itself as a *reaction* force on the body from the fluid, given by Lagrange's equation as:

$$\widetilde{F}_{app} = \frac{d}{dt}\frac{d\widetilde{T}}{dV} = Const \times \frac{dV}{dt}$$

This has the form of the classical Newtonian force-acceleration equation, so one may define the *Const* term to be $\widetilde{m}$, an "apparent mass", and the equation becomes:

$$\widetilde{F}_{app} = \widetilde{m}\,\dot{V}$$

In the author's experience, there has been considerable confusion about the concept of apparent mass. In some cases, it is looked upon as some sort of "aerodynamic snow" that resides upon the body. However, as this simple derivation shows, it is a matter of the change of *fluid* kinetic energy produced by the changing velocity of the body.

ASIDE

As another observation about apparent mass is a hypothetical situation of being in a canoe on a lake of potential fluid. Steady motions of the paddle would produce no propulsion because the fluid would not separate at the edges of the paddle (no separated-flow pressure differential) and there would be no friction drag on the paddle. However, if the paddle is accelerated during its stroke in the fluid, the apparent-mass force would produce propulsion. Afterwards, the canoe would continue at a constant speed across the lake (no drag) and a reverse action of the paddle would be required to avoid crashing into the shore. There is something uniquely Canadian about this example.

END ASIDE

This development may be generalized for a body whose body-fixed coordinates align with its principal axes. In the figure below, of an ellipsoid, observe that *cv* is the volumetric center:

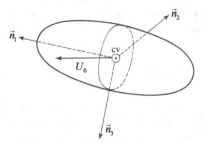

From "Hydrodynamics" by Horace Lamb (Dover Publications, 6th Edition, 1945, page 163), the fluid kinetic energy caused by this body's translational velocities and angular velocities about its principal axes is given by:

$$\widetilde{T} = \frac{m_1}{2}U^2 + \frac{m_2}{2}V^2 + \frac{m_3}{2}W^2 + \frac{i_1}{2}P^2 + \frac{i_2}{2}Q^2 + \frac{i_3}{2}R^2$$

where U, V and W are translational velocities along the $\vec{n}_1$, $\vec{n}_2$ and $\vec{n}_3$ principal axes, respectively; and P, Q and R are rotations about the $\vec{n}_1$, $\vec{n}_2$ and $\vec{n}_3$ axes, respectively.

Appendix D: Apparent-Mass Effects

Observe that, generally speaking, the values of the apparent-mass terms, m_i, depend on the direction of motion. This shows how they differ from the body's actual mass, m.

Now, again from Lamb, the reaction force in the $\vec{n}_1$ direction, X_{app}, is given by:

$$X_{app} = -\frac{d}{dt}\frac{\partial \tilde{T}}{\partial U} + R\frac{\partial \tilde{T}}{\partial V} - Q\frac{\partial \tilde{T}}{\partial W} \rightarrow X_{app} = -m_1 \dot{U} + m_2 RV - m_3 QW$$

Likewise, the Y_{app} and Z_{app} reaction forces in the $\vec{n}_2$ and $\vec{n}_3$ directions, respectively, are:

$$Y_{app} = -\frac{d}{dt}\frac{\partial \tilde{T}}{\partial V} + P\frac{\partial \tilde{T}}{\partial W} - R\frac{\partial \tilde{T}}{\partial U} \rightarrow Y_{app} = -m_2 \dot{V} + m_3 PW - m_1 RU$$

$$Z_{app} = -\frac{d}{dt}\frac{\partial \tilde{T}}{\partial W} + Q\frac{\partial \tilde{T}}{\partial U} - P\frac{\partial \tilde{T}}{\partial V} \rightarrow Z_{app} = -m_3 \dot{W} + m_1 QU - m_2 PV$$

The $\tilde{L}_{app}$ reaction moment about the $\vec{n}_1$ axis is given by:

$$\tilde{L}_{app} = -\frac{d}{dt}\frac{\partial \tilde{T}}{\partial P} + W\frac{\partial \tilde{T}}{\partial V} - V\frac{\partial \tilde{T}}{\partial W} + R\frac{\partial \tilde{T}}{\partial Q} - Q\frac{\partial \tilde{T}}{\partial R} \rightarrow$$

$$\tilde{L}_{app} = -i_1 \dot{P} + m_2 WV - m_3 WV + i_2 RQ - i_3 RQ \rightarrow$$

$$\tilde{L}_{app} = -i_1 \dot{P} + (m_2 - m_3)WV + (i_2 - i_3)RQ$$

Likewise, the M_{app} and N_{app} reaction moments about the $\vec{n}_2$ and $\vec{n}_3$ axes, respectively, are:

$$M_{app} = -\frac{d}{dt}\frac{\partial \tilde{T}}{\partial Q} + U\frac{\partial \tilde{T}}{\partial W} - W\frac{\partial \tilde{T}}{\partial U} + P\frac{\partial \tilde{T}}{\partial R} - R\frac{\partial \tilde{T}}{\partial P} \rightarrow$$

$$M_{app} = -i_2 \dot{Q} + (m_3 - m_1)UW + (i_3 - i_1)PR$$

$$N_{app} = -\frac{d}{dt}\frac{\partial \tilde{T}}{\partial R} + V\frac{\partial \tilde{T}}{\partial U} - U\frac{\partial \tilde{T}}{\partial V} + Q\frac{\partial \tilde{T}}{\partial P} - P\frac{\partial \tilde{T}}{\partial Q} \rightarrow$$

$$N_{app} = -i_3 \dot{R} - (m_2 - m_1)UV - (i_2 - i_1)PQ$$

Now, assume that the velocity component, U, is aligned with the equilibrium steady-state velocity, U_0. Further, for the small-perturbation assumptions, described in Chapter 6, where:

$$U = U_0 + u, \quad V = v, \quad W = w, \quad P = p, \quad Q = q, \quad R = r$$

$$u, v, w \ll U_0, \quad q \ll 2U_0/\bar{c}, \quad r, p \ll 2U_0/b, \quad \dot{u}, \dot{w} \ll 2U_0^2/\bar{c}, \quad \dot{v} \ll 2U_0^2/b$$

these equations become:

$$\Delta X_{app} = -m_1 \dot{u}, \quad \Delta Y_{app} = -m_2 \dot{v} - m_1 U_0 r, \quad \Delta Z_{app} = -m_3 \dot{w} + m_1 U_0 q$$

$$\Delta \tilde{L}_{app} = -i_1 \dot{p}, \quad \Delta M_{app} = -i_2 \dot{q} + (m_3 - m_1)U_0 w, \quad \Delta N_{app} = -i_3 \dot{r} - (m_2 - m_1)U_0 v$$

Note that one may use the apparent-mass coefficients, k_1, k_2, k_3, defined in Chapter 8 to obtain:

$$m_1 = \rho k_1 (Vol), \quad m_2 = \rho k_2 (Vol), \quad m_3 = \rho k_3 (Vol)$$

The force equations may then be non-dimensionalized by the definitions in Chapter 6. For ΔX_{app}, one obtains that:

$$\Delta X_{app} = (C_X)_{app} \frac{\rho V_{cm}^2 S}{2} = -\rho k_1 (Vol) \frac{2U_0}{\bar{c}} U_0 D\hat{u} = -\rho U_0^2 k_1 (Vol) \frac{2}{\bar{c}} D\hat{u}$$

For this application, $V_{cm} = U_0$, so this becomes:

$$(C_X)_{app} \frac{\rho U_0^2 S}{2} = -\rho U_0^2 k_1 (Vol) \frac{2}{\bar{c}} D\hat{u} \rightarrow (C_X)_{app} = -\frac{4 k_1 (Vol)}{S \bar{c}} D\hat{u}$$

Upon defining,

$$(C_{XDu})_{body} \equiv -\frac{4 k_1 (Vol)}{S \bar{c}}$$

One has:

$$(C_X)_{app} = (C_{XDu})_{body} D\hat{u}$$

Next, ΔY_{app} becomes:

$$\Delta Y_{app} = (C_Y)_{app} \frac{\rho V_{cm}^2 S}{2} = (C_Y)_{app} \frac{\rho U_0^2 S}{2} = -\rho k_2 (Vol) \frac{2 U_0}{b} U_0 D\hat{v} - \rho k_1 (Vol) U_0 \frac{2 U_0}{b} \hat{r} \rightarrow$$

$$(C_Y)_{app} \frac{\rho U_0^2 S}{2} = -\rho U_0^2 k_2 (Vol) \frac{2}{b} D\hat{v} - \rho U_0^2 k_1 (Vol) \frac{2}{b} \hat{r} \rightarrow$$

$$(C_Y)_{app} = -\frac{4 k_2 (Vol)}{S b} D\hat{v} - \frac{4 k_1 (Vol)}{S b} \hat{r}$$

Upon recognising, from Chapter 6, that $D\beta \equiv D\hat{v}$, one may define:

$$(C_{YD\beta})_{body} \equiv -\frac{4 k_2 (Vol)}{S b}$$

Thus, one obtains that:

$$(C_Y)_{app} = (C_{YD\beta})_{body} D\hat{v} + \frac{\bar{c}}{b} (C_{XDu})_{body} \hat{r}$$

Likewise, the ΔZ_{app} equation becomes:

$$(C_Z)_{app} = -\frac{4 k_3 (Vol)}{S \bar{c}} D\hat{w} + \frac{4 k_1 (Vol)}{S \bar{c}} \hat{q}$$

Again, upon recognizing from Chapter 6 that $D\alpha \equiv D\hat{w}$, one may define:

$$(C_{ZD\alpha})_{body} \equiv -\frac{4 k_3 (Vol)}{S \bar{c}}$$

which gives:

$$(C_Z)_{app} = (C_{ZD\alpha})_{body} D\hat{w} - (C_{XDu})_{body} \hat{q}$$

Notice how the body's x-direction acceleration stability derivative, $(C_{XDu})_{body}$, makes contributions to the body's C_{Yr} and C_{Zq} stability derivatives.

Next, the apparent moment-of-inertia terms may be expressed as:

$$i_1 = k_1' I_1, \quad i_2 = k_2' I_2, \quad i_3 = k_3' I_3,$$

Appendix D: Apparent-Mass Effects

where I_1, I_2 and I_3 are the principal moments of inertia of the ellipsoid, if its density is equal to that of the fluid through which it is moving. If l, d_y and d_z are the x, y and z values of the major axes of the ellipsoid (for an ellipsoid of revolution, l would be the total axial length and $d = d_y = d_z$ would be the maximum diameter), then:

$$I_1 = \frac{\rho(\text{Vol})}{20}\left(d_y^2 + d_z^2\right), \quad I_2 = \frac{\rho(\text{Vol})}{20}\left(l^2 + d_z^2\right), \quad I_3 = \frac{\rho(\text{Vol})}{20}\left(l^2 + d_y^2\right)$$

The non-dimensional form of $\Delta \tilde{L}_{\text{app}}$ is obtained from:

$$\Delta \tilde{L}_{\text{app}} = (C_{\tilde{L}})_{\text{app}} \frac{\rho V_{\text{cm}}^2 b S}{2} = (C_{\tilde{L}})_{\text{app}} \frac{\rho U_0^2 b S}{2} = -k_1' I_1 \frac{2U_0}{b} \frac{2U_0}{b} D\hat{p} = -k_1' I_1 \frac{4U_0^2}{b^2} D\hat{p} \rightarrow$$

$$(C_{\tilde{L}})_{\text{app}} = -\frac{2}{\rho U_0^2 b S} k_1' I_1 \frac{4U_0^2}{b^2} D\hat{p} = -k_1' \frac{2}{\rho b^3 S} \frac{\rho(\text{Vol})}{20}\left(d_y^2 + d_z^2\right) 4 D\hat{p} \rightarrow$$

$$(C_{\tilde{L}})_{\text{app}} = -k_1' \frac{2(\text{Vol})\left(d_y^2 + d_z^2\right)}{5 b^3 S} D\hat{p}$$

Upon defining:

$$\left(C_{\tilde{L} Dp}\right)_{\text{body}} \equiv -2 k_1' \frac{\left(d_y^2 + d_z^2\right)}{5 b^3 S}(\text{Vol})$$

One has:

$$(C_{\tilde{L}})_{\text{app}} = \left(C_{\tilde{L} Dp}\right)_{\text{body}} D\hat{p}$$

Observe that for a body of revolution, $k_1' = 0$, thus $\left(C_{\tilde{L} Dp}\right)_{\text{body}} = 0$.

Next, the non-dimensional form of ΔM_{app} is obtained from:

$$\Delta M_{\text{app}} = (C_M)_{\text{app}} \frac{\rho U_0^2 \bar{c} S}{2} = -k_2' I_2 \frac{4U_0^2}{\bar{c}^2} D\hat{q} + \rho(\text{Vol})(k_3 - k_1) U_0^2 \alpha \rightarrow$$

$$(C_M)_{\text{app}} = -\frac{2 k_2' I_2}{\rho U_0^2 \bar{c} S} \frac{4U_0^2}{\bar{c}^2} D\hat{q} + \frac{2\rho(\text{Vol})(k_3 - k_1) U_0^2}{\rho U_0^2 \bar{c} S} \alpha \rightarrow$$

$$(C_M)_{\text{app}} = -2 k_2' \frac{\left(l^2 + d_z^2\right)}{5 \bar{c}^3 S}(\text{Vol}) D\hat{q} + \frac{2(k_3 - k_1)(\text{Vol})}{\bar{c} S} \alpha$$

Define:

$$\left(C_{MDq}\right)_{\text{body}} \equiv -2 k_2' \frac{\left(l^2 + d_z^2\right)}{5 \bar{c}^3 S}(\text{Vol}), \quad \left(C_{M\alpha}\right)_{\text{body}} \equiv \frac{2(k_3 - k_1)(\text{Vol})}{\bar{c} S}$$

so that,

$$(C_M)_{\text{app}} = \left(C_{MDq}\right)_{\text{body}} D\hat{q} + \left(C_{M\alpha}\right)_{\text{body}} \alpha$$

Observe that the second right-hand term is the "Munk Moment", introduced in Chapter 8.

Finally, the non-dimensional form of $\Delta \tilde{N}$ is obtained in the same way as above, which gives:

$$(C_N)_{\text{app}} = \left(C_{NDr}\right)_{\text{body}} D\hat{r} + \left(C_{N\beta}\right)_{\text{body}} \beta$$

where,

$$\left(C_{\text{NDr}}\right)_{\text{body}} \equiv -2k_3' \frac{(l^2+d_y^2)}{5b^3 S}(Vol), \quad \left(C_{\text{N}\beta}\right)_{\text{body}} \equiv -\frac{2(k_2-k_1)(Vol)}{bS}$$

Again, the second term is the "Munk Moment", applied laterally.

The Munk Moment is usually associated with the steady-state motion of a body of revolution at a fixed angle of attack. Therefore, it is interesting that this seemingly static aerodynamic term appears in the course of deriving aerodynamic forces due to acceleration. An explanation is provided by observing that, besides a global increase in the fluid's kinetic energy, there is also an instantaneous redistribution of the field's kinetic energy due to the body's motion. This occurs even if the body's motion is steady. In that case, there is no global increase in the fluid field's kinetic energy, but the asymmetric redistribution of the local kinetic energy, due to the angle of attack, requires an asymmetric distribution of reaction forces on the body's surface.

The Munk Moment can be derived by steady-state fluid-dynamic analysis, with the pressure distribution obtained from Bernoulli's Equation:

$$P_0 = P + \rho V^2/2$$

However, note that this is an energy equation, where P is the specific stored energy (pressure energy) of the fluid and $\rho V^2/2$ is the specific kinetic energy of the fluid. In fact, all of fluid dynamics, steady state or accelerating, involves generation and/or redistribution of the kinetic energy throughout the field. Therefore, the Munk Moment terms are to be expected in this general derivation of the forces and moments on a body of revolution in motion through potential fluid.

Upon returning to the previous equations, observe that $(C_{M\alpha})_{\text{body}}$ and $(C_{N\beta})_{\text{body}}$ may be rewritten as:

$$\left(C_{M\alpha}\right)_{\text{body}} \equiv \frac{2(k_3-k_1)(Vol)}{\bar{c} S} = \frac{1}{2}\left[\left(C_{\text{XDu}}\right)_{\text{body}} - \left(C_{\text{ZD}\alpha}\right)_{\text{body}}\right]$$

$$\left(C_{N\beta}\right)_{\text{body}} \equiv -\frac{2(k_2-k_1)(Vol)}{bS} = \frac{1}{2}\left[\left(C_{\text{YD}\beta}\right)_{\text{body}} - \frac{\bar{c}}{b}\left(C_{\text{XDu}}\right)_{\text{body}}\right]$$

Finally, the following carpet plots give values for the apparent-mass coefficients for an ellipsoid with three unequal axes. Note the axes definitions on the plots, such that,

$$k_x \equiv k_1, \quad k_y \equiv k_2, \quad k_z \equiv k_3, \quad k_x' \equiv k_1', \quad k_y' \equiv k_2', \quad k_z' \equiv k_3'$$

It must be acknowledged that the source of these plots is unknown to the author. These were given to the author by a colleague in 1970, and they were in the form of multiply-photocopied generations past the originals (separated from the original report). These have been considerably cleaned up with Photoshop to put them in an acceptable form for this book.

Appendix D: Apparent-Mass Effects

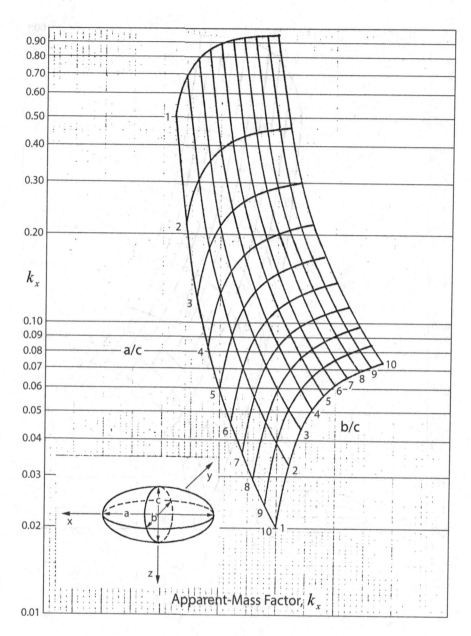

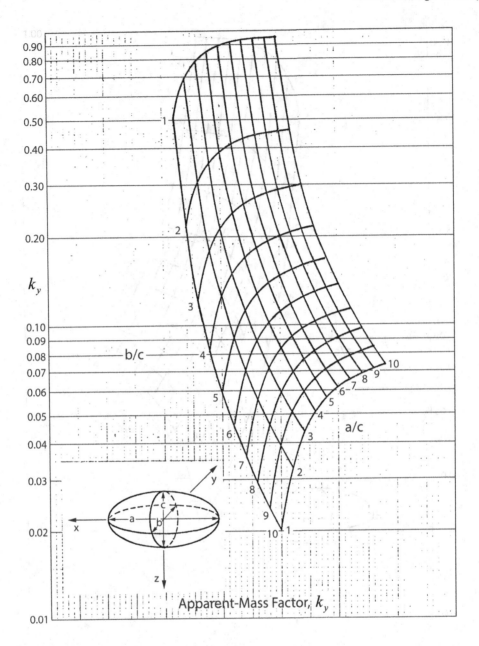

Appendix D: Apparent-Mass Effects

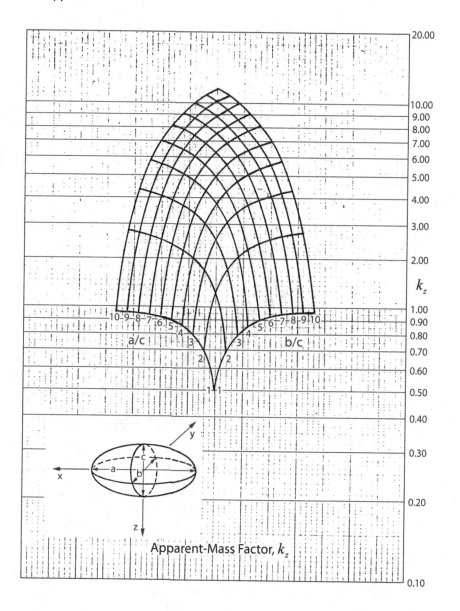

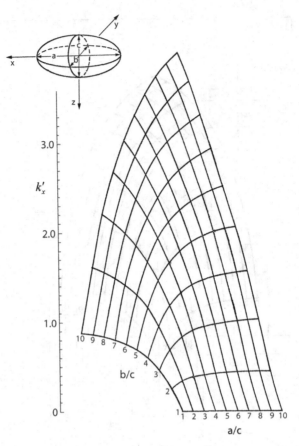

Apparent Inertia Factor, k'_x

Appendix D: Apparent-Mass Effects

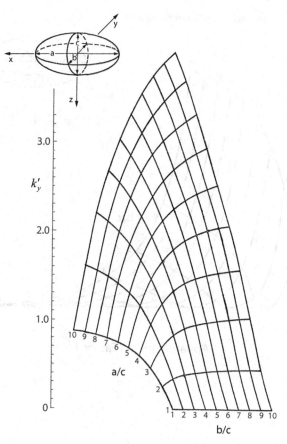

Apparent Inertia Factor, k'_y

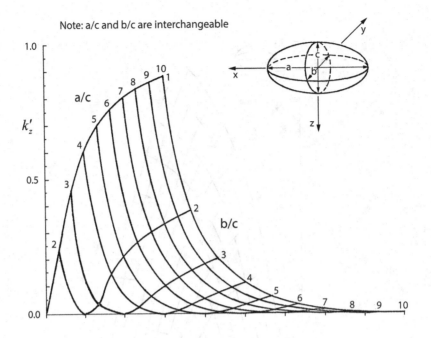

Apparent Inertia Factor, k'_z

Appendix E: Lift of Finite Wings Due To Oscillatory Plunging Acceleration

For an oscillating two-dimensional airfoil in incompressible flow, Garrick ("Propulsion of a Flapping and Oscillating Airfoil", NACA Report No. 567, 1936) gives the following equation for its lift force:

$$L = \rho \pi U (c/2)^2 \dot{\alpha}_{c/2} + 2\pi \frac{\rho U^2}{2} c\, C(k) \alpha_{3c/4}$$

where $\alpha_{c/2}$ is the incident flow angle at the mid-chord location, and $\alpha_{3c/4}$ is the incident flow angle at the three-quarters chord location. These two oscillating angles may be expressed as:

$$\alpha_{c/2} = A_{c/2}\, e^{i\omega t}, \quad \alpha_{3/4} = A_{3c/4}\, e^{i\omega t}$$

For *plunging* motion only, one obtains that:

$$\alpha_{c/2} = \alpha_{3c/4} \equiv \alpha \rightarrow L = \rho \pi U (c/2)^2 \dot{\alpha} + 2\pi \frac{\rho U^2}{2} c\, C(k) \alpha$$

where,

$$\alpha = A e^{i\omega t}$$

The first part of the lift equation's right-hand side is the force due to the apparent-mass effect, and will be denoted as L_{app}. The second term on the right-hand side is the force from the airfoil's circulation, and will be denoted as L_{cir}. Therefore, the lift force for the plunging airfoil may be expressed as $L = L_{app} + L_{cir}$, where,

$$L_{app} = \rho \pi U (c/2)^2 \dot{\alpha}, \quad L_{cir} = 2\pi \frac{\rho U^2}{2} c\, C(k) \alpha$$

The $C(k)$ term is known as the Theodorsen Function, which represents the aerodynamic effect of the bound and shed vorticity on an oscillating airfoil. It is a complex function and can be written as:

$$C(k) = F(k) + i G(k)$$

Therefore, L_{cir} may be written as:

$$L_{cir} = 2\pi \frac{\rho U^2}{2} c \left[F(k) + i G(k) \right] \alpha$$

Note that:

$$\dot{\alpha} = i\omega A e^{i\omega t} = i\omega \alpha \rightarrow \alpha = -i\dot{\alpha}/\omega$$

So, L_{cir} becomes:

$$L_{cir} = 2\pi \frac{\rho U^2}{2} c F(k)\alpha + 2\pi \frac{\rho U^2}{2} c \frac{G(k)}{\omega}\dot{\alpha} \equiv L_{cir}(\alpha) + L_{cir}(\dot{\alpha})$$

Further, the "reduced frequency", k, is given by:

$$k = \frac{\omega c}{2U}, \quad \text{which gives } \omega = \frac{2Uk}{c}$$

Thus,

$$L_{cir}(\dot{\alpha}) = 2\pi \frac{\rho U c^2}{4} \frac{G(k)}{k}\dot{\alpha}$$

Observe that 2π represents the lift-curve slope of a two-dimensional airfoil. This is seen from the equation for $L_{cir}(\alpha)$ when $k=0 \rightarrow F(k)=1$, which then gives the lift force for a two-dimensional airfoil. Therefore,

$$\left(C_{l\alpha}\right)_{2\text{-dim}} = 2\pi$$

For a *finite wing*, replace 2π with the lift-curve slope of a steady finite wing, $\left(C_{L\alpha}\right)_0$, and c with the wing area, S, so that L_{cir} becomes:

$$L_{cir} = \left(C_{L\alpha}\right)_0 \frac{\rho U^2 S}{2} F'(k)\alpha + \left(C_{L\alpha}\right)_0 \frac{\rho U S c}{4} \frac{G'(k)}{k}\dot{\alpha}$$

where $F'(k)$ and $G'(k)$ are terms from the Theodorsen function applied to a finite wing (more about this later).

The L_{app} term is also changed by finite-wing effects, and may more-generally be written as:

$$L_{app} = m_{app} U \dot{\alpha}$$

where m_{app} is the apparent mass for a finite wing. The value for an elliptical-planform disk is:

$$m_{app} = \rho b c_0^2 \frac{\pi}{6} \tilde{k}$$

The term, $\tilde{k}$, is the apparent-mass coefficient for the elliptical disk. When $b/c_0 \rightarrow \infty$, $\tilde{k} \rightarrow 1$, so that L_{app} reduces to the two-dimensional equation applied to a wing of elliptical planform.

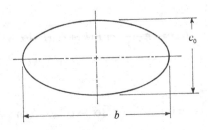

Appendix E: Lift of Finite Wings Due To Oscillatory Plunging Acceleration

Now, the aspect ratio, AR, of an elliptical-planform wing is given by:

$$AR = \frac{b^2}{S} = \frac{b^2}{\pi(b/2)(c_0/2)} = \frac{4b}{\pi c_0} \rightarrow c_0 = \frac{4b}{\pi AR}$$

Also,

$$S = \frac{\pi}{4} b c_0 \rightarrow b = \frac{4S}{\pi c_0} \rightarrow$$

$$m_{app} = \rho(bc_0)c_0 \frac{\pi}{6}\tilde{k} = \rho \frac{4S}{\pi} c_0 \frac{\pi}{6}\tilde{k} = \frac{2}{3}\rho S c_0 \tilde{k} = \frac{2}{3}\rho S \frac{4b}{\pi AR}\tilde{k} \rightarrow$$

$$m_{app} = \frac{8}{3}\rho S \frac{b}{\pi AR}\tilde{k}$$

Therefore, upon combining the circulation and apparent-mass terms, one has:

$$L = L_\alpha \alpha + L_{d\alpha} \dot{\alpha}, \quad \text{where}$$

$$L_\alpha = C_{L\alpha}\frac{\rho U^2 S}{2} F'(k), \quad L_{d\alpha} = \frac{8}{3}\rho S \frac{b}{\pi AR} U \tilde{k} + (C_{L\alpha})_0 \frac{\rho U S c}{4} \frac{G'(k)}{k}$$

Note that:

$$L = C_L \frac{\rho U^2 S}{2} \rightarrow \frac{\partial L}{\partial \alpha} = \frac{\partial C_L}{\partial \alpha}\frac{\rho U^2 S}{2} \rightarrow C_{L\alpha} = (C_{L\alpha})_0 F'(k)$$

And,

$$\frac{\partial L}{\partial \dot{\alpha}} = \frac{\partial C_L}{\partial \dot{\alpha}}\frac{\rho U^2 S}{2} = \frac{\partial C_L}{\partial (D\alpha)}\frac{\partial (D\alpha)}{\partial \dot{\alpha}}\frac{\rho U^2 S}{2}$$

Now, from the "Non-Dimensional Form of the Equations" portion of the "Aircraft Longitudinal Small-Perturbation Dynamic Equations" section of Chapter 6, one has that:

$$D\alpha = \frac{c}{2U}\dot{\alpha} \rightarrow \frac{\partial (D\alpha)}{\partial \dot{\alpha}} = \frac{c}{2U}$$

Further,

$$\frac{\partial C_L}{\partial (D\alpha)} \equiv C_{LD\alpha}$$

So, one has that:

$$L_{d\alpha} = C_{L D\alpha} \frac{c}{2U} \frac{\rho U^2 S}{2} = C_{L D\alpha} \frac{\rho U S c}{4} \rightarrow C_{L D\alpha} = \frac{4}{\rho U S c} L_{d\alpha}$$

When this is substituted into the previous equation for $L_{d\alpha}$, one obtains:

$$C_{L D\alpha} = \frac{4}{c} \times \frac{8}{3} \frac{b}{\pi AR} \tilde{k} + (C_{L\alpha})_0 \frac{G'(k)}{k}$$

Let $c = c_0 = 4b/(\pi AR)$, then this equation becomes:

$$C_{L D\alpha} = \frac{8}{3} \tilde{k} + (C_{L\alpha})_0 \frac{G'(k)}{k}$$

Now, the Theodorsen Function for elliptical-planform finite wings was derived by R. T. Jones ("The Unsteady Lift of a Wing of Finite Aspect Ratio", NACA Report No. 681, 1940), and the resulting expressions for $F'(k)$ and $G'(k)$ are fairly complicated Bessel Functions. However, J. Otto Scherer ("Experimental and Theoretical Investigation of Large Amplitude Oscillating Foil Propulsion Systems", Hydronautics company report, Laurel, Md, December, 1968) obtained simpler approximate curve-fit equations to these functions:

$$F'(k) \approx 1 - \frac{C_1 k^2}{k^2 + C_2^2}, \quad G'(k) = -\frac{C_1 C_2 k}{k^2 + C_2^2}$$

Where,

$$C_1 = \frac{0.5 AR}{AR + 2.32}, \quad C_2 = 0.181 + \frac{0.772}{AR}$$

When these functions are substituted into the equations for $C_{L\alpha}$ and $C_{L D\alpha}$, one obtains:

$$C_{L\alpha} = (C_{L\alpha})_0 \left(1 - \frac{C_1 k^2}{k^2 + C_2^2} \right), \quad C_{L D\alpha} = \frac{8}{3} \tilde{k} - (C_{L\alpha})_0 \left(\frac{C_1 C_2}{k^2 + C_2^2} \right)$$

Recall, from Chapter 6, that the aerodynamic terms are assumed to be quasi-steady. Namely, the assumption is that a typical aircraft's modes of motion are slow enough that the corresponding reduced frequencies are approximately zero. Therefore, in this case, as $k \rightarrow 0$, the above equations become:

$$C_{L\alpha} = (C_{L\alpha})_0, \quad C_{L D\alpha} = \frac{8}{3} \tilde{k} - (C_{L\alpha})_0 \frac{C_1}{C_2}$$

Values of $\tilde{k}$ vs. b/c_0 were found in a 1965 ASME publication by Kirk T. Patton ("Tables of Hydrodynamic Mass Factors for Translational Motion"). Also, for an elliptical planform wing, one has that:

$$(C_{L\alpha})_0 = \frac{2\pi}{1 + 2/AR}$$

Appendix E: Lift of Finite Wings Due To Oscillatory Plunging Acceleration

From this information, the following values for $C_{LD\alpha}$ are obtained:

AR	$(C_{L\alpha})_0$	$\tilde{k}$	$(C_{LD\alpha})_{app} \equiv 8\tilde{k}/3$	C_1	C_2	$(C_{LD\alpha})_{cir} \equiv -(C_{L\alpha})_0 C_1/C_2$	$C_{LD\alpha} = (C_{LD\alpha})_{app} + (C_{LD\alpha})_{cir}$
1.27	2.440	0.637	1.699	0.177	0.789	−0.547	1.152
2.55	3.521	0.826	2.203	0.262	0.484	−1.906	0.297
3.82	4.124	0.900	2.400	0.311	0.383	−3.349	−0.949
5.09	4.511	0.933	2.488	0.344	0.333	−4.657	−2.169
6.40	4.787	0.952	2.539	0.367	0.302	−5.824	−3.285
7.64	4.980	0.964	2.571	0.384	0.282	−6.753	−4.182
8.91	5.131	0.972	2.592	0.397	0.268	−7.606	−5.014
10.4	5.270	0.978	2.608	0.409	0.255	−8.442	−5.834
12.2	5.398	0.983	2.621	0.420	0.244	−9.284	−6.663
13.3	5.462	0.985	2.623	0.426	0.240	−9.692	−7.069
16.2	5.593	0.987	2.632	0.437	0.229	−10.698	−8.066
18.2	5.661	0.991	2.643	0.444	0.223	−11.237	−8.594

The curve fits to these tabular data are plotted below. It is seen that the apparent-mass and circulation terms act to oppose each other. In fact, $C_{LD\alpha} = 0$ at $AR \approx 2.8$. Beyond that value, $C_{LD\alpha}$ becomes increasingly negative. Now, it may seem rather curious that an aerodynamic force can be generated in the same direction as the wing's acceleration. This seems to be an opportunity for a perpetual-motion machine. The fact is, however, that this simply represents a deficit to the increasing lift force generated by α, through $(C_{L\alpha})_{cir}$:

$$C_L = (C_{L\alpha})_{cir} \alpha + (C_{LD\alpha})_{cir} D\alpha + (C_{LD\alpha})_{app} D\alpha$$

The equations for the fitted curves are:

$$(C_{LD\alpha})_{app} = 0.8905 + 0.8297 AR - 0.1638 AR^2 + 0.0158 AR^3 - 0.00073 AR^4 + 0.000013 AR^5$$

$$(C_{LD\alpha})_{cir} = 0.7749 - 0.9488 AR - 0.0810 AR^2 + 0.0155 AR^3 - 0.00085 AR^4 + 0.000016 AR^5$$

$$C_{LD\alpha} = 1.6654 - 0.1192 AR - 0.2447 AR^2 + 0.0313 AR^3 - 0.0016 AR^4 + 0.00003 AR^5$$

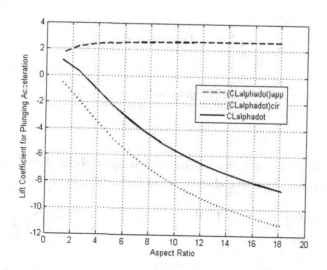

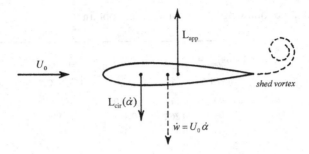

The concept of a lift deficit caused by a shed vortex is seen in the indicial-lift case, where an airfoil is given an instantaneous angle of attack in steady flow:

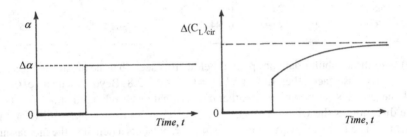

$$\Delta\alpha = 1(t)\alpha, \quad \Delta(C_L)_{cir} = 2\pi\varphi(t)\Delta\alpha$$

The $1(t)$ term is the Heavyside "step (or indicial)" function. Observe that the lift coefficient does not instantaneously reach the steady-state value associated with $\Delta\alpha$. Instead, it starts at half that value and then asymptotically achieves this with time, which illustrates the deficit in the steady-state lift caused by this motion. The expression $\varphi(t)$ in the above equation represents that behavior, and it is known as the Wagner Function.

For arbitrary α variations, this indicial-lift function may be used to find the resulting $(C_L)_{cir}$ behavior:

$$(C_L)_{cir} = 2\pi \int_0^t \varphi(t-\tau)\dot{\alpha}(\tau)d\tau$$

Assume a uniform acceleration for the plunging case:

$$\dot{\alpha}(\tau) = const. \equiv \dot{\alpha}$$

Also, the indicial function for a finite aspect-ratio wing is obtained from "Aeroelasticity" by Bisplinghoff, Ashley, and Halfman (Addison-Wesley Press, 1957, page 394):

$$\phi(s) = b_0 + b_1 \exp(-\beta_1 s) + b_2 \exp(-\beta_2 s) + b_3 \exp(-\beta_3 s)$$

Appendix E: Lift of Finite Wings Due To Oscillatory Plunging Acceleration

where $s = 2Ut/\bar{c}$. Upon substitution of this into the integral, and choosing $\beta_i \equiv \tilde{\beta}_i \bar{c}/(2U)$, one obtains:

$$(C_L)_{cir} = 2\pi b_0 \alpha + 2\pi \dot{\alpha} \sum_1^3 b_i \int_0^t \exp\left(-\tilde{\beta}_i(t-\tau)\right) d\tau, \quad i = 1,2,3$$

Now, the equation for $(C_L)_{cir}$ may be written as:

$$(C_L)_{cir} = C_{L\alpha}\alpha + \left(\frac{dC_L}{d\dot{\alpha}}\right)_{cir}\dot{\alpha}$$

Hence, in comparison with the integral, one sees that:

$C_{L\alpha} = 2\pi b_0$ (which is the steady-state lift coefficient derivative for a finite wing)

$$\left(\frac{dC_L}{d\dot{\alpha}}\right)_{cir} = 2\pi \sum_1^3 b_i \int_0^t \exp\left(-\tilde{\beta}_i(t-\tau)\right) d\tau$$

Further, note that the integral term becomes:

$$\int_0^t \exp\left(-\tilde{\beta}_i(t-\tau)\right) d\tau = \exp\left(-\tilde{\beta}_i t\right) \int_1^t \exp\tilde{\beta}_i \tau \, d\tau = \frac{\exp\left(-\tilde{\beta}_i t\right)}{\tilde{\beta}_i} \left|\exp\left(\tilde{\beta}_i \tau\right)\right|_0^t \to$$

$$\frac{1}{\tilde{\beta}_i} \exp\left(-\tilde{\beta}_i t\right) \left[\exp\left(\tilde{\beta}_i t\right) - 1\right] = \frac{1}{\tilde{\beta}_i}\left[1 - \exp\left(-\tilde{\beta}_i t\right)\right]$$

As $t \to \infty$, this represents the quasi steady-state behavior assumed for flight-dynamic analysis. Thus, the integral reduces to $1/\tilde{\beta}_i$ (note that all $\tilde{\beta}_i$ values are positive). One, therefore, obtains:

$$\left(\frac{dC_L}{d\dot{\alpha}}\right)_{cir} = 2\pi \sum_1^3 \frac{b_i}{\tilde{\beta}_i}$$

Further, $\dot{\alpha} = \frac{2U}{\bar{c}} D\alpha$, which gives:

$$\left(C_{LD\alpha}\right)_{cir} = 2\pi \sum_1^3 \frac{b_i}{\tilde{\beta}_i}$$

The table below is obtained from the previously cited reference:

AR	b_0	b_1	b_2	b_3	β_1	β_2	β_3
3	0.600	−0.170	0	0	0.540	−	−
6	0.740	−0.267	0	0	0.381	−	−
∞	1.000	−0.165	−0.335	0	0.0455	0.300	−

Therefore, the above equation gives the following values for $(C_{LD\alpha})_{cir}$. These are compared with corresponding values calculated from the previous analysis:

AR	Current $(C_{LD\alpha})_{cir}$ values	Previous $(C_{LD\alpha})_{cir}$ values
3	−1.978	−2.425
6	−4.403	−5.487
∞	−29.801	−17.357

There are quantitative differences between the results from the two methods because different curve-fit approximate expressions were used for each method. As stated before, Scherer's equations are approximations to a complicated expression involving Bessel Functions, and the second analysis is based on numerical curve fits to the indicial functions. The author can't authoritatively state which method is more accurate. However, it is seen that the two methods give approximately-corresponding values and, most importantly, they both confirm the lift-deficit behavior of $(C_{LD\alpha})_{cir}$.

As a practical matter, the author has used the first method since the 1970s for the flight-dynamic analyses of tethered aerostats and airships. Where experimental data has been available, the comparisons with theory have been very good.

Finally, recall that the analysis assumes an elliptical-planform wing. However, this methodology should still be applicable to tapered planforms and even rectangular planforms of higher aspect ratio. Further, it is important to note that the apparent-mass component of the plunging wing's lift acts at the planform's centroid; and the circulation component of the lift acts at the aerodynamic center. This fact has consequence for calculating the moment generated by $C_{LD\alpha}$ about a fixed point (such as an aircraft's mass center).

Index

A

Acceleration derivatives, 261, 329
Acceleration equation for take-off, 428
ac, see Aerodynamic center
Actuator-disk momentum theory, 82–84
Aerodynamic center (*ac*), 25, 29, 36, 38, 121, 130, 213, 214, 247, 268, 271, 318, 320, 475, 482, 488
Aerodynamic corrections for inflated fins, 494–496
Aerodynamics, 25
  airfoils, 25–35
  bodies, 40–46
  complete-aircraft aerodynamics, 49
    drag, 53–55
    lift, 50–53
    pitching moment (with extended undercarriage), 60
    pitching moment (with non-extended undercarriage), 55–60
  numerical example, 60
    body, 70–74
    complete airplane lift and drag coefficients, 75–76
    complete airplane moment characteristics, 76–77
    propeller, 74
    tail, 64–69
    undercarriage, 74–75
    wing, 61–64
  undercarriage, 46–48
  wing-body combination, 48–49
  wings, 36–40
"*Aerodynamics of The Airplane*" (Hermann Schlichting and Erich Truckenbrodt), 49, 213, 416
Aerostats, 493, 494, 496–497
AeroVironment Inc., 110, 380
Airfoils, 25–35
  angle-of-attack of, 25
  camber, 25
  chord, 25
  Clark-Y airfoil, 62
  flat-plate airfoil, 69
  ideal angle-of-attack, 30–31
  leading edge suction, 30
  moment about the *ac*, 27, 29, 34
  NACA 0009 symmetrical airfoil, 132
  NACA 0018 symmetrical airfoil, 495
  NACA 2412 airfoil, 132
  profile drag coefficient, 35
  propeller airfoils, 97
  sectional drag coefficient, 189
  Selig SD8020 airfoil, 65
  standard parameters, 25
  zero-lift line, 25, 34, 270
Airships, 446–461
Airshipwreck, 451
Airship ZRS-4 "Akron" 5.98m wind-tunnel model (example), 479–485
Akron airship, 450, 464
Allen, H.J., 464
All-metal ZMC-2 airship (1929), 455

All-wood "boxspar" design, 22
Alternative swept-wing analysis, 519–525
Aluminum alloys, evolution of, 4
Aluminum column, buckling load for, 7
American Blimp Corporation, 381, 459
Anderson, John, 40
Angle-of-attack, 25, 30, 37, 117, 464
Antoinette (1909), 175
Apparent-mass coefficients, 41
Apparent-mass effects, 271–272, 539–550
"Applied Wing Theory" (Elliott G. Reid), 33
Argand diagram, 364
Armstrong-Whitworth AW-52, 114, 136
Armstrong-Whitworth FK-10, 209
Army Air Corps, 199
Arnoux, Rene, 101
"Arup" design, 159
A-60 design, 464
Aspect Ratio, 36
Avanti, 12
Avdicevic, Benjamin, 406
Average sweep angle, 525
Axial drag force, 29
Axial force equation, 465
Axial-momentum theory, 85

B

Ballonet, 444, 455, 458
Balloons and airships, 431
  aerodynamics of finned axisymmetric bodies, 461
    Airship ZRS-4 "Akron" 5.98m wind-tunnel model (example), 479–485
    Goodyear "Wingfoot2" Airship (Zeppelin LZ N07) (example), 485–492
    longitudinal static aerodynamic coefficients, method for calculating, 464–472
    small aerostat (example), 472–479
    TCOM CBV-71 Aerostat (example), 493–494
  aerostats, observations about, 496–497
  airship aerodynamic mystery, 502–503
  airships, 446–461
  buoyancy, physics of, 439–441
  free balloons, 431–439
  inflated fins, aerodynamic corrections for, 494–496
  lateral force and yawing moment calculation, 498–502
  Matlab program, 503–516
  tethered balloons, 441–446
Basic lift distribution, 118, 519–522
Beech "Starship," 11
Bernoulli's equation, 82
Bilton, Amy, 406
Biplane Glider, 241
Biplanes, 197
  biplane analysis, considerations of, 229–232
  biplane glider (example), 232–243
  decalage, 203, 243
  estimating the aerodynamic characteristics of, 210–212

559

gap, 203
"Slow SHARP" biplane (example), 212–228
stagger, 203
total induced drag coefficient of, 201
Bismarck, 198
Bituminous compound, 455
"Blended" compound-delta wing, 159
Bleriot tail-aft (1913), 175
Blimp fabric, 449
Blimps, 446
Blower-fan system, 444
Bodies, 40–46, 70–74, 272–273, 276–277, 285–286, 298–299, , 306–308, 314, 329–331
Body-fixed axes system, 29, 251
Boeing Company, 31
Borst, H. V., 41
Boundary layer, 26
Boxwing, 208
Brazilian CEA-308 airplane, 3
British airship R100, 503
British "Lancaster" bomber, 32
Bucco-Lechat, Clement, 109
Buoyancy, 435
  physics of, 439–441
Buoyant flying-wing design, 458

C

Cameron Balloons, 436
Cameron GB1000 gas balloon, 432
Canadian Communications and Research Centre (CRC), 205
Canadian flying-wing glider, 114
Canadian NRC Tailless Glider, 114
Canard airplanes, 167
  canard glider (example), 185–197
  estimating the aerodynamic characteristics of, 179–183
  rectangular wing (example), 183–184
Canard glider (example), 185–197
Caproni, Gianni, 209
Caproni Ca-60 "Transaereo" (1921), 209
"Carbon-fiber-boxspar" wing, 21
Carbon-fiber/epoxy composite tubes, 11
CargoLifter airship, 457
CBV-71 aerostat, 494
Center-of-pressure (cp), 27
Chance-Vought XF5U-1, 160
Charles, Jacques Alexandre, 431, 435
Chord, 25
Chord-distribution function, 118
Chord line, 25
Clark-Y drag polar plots, 63
Coalition Army, 441
Coefficient of pitching moment, 29
Column buckling, different materials in, 6–11
Complete-aircraft aerodynamics, 49
  drag, 53–55
  lift, 50–53
  pitching moment
    with extended undercarriage, 60
    with non-extended undercarriage, 55–60
Complete airplane lift and drag coefficients
  numerical example, 75–76
Complete airplane moment characteristics
  numerical example, 76–77
Compressive failure load, 9
"Concorde" supersonic airliner, 159
Constant-thrust/zero-thrust aircraft, 377
Controls-fixed stability, 360
Convair 1955 F-102A "Delta Dagger" jet-propelled fighter, 155
Cotton/Neoprene/silver paint laminates, 444
cp, see Center-of-pressure
CRC, see Canadian Communications and Research Centre
Critical damping, 377
Cross-flow drag equations, 481
Cruising-flight Reynolds number, 217
Cube-square law, 460
Culick, Fred E. C., 169
Curtis, Howard "Pat," 461

D

Dacron, 456
Dacron/Mylar/Tedlar laminates, 443–444
Damping ratio, 377
Damping term, 377
D'Arlandes, Marquis Francois, 435
DATCOM method, 303–304, 329, 333, 337
Dayton-Wright Racer, 1
Decalage, 243
Deighton, Len, 451
"Delta 1" (1931), 154
De Rozier, Jean-Francois Pilatre, 434, 435
Design features, of aircraft, 1–5
Design of Light Aircraft (Richard Hiscocks), 202
Design point, 31
Diederich, Franklin, 519
Dihedral angle, 305
Dimensional stability derivatives, 258
Directional stability, 360, 371, 372
Distributed profile-drag coefficients, 322
Distribution function, 182
Double-swept and double-tapered wing (example), 129
Douglas DC-2, 15
Downwash, 51–52
Downwash-lag effect, 287
"Downwash" angle, 87
Drag, 53–55
Drag-coefficient equation, 28, 264
Drag coefficient of segment, 81
Drzewiecki, Stefan, 79
Dunne D8 biplane Flying Wing (1914), 107
Duralumin, 450
Dutch-Roll Mode, 386, 392, 395–396
Dynamic instability, 463
Dynamic neutral stability, 380

E

Easy-to-fly model airplane, 403
Efficiency factor, 48
"Elements of Practical Aerodynamics" (Bradley Jones), 207
Elliptical function, 182
Engine power, 459
Equilibrium flight

Index

full solution, 412–419
  performance parameters, 419–428
  trim state, 408–412
Equivalent monoplane, 207
Euler Angles, 536
Euler buckling, 6
Example wing structures, 17–23
Excrescence drag, 45

F

Fairey "Swordfish," 197
Farashahi, Hadi, 406
Farman design (1910), 175
Fatness factor, 486
Fin contribution, 341, 343, 353, 354, 355, 356
Fineness ratio, 470, 480, 486
Finned axisymmetric bodies, aerodynamics of, 461
  Airship ZRS-4 "Akron" 5.98m wind-tunnel model (example), 479–485
  Goodyear "Wingfoot2" Airship (Zeppelin LZ N07) (example), 485–492
  longitudinal static aerodynamic coefficients, method for calculating, 464–472
  small aerostat (example), 472–479
  TCOM CBV-71 Aerostat (example), 493–494
Fins, 463, 464
  drag coefficient, 471
  inflated, 494–496
Flat-plate friction-drag coefficient, 470
Flight dynamics, 245
  aircraft lateral small-perturbation dynamic equations, 294
    estimation of the lateral stability derivatives, 297–329
    non-dimensional form of the equations, 295–297
    "Scholar" tail-aft monoplane, example lateral stability derivatives for, 329–359
  aircraft longitudinal small-perturbation dynamic equations, 256
    estimation of the longitudinal stability derivatives, 264–279
    longitudinal numerical example, 279–294
    non-dimensional form of the equations, 259–264
  lateral dynamic stability, 382
    Dutch-roll mode, 386, 395–396
    flight paths, 386–392
    rolling mode, 385, 394–395
    roots-locus plots, 396–397
    spiral mode, 385, 393–394
    stability-boundary plot, 398
  longitudinal dynamic stability, 362
    flight paths, 369–374
    numerical example, 366–368
    phugoid mode, 368, 376–378
    roots-locus plots, 378–382
    short-period mode, 368, 374–375
  radii-of-gyration values for representative airplanes, 359
  stability, definitions of, 360
    directional stability, 360
    path-keeping stability, 360–362
    straight-line stability, 360
Flight paths, 386–392

"*Fluid-Dynamic Drag*" (Sighard Hoerner), 207
Flying Plank, 101, 110
Flying-wing hang gliders, 109
Flying Wings, 108, 154, 164, 165
Flying-Wing type airplanes, 101
  Canadian flying-wing glider, 114–115
  "Delta" tailless aircraft, 154–162
  estimating the aerodynamic characteristics of wings, 115
    double-swept and double-tapered wing (example), 129
    Matlab computer program, 143–154
    straight-tapered linear-twisted wing (example), 118
  "flying planks," 101–105
  paragliders, 108
  Rogallo-type hang gliders, 109–110
  span-loader flying wings, 110–113
  swept flying wings, 105–108
Fokker Company, 197
Fokker D-VII fighter, 197
Fokker D-VIII fighter, 12
Fokker Triplane, 200
Fokker wooden-wing structural-design concept, 14
Force-acceleration equation, 530
Form drag, 45
Forward-facing scoop, 442
Free balloons, 431–439
Free flight, 426
French Caquot observation balloon, 442
Friction-drag coefficient, 35
Froude, Robert Edmond, 82
Froude actuator-disk theory, 82–84

G

Gamma, 1, 3
German radars, 5
Gigantism, 457, 458
Glauert correction factor, 37
Glide Tests, 405–407
Gloster "Gladiator," 203
Goodyear Airship Operations, 485
Goodyear Corporation, 455
Goodyear "Inflatoplane," 17
Goodyear "Wingfoot2" Airship (Zeppelin LZ N07) (example), 485–492
Gordon, James Edward, 5
Gossamer Albatross, 172
Gossamer Condor, 172, 380
Göttingen wind tunnel, 14
Graf Zeppelin, 452
"Great Aerodrome" of 1903, 179
Gruelling never-ending chore, *see* Proof-Reading
Grumman F2F carrier-based Navy fighter, 199

H

Hage, R., 415
Hall, George, 447
Harpoothian, Edward, 87
Heavy-Lift Airship (HLA) concept, 460, 461
Hillsboro Oregon, 459
Hiscocks, Richard, 202

HLA concept, *see* Heavy-Lift Airship concept
Hoerner, Sighard, 35, 37, 41, 207
Hoerner's plot, 208
Hoffman, Raoul, 159
Hoop-stress limitations, 455
Horten Ho-VI sailplane, 162
Hot-air balloons, 434, 435
Hughes H-4 "Hercules," 16
Hull drag coefficient, 474
Hull zero-angle drag coefficient, 470
Hull's "fatness factor," 480, 486
Hull's reference area, 471
Human-carrying free balloon, 434, 436

I

Ideal angle of attack, 426
    airfoil at, 30–31
Imperial German Air Force, 197
Induced drag, 37, 67,141
Induced drag (biplanes), 210–211
Induced drag coefficient, 198, 201, 208, 210
Inflated fins, aerodynamic corrections for, 494–496
Italia, 454

J

"Japanese-lantern" effect, 459
Jex, Henry R., 169
Johansson, P., 156
Jones, Bradley, 207
Jones, Brian, 436
Jones, Samuel P., 464, 285

K

Kevlar, 23, 456

L

Lagrange's equation, 540
Lakehurst New Jersey, 452
Lam, Andrew, 406
Lanchester, Frederick, 368
Langley, Samuel P., 178–179
Larson, George, 447
Lateral-displacement modal vector, 391
Lateral dynamic stability, 382
    approximate equations, 392
        Dutch-roll mode, 395–396
        rolling mode, 394–395
        spiral mode, 393–394
    flight paths, 386–392
    roots-locus plots, 396–397
    stability-boundary plot, 398
Lateral force and yawing moment calculation, 498–502
Lateral lift-coefficient, 334
Lateral small-perturbation dynamic equations, 294
    estimation of lateral stability derivatives, 297
        β derivatives, 298–314
        $D\hat{p}$ derivatives, 327–329
        $D\hat{r}$ derivatives, 326–327
        $D\beta$ derivatives, 324–326
        $\hat{p}$ derivatives, 319–324
        $\hat{r}$ derivatives, 314–319
    non-dimensional form of the equations, 295–297
    "Scholar" tail-aft monoplane, lateral stability derivatives for, 329–359
Leading-edge separation bubble, 97
Leading-edge spar, 23
Leading-edge suction, 29, 30
Leading-edge suction efficiency, 35, 472, 477, 489
Le Moigne, T., 156
Lift, 50–53
Lift-coefficient data for different values of RN, 62
Lift-coefficient equations, 36, 140, 185, 211, 229, 231, 405, 478
Lift-curve slope, 36, 62
Lift/Drag ratio, 76, 108, 142, 192, 220, 239, 377–378, 421, 425
Lift equation, 28, 39, 50, 117, 202, 405, 408
Lifting gas, 431, 432, 434, 440, 444–446
Lift-loading contribution, 40
"Lightness" of the aircraft, 174
Lippisch, Alexander, 154
Local chord, of wing, 182
Lockheed 9D-2 Orion, 3
Loft line, 49, 302, 309
Longitudinal dihedral, 243
Longitudinal dynamic stability, 362
    approximate equations, 374
        phugoid mode, 376–378
        short-period mode, 374–375
        comments on α and θ, 368–369
    flight paths, 369–374
    numerical example, 366–368
    roots-locus plots, 378–382
Longitudinal small-perturbation dynamic equations, 256
    estimation of longitudinal stability derivatives, 264
        $(\ )_0$ terms, 264
        α derivatives, 266–267
        $D\hat{q}$ derivatives, 276–277
        $D\hat{u}$ derivatives, 271–273
        $D\alpha$ derivatives, 273–276
        propulsion derivatives, 277–279
        $\hat{u}$ derivatives, 265
        $\hat{q}$ derivatives, 267–271
    longitudinal numerical example, 279–294
    non-dimensional form of the equations, 259–264
Longitudinal static aerodynamic coefficients, method for calculating, 464–472
Lowe, Thaddeus, 441
Low-flying enemy airplanes, 443
Lowry and Polhamus equation, 36
Luftschifftechnik, Zeppelin, 454
"Lysander" design, 178
LZ 127 "Graf Zeppelin" (1928), 452
LZ 129 "Hindenburg" (1936), 452, 457
LZ 130 "Graf Zeppelin II" (1938), 454

M

Mach Number, 265, 269
Makhonine designs, 422, 424
Marske, Jim, 104, 128

Index

Marske *Monarch* ultra-light glider (1974), 104
Marske *Pioneer 3* (2009), 107
Mass-center of airplane, 251, 295, 306, 380, 381, 530, 535
Matlab program, 97–100, 143–154, 225–228, 417–419, 503–516
Matlab solution, 250, 367, 384
MAV, *see* Micro Air Vehicle
Maximum achievable lift coefficient, 426
Maximum endurance parameter, 426
Mayfly, 452
Mayrhofer, Koloman, 12
McCoubrey, Ryan, 406
Mean aerodynamic chord, 36
Michael's equations, 321
Micro Air Vehicle (MAV), 162
Miller, William H., 87
Modal vector diagram, 249, 364, 384
Model-train mode, 464
Modern tethered balloon, 444
Moment-dynamic equations, 533
Moment-of-inertia, 245, 260, 359, 531
Montgolfier brothers, 434–435
Mosquito, de Havilland, 5
Multhopp, Hans, 127
Multhropp body-moment method, 71, 74, 517
Munk, Max, 462, 464
Munk equations, 285
Munk Moment, 462, 463, 544
Munk stagger-lift factor, 233
Murray, Graham, 406
Mylar film, 456

N

NACA 0018 symmetrical airfoil, 495
Natural-shape free balloon, 456
Neoprene, 449
Neutral point, 57, 60
  pitching-moment coefficient about, 59–60, 194, 224, 241
  solution for, 57–60, 76, 193–195, 224, 241
Neutral stability, 57, 380
Newtonian force-acceleration equation, 540
Nickel, Karl, 115
Nobile, General Umberto, 453
Non-dimensional lateral displacement, 392
Non-rigid airships, 446, 456
Norge, 453
Normal Force, 29
Normal-force coefficient, 475
Normal-Force equation, 465
Northrop, Jack, 1, 3, 164, 165
Northrop B-2 bomber, 163
Northrop B-35 Flying-Wing bomber, 112
Northrop bombers, 112
Northrop Gamma, 1, 3
Number of cycles to half (or double) amplitude, 246, 363, 383

O

Ornithopter, 23
Oscillatory plunging acceleration, lift of finite wings due to, 551–558

P

Paragliders, 108
Path-keeping stability, 360–361
Patton, Kirk T., 554
Paul Matt Scale Airplane Drawings, Vol. 1, 31
Performance, 405
  equilibrium flight
    full solution, 412–419
    performance parameters, 419–428
    trim state, 408–412
  glide tests, 405–407
  take-off dynamics, 428
    Wood's methodology, 428–430
    final comments, 430
Perkins, C., 415
Perkins, E.W., 464
Pessimistic coordinate system, 371
Phillips, Horatio, 200
Photographic stability, 360
Phugoid Mode, 368, 369, 372, 376–378, 379
Phugoid Spiral, 372–373, 374
Piccard, Betrand, 436
Pitching moment, 29, 55, 125, 271, 465
  with extended undercarriage, 60
  with non-extended undercarriage, 55–60
Pitching-moment coefficient, 29, 38, 41, 194, 222
Pitching-moment equation, 56
Platz, Reinhold, 12
"Plunging" velocity, 369
Plywood shear web, 23
Polhamus, Edward C., 158
Polynomial curve-fit equations, 63, 65, 192, 220, 238, 490
Positive decalage angle, 203, 243
Positive stagger, 203
Potential-fluid field, 539
Potomac River, 179
Pressure altitude, 445
Product-of-inertia, 532
Profile-drag equation, 35, 139–140, 142
Proof-Reading, *see* Gruelling never-ending chore
Propeller, 52, 74, 79, 305
  actuator-disk momentum theory, 82–84
  application of the analysis, 94–96
  Matlab program, 97–100
  method of calculation, 87–89
  numerical example, 89–94
  propeller airfoils, 97
  simple blade-element analysis, 79–82
  simple blade-element analysis, extensions to, 84–87
Propeller-driven aircraft, 377
Propulsive efficiency equation, 82, 87
Pulitzer Trophy Race (1924), 3
"Pure" aircraft, 165
Putman, William F., 461

R

Radii-of-gyration values for representative airplanes, 359
Radio-controlled flying, 403
Rate of climb, 426
Rectangular wing (example), 183–184
The Red Tent, 454
Reduced frequency, 552

Reid method, 33
Resultant force of segment, 81
Reynolds Number (*RN*), 13, 26, 28–29, 32, 46–47, 62, 65, 102, 143, 188, 265, 269, 308, 337, 470, 480, 491
Ribner, Herbert, 52
Rigid-body equations of motion, 527
    dynamic equations, 530–535
    kinematics, 527–529
    linearized versions of the equations, 537–538
    orientation, 535–537
"Rigid" airships, 450
Rip panel, 432
Rizzi, A., 156
RN, *see* Reynolds Number
Robert, Nicolas-Louis, 435
Rogallo-type hang gliders, 109–110
Rogallo-Wing configuration, 109
"Roller coaster" type of motion, 372
Rolling Mode, 385, 394–395, 402
Rolling-moment equation, 329
Roma, 454
Roots-locus plots, 378–382, 396–397
Rotational interference factor, 85

S

SAAB Viggen fighter airplane, 176–177
SAS, *see* Stability Augmentation Systems
Scherer's equations, 558
Schirmer, Eng. M., 502
Schlichting, Hermann, 213, 416
Scholar model airplane, 400
"Scholar" tail-aft monoplane, 61–77, 279–294, 329–359, 399–403
Schrenk Approximation, 181
Schrenk model, 115–118, 121–123, 127, 143, 181–182
Schwartzman, Arnold, 451
Sea-level Reynolds Number, 26, 337
Sectional drag coefficient, of airfoil, 188–189
Segmental thrust, 86
Segment's drag coefficient, 81
Segment's resultant force, 81
Selig, Michael, 40
Semi-graphical method, 33
Semi-monocoque construction, 1, 4, 5, 16
Shell buckling, 10
SHARP, *see* Stationary High Altitude Relay Platform
Short-Period Mode, 173, 368, 371, 374–375, 378, 380
Short Take-Off and Landing (STOL) ability, 160
Sikorsky, Igor, 164
Sikorsky "Skycrane," 460
"Silkspan-on-foam" wing, 20, 21
Simple blade-element analysis, 79–82
Simple blade-element analysis, extensions to, 84–87
Simplex-Arnoux racer (1922), 101
Single-design-point aircraft, 206
Skin-friction coefficient, 26, 476, 482
Skin-friction drag, 45, 470
Skin-friction drag coefficient, 26, 35, 471
Skyship 600, 464
"Slow SHARP" biplane (example), 212–228
Small aerostat (example), 472–479
Small-perturbation dynamic equations for aircraft, longitudinal 256
Small-perturbation dynamic equations for aircraft, lateral 294
Small-perturbation stability analysis, 361
Smetana, Frederick, 318
"Snoopy Two" blimp, 459
Snyder, C. L., 159
Soltani, Tahoura, 406
Soondarsingh, Devi, 406
Space Shuttle, 156
Span-loader flying wings, 110–113
Spiral Mode, 385, 388, 393–394, 397
Square-cube law, 460
Stability, definitions of, 360
    directional stability, 360
    path-keeping stability, 360–362
    straight-line stability, 360
Stability Augmentation Systems (SAS), 171
Stability-boundary plot, 398
Stabilizer, 49, 213–214, 415–416, 464
"Stagger" ratio, 202–203
Stagger SHARP, 205–206, 212
Staggerwing, 204
Static stability/instability, 59, 60, 102, 106, 170–174, 194, 224, 378, 379, 380–382, 463
Static margin, 59, 127–128, 134–135, 142, 194, 196, 242, 380
Stationary High Altitude Relay Platform (SHARP), 17, 205
STOL ability, *see* Short Take-Off and Landing ability
Straight-line stability, 360
Straight-tapered linear-twisted wing (example), 118
Stratoliner, 31
Structural-weight curve, 460
Super-pressure free balloons, 438–439, 456
"Surging" velocity, 369
Swedish SAAB "Draken," 156
Sweet spot, 48
Swept flying wings, 105–108
Swordfish, 197–198

T

Tail, 49–50, 51, 54–55, 64–69, 275, 464
    boat tail, 45, 470–471
    numerical example, 64–69
    V-Tail, 49, 321
Tail-body efficiency factor, 67
Tail drag coefficient, 54–55
Tail draggers, 47, 430
Tailless Aircraft, 101; *see also* Flying-Wing type airplanes
    Canadian NRC Tailless Glider, 114–115
    "Delta" tailless aircraft, 154–162
"Tailless Aircraft in Theory and Practice" (Karl Nickel and Michael Wohlfahrt), 115
Take-off dynamics, 428
    Wood's methodology, 428–430
"Tandem" configuration, 178
Taylor series expansion, 257
TCOM aerostat, 463, 464
TCOM CBV-71 Aerostat (example), 493–494
Tedlar, 457
Telescoping-wing concept, 423–424
Tethered aerostat lateral dynamic stability, 496
Tethered aircraft, 496

Index

Tethered balloons, 441–446
Theodorsen Function, 551
"Theory of Flight" (Richard von Mises), 210
Thickness correction factor, 35, 471, 477
Thrust coefficient, 57–58, 84
Thrust derivatives, 58, 74, 278–279
Thwaites, Brian, 298
Time-to-double-amplitude, spiral instability (example), 389
Time to half (or double) amplitude, 246, 363, 383
Torque factor, 91
Total drag coefficient (undercarriage), 47
Total drag coefficient ("Slow SHARP" biplane example), 219
Total drag of the airplane, 53
Total induced drag, 199
Total induced drag coefficient, 199, 201, 231
Total lift-curve slope, 57, 183
Total lift distribution, 117
Trailing-edge spar, 218
Truckenbrodt, Erich, 213, 416
Twisted lift distribution, 118, 522–525

U

Undercarriage, 46–48, 60, 74–75, 430
Union Balloon Corps, 441
Un-staggered lift coefficients, 211, 233

V

Velocity canard airplane, 168
Verville-Sperry R3 monoplane racer, 2
Vincenti, Walter G., 3
Von Karman, Theodore, 371
Von Mises, Richard, 210
Von Richthofen, Baron Manfred, 200
Vought V-173 "Flying Flapjack," 160
V-Tail, 49, 321

W

Wagner Function, 556
Wardlaw's cross-flow model, 469, 501
Wavy-pavement model, 362
Weather-vane model, 245
Weather-vane motion, 392
Weather-vane stability, 381
Westland P.12 "Wendover" airplane, 178
Wings, 36–40, 115–154
    aerodynamic-center point, 36, 121–124, 130–131, 522
    angle of attack, 37, 50
    drag-coefficient, 37–38, 53–54, 139–142
    estimating the aerodynamic characteristics of, 115
    double-swept and double-tapered wing (example), 129
        Matlab computer program, 143–154
        straight-tapered linear-twisted wing (example), 118
    lift coefficient, 36, 50, 53–54, 117, 138
    lift-coefficient equations (biplane), 202, 211, 229, 231
    lift-coefficient equation (with canard), 185
    lift-curve slope, 36, 136, 183
    local chord, 116, 182
    mean aerodynamic chord, 36, 116, 134
    numerical example, 39, 61–64, 118–128, 129–143
    pitching-moment coefficient, 38, 125–126, 132–134, 523–524
    zero-lift line, 37, 117, 119–120, 133, 523
Wing-body combination, 48–49
Wing-body lift-curve slope, 44
Wing loading, 348
Wohlfahrt, Michael, 115
Wolman, Baron, 447
Wood, Karl, 243, 428
Wood, metal and composite airplanes, 11–17
Wood's methodology, 428–430
Wright Flyer, 168, 169, 171
Wright glider (1902), 173
Wright Model A (1908), 167

X

X-29 experimental airplane, 171
XFOIL, 32
X-force equation, 263

Y

Yates, A. H., 309

Z

Zdunich, Patrick, 423
Zeppelin Corporation, 452, 485, 502
Zeppelin NT, 459
Zero-angle drag coefficient of stabilizer, 69, 489
Zero-angle lift coefficient, 39
Zero-Lift Line (ZLL), 25, 117, 119, 133, 270
"Zero-pressure" design, 433
Zero-thrust aircraft, 377
Zimmerman, Charles, 160
Zimmer Skimmer, 160
ZLL, *see* Zero-Lift Line
ZMC-2, 455
ZPG-3W blimp, 449
ZRS-4 "Akron," 450, 479, 497, 503

Printed in the United States
by Baker & Taylor Publisher Services

Printed in the United States
by Baker & Taylor Publisher Services